Student Solutions Man

Finite Mathematics

SEVENTH EDITION

Stefan Waner
Hofstra University

Steven R. Costenoble
Hofstra University

Prepared by

Stefan Waner
Hofstra University

Steven R. Costenoble
Hofstra University

CENGAGE
Learning

Australia • Brazil • Mexico • Singapore • United Kingdom • United States

ISBN: 978-1-337-28047-1

Cengage Learning
20 Channel Center Street
Boston, MA 02210
USA

Cengage Learning is a leading provider of customized learning solutions with office locations around the globe, including Singapore, the United Kingdom, Australia, Mexico, Brazil, and Japan. Locate your local office at: **www.cengage.com/global**.

Cengage Learning products are represented in Canada by Nelson Education, Ltd.

To learn more about Cengage Learning Solutions, visit **www.cengage.com**.

Purchase any of our products at your local college store or at our preferred online store **www.cengagebrain.com**.

Printed in the United States of America
Print Number: 01   Print Year: 2017

Table of Contents

Section 0.1

1. $2(4 + (-1))(2 \cdot -4)$

$= 2(3)(-8) = (6)(-8) = -48$

3. $20/(3*4) - 1$

$= \frac{20}{12} - 1 = \frac{5}{3} - 1 = \frac{2}{3}$

5. $\dfrac{3 + ([3 + (-5)])}{3 - 2 \times 2}$

$= \dfrac{3 + (-2)}{3 - 4} = \dfrac{1}{-1} = -1$

7. $(2 - 5*(-1))/1 - 2*(-1)$

$= \dfrac{2 - 5 \cdot (-1)}{1} - 2 \cdot (-1)$

$= \dfrac{2+5}{1} + 2 = 7 + 2 = 9$

9. $2 \cdot (-1)^2/2 = \dfrac{2 \times (-1)^2}{2} = \dfrac{2 \times 1}{2} = \dfrac{2}{2} = 1$

11. $2 \cdot 4^2 + 1$

$= 2 \times 16 + 1 = 32 + 1 = 33$

13. $3\char`^2+2\char`^2+1$

$= 3^2 + 2^2 + 1 = 9 + 4 + 1 = 14$

15. $\dfrac{3 - 2(-3)^2}{-6(4 - 1)^2}$

$= \dfrac{3 - 2 \times 9}{-6(3)^2} = \dfrac{3 - 18}{-6 \times 9}$

$= \dfrac{-15}{-54} = \dfrac{5}{18}$

17. $10*(1+1/10)\char`^3$

$= 10\left(1 + \dfrac{1}{10}\right)^3 = 10(1.1)^3$

$= 10 \times 1.331 = 13.31$

19. $3\left[\dfrac{-2 \cdot 3^2}{-(4 - 1)^2}\right]$

$= 3\left[\dfrac{-2 \times 9}{-3^2}\right] = 3\left[\dfrac{-18}{-9}\right]$

$= 3 \times 2 = 6$

21. $3\left[1 - \left(-\dfrac{1}{2}\right)^2\right]^2 + 1$

$= 3\left[1 - \dfrac{1}{4}\right]^2 + 1$

$= 3\left[\dfrac{3}{4}\right]^2 + 1 = 3\left[\dfrac{9}{16}\right] + 1$

$= \dfrac{27}{16} + 1 = \dfrac{43}{16}$

23. $(1/2)\char`^2-1/2\char`^2$

$= \left[\dfrac{1}{2}\right]^2 - \dfrac{1}{2^2} = \dfrac{1}{4} - \dfrac{1}{4} = 0$

25. $3 \times (2 - 5) = 3*(2-5)$

27. $\dfrac{3}{2 - 5} = 3/(2-5)$

Note $3/2-5$ is wrong, as it corresponds to $\dfrac{3}{2} - 5$.

29. $\dfrac{3-1}{8+6}$ = (3-1)/(8+6)

Note 3-1/8-6 is wrong, as it corresponds to

$3 - \dfrac{1}{8} - 6.$

31. $3 - \dfrac{4+7}{8}$ = 3-(4+7)/8

33. $\dfrac{2}{3+x} - xy^2$ = 2/(3+x)-x*y^2

35. $3.1x^3 - 4x^{-2} - \dfrac{60}{x^2-1}$

= 3.1x^3-4x^(-2)-60/(x^2-1)

37. $\dfrac{\left(\frac{2}{3}\right)}{5}$

= (2/3)/5

39. $3^{4-5} \times 6$

= 3^(4-5)*6

Note that the entire exponent is in parentheses.

41. $3\left(1 + \dfrac{4}{100}\right)^{-3}$

= 3*(1+4/100)^(-3)

43. $3^{2x-1} + 4^x - 1$

= 3^(2*x-1)+4^x-1

45. 2^{2x^2-x+1}

= 2^(2x^2-x+1)

Note that the entire exponent is in parentheses.

47. $\dfrac{4e^{-2x}}{2 - 3e^{-2x}}$

= 4*e^(-2*x)/(2-3e^(-2*x))
or 4(*e^(-2*x))/(2-3e^(-2*x))
or (4*e^(-2*x))/(2-3e^(-2*x))

49. $3\left(1 - \left(-\dfrac{1}{2}\right)^2\right)^2 + 1$

= 3(1-(-1/2)^2)^2+1

2

Section 0.2

1. $3^3 = 27$

3. $-(2 \cdot 3)^2 = -(2^2 \cdot 3^2) = -(4 \cdot 9) = -36$

or

$-(2 \cdot 3)^2 = -6^2 = -36$

5. $\left(\dfrac{-2}{3}\right)^2 = \dfrac{(-2)^2}{3^2} = \dfrac{4}{9}$

7. $(-2)^{-3} = \dfrac{1}{(-2)^3} = \dfrac{1}{-8} = -\dfrac{1}{8}$

9. $\left(\dfrac{1}{4}\right)^{-2} = \dfrac{1}{(1/4)^2} = \dfrac{1}{1/16} = 16$

11. $2 \cdot 3^0 = 2 \cdot 1 = 2$

13. $2^3 2^2 = 2^{3+2} = 2^5 = 32$

or

$2^3 2^2 = 8 \cdot 4 = 32$

15. $2^2 2^{-1} 2^4 2^{-4} = 2^{2-1+4-4} = 2^1 = 2$

17. $x^3 x^2 = x^{3+2} = x^5$

19. $-x^2 x^{-3} y = -x^{2-3} y = -x^{-1} y = -\dfrac{y}{x}$

21. $\dfrac{x^3}{x^4} = x^{3-4} = x^{-1} = \dfrac{1}{x}$

23. $\dfrac{x^2 y^2}{x^{-1} y} = x^{2-(-1)} y^{2-1} = x^3 y$

25. $\dfrac{(xy^{-1}z^3)^2}{x^2 y z^2} = \dfrac{x^2 (y^{-1})^2 (z^3)^2}{x^2 y z^2}$

$= \dfrac{x^2 y^{-2} z^6}{x^2 y z^2} = x^{2-2} y^{-2-1} z^{6-2}$

$= x^0 y^{-3} z^4 = \dfrac{z^4}{y^3}$

27. $\left(\dfrac{xy^{-2}z}{x^{-1}z}\right)^3 = \dfrac{(xy^{-2}z)^3}{(x^{-1}z)^3}$

$= \dfrac{x^3 y^{-6} z^3}{x^{-3} z^3} = x^{3-(-3)} y^{-6} z^{3-3}$

$= x^6 y^{-6} z^0 = \dfrac{x^6}{y^6}$

29. $\left(\dfrac{x^{-1} y^{-2} z^2}{xy}\right)^{-2} = (x^{-1-1} y^{-2-1} z^2)^{-2}$

$= (x^{-2} y^{-3} z^2)^{-2} = x^4 y^6 z^{-4}$

$= \dfrac{x^4 y^6}{z^4}$

31. $3x^{-4} = \dfrac{3}{x^4}$

33. $\dfrac{3}{4} x^{-2/3} = \dfrac{3}{4x^{2/3}}$

35. $1 - \dfrac{0.3}{x^{-2}} - \dfrac{6}{5} x^{-1} = 1 - 0.3x^2 - \dfrac{6}{5x}$

37. $\sqrt{4} = 2$

39. $\sqrt{\dfrac{1}{4}} = \dfrac{\sqrt{1}}{\sqrt{4}} = \dfrac{1}{2}$

3

41. $\sqrt{\dfrac{16}{9}} = \dfrac{\sqrt{16}}{\sqrt{9}} = \dfrac{4}{3}$

43. $\dfrac{\sqrt{4}}{5} = \dfrac{2}{5}$

45. $\sqrt{9} + \sqrt{16} = 3 + 4 = 7$

47. $\sqrt{9+16} = \sqrt{25} = 5$

49. $\sqrt[3]{8-27} = \sqrt[3]{-19} \approx -2.668$

51. $\sqrt[3]{27/8} = \dfrac{\sqrt[3]{27}}{\sqrt[3]{8}} = \dfrac{3}{2}$

53. $\sqrt{(-2)^2} = \sqrt{4} = 2$

55. $\sqrt{\dfrac{1}{4}(1+15)} = \sqrt{\dfrac{16}{4}} = \dfrac{\sqrt{16}}{\sqrt{4}} = \dfrac{4}{2} = 2$

57. $\sqrt{a^2 b^2} = \sqrt{a^2}\sqrt{b^2} = ab$

59. $\sqrt{(x+9)^2} = x+9$

$(x+9 > 0$ because x is positive.$)$

61. $\sqrt[3]{x^3(a^3+b^3)} = \sqrt[3]{x^3}\,\sqrt[3]{(a^3+b^3)}$

$= x\,\sqrt[3]{(a^3+b^3)}\quad$ *(Not $x(a+b)$)*

63. $\sqrt{\dfrac{4xy^3}{x^2 y}} = \sqrt{\dfrac{4y^2}{x}} = \dfrac{\sqrt{4}\sqrt{y^2}}{\sqrt{x}} = \dfrac{2y}{\sqrt{x}}$

65. $\sqrt{3} = 3^{1/2}$

67. $\sqrt{x^3} = x^{3/2}$

69. $\sqrt[3]{xy^2} = (xy^2)^{1/3}$

71. $\dfrac{x^2}{\sqrt{x}} = \dfrac{x^2}{x^{1/2}} = x^{2-1/2} = x^{3/2}$

73. $\dfrac{3}{5x^2} = \dfrac{3}{5}x^{-2}$

75. $\dfrac{3x^{-1.2}}{2} - \dfrac{1}{3x^{2.1}} = \dfrac{3}{2}x^{-1.2} - \dfrac{1}{3}x^{-2.1}$

77. $\dfrac{2x}{3} - \dfrac{x^{0.1}}{2} + \dfrac{4}{3x^{1.1}} = \dfrac{2}{3}x - \dfrac{1}{2}x^{0.1} + \dfrac{4}{3}x^{-1.1}$

79. $\dfrac{3\sqrt{x}}{4} - \dfrac{5}{3\sqrt{x}} + \dfrac{4}{3x\sqrt{x}}$

$= \dfrac{3x^{1/2}}{4} - \dfrac{5}{3x^{1/2}} + \dfrac{4}{3x \cdot x^{1/2}}$

$= \dfrac{3}{4}x^{1/2} - \dfrac{5}{3}x^{-1/2} + \dfrac{4}{3}x^{-3/2}$

81. $\dfrac{3\sqrt[5]{x^2}}{4} - \dfrac{7}{2\sqrt{x^3}} = \dfrac{3x^{2/5}}{4} - \dfrac{7}{2x^{3/2}}$

$= \dfrac{3}{4}x^{2/5} - \dfrac{7}{2}x^{-3/2}$

83. $\dfrac{1}{(x^2+1)^3} - \dfrac{3}{4\sqrt[3]{x^2+1}}$

$= \dfrac{1}{(x^2+1)^3} - \dfrac{3}{(x^2+1)^{1/3}}$

$= (x^2+1)^{-3} - \dfrac{3}{4}(x^2+1)^{-1/3}$

85. $2^{2/3} = \sqrt[3]{2^2}$

87. $x^{4/3} = \sqrt[3]{x^4}$

89. $\left(x^{1/2}y^{1/3}\right)^{1/5} = \sqrt[5]{\sqrt{x}\,\sqrt[3]{y}}$

91. $-\dfrac{3}{2}x^{-1/4} = -\dfrac{3}{2x^{1/4}} = -\dfrac{3}{2\sqrt[4]{x}}$

93. $0.2x^{-2/3} + \dfrac{3}{7x^{-1/2}} = \dfrac{0.2}{x^{2/3}} + \dfrac{3x^{1/2}}{7}$

$\qquad = \dfrac{0.2}{\sqrt[3]{x^2}} + \dfrac{3\sqrt{x}}{7}$

95. $\dfrac{3}{4(1-x)^{5/2}} = \dfrac{3}{4\sqrt{(1-x)^5}}$

97. $4^{-1/2}4^{7/2} = 4^{-1/2+7/2} = 4^{6/2} = 4^3 = 64$

99. $3^{2/3}3^{-1/6} = 3^{2/3-1/6} = 3^{1/2} = \sqrt{3}$

101. $\dfrac{x^{3/2}}{x^{5/2}} = x^{3/2-5/2} = x^{-1} = \dfrac{1}{x}$

103. $\dfrac{x^{1/2}y^2}{x^{-1/2}y} = x^{1/2-(-1/2)}y^{2-1} = xy$

105. $\left(\dfrac{x}{y}\right)^{1/3}\left(\dfrac{y}{x}\right)^{2/3} = \dfrac{x^{1/3}}{y^{1/3}} \cdot \dfrac{y^{2/3}}{x^{2/3}}$

$\qquad = x^{1/3-2/3}y^{-1/3+2/3} = x^{-1/3}y^{1/3} \text{ or } \left(\dfrac{y}{x}\right)^{1/3}$

107. $x^2 - 16 = 0 \Rightarrow x^2 = 16$

$\qquad \Rightarrow x = \pm\sqrt{16} \Rightarrow x = \pm 4$

109. $x^2 - \dfrac{4}{9} = 0 \Rightarrow x^2 = \dfrac{4}{9}$

$\qquad \Rightarrow x = \pm\sqrt{\dfrac{4}{9}} \Rightarrow x = \pm\dfrac{2}{3}$

111. $x^2 - (1+2x)^2 = 0 \Rightarrow x^2 = (1+2x)^2$

$\qquad \Rightarrow x = \pm 1 + 2x$

If $x = 1 + 2x$, then $-x = 1 \Rightarrow x = -1$.

If $x = -(1+2x)$, then $3x = -1 \Rightarrow x = -\dfrac{1}{3}$.

So, $x = 1 \text{ or } -\dfrac{1}{3}$.

113. $x^5 + 32 = 0 \Rightarrow x^5 = -32$

$\qquad \Rightarrow x = \sqrt[5]{-32} = -2$

115. $x^{1/2} - 4 = 0 \Rightarrow x^{1/2} = 4$

$\qquad \Rightarrow x = 4^2 = 16$

117. $1 - \dfrac{1}{x^2} = 0 \Rightarrow 1 = \dfrac{1}{x^2}$

$\qquad \Rightarrow x^2 = 1 \Rightarrow x = \pm\sqrt{1} = \pm 1$

119. $(x-4)^{-1/3} = 2 \Rightarrow x - 4 = 2^{-3} = \dfrac{1}{8}$

$\qquad \Rightarrow x = 4 + \dfrac{1}{8} = \dfrac{33}{8}$

Section 0.3

1. $x(4x + 6) = 4x^2 + 6x$

3. $(2x - y)y = 2xy - y^2$

5. $(x + 1)(x - 3) = x^2 + x - 3x - 3$
$\quad = x^2 - 2x - 3$

7. $(2y + 3)(y + 5) = 2y^2 + 3y + 10y + 15$
$\quad = 2y^2 + 13y + 15$

9. $(2x - 3)^2 = 4x^2 - 12x + 9$

11. $\left(x + \dfrac{1}{x}\right)^2 = x^2 + 2 + \dfrac{1}{x^2}$

13. $(2x - 3)(2x + 3) = (2x)^2 - 3^2 = 4x^2 - 9$

15. $\left(y - \dfrac{1}{y}\right)\left(y + \dfrac{1}{y}\right) = y^2 - \left(\dfrac{1}{y}\right)^2 = y^2 - \dfrac{1}{y^2}$

17. $(x^2 + x - 1)(2x + 4)$
$\quad = x^2(2x + 4) + x(2x + 4) - 1(2x + 4)$
$\quad = 2x^3 + 4x^2 + 2x^2 + 4x - 2x - 4$
$\quad = 2x^3 + 6x^2 + 2x - 4$

19. $(x^2 - 2x + 1)^2 = (x^2 - 2x + 1)(x^2 - 2x + 1)$
$\quad = x^2(x^2 - 2x + 1) - 2x(x^2 - 2x + 1) + 1(x^2 - 2x + 1)$
$\quad = x^4 - 2x^3 + x^2 - 2x^3 + 4x^2 - 2x + x^2 - 2x + 1$
$\quad = x^4 - 4x^3 + 6x^2 - 4x + 1$

21. $(y^3 + 2y^2 + y)(y^2 + 2y - 1)$
$\quad = y^3(y^2 + 2y - 1) + 2y^2(y^2 + 2y - 1) + y(y^2 + 2y - 1)$
$\quad = y^5 + 2y^4 - y^3 + 2y^4 + 4y^3 - 2y^2 + y^3 + 2y^2 - y$
$\quad = y^5 + 4y^4 + 4y^3 - y$

23. $(x + 1)(x + 2) + (x + 1)(x + 3)$
$\quad = (x + 1)(x + 2 + x + 3)$
$\quad = (x + 1)(2x + 5)$

25. $(x^2 + 1)^5(x + 3)^4 + (x^2 + 1)^6(x + 3)^3$
$\quad = (x^2 + 1)^5(x + 3)^3(x + 3 + x^2 + 1)$
$\quad = (x^2 + 1)^5(x + 3)^3(x^2 + x + 4)$

27. $(x^3 + 1)\sqrt{x + 1} - (x^3 + 1)^2\sqrt{x + 1}$
$\quad = (x^3 + 1)\sqrt{x + 1} \cdot [1 - (x^3 + 1)]$
$\quad = -x^3(x^3 + 1)\sqrt{x + 1}$

29. $\sqrt{(x + 1)^3} + \sqrt{(x + 1)^5}$
$\quad = \sqrt{(x + 1)^3} \cdot [1 + \sqrt{(x + 1)^2}]$
$\quad = \sqrt{(x + 1)^3} \cdot (1 + x + 1)$
$\quad = (x + 2)\sqrt{(x + 1)^3}$

31. a. $2x + 3x^2 = x(2 + 3x)$

b. $x(2 + 3x) = 0$

    $x = 0$ or $2 + 3x = 0$

    $x = 0$ or $-2/3$

33. a. $6x^3 - 2x^2 = 2x^2(3x - 1)$

b. $2x^2(3x - 1) = 0$

    $x^2 = 0$ or $3x - 1 = 0$

    $x = 0$ or $1/3$

35. a. $x^2 - 8x + 7 = (x - 1)(x - 7)$

b. $(x - 1)(x - 7) = 0$

    $x - 1 = 0$ or $x - 7 = 0$

    $x = 1$ or 7

37. a. $x^2 + x - 12 = (x - 3)(x + 4)$

b. $(x - 3)(x + 4) = 0$

    $x - 3 = 0$ or $x + 4 = 0$

    $x = 3$ or -4

39. a. $2x^2 - 3x - 2 = (2x + 1)(x - 2)$

b. $(2x + 1)(x - 2) = 0$

    $2x + 1 = 0$ or $x - 2 = 0$

    $x = -1/2$ or 2

41. a. $6x^2 + 13x + 6 = (2x + 3)(3x + 2)$

b. $(2x + 3)(3x + 2) = 0$

    $2x + 3 = 0$ or $3x + 2 = 0$

    $x = -3/2$ or $-2/3$

43. a. $12x^2 + x - 6 = (3x - 2)(4x + 3)$

b. $(3x - 2)(4x + 3) = 0$

    $3x - 2 = 0$ or $4x + 3 = 0$

    $x = 2/3$ or $-3/4$

45. a. $x^2 + 4xy + 4y^2 = (x + 2y)^2$

b. $(x + 2y)^2 = 0$

    $x + 2y = 0$

    $x = -2y$

47. a. $x^4 - 5x^2 + 4 = (x^2 - 1)(x^2 - 4)$

    $= (x + 1)(x - 1)(x + 2)(x - 2)$

b. $(x + 1)(x - 1)(x + 2)(x - 2) = 0$

    $x + 1 = 0$ or $x - 1 = 0$ or

        $x + 2 = 0$ or $x - 2 = 0$

    $x = \pm 1$ or ± 2

Section 0.4

1. $\dfrac{x-4}{x+1} \cdot \dfrac{2x+1}{x-1} = \dfrac{(x-4)(2x+1)}{(x+1)(x-1)}$

   $= \dfrac{2x^2 - 7x - 4}{x^2 - 1}$

3. $\dfrac{x-4}{x+1} + \dfrac{2x+1}{x-1} = \dfrac{(x-4)(x-1)+(x+1)(2x+1)}{(x+1)(x-1)}$

   $= \dfrac{3x^2 - 2x + 5}{x^2 - 1}$

5. $\dfrac{x^2}{x+1} - \dfrac{x-1}{x+1} = \dfrac{x^2 - (x-1)}{x+1}$

   $= \dfrac{x^2 - x + 1}{x+1}$

7. $\dfrac{1}{\left(\dfrac{x}{x-1}\right)} + x - 1 = \dfrac{x-1}{x} + x - 1$

   $= \dfrac{x - 1 + x(x-1)}{x} = \dfrac{x^2 - 1}{x}$

9. $\dfrac{1}{x}\left[\dfrac{x-3}{xy} + \dfrac{1}{y}\right] = \dfrac{1}{x}\left[\dfrac{x-3+x}{xy}\right]$

   $= \dfrac{2x-3}{x^2 y}$

11. $\dfrac{(x+1)^2(x+2)^3 - (x+1)^3(x+2)^2}{(x+2)^6}$

    $= \dfrac{(x+1)^2(x+2)^2[(x+2)-(x+1)]}{(x+2)^6} = \dfrac{(x+1)^2}{(x+2)^4}$

13. $\dfrac{(x^2-1)\sqrt{x^2+1} - \dfrac{x^4}{\sqrt{x^2+1}}}{x^2+1}$

    $= \dfrac{(x^2-1)(x^2+1) - x^4}{(x^2+1)\sqrt{x^2+1}} = \dfrac{-1}{\sqrt{(x^2+1)^3}}$

15. $\dfrac{\dfrac{1}{(x+y)^2} - \dfrac{1}{x^2}}{y} = \dfrac{x^2 - (x+y)^2}{yx^2(x+y)^2}$

    $= \dfrac{x^2 - x^2 - 2xy - y^2}{yx^2(x+y)^2} = \dfrac{-y(2x+y)}{yx^2(x+y)^2} = \dfrac{-(2x+y)}{x^2(x+y)^2}$

8

Section 0.5

1. $x + 1 = 0 \Rightarrow x = 0 - 1 \Rightarrow x = -1$

3. $-x + 5 = 0 \Rightarrow -x = -5 \Rightarrow x = 5$

5. $4x - 5 = 8 \Rightarrow 4x = 13 \Rightarrow x = \dfrac{13}{4}$

7. $7x + 55 = 98 \Rightarrow 7x = 43 \Rightarrow x = -\dfrac{43}{7}$

9. $x + 1 = 2x + 2 \Rightarrow -x = 1 \Rightarrow x = -1$

11. $ax + b = c \Rightarrow ax = c - b \Rightarrow x = \dfrac{c - b}{a}$

13. $2x^2 + 7x - 4 = 0 \Rightarrow (2x - 1)(x + 4) = 0$

$\Rightarrow x = -4, \dfrac{1}{2}$

15. $x^2 - x + 1 = 0 \Rightarrow \Delta = b^2 - 4ac = -3 < 0$,
so this equation has no real solutions.

17. $2x^2 - 5 = 0 \Rightarrow x^2 = \dfrac{5}{2} \Rightarrow x = \pm\sqrt{\dfrac{5}{2}}$

19. $-x^2 - 2x - 1 = 0 \Rightarrow -(x + 1)^2 = 0$

$\Rightarrow x = -1$

21. $\dfrac{1}{2}x^2 - x - \dfrac{3}{2} = 0 \Rightarrow x^2 - 2x - 3 = 0$

$\Rightarrow (x + 1)(x - 3) = 0 \Rightarrow x = -1, 3$

23. $x^2 - x = 1 \Rightarrow x^2 - x - 1 = 0$

$\Rightarrow x = \dfrac{-b \pm \sqrt{b^2 - 4ac}}{2a} = \dfrac{1 \pm \sqrt{5}}{2}$

25. $x = 2 - \dfrac{1}{x} \Rightarrow x^2 = 2x - 1$

$\Rightarrow x^2 - 2x + 1 = 0 \Rightarrow (x - 1)^2 = 0$

$\Rightarrow x = 1$

27. $x^4 - 10x^2 + 9 = 0 \Rightarrow (x^2 - 1)(x^2 - 9) = 0$

$\Rightarrow x^2 = 1 \text{ or } x^2 = 0 \Rightarrow x = \pm 1, \pm 3$

29. $x^4 + x^2 - 1 = 0 \Rightarrow x^2 = \dfrac{-b \pm \sqrt{b^2 - 4ac}}{2a}$

$\Rightarrow x^2 = \dfrac{-1 \pm \sqrt{5}}{2}$

$\Rightarrow x = \pm\sqrt{\dfrac{-1 \pm \sqrt{5}}{2}}$

31. $x^3 + 16x^2 + 11x + 6 = 0$

$\Rightarrow (x + 1)(x + 2)(x + 3) = 0 \Rightarrow x = -1, -2, -3$

33. $x^3 + 4x^2 + 4x + 3 = 0$

$\Rightarrow (x + 3)(x^2 + x + 1) = 0 \Rightarrow x = -3$

(For $x^2 + x + 1 = 0$, $\Delta = b^2 - 4ac = -3 < 0$, so
there are no real solutions to this quadratic equation.)

35. $x^3 - 1 = 0$

$\Rightarrow x^3 = 1 \Rightarrow x = \sqrt[3]{1} = 1$

37. $y^3 + 3y^2 + 3y + 2 = 0$

$\Rightarrow (y + 2)(y^2 + y + 1) = 0 \Rightarrow y = -2$

(For $y^2 + y + 1 = 0$, $\Delta = b^2 - 4ac = -3 < 0$, so
there are no real solutions to this quadratic equation.)

39. $x^3 - x^2 - 5x + 5 = 0$

$\Rightarrow (x - 1)(x^2 - 5) = 0 \Rightarrow x = 1, \pm\sqrt{5}$

9

41. $2x^6 - x^4 - 2x^2 + 1 = 0$

$\Rightarrow (2x^2 - 1)(x^4 - 1) = 0$

$\Rightarrow (2x^2 - 1)(x^2 - 1)(x^2 + 1) = 0$

(Think of the cubic you get by substituting y for x^2)

$\Rightarrow x = \pm 1, \ \pm \dfrac{1}{\sqrt{2}}$

43. $(x^2 + 3x + 2)(x^2 - 5x + 6) = 0$

$\Rightarrow (x + 2)(x + 1)(x - 2)(x - 3) = 0$

$\Rightarrow x = -2, -1, 2, 3$

Section 0.6

1. $x^4 - 3x^3 = 0$, $x^3(x - 3) = 0$, $x = 0, 3$

3. $x^4 - 4x^2 = -4$, $x^4 - 4x^2 + 4 = 0$,

$(x^2 - 2)^2 = 0, x = \pm\sqrt{2}$

5. $(x + 1)(x + 2) + (x + 1)(x + 3) = 0$,

$(x + 1)(x + 2 + x + 3) = 0$,

$(x + 1)(2x + 5) = 0, x = -1, -5/2$

7. $(x^2 + 1)^5(x + 3)^4 + (x^2 + 1)^6(x + 3)^3 = 0$,

$(x^2 + 1)^5(x + 3)^3(x + 3 + x^2 + 1) = 0$,

$(x^2 + 1)^5(x + 3)^3(x^2 + x + 4) = 0, x = -3$

(Neither $x^2 + 1 = 0$ nor $x^2 + x + 4 = 0$ has a real solution.)

9. $(x^3 + 1)\sqrt{x + 1} - (x^3 + 1)^2\sqrt{x + 1} = 0$,

$(x^3 + 1)\sqrt{x + 1}\,[1 - (x^3 + 1)] = 0$,

$-x^3(x^3 + 1)\sqrt{x + 1} = 0, \quad x = 0, -1$

11. $\sqrt{(x + 1)^3} + \sqrt{(x + 1)^5} = 0$,

$\sqrt{(x + 1)^3}\,(1 + x + 1) = 0$,

$(x + 2)\sqrt{(x + 1)^3} = 0, \quad x = -1$

($x = -2$ is not a solution because $\sqrt{(x + 1)^3}$ is not defined for $x = -2$.)

13. $(x + 1)^2(2x + 3) - (x + 1)(2x + 3)^2 = 0$,

$(x + 1)(2x + 3)(x + 1 - 2x - 3) = 0$,

$(x + 1)(2x + 3)(-x - 2) = 0$,

$x = -2, -3/2, -1$

15. $\dfrac{(x + 1)^2(x + 2)^3 - (x + 1)^3(x + 2)^2}{(x + 2)^6} = 0$,

$\dfrac{(x + 1)^2(x + 2)^2[(x + 2) - (x + 1)]}{(x + 2)^6} = 0$,

$\dfrac{(x + 1)^2}{(x + 2)^4} = 0, \ (x + 1)^2 = 0$,

$x = -1$

17. $\dfrac{2(x^2 - 1)\sqrt{x^2 + 1} - \dfrac{x^4}{\sqrt{x^2 + 1}}}{x^2 + 1} = 0$,

$\dfrac{2(x^2 - 1)(x^2 + 1) - x^4}{(x^2 + 1)\sqrt{x^2 + 1}} = 0$,

$\dfrac{x^4 - 2}{(x^2 + 1)\sqrt{x^2 + 1}} = 0$,

$x^4 - 2 = 0, x = \pm\sqrt[4]{2}$

19. $x - \dfrac{1}{x} = 0$, $x^2 - 1 = 0, x = \pm 1$

21. $\dfrac{1}{x} - \dfrac{9}{x^3} = 0$, $x^2 - 9 = 0$, $x = \pm 3$

23. $\dfrac{x-4}{x+1} - \dfrac{x}{x-1} = 0$,

$\dfrac{(x-4)(x-1) - x(x+1)}{(x+1)(x-1)} = 0$,

$\dfrac{-6x+4}{(x+1)(x-1)} = 0$,

$-6x + 4 = 0$, $x = 2/3$

25. $\dfrac{x+4}{x+1} + \dfrac{x+4}{3x} = 0$,

$\dfrac{3x(x+4) + (x+1)(x+4)}{3x(x+1)} = 0$,

$\dfrac{(x+4)(3x + x + 1)}{3x(x+1)} = 0$,

$\dfrac{(x+4)(4x+1)}{3x(x+1)} = 0$,

$(x+4)(4x+1) = 0$, $x = -4, -1/4$

Section 0.7

1. $P(0, 2), Q(4, -2), R(-2, 3), S(-3.5, -1.5), T(-2.5, 0), U(2, 2.5)$

3.

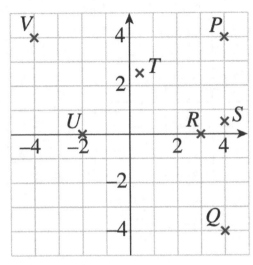

5. Solve the equation $x + y = 1$ for y to get $y = 1 - x$. Then plot some points:

Graph:

x	-2	-1	0	1	2
$y = 1 - x$	3	2	1	0	-1

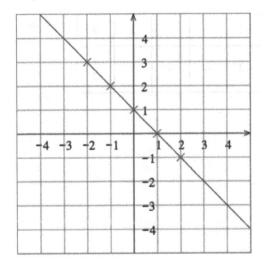

7. Solve the equation $2y - x^2 = 1$ for y to get $y = \dfrac{1 + x^2}{2}$. Then plot some points:

Graph:

x	−2	−1	0	1	2
$y = \dfrac{1 + x^2}{2}$	2.5	1	0.5	1	2.5

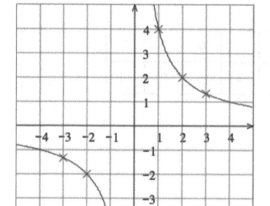

9. Solve the equation $xy = 4$ for y to get $y = \dfrac{4}{x}$. Then plot some points:

Graph:

x	−3	−2	−1	1	2	3
$y = \dfrac{4}{x}$	$-\dfrac{4}{3}$	−2	−4	4	2	$\dfrac{4}{3}$

11. Solve the equation $xy = x^2 + 1$ for y to get $y = x + \dfrac{1}{x}$. Then plot some points:

Graph:

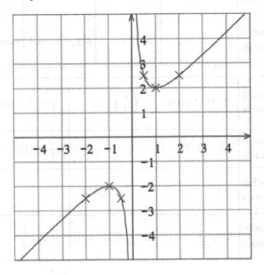

x	-2	-1	$-\dfrac{1}{2}$	$\dfrac{1}{2}$	1	2
$y = x + \dfrac{1}{x}$	$-\dfrac{5}{2}$	-2	$-\dfrac{5}{2}$	$\dfrac{5}{2}$	2	$\dfrac{5}{2}$

13. $\sqrt{(2-1)^2 + (-2+1)^2} = \sqrt{2}$

15. $\sqrt{(0-a)^2 + (b-0)^2} = \sqrt{a^2 + b^2}$

17. Set the two distances equal and solve:

$$\sqrt{(1-0)*2 + (k-0)^2} = \sqrt{(1-2)^2 + (k-1)^2}$$

$$\Rightarrow \sqrt{1 + k^2} = \sqrt{2 - 2k + k^2}$$

$$\Rightarrow 1 + k^2 = 2 - 2k + k^2$$

$$\Rightarrow 2k = 1 \Rightarrow k = \dfrac{1}{2}$$

19. Circle with center $(0, 0)$ and radius 3.

Section 0.8

1.

Exponential Form	$10^2 = 100$	$4^3 = 64$	$4^4 = 256$	$0.45^0 = 1$	$8^{1/2} = 2\sqrt{2}$	$4^{-3} = \dfrac{1}{64}$
Logarithmic Form	$\log_{10} 100 = 2$	$\log_4 64 = 3$	$\log_4 256 = 4$	$\log_{0.45} 1 = 0$	$\log_8 2\sqrt{2} = \dfrac{1}{2}$	$\log_4\left(\dfrac{1}{64}\right) = -3$

3.

Exponential Form	$0.3^2 = 0.09$	$\left(\dfrac{1}{2}\right)^0 = 1$	$10^{-3} = 0.001$
Logarithmic Form	$\log_{0.3} 0.09 = 2$	$\log_{1/2} 1 = 0$	$\log_{10} 0.001 = -3$
Exponential Form	$9^{-2} = \dfrac{1}{81}$	$2^{10} = 1{,}024$	$64^{-1/3} = \dfrac{1}{4}$
Logarithmic Form	$\log_9 \dfrac{1}{81} = -2$	$\log_2 1{,}024 = 10$	$\log_{64} \dfrac{1}{4} = -\dfrac{1}{3}$

5. $\log_4 16 =$ the power to which we need to raise 4 in order to get 16. Because $4^{\boxed{2}} = 16$, this power is 2, so $\log_4 16 = 2$.

7. $\log_5 \dfrac{1}{25} =$ the power to which we need to raise 5 in order to get $\dfrac{1}{25}$. Because $5^{\boxed{-2}} = \dfrac{1}{25}$, this power is -2, so $\log_5 \dfrac{1}{25} = -2$.

9. $\log 100{,}000 = \log_{10} 100{,}000 =$ the power to which we need to raise 10 in order to get 100,000. Because $10^{\boxed{5}} = 100{,}000$, this power is 5, so $\log 100{,}000 = 5$.

11. $\log_{16} 16 =$ the power to which we need to raise 16 in order to get 16. Because $16^{\boxed{1}} = 16$, this power is 1, so $\log_{16} 16 = 1$.

13. $\log_4 \dfrac{1}{16} =$ the power to which we need to raise 4 in order to get $\dfrac{1}{16}$. Because $4^{\boxed{-2}} = \dfrac{1}{16}$, this power is -2, so $\log_4 \dfrac{1}{16} = -2$.

15. $\log_2 \sqrt{2} =$ the power to which we need to raise 2 in order to get $\sqrt{2}$. Because $2^{\boxed{1/2}} = \sqrt{2}$, this power is $\dfrac{1}{2}$, so $\log_2 \sqrt{2} = \dfrac{1}{2}$.

17. By Identity (1), $\log_b 3 + \log_b 4 = \log_b(3 \times 4) = \log_b \boxed{12}$.

19. $\log_b 2 - \log_b 5 - \log_b 4 = \log_b 2 - (\log_b 5 + \log_b 4) = \log_b 2 - (\log_b(5 \times 4))$ (Identity 1)

$= \log_b\left(\dfrac{2}{5 \times 4}\right)$ (Identity 2) $= \log_b \boxed{\dfrac{1}{10}}$

21. $\log_b 3 - 3\log_b 2 = \log_b 3 - \log_b 2^3$ (Identity 3) $= \log_b\left(\dfrac{3}{2^3}\right)$ (Identity 2) $= \log_b\boxed{\dfrac{3}{8}}$

23. $4\log_b x + 5\log_b y = \log_b x^4 + \log_b y^5$ (Identity 3) $= \log_b\boxed{x^4 y^5}$ (Identity 1)

25. $2\log_b x + 3\log_b y - 4\log_b z = \log_b x^2 + \log_b y^3 - \log_b z^4$ (Identity 3)

$= \log_b(x^2 y^3) - \log_b z^4$ (Identity 1) $= \log_b\boxed{\dfrac{x^2 y^3}{z^4}}$ (Identity 2)

27. $x\log_b 2 - 2\log_b x = \log_b 2^x - \log_b x^2$ (Identity 3) $= \log_b\boxed{\dfrac{2^x}{x^2}}$ (Identity 2)

29. $\log 21 = \log(3 \times 7) = \log 3 + \log 7 = b + c$ (Identity 1)

31. $\log 42 = \log(2 \times 3 \times 7) = \log 2 + \log 3 + \log 7 = a + b + c$ (Identity 1)

33. $\log\left(\dfrac{1}{7}\right) = -\log 7 = -c$ (Identity 5)

35. $\log\left(\dfrac{2}{3}\right) = \log 2 - \log 3 = a - b$ (Identity 2)

37. $\log\left(\dfrac{4}{7}\right) = \log 4 - \log 7$ (Identity 2) $= \log 2 + \log 2 - \log 7$ (Identity 1) $= 2a - c$

39. $\log 16 = \log 2^4 = 4\log 2 = 4a$ (Identity 3)

41. $\log 0.03 = \log\left(\dfrac{3}{10^2}\right) = \log 3 - \log 10^2 = \log 3 - 2\log 10 = b - 2$

43. $\log 5 = \log\left(\dfrac{10}{2}\right) = \log 10 - \log 2 = 1 - a$ (Identity 2)

45. $\log\sqrt{7} = \log(7^{1/2}) = \dfrac{1}{2}\log 7 = \dfrac{c}{2}$

47. $4 = 2^x$ is exactly the equation we solve in order to calculate $\log_2 4$; answer: 2. alternatively, write the equation $4 = 2^x$ in logarithmic form to get $x = \log_2 4 = 2$.

49. Take the base 3 logarithm of both sides of the given equation $27 = 3^{2x-1}$ to get

$\log_3 27 = \log_3 3^{2x-1}$

$3 = (2x - 1)\log_3 3$ By Identity 3

$3 = 2x - 1$ By Identity 4

$x = \dfrac{3 + 1}{2} = 2$ Solve for x.

51. Take the base 5 logarithm of both sides of the given equation $5^{-x+1} = \dfrac{1}{125}$ to get

$\log_5 5^{-x+1} = \log_5\left(\dfrac{1}{125}\right)$

$(-x + 1)\log_5 5 = -\log_5 125$ By Identities 3 and 5

$-x + 1 = -3$ By Identity 4

$x = 1 + 3 = 4$ Solve for x.

53. First divide both sides of the given equation by 50 to get $2.4 = 2^{3t}$. Take the common logarithm of both sides:

$\log 2.4 = \log(2^{3t})$

$\log 2.4 = 3t \log 2$ By Identity 3

$3t = \dfrac{\log 2.4}{\log 2}$ Divide.

$t = \dfrac{\log 2.4}{3\log 2} \approx 0.4210$ Solve for t.

If instead you take the base-2 logarithm, the answer is represented as $\dfrac{\log_2(2.4)}{3}$.

55. First divide both sides of the given equation by 300 to get $\dfrac{10}{3} = 1.3^{4t-1}$. Take the common logarithm of both sides:

$\log\left(\dfrac{10}{3}\right) = \log(1.3^{4t-1})$

$\log 10 - \log 3 = (4t - 1)\log 1.3$ By Identities 2 and 3

$4t - 1 = \dfrac{\log 10 - \log 3}{\log 1.3}$ Divide.

$t = \dfrac{1}{4}\left(\dfrac{\log 10 - \log 3}{\log 1.3} + 1\right) \approx 1.3972$ Solve for t.

Section 1.1

1. Using the table: **a.** $f(0) = 2$ **b.** $f(2) = -0.5$

3. Using the table: **a.** $f(2) - f(-2) = -0.5 - 2 = -2.5$ **b.** $f(-1)f(-2) = (4)(2) = 8$
    c. $-2f(-1) = -2(4) = -8$

5. From the graph, we estimate: **a.** $f(1) = 20$ **b.** $f(2) = 30$

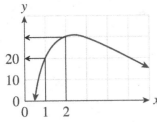

In a similar way, we find: **c.** $f(3) = 30$ **d.** $f(5) = 20$ **e.** $f(3) - f(2) = 30 - 30 = 0$ **f.**
$f(3 - 2) = f(1) = 20$

7. From the graph, we estimate: **a.** $f(-1) = 0$ **b.** $f(1) = -3$ since the solid dot is on $(1, -3)$.

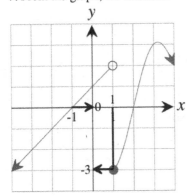

In a similar way, we estimate **c.** $f(3) = 3$ **d.** Since $f(3) = 3$ and $f(1) = -3$, $\dfrac{f(3) - f(1)}{3 - 1} = \dfrac{3 - (-3)}{3 - 1} = 3.$

9. $f(x) = x - \dfrac{1}{x^2}$, with its natural domain.

The natural domain consists of all x for which $f(x)$ makes sense: all real numbers other than 0.

a. Since 4 is in the natural domain, $f(4)$ is defined, and $f(4) = 4 - \dfrac{1}{4^2} = 4 - \dfrac{1}{16} = \dfrac{63}{16}$.

b. Since 0 is not in the natural domain, $f(0)$ is not defined.

c. Since -1 is in the natural domain, $f(-1) = -1 - \dfrac{1}{(-1)^2} = -1 - \dfrac{1}{1} = -2.$

11. $f(x) = \sqrt{x + 10}$, with domain $[-10, 0)$

a. Since 0 is not in $[-10, 0)$, $f(0)$ is not defined.   **b.** Since 9 is not in $[-10, 0)$, $f(9)$ is not defined.

c. Since -10 is in $[-10, 0)$, $f(-10)$ is defined, and $f(-10) = \sqrt{-10 + 10} = \sqrt{0} = 0$

13. $f(x) = 4x - 3$

a. $f(-1) = 4(-1) - 3 = -4 - 3 = -7$ **b.** $f(0) = 4(0) - 3 = 0 - 3 = -3$

c. $f(1) = 4(1) - 3 = 4 - 3 = 1$ **d.** Substitute y for x to obtain $f(y) = 4y - 3$

e. Substitute $(a + b)$ for x to obtain $f(a + b) = 4(a + b) - 3$.

15. $f(x) = x^2 + 2x + 3$

a. $f(0) = (0)^2 + 2(0) + 3 = 0 + 0 + 3 = 3$ **b.** $f(1) = 1^2 + 2(1) + 3 = 1 + 2 + 3 = 6$

c. $f(-1) = (-1)^2 + 2(-1) + 3 = 1 - 2 + 3 = 2$ **d.** $f(-3) = (-3)^2 + 2(-3) + 3 = 9 - 6 + 3 = 6$

e. Substitute a for x to obtain $f(a) = a^2 + 2a + 3$.   **f.** Substitute $(x + h)$ for x to obtain

$f(x + h) = (x + h)^2 + 2(x + h) + 3$.

17. $g(s) = s^2 + \dfrac{1}{s}$

a. $g(1) = 1^2 + \dfrac{1}{1} = 1 + 1 = 2$ **b.** $g(-1) = (-1)^2 + \dfrac{1}{-1} = 1 - 1 = 0$

c. $g(4) = 4^2 + \dfrac{1}{4} = 16 + \dfrac{1}{4} = \dfrac{65}{4}$ or 16.25 **d.** Substitute x for s to obtain $g(x) = x^2 + \dfrac{1}{x}$

e. Substitute $(s + h)$ for s to obtain $g(s + h) = (s + h)^2 + \dfrac{1}{s + h}$

f. $g(s + h) - g(s) = $ Answer to part (e) $-$ Original function $= \left((s + h)^2 + \dfrac{1}{s + h} \right) - \left(s^2 + \dfrac{1}{s} \right)$

19. $f(x) = -x^3$ (domain $(-\infty, +\infty)$)
Technology formula: $-(x^3)$

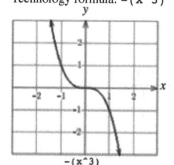

21. $f(x) = x^4$ (domain $(-\infty, +\infty)$)
Technology formula: x^4

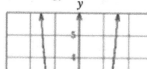
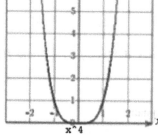

23. $f(x) = \dfrac{1}{x^2}$ $(x \neq 0)$

Technology formula: $1/(x^2)$

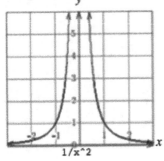

1/x^2

25. a. $f(x) = x$ $(-1 \leq x \leq 1)$

Since the graph of $f(x) = x$ is a diagonal 45° line through the origin inclining up from left to right, the correct graph is (A).

b. $f(x) = -x$ $(-1 \leq x \leq 1)$

Since the graph of $f(x) = -x$ is a diagonal 45° line through the origin inclining down from left to right, the correct graph is (D).

c. $f(x) = \sqrt{x}$ $(0 < x < 4)$

Since the graph of $f(x) = \sqrt{x}$ is the top half of a sideways parabola, the correct graph is (E).

d. $f(x) = x + \dfrac{1}{x} - 2$ $(0 < x < 4)$

If we plot a few points like $x = 1/2,\ 1, 2,$ and 3, we find that the correct graph is (F).

e. $f(x) = |x|$ $(-1 \leq x \leq 1)$

Since the graph of $f(x) = |x|$ is a "V"-shape with its vertex at the origin, the correct graph is (C).

f. $f(x) = x - 1$ $(-1 \leq x \leq 1)$

Since the graph of $f(x) = x - 1$ is a straight line through $(0, -1)$ and $(1, 0)$, the correct graph is (B).

27. Technology formula: $0.1*x^2 - 4*x+5$
Table of values:

x	0	1	2	3	4	5	6	7	8	9	10
$f(x)$	5	1.1	−2.6	−6.1	−9.4	−12.5	−15.4	−18.1	−20.6	−22.9	−25

29. Technology formula: $(x^2-1)/(x^2+1)$
Table of values:

x	0.5	1.5	2.5	3.5	4.5	5.5	6.5	7.5	8.5	9.5	10.5
$h(x)$	−0.6000	0.3846	0.7241	0.8491	0.9059	0.9360	0.9538	0.9651	0.9727	0.9781	0.9820

31. $f(x) = \begin{cases} x & \text{if } -4 \leq x < 0 \\ 2 & \text{if } 0 \leq x \leq 4 \end{cases}$

Technology formula: $x*(x<0)+2*(x>=0)$ (For a graphing calculator, use $\geq$ instead of >=.)

21

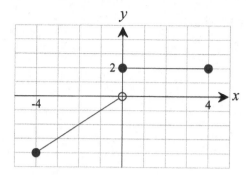

a. $f(-1) = -1$. We used the first formula, since -1 is in $[-4, 0)$.

b. $f(0) = 2$. We used the second formula, since 0 is in $[0, 4]$.

c. $f(1) = 2$. We used the second formula, since 1 is in $[0, 4]$.

33. $f(x) = \begin{cases} x^2 & \text{if } -2 < x \le 0 \\ 1/x & \text{if } 0 < x \le 4 \end{cases}$

Technology formula: $(x^2)*(x<=0)+(1/x)*(0<x)$ (For a graphing calculator, use $\le$ instead of $<=$.)

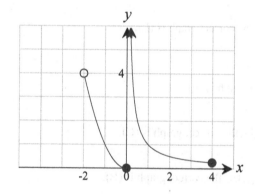

a. $f(-1) = 1^2 = 1$. We used the first formula, since -1 is in $(-2, 0]$.

b. $f(0) = 0^2 = 0$. We used the first formula, since 0 is in $(-2, 0]$.

c. $f(1) = 1/1 = 1$. We used the second formula, since 1 is in $(0, 4]$.

35. $f(x) = \begin{cases} x & \text{if } -1 < x \le 0 \\ x+1 & \text{if } 0 < x \le 2 \\ x & \text{if } 2 < x \le 4 \end{cases}$

Technology formula: $x*(x<=0)+(x+1)*(0<x)*(x<=2)+x*(2<x)$ (For a graphing calculator, use $\le$ instead of $<=$.)

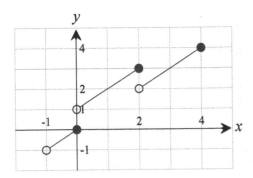

a. $f(0) = 0$. We used the first formula, since 0 is in $(-1, 0]$.

b. $f(1) = 1 + 1 = 2$. We used the second formula, since 1 is in $(0, 2]$.

c. $f(2) = 2 + 1 = 3$. We used the second formula, since 2 is in $(0, 2]$.

d. $f(3) = 3$. We used the third formula, since 3 is in $(2, 4]$.

37. $f(x) = x^2$

a. $f(x + h) = (x + h)^2$ Therefore,

$$
\begin{aligned}
f(x + h) - f(x) \quad &= (x + h)^2 - x^2 \\
&= x^2 + 2xh + h^2 - x^2 \\
&= 2xh + h^2 = h(2x + h)
\end{aligned}
$$

b. Using the answer to part (a),

$$\frac{f(x + h) - f(x)}{h} = \frac{h(2x + h)}{h} = 2x + h$$

39. $f(x) = 2 - x^2$

a. $f(x + h) = 2 - (x + h)^2$ Therefore,

$$
\begin{aligned}
f(x + h) - f(x) \quad &= 2 - (x + h)^2 - (2 - x^2) \\
&= 2 - x^2 - 2xh - h^2 - 2 + x^2 \\
&= -2xh - h^2 = -h(2x + h)
\end{aligned}
$$

b. Using the answer to part (a),

$$\frac{f(x + h) - f(x)}{h} = \frac{-h(2x + h)}{h} = -(2x + h)$$

Applications

41. From the table,

a. $p(2) = 2.95$; Pemex produced 2.95 million barrels of crude oil per day in 2010 $(t = 2)$.

$p(3) = 2.94$; Pemex produced 2.94 million barrels of crude oil per day in 2011 $(t = 3)$.

$p(6) = 2.79$; Pemex produced 2.79 million barrels of crude oil per day in 2014 $(t = 6)$.

b. $p(4) - p(2) = 2.91 - 2.95 = -0.04$; Crude oil production by Pemex decreased by 0.04 million barrels/day from 2010 $(t = 2)$ to 2012 $(t = 4)$.

43. a. Graph of p (below left):

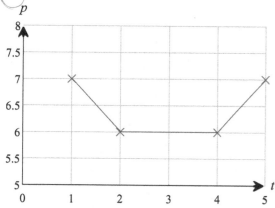 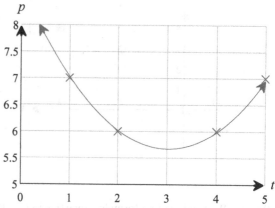

From the graph, $p(4.5) \approx 6.5$. Interpretation: $t = 4.5$ represents 4.5 years since the start of 2008, or midway through 2012.
Thus, we interpret the answer as follows: The popularity of Twitter midway through 2012 was about 6.5%.
b. The four points suggest a "u"-shaped curve such as a parabola, and only Choice (D) is of this type. (Choice (B) gives an "upside-down" (concave down) parabola.)

45. From the graph, $f(7) \approx 1,000$. Because f is the number of thousands of housing starts in year t, we interpret the result as follows: Approximately 1,000,000 homes were started in 2007.

Similarly, $f(14) \approx 600$: Approximately 600,000 homes were started in 2014.

Also, we estimate $f(9.5) \approx 450$. Because $t = 9.5$ is midway between 2009 and 2010, we interpret the result as follows: 450,000 homes were started in the year beginning July 2009.

47. $f(7 - 3) = f(4) \approx 1,600$ Interpretation: 1,600,000 homes were started in 2004 ($t = 4$).

$f(7) - f(3) = 1,000 - 1,500 = -500$
Interpretation:

$f(7) - f(3)$ is the change in the number of housing starts (in thousands) from 2003 to 2007; there were 500,000 fewer housing starts in 2007 than in 2003.

49. $f(t + 5) - f(t)$ measures the change from year t to the year five years later. It is greatest when the the line segment from year t to year $t + 5$ is steepest upward-sloping. From the graph, this occurs when $t = 0$, for a change of $1,700 - 1,200 = 500$.
Interpretation: The greatest five-year increase in the number of housing starts occurred in 2000–2005.

51. a. From the graph, $n(2) \approx 400$, $n(4) \approx 400$, $n(4.5) \approx 350$. Because $n(t)$ is Abercrombie's net income in the year ending $t + 2004$, we interpret the results as follows:
Abercrombie's net income was $400 million in 2006.
Abercrombie's net income was $400 million in 2008.
Abercrombie's net income was $350 million in the year ending June 2009 (because $t = 4.5$ represents June 2009).

b. Increasing most rapidly at $t \approx 8$ (over the interval $[3, 8]$ the graph is steepest upward-sloping at around $t = 8$.)
Interpretation: Between Dec. 2007 and Dec. 2012 Abercrombie's net income was increasing most rapidly in December 2012.

c. Decreasing most rapidly at $t \approx 5$ (over the interval $[3, 8]$ the graph is steepest downward-sloping at around $t = 5$.)
Interpretation: Between Dec. 2007 and Dec. 2012, Abercrombie's net income was decreasing most rapidly in Dec. 2009.

53. a. The model is valid for the range 1958 ($t = 0$) through 1966 ($t = 8$). Thus, an appropriate domain is $[0, 8]$. $t \geq 0$ is not an appropriate domain because it would predict federal funding of NASA beyond 1966, whereas the model is based only on data up to 1966.

b. $p(t) = \dfrac{4.5}{1.07^{(t-8)^2}}$

$\Rightarrow \qquad p(5) = \dfrac{4.5}{1.07^{(5-8)^2}} \approx 2.4 \qquad$ Technology formula: `4.5/(1.07^((t-8)^2))`

$t = 5$ represents $1958 + 5 = 1963$, and therefore we interpret the result as follows: In 1963, 2.4% of the U.S. federal budget was allocated to NASA.

c. $p(t)$ is increasing most rapidly when the graph is steepest upward-sloping from left to right, and, among the given values of t, this occurs when $t = 5$. Thus, the percentage of the budget allocated to NASA was increasing most rapidly in 1963.

55. $p(t) = 100\left(1 - \dfrac{12,200}{t^{4.48}}\right) \qquad (t \ge 8.5)$ **a.** Technology formula: `100*(1-12200/t^4.48)`

b. Graph:

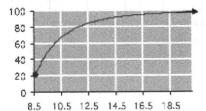

c. Table of values:

t	9	10	11	12	13	14	15	16	17	18	19	20
$p(t)$	35.2	59.6	73.6	82.2	87.5	91.1	93.4	95.1	96.3	97.1	97.7	98.2

d. From the table, $p(12) = 82.2$, so that 82.2% of children are able to speak in at least single words by the age of 12 months. **e.** We seek the first value of t such that $p(t)$ is at least 90. Since $t = 14$ has this property ($p(14) = 91.1$) we conclude that, at 14 months, 90% or more children are able to speak in at least single words.

57. $v(t) = \begin{cases} 8(1.22)^t & \text{if } 0 \le t < 16 \\ 400t - 6{,}200 & \text{if } 16 \le t < 25 \\ 3800 & \text{if } 25 \le t \le 30. \end{cases}$

a. $v(10) = 8(1.22)^{10} \approx 58$. We used the first formula, since 10 is in $[0, 16)$.

$v(16) = 400(16) - 6{,}200 = 200$. We used the second formula, since 16 is in $[16, 25)$.

$v(28) = 3{,}800$. We used the third formula, since 28 is in $[25, 30]$.

Interpretation: Processor speeds were about 58 MHz in 1990, 200 MHz in 1996, and 3800 MHz in 2008.

b. Technology formula (using x as the independent variable):

`(8*(1.22)^x)*(x<16)+(400*x-6200)*(x>=16)*(x<25)+3800*(x>=25)`

(For a graphing calculator, use ≤ instead of <=.)

c. Using the above technology formula (for instance, on the Function Evaluator and Grapher on the Web site) we obtain the graph and table of values. Graph:

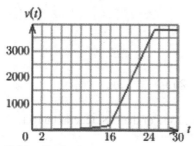

Table of values:

t	0	2	4	6	8	10	12	14	16	18	20	22	24	26	28	30
$v(t)$	8	12	18	26	39	58	87	129	200	1,000	1,800	2,600	3,400	3,800	3,800	3,800

d. From either the graph or the table, we see that the speed reached 3,000 MHz around $t = 23$. We can obtain a more precise answer algebraically by using the formula for the corresponding portion of the graph:

$$3,000 = 400t - 6,200$$

giving

$$t = \frac{9,200}{400} = 23$$

Since t is time since 1980, $t = 23$ corresponds to 2003.

59. a. Each row of the table gives us a formula with a condition:
First row in words: 10% of the amount over \$0 if your income is over \$0 and not over \$9,225.
Translation to formula:

$$0.10x \quad \text{if } 0 < x \le 9,225.$$

Second row in words: \$922.50 + 15% of the amount over \$9,225 if your income is over \$9,225 and not over \$37,450.
Translation to formula:

$$922.50 + 0.15(x - 9,225) \quad \text{if } 9,225 < x \le 37,450.$$

Continuing in this way leads to the following piecewise-defined function:

$$T(x) = \begin{cases} 0.10x & \text{if } 0 < x \le 9,225 \\ 922.50 + 0.15(x - 9,225) & \text{if } 9,225 < x \le 37,450 \\ 5,156.25 + 0.25(x - 37,450) & \text{if } 37,450 < x \le 90,750 \\ 18,481.25 + 0.28(x - 90,750) & \text{if } 90,750 < x \le 189,300 \\ 46,075.25 + 0.33(x - 189,300) & \text{if } 189,300 < x \le 411,500 \\ 119,401.25 + 0.35(x - 411,500) & \text{if } 411,500 < x \le 413,200 \\ 119,996.25 + 0.396(x - 413,200) & \text{if } 413,200 < x \end{cases}$$

b. A taxable income of \$45,000 falls in the bracket $37,450 < x \le 90,750$ and so we use the formula

$5,156.25 + 0.25(x - 37,450)$:

$$5,156.25 + 0.25(45,000 - 37,450) = 5,156.25 + 0.25(7,550) = \$7,043.75.$$

Communication and reasoning exercises

61. The dependent variable is a function of the independent variable. Here, the market price of gold m is a function of time *t*. Thus, the independent variable is *t* and the dependent variable is *m*.

63. To obtain the function notation, write the dependent variable as a function of the independent variable. Thus $y = 4x^2 - 2$ can be written as

$$f(x) = 4x^2 - 2 \text{ or } y(x) = 4x^2 - 2$$

65. False. A graph usually gives infinitely many values of the function while a numerical table will give only a finite number of values.

67. False. In a numerically specified function, only certain values of the function are specified so we cannot know its value on every real number in [0, 10], whereas an algebraically specified function would give values for every real number in [0, 10].

69. Functions with infinitely many points in their domain (such as $f(x) = x^2$) cannot be specified numerically. So, the assertion is false.

71. As the text reminds us: to evaluate *f* of a quantity (such as $x + h$) replace *x* everywhere by the *whole quantity* $x + h$:

$$f(x) \quad = x^2 - 1$$
$$f(x + h) = (x + h)^2 - 1.$$

73. If two functions are specified by the same formula $f(x)$, say, their graphs must follow the same curve $y = f(x)$. However, it is the domain of the function that specifies what portion of the curve appears on the graph. Thus, if the functions have different domains, their graphs will be different portions of the curve $y = f(x)$.

75. Suppose we already have the graph of *f* and want to construct the graph of g. We can plot a point of the graph of g as follows: Choose a value for *x* ($x = 7$, say) and then "look back" 5 units to read off $f(x - 5)$ ($f(2)$ in this instance). This value gives the *y*-coordinate we want. In other words, points on the graph of g are obtained by "looking back 5 units" to the graph of *f* and then copying that portion of the curve. Put another way, the graph of g is the same as the graph of *f*, but shifted 5 units to the right:

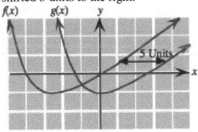

Section 1.2

1. $f(x) = x^2 + 1$ with domain $(-\infty, +\infty)$

$g(x) = x - 1$ with domain $(-\infty, +\infty)$

a. $s(x) = f(x) + g(x) = (x^2 + 1) + (x - 1) = x^2 + x$

b. Since both functions are defined for every real number x, the domain of s is the set of all real numbers: $(-\infty, +\infty)$.

c. $s(-3) = (-3)^2 + (-3) = 9 - 3 = 6$

3. $g(x) = x - 1$ with domain $(-\infty, +\infty)$

$u(x) = \sqrt{x + 10}$ with domain $[-10, 0)$

a. $p(x) = g(x)u(x) = (x - 1)\sqrt{x + 10}$

b. The domain of p consists of all real numbers x simultaneously in the domains of g and u; that is, $[-10, 0)$.

c. $p(-6) = (-6 - 1)\sqrt{-6 + 10} = (-7)(2) = -14$

5. $g(x) = x - 1$ with domain $(-\infty, +\infty)$

$v(x) = \sqrt{10 - x}$ with domain $[0, 10]$

a. $q(x) = \dfrac{v(x)}{g(x)} = \dfrac{\sqrt{10 - x}}{x - 1}$

b. The domain of q consists of all real numbers x simultaneously in the domains of v and g such that $g(x) \neq 0$. Since

$g(x) = 0$ when $x - 1 = 0$, or $x = 1$

we exclude $x = 1$ from the domain of the quotient. Thus, the domain consists of all x in $[0, 10]$ excluding $x = 1$ (since $g(1) = 0$), or $0 \leq x \leq 10$; $x \neq 1$.

c. As 1 is not in the domain of q, $q(1)$ is not defined.

7. $f(x) = x^2 + 1$ with domain $(-\infty, +\infty)$

a. $m(x) = 5f(x) = 5(x^2 + 1)$

b. The domain of m is the same as the domain of f: $(-\infty, +\infty)$.

c. $m(1) = 5f(1) = 5(1^2 + 1) = 10$

Applications

9. Number of music files = Starting number + New files = $200 + 10 \times$ Number of days

So, $N(t) = 200 + 10t$ (N = number of music files, t = time in days)

11. Take y to be the width. Since the length is twice the width,

$x = 2y$, so $y = x/2$.

The area is therefore

$A(x) = xy = x(x/2) = x^2/2$.

13. Since the patch is square the width and length are both equal to x. The costs are:

East and West sides: $4x + 4x = 8x$

North and South Sides: $2x + 2x = 4x$

Total cost $C(x) = 8x + 4x = 12x$

15. The number of hours you study, $h(n)$, equals 4 on Sunday through Thursday and equals 0 on the remaining days.

Since Sunday corresponds to $n = 1$ and Thursday to $n = 5$, we get

$$h(n) = \begin{cases} 4 & \text{if } 1 \leq n \leq 5 \\ 0 & \text{if } n > 5 \end{cases}.$$

17. For a linear cost function, $C(x) = mx + b$. Here, m = marginal cost = $1,500 per piano, b = fixed cost = $1,000.
Thus, the daily cost function is

$C(x) = 1,500x + 1,000.$

a. The cost of manufacturing 3 pianos is

$C(3) = 1,500(3) + 1,000 = 4,500 + 1,000 = \$5,500.$

b. The cost of manufacturing each additional piano (such as the third one or the 11th one) is the marginal cost,
$m = \$1,500.$

c. Same answer as (b).

d. Variable cost = part of the cost function that depends on $x = \$1,500x$

Fixed cost = constant summand of the cost function = $1,000
Marginal cost = slope of the cost function = $1,500 per piano

e. Graph:

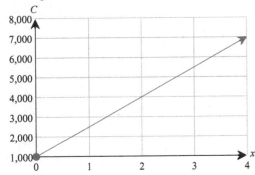

19. a. For a linear cost function, $C(x) = mx + b$. Here, m = marginal cost = $0.40 per copy, b = fixed cost = $70.

Thus, the cost function is $C(x) = 0.4x + 70.$

The revenue function is $R(x) = 0.50x$. (x copies at 50¢ per copy)
The profit function is

$P(x) = R(x) - C(x)$

$= 0.5x - (0.4x + 70)$

$= 0.5x - 0.4x - 70$

$= 0.1x - 70$

b. $P(500) = 0.1(500) - 70 = 50 - 70 = -20$

Since P is negative, this represents a loss of $20.

c. For breakeven, $P(x) = 0$:

$0.1x - 70 = 0$

$0.1x = 70$

$x = \dfrac{70}{0.1} = 700 \text{ copies}$

21. The revenue per jersey is $100. Therefore, Revenue $R(x) = \$100x$.

Profit = Revenue − Cost

$$P(x) = R(x) - C(x)$$

$$= 100x - (2,000 + 10x + 0.2x^2)$$

$$= -2,000 + 90x - 0.2x^2$$

To break even, $P(x) = 0$, so $-2,000 + 90x - 0.2x^2 = 0$.

This is a quadratic equation with $a = -0.2$, $b = 90$, $c = -2,000$ and solution

$$x = \frac{-b \pm \sqrt{b^2 - 4ac}}{2a}$$

$$= \frac{-90 \pm \sqrt{(90)^2 - 4(-2,000)(-0.2)}}{2(-0.2)} \approx 23.44 \text{ or } 426.56 \text{ jerseys.}$$

Since the second value is outside the domain, we use the first: $x = 23.44$ jerseys. To make a profit, x should be larger than this value: at least 24 jerseys.

23. The revenue from one thousand square feet ($x = 1$) is $0.1 million. Therefore, Revenue $R(x) = \$0.1x$. Profit = Revenue − Cost

$$P(x) = R(x) - C(x)$$

$$= 0.1x - (1.7 + 0.12x - 0.0001x^2)$$

$$= -1.7 - 0.02x + 0.0001x^2$$

To break even, $P(x) = 0$, so $-1.7 - 0.02x + 0.0001x^2 = 0$.

This is a quadratic equation with $a = 0.0001, b = -0.02, c = -1.7$ and solution

$$x = \frac{-b \pm \sqrt{b^2 - 4ac}}{2a}$$

$$= \frac{0.02 \pm \sqrt{(-0.02)^2 - 4(0.0001)(-1.7)}}{2(0.0001)} = \frac{0.02 \pm 0.03286}{0.0002} \approx 264 \text{ thousand square feet}$$

25. The hourly profit function is given by

Profit = Revenue − Cost

$$P(x) = R(x) - C(x)$$

(Hourly) cost function: This is a fixed cost of $5,132 only:

$$C(x) = 5,132$$

(Hourly) revenue function: This is a variable of $100 per passenger cost only:

$$R(x) = 100x$$

Thus, the profit function is

$$P(x) = R(x) - C(x)$$

$$P(x) = 100x - 5,132$$

For the domain of $P(x)$, the number of passengers x cannot exceed the capacity: 405. Also, x cannot be negative.

Thus, the domain is given by $0 \le x \le 405$, or $[0, 405]$.

For breakeven, $P(x) = 0$

$$100x - 5,132 = 0$$

$100x = 5{,}132$, or $x = \dfrac{5{,}132}{100} = 51.32$

If x is larger than this, then the profit function is positive, and so there should be at least 52 passengers (note that x must be a whole number); $x \geq 52$, for a profit.

27. To compute the break-even point, we use the profit function: Profit = Revenue – Cost

$\quad P(x) = R(x) - C(x)$

$\quad R(x) = 2x \qquad \$2$ per unit

$\quad C(x) =$ Variable Cost + Fixed Cost

$\qquad = 40\%$ of Revenue $+ 6{,}000 = 0.4(2x) + 6{,}000 = 0.8x + 6{,}000$

Thus, $P(x) = R(x) - C(x)$

$\quad P(x) = 2x - (0.8x + 6{,}000) = 1.2x - 6{,}000$

For breakeven, $P(x) = 0$

$\quad 1.2x - 6{,}000 = 0$

$\quad 1.2x = 6{,}000$, so $x = \dfrac{6{,}000}{1.2x} = 5{,}000$

Therefore, 5,000 units should be made to break even.

29. To compute the break-even point, we use the revenue and cost functions:

$\quad R(x) =$ Selling price $\times$ Number of units $= SPx$

$\quad C(x) =$ Variable Cost + Fixed Cost $= VCx + FC$

(Note that "variable cost per unit" is marginal cost.) For breakeven

$\quad R(x) = C(x)$

$\quad SPx = VCx + FC$

Solve for x:

$\quad SPx - VCx = FC$ n $\qquad x(SP - VC) = FC$, so $x = \dfrac{FC}{SP - VC}$.

31. Take x to be the number of grams of perfume he buys and sells. The profit function is given by Profit = Revenue – Cost: that is, $P(x) = R(x) - C(x)$

Cost function $C(x)$:

$\quad$ Fixed costs: $\qquad\qquad\qquad$ 20,000

$\quad$ Cheap perfume @ \$20 per g: $\quad 20x$

$\quad$ Transportation @ \$30 per 100 g: $0.3x$

Thus the cost function is

$\quad C(x) = 20x + 0.3x + 20{,}000 = 20.3x + 20{,}000$

Revenue function $R(x)$

$\quad R(x) = 600x \qquad \$600$ per gram

Thus, the profit function is

$\quad P(x) = R(x) - C(x)$

$\quad P(x) = 600x - (20.3x + 20{,}000) = 579.7x - 20{,}000$, with domain $x \geq 0$.

For breakeven, $P(x) = 0$

$\quad 579.7x - 20{,}000 = 0$

$$579.7x = 20,000, \text{ so } x = \frac{20,000}{579.7} \approx 34.50$$

Thus, he should buy and sell 34.50 grams of perfume per day to break even.

33. a. To graph the demand function we use technology with the formula

$$760/x-1$$

and with xMin = 60 and xMax = 400.
Graph:

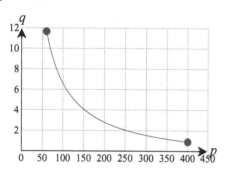

b. $q(100) = \dfrac{760}{100} - 1 = 7.6 - 1 = 6.6,$ $q(200) = \dfrac{760}{200} - 1 = 3.8 - 1 = 2.8$

So, the change in demand is $2.8 - 6.6 = -3.8$ million units, which means that demand decreases by about 3.8 million units per year.

c. The value of q on the graph decreases by smaller and smaller amounts as we move to the right on the graph, indicating that the demand decreases at a smaller and smaller rate (Choice (D)).

35. $q(p) = 0.17p^2 - 63p + 5,900$

a. Setting $p = 110$ gives

$$q(110) = 0.17(110)^2 - 63(110) + 5,900 = 1,027 \text{ million units.}$$

b. Setting $p = 90$ gives

$$q(90) = 0.17(90)^2 - 63(90) + 5,900 = 1,607 \text{ million units.}$$

c. Revenue = Price × Quantity

$$R(p) \quad = pq(p) = p(0.17p^2 - 63p + 5,900)$$

$$= 0.17p^3 - 63p^2 + 5,900p \text{ million dollars/year} \qquad \text{Revenue function}$$

$$R(110) = 0.17(110)^3 - 63(110)^2 + 5,900(110)$$

$$= 112,970 \approx 113,000 \text{ million dollars/year, or \$113 billion/year}$$

d. Graph: Technology formula: `0.17x^3-63x^2+5900x`

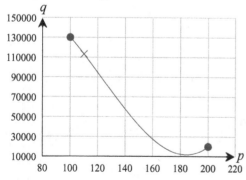

The graph shows revenue increasing as p decreases past \$110.

37. The price at which there is neither a shortage nor surplus is the equilibrium price, which occurs when demand = supply:

$$-3p + 700 = 2p - 500$$

$$5p = 1200$$

$$p = \$240 \text{ per skateboard}$$

39. a. The equilibrium price occurs when demand = supply:

$$-p + 156 = 4p - 394$$

$$5p = 550$$

$p = \$110$ per phone. **b.** Since \$105 is below the equilibrium price, there would be a shortage at that price. To calculate it, compute demand and supply:

Demand: $q = -105 + 156 = 51$ million phones

Supply: $q = 4(105) - 394 = 26$ million phones

Shortage = Demand − Supply = 51 − 26 = 25 million phones

41. a. For equilibrium, Demand = Supply:

$$\frac{760}{p} - 1 = 0.019p - 1$$

$$\frac{760}{p} = 0.019p$$

Cross-multiply:

$$0.019p^2 = 760 \quad \Rightarrow \quad p^2 = \frac{760}{0.019} = 40{,}000$$

So $\quad p = \sqrt{40{,}000} = \$200.$

Thus, the equilibrium price is \$200, and the equilibrium demand (or supply) is $760/200 - 1 = 2.8$ million e-readers

b. Graph:

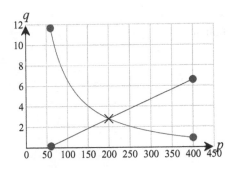

Technology formulas:
Demand: `y = 760/x-1`
Supply: `y = 0.019x-1`
The graphs cross at $(200, 2.8)$ confirming the calclation in part (a)

c. Since \$72 is below the equilibrium price, there would be a shortage at that price. To calculate it, compute demand and supply:

Demand: $q = \dfrac{760}{72} - 1 \approx 9.556$ million e-readers

Supply: $0.019(72) - 1 = 0.368$ million e-readers

Shortage = Demand – Supply $\approx 9.556 - 0.368 \approx 9.2$ million e-readers, or 9,200,000 e-readers.

43. $C(q) = 2,000 + 100q^2$

a. $C(10) = 2,000 + 100(10)^2 = 2,000 + 10,000 = \$12,000$

b. $N = C - S$, so

$$N(q) = C(q) - S(q) = 2,000 + 100q^2 - 500q$$

This is the cost of removing q lb of PCPs per day after the subsidy is taken into account.

c. $N(20) = 2,000 + 100(20)^2 - 500(20) = 2,000 + 40,000 - 10,000 = \$32,000$

45. The technology formulas are:
(A) `-0.2*t^2+t+16`
(B) `0.2*t^2+t+16`
(C) `t+16`
The following table shows the values predicted by the three models:

t	0	2	4	6	7
$S(t)$	16	18	22	28	30
(A)	16	17.2	16.8	14.8	13.2
(B)	16	18.8	23.2	29.2	32.8
(C)	16	18	20	22	23

As shown in the table, the values predicted by model (B) are much closer to the observed values $S(t)$ than those predicted by the other models.

b. Since 1998 corresponds to $t = 8$,

$$S(t) = 0.2t^2 + t + 16$$

$$S(8) = 0.2(8)^2 + 8 + 16 = 36.8$$

So the spending on corrections in 1998 was predicted to be approximately \$37 billion.

47. The technology formulas are:
(A) `0.005*x+2.75`
(B) `0.01*x+20+25/x`
(C) `0.0005*x^2-0.07*x+23.25`
(D) `25.5*1.08^(x-5)`
The following table shows the values predicted by the four models:

x	5	25	40	100	125
$A(x)$	22.91	21.81	21.25	21.25	22.31
(A)	20.775	20.875	20.95	21.25	21.375
(B)	25.05	21.25	21.025	21.25	21.45
(C)	22.913	21.813	21.25	21.25	22.313
(D)	25.5	118.85	377.03	38,177	261,451

Model (C) fits the data almost perfectly—more closely than any of the other models.

b. Graph of model (C):

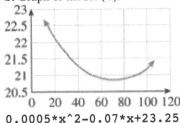

`0.0005*x^2-0.07*x+23.25`

The lowest point on the graph occurs at $x = 70$ with a y-coordinate of 20.8. Thus, the lowest cost per shirt is \$20.80, which the team can obtain by buying 70 shirts.

49. Here are the technology formulas as entered in the online Function Evaluator and Grapher at the Web Site:
(A) `6.3*0.67^x`
(B) `0.093*x^2-1.6*x+6.7`
(C) `4.75-0.50*x`
(D) `12.8/(x^1.7+1.7)`
The following graph shows all three curves together with the plotted points (entered as shown in the margin technology note with Example 5).

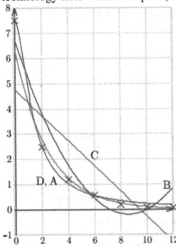

As shown in the graph, the values predicted by models (A) and (D) are much closer to the observed values than those predicted by the other models.

b. Since 2020 corresponds to $t = 20$,

$$\text{Model (A): } c(20) = 6.3(0.67)^{20} \approx \$0.0021$$

$$\text{Model (D): } c(20) = \frac{12.8}{10^{1.7} + 1.7} \approx \$0.0778$$

So model (A) gives the lower price: approximately \$0.0021.

51. A plot of the given points gives a straight line (Option (A)). Options (B) and (C) give curves, so (A) is the best choice.

53. A plot of the given data suggests a concave-down curve that becomes steeper downward as the price p increases, suggesting Model (D). Model (A) would predict increasing demand with increasing price, Model (B) would correspond to a descending curve that becomes less steep as p increases (a concave up curve), and Model (C) would give a concave-up parabola.

55. Apply the formula

$$A(t) = P\left(1 + \frac{r}{m}\right)^{mt}$$

with $P = 5{,}000$, $r = 0.05/100 = 0.0005$, and $m = 12$. We get the model

$$A(t) = 5{,}000(1 + 0.0005/12)^{12t}$$

In August 2020 ($t = 7$), the deposit would be worth $5{,}000(1 + 0.0005/12)^{12(7)} \approx \$5{,}018$.

57. From the answer to Exercise 55, the value of the investment after t years is

$$A(t) = 5{,}000(1 + 0.0005/12)^{12t}.$$

TI-83/84 Plus: Enter

`Y1 = 5000*(1+0.0005/12)^(12*X)`, Press [2nd] [TBLSET], and set `Indpnt` to Ask. (You do this once and for all; it will permit you to specify values for x in the table screen.) Then, press [2nd] [TABLE], and you will be able to evaluate the function at several values of x. Here are some values of x and the resulting values of Y1.

x	18	19	20	21
Y1	5045.20	5047.73	5050.25	5052.78

Notice that Y1 first exceeds 5,050 when $x = 20$. Since $x = 0$ represents August 2013, $x = 20$ represents August 2033, so the investment will first exceed $5,050 in August 2033.

59. $C(t) = 104(0.999879)^t$, so $C(10{,}000) \approx 31.0$ g, $C(20{,}000) \approx 9.25$ g, and $C(30{,}000) \approx 2.76$ g.

61. We are looking for t such that $4.06 = C(t) = 46(0.999879)^t$. Among the values suggested we find that $C(15{,}000) \approx 7.5$, $C(20{,}000) \approx 4.09$ and $C(25{,}000) \approx 2.23$. Thus, the answer is 20,000 years to the nearest 5,000 years.

63. a. Amount left after 1,000 years: $C(1{,}000) = A(0.999567)^{1{,}000} \approx 0.6485A$, or about 65% of the original amount.

Amount left after 2,000 years: $C(2{,}000) = A(0.999567)^{2{,}000} \approx 0.4206A$, or about 42% of the original amount.

Amount left after 3,000 years: $C(3{,}000) = A(0.999567)^{3{,}000} \approx 0.2727A$, or about 27% of the original amount.

b. For a sample of 100 g, $C(t) = 100(0.999567)^t$.

Here is the graph, together with the line $y = 50$ (one half the original sample):
Technology: `100*(0.999567)^x`

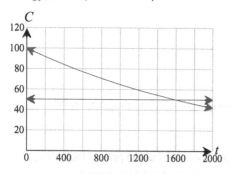

Since the graphs intersect close to $x = 1{,}600$, we conclude that half the sample will have decayed after about 1,600 years.

Communication and reasoning exercises

65. $P(0) = 200,\ P(1) = 230,\ P(2) = 260, \dots$ and so on. Thus, the population is increasing by 30 per year.

67. Curve fitting. The model is based on fitting a curve to a given set of observed data.

69. The given model is $c(t) = 4 - 0.2t$. This tells us that c is \$4 at time $t = 0$ (January) and is decreasing by \$0.20 per month. So, the cost of downloading a movie was \$4 in January and is decreasing by 20¢ per month.

71. In a linear cost function, the <u>variable</u> cost is x times the <u>marginal</u> cost.

73. Yes, as long as the supply is going up at a faster rate, as illustrated by the following graph:

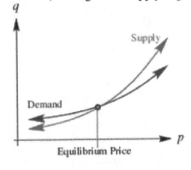

75. Extrapolate both models and choose the one that gives the most reasonable predictions.

77. The value of $f - g$ at x is $f(x) - g(x)$. Since $f(x) \geq g(x)$ for every x, it follows that $f(x) - g(x) \geq 0$ for every x.

79. Since the values of $\dfrac{f}{g}$ at x are ratios $\dfrac{f(x)}{g(x)}$, it follows that the units of measurement of $\dfrac{f}{g}$ are units of f per unit of g; that is, books per person.

Section 1.3

1.

x	−1	0	1
y	5	8	

We calculate the slope m first. The first two points shown give changes in x and y of

$$\Delta x = 0 - (-1) = 1$$

$$\Delta y = 8 - 5 = 3 \text{ This gives a slope of}$$

$$m = \frac{\Delta y}{\Delta x} = \frac{3}{1} = 3.$$

Now look at the second and third points: The change in x is again

$$\Delta x = 1 - 0 = 1$$

and so Δy must be given by the formula

$$\Delta y = m\Delta x$$

$$\Delta y = 3(1) = 3$$

This means that the missing value of y is

$$8 + \Delta y = 8 + 3 = 11.$$

3.

x	2	3	5
y	−1	−2	

We calculate the slope m first. The first two points shown give changes in x and y of

$$\Delta x = 3 - 2 = 1$$

$$\Delta y = -2 - (-1) = -1$$

This gives a slope of

$$m = \frac{\Delta y}{\Delta x} = \frac{-1}{1} = -1.$$

Now look at the second and third points: The change in x is

$$\Delta x = 5 - 3 = 2$$

and so Δy must be given by the formula

$$\Delta y = m\Delta x$$

$$\Delta y = (-1)(2) = -2$$

This means that the missing value of y is

$$-2 + \Delta y = -2 + (-2) = -4.$$

5.

x	−2	0	2
y	4		10

We calculate the slope m first. The first and third points shown give changes in x and y of

$$\Delta x = 2 - (-2) = 4$$

$$\Delta y = 10 - 4 = 6$$

This gives a slope of

$$m = \frac{\Delta y}{\Delta x} = \frac{6}{4} = \frac{3}{2}.$$

Now look at the first and second points: The change in x is

$$\Delta x = 0 - (-2) = 2$$

and so Δy must be given by the formula

$$\Delta y = m\Delta x$$

$$\Delta y = \left(-\frac{3}{2}\right)(2) = 3$$

This means that the missing value of y is

$$4 + \Delta y = 4 + 3 = 7.$$

7. From the table, $b = f(0) = -2$.
The slope (using the first two points) is

$$m = \frac{y_2 - y_1}{x_2 - x_1} = \frac{-2 - (-1)}{0 - (-2)} = \frac{-1}{2} = -\frac{1}{2}.$$

Thus, the linear equation is

$$f(x) = mx + b = -\frac{1}{2}x - 2, \text{ or } f(x) = -\frac{x}{2} - 2.$$

9. The slope (using the first two points) is

$$m = \frac{y_2 - y_1}{x_2 - x_1} = \frac{-2 - (-1)}{-3 - (-4)} = \frac{-1}{1} = -1.$$

To obtain $f(0) = b$, use the formula for b:

$$f(0) = b = y_1 - mx_1 = -1 - (-1)(-4) = -5 \qquad \text{Using the point } (x_1, y_1) = (-4, -1)$$

This gives

$$f(x) = mx + b = -x - 5.$$

11. In the table, x increases in steps of 1 and f increases in steps of 4, showing that f is linear with slope

$$m = \frac{\Delta y}{\Delta x} = \frac{4}{1} = 4$$

and intercept

$$b = f(0) = 6$$

giving

$$f(x) = mx + b = 4x + 6.$$

The function g does not increase in equal steps, so g is not linear.

13. In the first three points listed in the table, x increases in steps of 3, but f does not increase in equal steps, whereas g increases in steps of 6. Thus, based on the first three points, only g could possibly be linear, with slope

$$m = \frac{\Delta y}{\Delta x} = \frac{6}{3} = 2$$

and intercept

$$b = g(0) = -1$$

giving

$$g(x) = mx + b = 2x - 1.$$

We can now check that the remaining points in the table fit the formula $g(x) = 2x - 1$, showing that g is indeed linear.

15. Slope = coefficient of $x = -\dfrac{3}{2}$

17. Slope = coefficient of $x = \dfrac{1}{6}$

19. If we solve for x we find that the given equation represents the vertical line $x = -1/3$, and so its slope is infinite (undefined).

21. $3y + 1 = 0$. Solving for y:

$$3y = -1$$

$$y = -\dfrac{1}{3}$$

Slope = coefficient of $x = 0$

23. $4x + 3y = 7$. Solve for y:

$$3y = -4x + 7$$

$$y = -\dfrac{4}{3}x + \dfrac{7}{3}$$

Slope = coefficient of $x = -\dfrac{4}{3}$

25. $y = 2x - 1$

y-intercept $= -1$, slope $= 2$

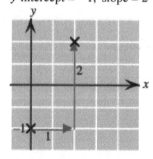
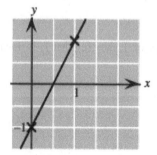

27. y-intercept $= 2$, slope $= -\dfrac{2}{3}$

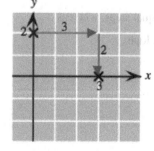
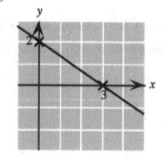

29. $y + \dfrac{1}{4}x = -4$. Solve for y to obtain $y = -\dfrac{1}{4}x - 4$

y-intercept $= -4$, slope $= -\dfrac{1}{4}$

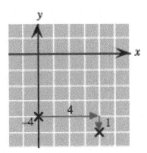
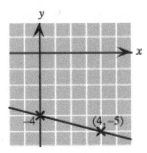

31. $7x - 2y = 7$. Solve for y:

$$-2y = -7x + 7, \text{ so } y = \dfrac{7}{2}x - \dfrac{7}{2}$$

y-intercept $= -\dfrac{7}{2} = -3.5$, slope $= \dfrac{7}{2} = 3.5$

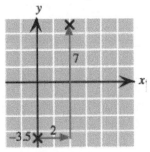
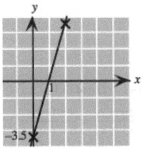

33. $3x = 8$. Solve for x to obtain $x = \dfrac{8}{3}$.

The graph is a vertical line:

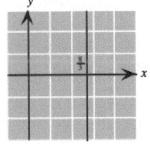

35. $6y = 9$. Solve for y to obtain $y = \dfrac{9}{6} = \dfrac{3}{2} = 1.5$

y-intercept $= \dfrac{3}{2} = 1.5$, slope $= 0$. The graph is a horizontal line:

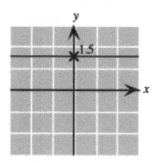

37. $2x = 3y$. Solve for y to obtain $y = \dfrac{2}{3}x$

y-intercept $= 0$, slope $= \dfrac{2}{3}$

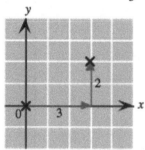

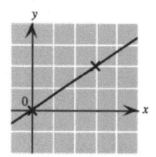

39. $(0, 0)$ and $(1, 2)$

$m = \dfrac{y_2 - y_1}{x_2 - x_1} = \dfrac{2 - 0}{1 - 0} = 2$

41. $(-1, -2)$ and $(0, 0)$

$m = \dfrac{y_2 - y_1}{x_2 - x_1} = \dfrac{0 - (-2)}{0 - (-1)} = 2$

43. $(4, 3)$ and $(5, 1)$

$m = \dfrac{y_2 - y_1}{x_2 - x_1} = \dfrac{1 - 3}{5 - 4} = \dfrac{-2}{1} = -2$

45. $(1, -1)$ and $(1, -2)$

$m = \dfrac{y_2 - y_1}{x_2 - x_1} = \dfrac{-2 - (-1)}{1 - 1}$ Undefined

47. $(2, 3.5)$ and $(4, 6.5)$

$m = \dfrac{y_2 - y_1}{x_2 - x_1} = \dfrac{6.5 - 3.5}{4 - 2} = \dfrac{3}{2} = 1.5$

49. $(300, 20.2)$ and $(400, 11.2)$

$m = \dfrac{y_2 - y_1}{x_2 - x_1} = \dfrac{11.2 - 20.2}{400 - 300} = \dfrac{-9}{100} = -0.09$

51. $(0, 1)$ and $\left(-\dfrac{1}{2}, \dfrac{3}{4}\right)$

$m = \dfrac{y_2 - y_1}{x_2 - x_1} = \dfrac{3/4 - 1}{-1/2 - 0} = \dfrac{-1/4}{-1/2} = \dfrac{2}{4} = \dfrac{1}{2}$

53. (a, b) and (c, d) $(a \neq c)$

$m = \dfrac{y_2 - y_1}{x_2 - x_1} = \dfrac{d - b}{c - a}$

55. (a, b) and (a, d) $(b \neq d)$

$m = \dfrac{y_2 - y_1}{x_2 - x_1} = \dfrac{d - b}{a - a}$ Undefined

57. $(-a, b)$ and $(a, -b)$ $(a \neq 0)$

$m = \dfrac{y_2 - y_1}{x_2 - x_1} = \dfrac{-b - b}{a - (-a)} = \dfrac{-2b}{2a} = -\dfrac{b}{a}$

59. a.

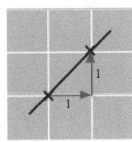

$$m = \frac{\Delta y}{\Delta x} = \frac{1}{1} = 1$$

b.

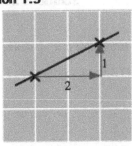

$$m = \frac{\Delta y}{\Delta x} = \frac{1}{2}$$

c.

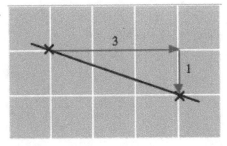

$$m = \frac{\Delta y}{\Delta x} = \frac{0}{1} = 0$$

d.

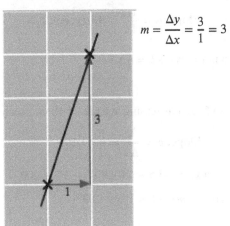

$$m = \frac{\Delta y}{\Delta x} = \frac{3}{1} = 3$$

e.

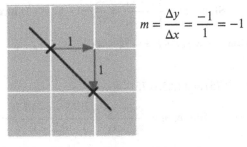

$$m = \frac{\Delta y}{\Delta x} = \frac{-1}{3} = -\frac{1}{3}$$

f.

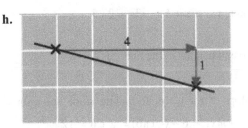

$$m = \frac{\Delta y}{\Delta x} = \frac{-1}{1} = -1$$

g.

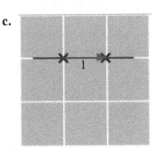

Vertical line; undefined slope

h.

$$m = \frac{\Delta y}{\Delta x} = \frac{-1}{4} = -\frac{1}{4}$$

i.

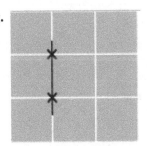

$$m = \frac{\Delta y}{\Delta x} = \frac{-2}{1} = -2$$

61. Through $(1, 3)$ with slope 3

Point: $(1, 3)$ **Slope:** $m = 3$ **Intercept:** $b = y_1 - mx_1 = 3 - 3(1) = 0$

Thus, the equation is $y = mx + b = 3x + 0$, or $y = 3x$

63. Through $\left(1, -\dfrac{3}{4}\right)$ with slope $\dfrac{1}{4}$

Point: $\left(1, -\dfrac{3}{4}\right)$ **Slope:** $m = \dfrac{1}{4}$ **Intercept:** $b = y_1 - mx_1 = -\dfrac{3}{4} - \dfrac{1}{4}(1) = -1$

Thus, the equation is $y = mx + b = \dfrac{1}{4}x - 1$

65. Through $(20, -3.5)$ and increasing at a rate of 10 units of y per unit of x

Point: $(20, -3.5)$ **Slope:** $m = \dfrac{\Delta y}{\Delta x} = \dfrac{10}{1} = 10$

Intercept: $b = y_1 - mx_1 = -3.5 - (10)(20) = -3.5 - 200 = -203.5$

Thus, the equation is $y = mx + b = 10x - 203.5$

67. Through $(2, -4)$ and $(1, 1)$

Point: $(2, -4)$ **Slope:** $m = \dfrac{y_2 - y_1}{x_2 - x_1} = \dfrac{5}{-1} = -5$ **Intercept:** $b = y_1 - mx_1 = -4 - (-5)(2) = 6$

Thus, the equation is $y = mx + b = -5x + 6$

69. Through $(1, -0.75)$ and $(0.5, 0.75)$

Point: $(1, -0.75)$ **Slope:** $m = \dfrac{y_2 - y_1}{x_2 - x_1} = \dfrac{0.75 - (-0.75)}{0.5 - 1} = \dfrac{1.5}{-0.5} = -3$

Intercept: $b = y_1 - mx_1 = -0.75 - (-3)(1) = -0.75 + 3 = 2.25$

Thus, the equation is $y = mx + b = -3x + 2.25$

71. Through $(6, 6)$ and parallel to the line $x + y = 4$

Point: $(6, 6)$

Slope: Same as slope of $x + y = 4$. To find the slope, solve for y, getting $y = -x + 4$. Thus, $m = -1$.

Intercept: $b = y_1 - mx_1 = 6 - (-1)(6) = 6 + 6 = 12$

Thus, the equation is $y = mx + b = -x + 12$

73. Through $(0.5, 5)$ and parallel to the line $4x - 2y = 11$

Point: $(0.5, 5)$

Slope: Same as slope of $4x - 2y = 11$. To find the slope, solve for y, getting $y = 2x - \dfrac{11}{2}$. Thus, $m = 2$.

Intercept: $b = y_1 - mx_1 = 5 - (2)(0.5) = 5 - 1 = 4$

Thus, the equation is $y = mx + b = 2x + 4$

75. Through $(0, 0)$ and (p, q)

Point: $(0, 0)$ **Slope:** $m = \dfrac{y_2 - y_1}{x_2 - x_1} = \dfrac{q - 0}{p - 0} = \dfrac{q}{p}$ **Intercept:** $b = y_1 - mx_1 = 0 - \dfrac{q}{p}(0) = 0$

Thus, the equation is $y = mx + b = \dfrac{q}{p}x$

77. Through (p, q) and (r, q) $(p \neq r)$

Point: (p, q) **Slope:** $m = \dfrac{y_2 - y_1}{x_2 - x_1} = \dfrac{q - q}{r - p} = 0$

Intercept: $b = y_1 - mx_1 = q - (0)p = q$

Thus, the equation is $y = mx + b$; that is, $y = q$

79. Through $(-p, q)$ and $(p, -q)$ $(p \neq 0)$

Point: $(-p, q)$ **Slope:** $m = \dfrac{y_2 - y_1}{x_2 - x_1} = \dfrac{-q - q}{p - (-p)} = \dfrac{-2q}{2p} = -\dfrac{q}{p}$

Intercept: $b = y_1 - mx_1 = q - \left(-\dfrac{q}{p}\right)(-p) = q - q = 0$

Thus, the equation is $y = mx + b = -\dfrac{q}{p}x$

Applications

81. We are given two points on the graph of the linear cost function: $(100, 10{,}500)$ and $(120, 11{,}000)$ (x is the number of bicycles, and the second coordinate is the cost C).

Marginal cost:

$$m = \frac{C_2 - C_1}{x_2 - x_1} = \frac{11{,}000 - 10{,}500}{120 - 100} = \frac{500}{20} = \$25 \text{ per bicycle}$$

Fixed cost:

$$b = C_1 - mx_1 = 10{,}500 - (25)(100) = 10{,}500 - 2{,}500 = \$8{,}000$$

83. We are given two points on the graph of the linear cost function: $(10, 2{,}070)$ and $(20, 4{,}120)$ (x is the number of iPhones, and the second coordinate is the cost C).

Marginal cost:

$$m = \frac{C_2 - C_1}{x_2 - x_1} = \frac{4{,}120 - 2{,}070}{20 - 10} = \frac{2{,}050}{10} = \$205 \text{ per iPhone}$$

Fixed cost:

$$b = C_1 - mx_1 = 2{,}070 - (205)(10) = \$20$$

Thus, the cost equation is $C = mx + b = 205x + 20$.

The cost to manufacture each additional iPhone is the marginal cost: 205.

The cost to manufacture 40 iPhones is obtained by setting $x = 40$ in the cost equation:

$$C(40) = 205(40) + 20 = \$8{,}220$$

85. A linear demand function has the form

$$q = mp + b$$

45

(p is the price, and q is the demand). We are given two points on its graph: $(1, 1,960)$ and $(5, 1,800)$.

Slope:

$$m = \frac{q_2 - q_1}{p_2 - p_1} = \frac{1,800 - 1,960}{5 - 1} = \frac{-160}{4} = -40$$

Intercept:

$$b = q_1 - mp_1 = 1,960 - (-40)(1) = 1,960 + 40 = 2,000$$

Thus, the demand equation is

$$q = mp + b = -40p + 2,000$$

87. a. A linear demand function has the form $q = mp + b$. (p is the price, and q is the demand). We are given two points on its graph:

2012: $(385, 720)$

2013: $(335, 1,010)$

Slope: $m = \dfrac{q_2 - q_1}{p_2 - p_1} = \dfrac{1,010 - 720}{335 - 385} = \dfrac{290}{-50} = -5.8$

Intercept: $b = q_1 - mp_1 = 720 - (-5.8)385 = 2,953$

Thus, the demand equation is $q = mp + b = -5.8p + 2,953$.

If $p = \$265$, then $q = -5.8(265) + 2,953 = 1,416$ million phones.

b. Since the slope is -5.8 million phones per unit increase in price, we interpret of the slope as follows: For every $\underline{\$1}$ increase in price, sales of smartphones decrease by $\underline{5.8\ \text{million}}$ units.

89. a. A linear demand function has the form $q = mp + b$. (p is the price, and q is the demand).

We are given two points on its graph: $(3, 28,000)$ and $(5, 19,000)$.

Slope: $m = \dfrac{q_2 - q_1}{p_2 - p_1} = \dfrac{19,000 - 28,000}{5 - 3} = \dfrac{-9,000}{2} = -4,500$

Intercept: $b = q_1 - mp_1 = 28,000 - (-4,500)3 = 28,000 + 13,500 = 41,500$

Thus, the demand equation is $q = mp + b = -4,500p + 41,500$.

b. The units of measurement of the slope are generally units of y per unit of x. In this case: Units of q per unit of p. That is,

Rides per day per \$1 increase in the fare.

Since the slope is $-4,500$ rides/day per \$1 increase in the price, we interpret it as saying that ridership decreases by 4,500 rides per day for every \$1 increase in the fare.

c. From part (a), the demand equation is

$$q = -4,500p + 41,500$$

If the fare is \$6, we have $p = 6$, so

$$q = -4500(6) + 41,500 = -27,000 + 41,500 = 14,500 \text{ rides/day.}$$

91. a. In a linear model of y versus time t, the slope is the number of units of y per unit time, and we are given this quantity: 40 million pounds/year. Thus, working in millions of pounds, we can take $m = 40$. We are also given the y-intercept (the value of y at $t = 0$) as $b = 290$. Thus, the model is

$$y = 40t + 290 \text{ million pounds of pasta}$$

b. In 2005, $t = 15$, and so $y(15) = 40(15) + 290 = 890$ million pounds.

93. a. The desired linear model has the form $N = mt + b$, where t is time in years since 2010. We are given two points on its graph: 2011 data: $(1, 0.63)$; 2014 data: $(4, -0.24)$

Slope: $m = \dfrac{N_2 - N_1}{t_2 - t_1} = \dfrac{-0.24 - 0.63}{4 - 1} = \dfrac{-0.87}{3} = -0.29$

Intercept: $b = N_1 - mt_1 = 0.63 - (-0.29)(1) = 0.92$

Thus, the linear model is $N = mp + b = -0.29t + 0.92$.

b. The units of measurement of the slope are units of N per unit of t; that is, billions of dollars per year. Amazon's net income decreased at a rate of $0.29 billion per year.

c. The year 2013 corresponds to $t = 3$, and so

$N = -0.29(3) + 0.92 = \$0.05$ billion,

which differs quite significantly from the actual net income.

95. $s(t) = 2.5t + 10$

a. Velocity = slope = 2.5 feet/sec.

b. After 4 seconds, $t = 4$, so

$s(4) = 2.5(4) + 10 = 10 + 10 = 20$

Thus the model train has moved 20 feet along the track.

c. The train will be 25 feet along the track when $s = 25$. Substituting gives

$25 = 2.5t + 10$

Solving for time t gives

$2.5t = 25 - 10 = 15$

$t = \dfrac{15}{2.5} = 6$ seconds

97. a. Take s to be displacement from Jones Beach, and t to be time in hours. We are given two points:

$(t, s) = (10, 0)$ $s = 0$ for Jones Beach.

$(t, s) = (10.1, 13)$ 6 minutes = 0.1 hours

We are asked for the speed, which equals the magnitude of the slope.

$m = \dfrac{s_2 - s_1}{t_2 - t_1} = \dfrac{13 - 0}{10.1 - 10} = \dfrac{13}{0.1} = 130$

Units of slope = units of s per unit of t = miles per hour

Thus, the police car was traveling at 130 mph.

b. For the displacement from Jones Beach at time t, we want to express s as a linear function of t; namely, $s = mt + b$.

We already know $m = 130$ from part (a). For the intercept, use

$b = s_1 - mt_1 = 0 - 130(10) = -1,300$

Therefore, the displacement at time t is

$s = mt + b = 130t - 1,300$

99. a. The desired linear model has the form $L = mn + b$. We are given two points on its graph: Second edition data:

$(2, 585)$; Sixth edition data: $(6, 755)$

Slope: $m = \dfrac{L_2 - L_1}{n_2 - n_1} = \dfrac{755 - 585}{6 - 2} = \dfrac{170}{4} = 42.5$

Intercept: $b = L_1 - mn_1 = 585 - (42.5)(2) = 500$

Thus, the linear model is $L = mn + b = 42.5n + 500$.

b. The units of measurement of the slope are units of L per unit of n; that is, pages per edition; *Applied Calculus* is growing at a rate of 42.5 pages per edition.

c. The length L will equal 1,500 when $42.5n + 500 = 1,500$. Solving for n gives

$$42.5n = 1,500 - 500 = 1,000$$

$$n = \frac{1,000}{42.5} \approx 23.5$$

Thus, by the 24th edition, the book will be over 1,500 pages long.

101. F = Fahrenheit temperature, C = Celsius temperature, and we want F as a linear function of C. That is,

$$F = mC + b.$$

(F plays the role of y and C plays the role of x.)
We are given two points:

$\quad (C, F) = (0, 32) \qquad$ Freezing point

$\quad (C, F) = (100, 212) \quad$ Boiling point

Slope: $m = \dfrac{F_2 - F_1}{C_2 - C_1} = \dfrac{212 - 32}{100 - 0} = \dfrac{180}{100} = 1.8$

Intercept: $b = F_1 - mC_1 = 32 - 1.8(0) = 32$
Thus, the linear relation is

$\quad F = mC + b = 1.8C + 32$

When $C = 30°$

$\quad F = 1.8(30) + 32 = 54 + 32 = 86°$

When $C = 22°$

$\quad F = 1.8(22) + 32 = 39.6 + 32 = 71.6°$. Rounding to the nearest degree gives 72°F.

When $C = -10°$, $F = 1.8(-10) + 32 = -18 + 32 = 14°$.

When $C = -14°$, $F = 1.8(-14) + 32 = -25.2 + 32 = 6.8°$. Rounding to the nearest degree gives 7°F.

103. a. S = Southwest Airlines net income (in $ millions), J = JetBlue Airways net income (in $ millions), and we want J as a linear function of S. That is,

$\quad J = mS + b \qquad$ *J plays the role of y and S plays the role of x.*
We are given two points:

$\quad (S, J) = (400, 130) \quad$ 2012 data

$\quad (S, J) = (900, 400) \quad$ 2014 data

Slope: $m = \dfrac{J_2 - J_1}{S_2 - S_1} = \dfrac{400 - 130}{900 - 400} = \dfrac{270}{500} = 0.54$

Intercept: $b = J_1 - mS_1 = 130 - (0.54)(400) = -86$
Thus, the linear relation is

$\quad J = mS + b = 0.54S - 86.$

b. In 2010, Southwest Airlines' net income was $S = 450$, so

$\quad J = 0.54(450) - 86 = 157,$

predicting a $157 million net income for JetBlue, $57 million higher than the actual $100 million net income JetBlue earned in 2010.

c. The units of measurement of the slope are units of J per unit of S; that is, millions of dollars of JetBlue Airways net income per million dollars of Southwest Airlines net income; JetBlue Airways earned an additional net income of $0.54 per $1 additional net income earned by Southwest Airlines.

105. Income = royalties + screen rights

$I = 5\%$ of net profits + 50,000

$I = 0.05N + 50,000$ Equation notation

$I(N) = 0.05N + 50,000$ Function notation

For an income of $100,000,

$100,00 = 0.05N + 50,000$

$0.05N = 50,000$

$N = \dfrac{50,000}{0.05} = \$1,000,000$

Her marginal income is her increase in income per $1 increase in net profit. This is the slope, $m = 0.05$ dollars of income per dollar of net profit, or $5¢$ per dollar of net profit.

107. The year 2000 corresponds to $t = 10$, which is in the range $6 \leq t < 15$, so we use the first equation:

$v(t) = 400t - 2,200$. The slope is 400 MHz/year, telling us that the speed of a processor was increasing by 400 MHz/year.

109. The data are

t	0	20	40
y	78	2,100	2,950

a. 1970–1990 (first two data points):

Slope: $m = \dfrac{y_2 - y_1}{t_2 - t_1} = \dfrac{700 - 78}{20 - 0} = 31.1$

Intercept: $b = 78$, specified in first data point

Thus, the linear model is

$y = mt + b = 31.1t + 78$.

b. 1990–2010 (second and third data points):

Slope: $m = \dfrac{y_2 - y_1}{t_2 - t_1} = \dfrac{2,950 - 700}{40 - 20} = 112.5$

Intercept: $b = y_1 - mt_1 = 700 - (112.5)20 = -1,550$

Thus, the linear model is

$y = mt + b = 112.5t - 1,550$.

c. Since the first model is valid for $0 \leq t \leq 20$ and the second one for $20 \leq t \leq 40$, we put them together as

$y = \begin{cases} 31.1t + 78 & \text{if } 0 \leq t < 20 \\ 112.5t - 1,550 & \text{if } 20 \leq t \leq 40. \end{cases}$

Notice that, since both formulas agree at $t = 20$, we can also say

$y = \begin{cases} 31.1t + 78 & \text{if } 0 \leq t \leq 20 \\ 112.5t - 1,550 & \text{if } 20 < t \leq 40. \end{cases}$

d. Since 2004 is represented by $t = 34$, we use the second formula to obtain

$$y = 112.5(34) - 1{,}550 = 2{,}275,$$

or $2,275,000$, in good agreement with the actual value shown in the graph.

111. 1995–2000: Points:

1995 data: $(t, N) = (0, 3)$

2000 data: $(t, N) = (5, 4.1)$

Slope: $m = \dfrac{N_2 - N_1}{t_2 - t_1} = \dfrac{4.1 - 3}{5 - 0} = 0.22$

Intercept: $b = 3$, specified in first data point

Thus, the linear model is

$$N = mt + b = 0.22t + 3.$$

2000–2004: Points:

2000 data: $(t, N) = (5, 4.1)$

2004 data: $(t, N) = (9, 3.5)$

Slope: $m = \dfrac{N_2 - N_1}{t_2 - t_1} = \dfrac{3.5 - 4.1}{9 - 5} = -0.15$

Intercept: $b = N_1 - mt_1 = 4.1 - (-0.15)(5) = 4.85$

Thus, the linear model is

$$N = mt + b = -0.15t + 4.85.$$

Putting them together gives

$$N = \begin{cases} 0.22t + 3 & \text{if } 0 \leq t \leq 5 \\ -0.15t + 4.85 & \text{if } 5 < t \leq 9 \end{cases}$$

The number of manufacturing jobs in Mexico in 2002 is $N(7)$, so we use the second formula to obtain

$$N(7) = -0.15(7) + 4.85 = 3.8 \text{ million jobs}$$

Communication and reasoning exercises

113. Compute the corresponding successive changes Δx in x and Δy in y, and compute the ratios $\Delta y / \Delta x$. If the answer is always the same number, then the values in the table come from a linear function.

115. To find the linear function, solve the equation $ax + by = c$ for y:

$$by = -ax + c$$

$$y = -\frac{a}{b}x + \frac{c}{b}$$

Thus, the desired function is $f(x) = -\dfrac{a}{b}x + \dfrac{c}{b}$.

If $b = 0$, then $\dfrac{a}{b}$ and $\dfrac{c}{b}$ are undefined, and y cannot be specified as a function of x. (The graph of the resulting equation would be a vertical line.)

117. The slope of the line is $m = \dfrac{\Delta y}{\Delta x} = \dfrac{3}{1} = 3$. Therefore, if, in a straight line, y is increasing three times as fast as x, then its <u>slope</u> is <u>3</u>.

119. If m is positive, then y will increase as x increases; if m is negative, then y will decrease as x increases; if m is zero, then y will not change as x changes.

121.

◇	A	B	C	D
1	x	y	m	b
2		1	2	=(B3-B2)/(A3-A2)	=B2-C2*A2
3		3	-1	Slope	Intercept

The slope computed in cell C2 is given by

$$m = \frac{y_2 - y_1}{x_2 - x_1} = \frac{-1 - 2}{3 - 1} = -1.5$$

If we increase the y-coordinate in cell B3, this increases y_2, and thus increases the numerator $\Delta y = y_2 - y_1$ without affecting the denominator Δx. Thus the slope will increase.

123. The units of the slope m are units of y (bootlags) per unit of x (zonars). The intercept b is on the y-axis, and is thus measured in units of y (bootlags). Thus, m is measured in <u>bootlags per zonar</u> and b is measured in <u>bootlags</u>.

125. If a quantity changes linearly with time, it must change by the same amount for every unit change in time. Thus, since it increases by 10 units in the first day, it must increase by 10 units each day, including the third.

127. Since the slope is 0.1, the velocity is increasing at a rate of 0.1 m/sec every second. Since the velocity is increasing, the object is accelerating (choice B).

129. Write $f(x) = mx + b$ and $g(x) = nx + c$. Then

$$f(x) + g(x) = mx + b + (nx + c) = (m + n)x + (b + c),$$

also a linear function with slope $m + n$.

131. Answers may vary. For example, $f(x) = x^{1/3}, g(x) = x^{2/3}$ gives

$$f(x)g(x) = x^{1/3}x^{2/3} = x.$$

133. Increasing the number of items from the break-even number results in a profit: Because the slope of the revenue graph is larger than the slope of the cost graph, it is higher than the cost graph to the right of the point of intersection, and hence corresponds to a profit.

Section 1.4

1. $(1, 1)$, $(2, 2)$, $(3, 4)$; $y = x - 1$

x	y	$\hat{y} = x - 1$	$y - \hat{y}$	$(y - \hat{y})^2$
1	1	0	1	1
2	2	1	1	1
3	4	2	2	4

SSE = Sum of squares of residuals (last column) = $1 + 1 + 4 = 6$

3. $(0, -1)$, $(1, 3)$, $(4, 6)$, $(5, 0)$; $y = -x + 2$

x	y	$\hat{y} = -x + 2$	$y - \hat{y}$	$(y - \hat{y})^2$
0	−1	2	−3	9
1	3	1	2	4
4	6	−2	8	64
5	0	−3	3	9

SSE = Sum of squares of residuals (last column) = $9 + 4 + 64 + 9 = 86$

5. $(1, 1)$, $(2, 2)$, $(3, 4)$

a. $y = 1.5x - 1$

x	y	$\hat{y} = 1.5x - 1$	$y - \hat{y}$	$(y - \hat{y})^2$
1	1	0.5	0.5	0.25
2	2	2	0	0
3	4	3.5	0.5	0.25

SSE = Sum of squares of residuals = 0.5

b. $y = 2x - 1.5$

x	y	$\hat{y} = 2x - 1.5$	$y - \hat{y}$	$(y - \hat{y})^2$
1	1	0.5	0.5	0.25
2	2	2.5	−0.5	0.25
3	4	4.5	−0.5	0.25

SSE = Sum of squares of residuals = 0.75
The model that gives the better fit is (a) because it gives the smaller value of SSE.

7. $(0,-1)$, $(1,3)$, $(4,6)$, $(5,0)$

a. $y = 0.3x + 1.1$

x	y	$\hat{y} = 0.3x + 1.1$	$y - \hat{y}$	$(y - \hat{y})^2$
0	-1	1.1	-2.1	4.41
1	3	1.4	1.6	2.56
4	6	2.3	3.7	13.69
5	0	2.6	-2.6	6.76

SSE = Sum of squares of residuals = 27.42

b. $y = 0.4x + 0.9$

x	y	$\hat{y} = 0.4x + 0.9$	$y - \hat{y}$	$(y - \hat{y})^2$
0	-1	0.9	-1.9	3.61
1	3	1.3	1.7	2.89
4	6	2.5	3.5	12.25
5	0	2.9	-2.9	8.41

SSE = Sum of squares of residuals = 27.16
The model that gives the better fit is (b) because it gives the smaller value of SSE.

9. Data points (x, y): $(1,1)$, $(2,2)$, $(3,4)$

x	y	xy	x^2
1	1	1	1
2	2	4	4
3	4	12	9
6	7	17	14

(The bottom row contains the column sums.)

$$n = 3 \text{ (number of data points)}$$

Slope: $\quad m = \dfrac{n(\sum xy) - (\sum x)(\sum y)}{n(\sum x^2) - (\sum x)^2} = \dfrac{3(17) - (6)(7)}{3(14) - 6^2} = \dfrac{9}{6} = 1.5$

Intercept: $b = \dfrac{\sum y - m(\sum x)}{n} = \dfrac{7 - 1.5(6)}{3} = \dfrac{-2}{3} \approx -0.6667$

Thus, the regression line is

$$y = mx + b \approx 1.5x - 0.6667.$$

Graph:

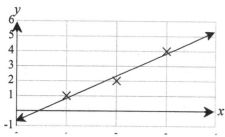

11. Data points (x, y): $(0, -1)$, $(1, 3)$, $(3, 6)$, $(4, 1)$

x	y	xy	x^2
0	−1	0	0
1	3	3	1
3	6	18	9
4	1	4	16
8	9	25	26

(The bottom row contains the column sums.)

$$n = 4 \text{ (number of data points)}$$

Slope: $m = \dfrac{n(\sum xy) - (\sum x)(\sum y)}{n(\sum x^2) - (\sum x)^2} = \dfrac{4(25) - (8)(9)}{4(26) - 8^2} = \dfrac{28}{40} = 0.7$

Intercept: $b = \dfrac{\sum y - m(\sum x)}{n} = \dfrac{9 - 0.7(8)}{4} = \dfrac{3.4}{4} = 0.85$

Thus, the regression line is

$$y = mx + b = 0.7x + 0.85.$$

Graph:

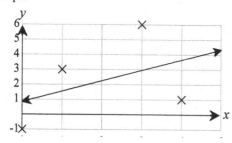

54

13. a. $(1, 3)$, $(2, 4)$, $(5, 6)$

x	y	xy	x^2	y^2
1	3	3	1	9
2	4	8	4	16
5	6	30	25	36
8	13	41	30	61

(The bottom row contains the column sums.)

$n = 3$ (number of data points)

$$r = \frac{n(\sum xy) - (\sum x)(\sum y)}{\sqrt{n(\sum x^2) - (\sum x)^2} \cdot \sqrt{n(\sum y^2) - (\sum y)^2}} = \frac{3(41) - (8)(13)}{\sqrt{3(30) - 8^2}\sqrt{3(61) - 13^2}} \approx \frac{19}{19.078784} \approx 0.9959$$

b. $(0, -1)$, $(2, 1)$, $(3, 4)$

x	y	xy	x^2	y^2
0	−1	0	0	1
2	1	2	4	1
3	4	12	9	16
5	4	14	13	18

(The bottom row contains the column sums.)

$n = 3$ (number of data points)

$$r = \frac{n(\sum xy) - (\sum x)(\sum y)}{\sqrt{n(\sum x^2) - (\sum x)^2} \cdot \sqrt{n(\sum y^2) - (\sum y)^2}} = \frac{3(14) - (5)(4)}{\sqrt{3(13) - 5^2}\sqrt{3(18) - 4^2}} \approx \frac{22}{23.0651252} \approx 0.9538$$

c. $(4, -3)$, $(5, 5)$, $(0, 0)$

x	y	xy	x^2	y^2
4	−3	−12	16	9
5	5	25	25	25
0	0	0	0	0
9	2	13	41	34

(The bottom row contains the column sums.)

$n = 3$ (number of data points)

$$r = \frac{n(\sum xy) - (\sum x)(\sum y)}{\sqrt{n(\sum x^2) - (\sum x)^2} \cdot \sqrt{n(\sum y^2) - (\sum y)^2}} = \frac{3(13) - (9)(2)}{\sqrt{3(41) - 9^2}\sqrt{3(34) - 2^2}} \approx \frac{21}{64.1560597} \approx 0.3273$$

The value of r in part (a) has the largest absolute value. Therefore, the regression line for the data in part (a) is the best fit.

The value of r in part (c) has the smallest absolute value. Therefore, the regression line for the data in part (c) is the worst fit.

Since r is not ± 1 for any of these lines, none of them is a perfect fit.

Applications

15. The entries in the xy column are obtained by multiplying the entries in the x column by the corresponding entries in the y column. The entries in the x^2 column are the squares of the entries in the x column. The entries in the last row are the sums of the respective columns.

Data points (x, y): $(0, 800)$, $(2, 1{,}600)$, $(4, 2{,}300)$

x	y	xy	x^2
0	800	0	0
2	1,600	3,200	4
4	2,300	9,200	16
6	4,700	12,400	20

(The bottom row contains the column sums.)

$$n = 3 \text{ (number of data points)}$$

Slope: $m = \dfrac{n(\sum xy) - (\sum x)(\sum y)}{n(\sum x^2) - (\sum x)^2} = \dfrac{3(12{,}400) - (6)(4{,}700)}{3(20) - 6^2} = \dfrac{9{,}000}{24} = 375$

Intercept: $b = \dfrac{\sum y - m(\sum x)}{n} = \dfrac{4{,}700 - 375(6)}{3} = \dfrac{2{,}450}{3} \approx 816.7$

Thus, the regression line is

$$y = mx + b \approx 375x + 816.7.$$

Since 2016 corresponds to $x = 6$, the prediction for 2016 is

$$y = 375(6) + 816.7 \approx 3{,}066.7 \text{ million subscribers.}$$

17. A linear demand function has the form $q = mp + b$. (p is the price, and q is the demand).

Data points (p, q): $(4, 0.7)$, $(3, 1)$, $(2, 2)$

p	q	pq	p^2
4	0.7	2.8	16
3	1	3	9
2	2	4	4
9	3.7	9.8	29

(The bottom row contains the column sums.)

$$n = 3 \text{ (number of data points)}$$

Slope: $m = \dfrac{n(\sum pq) - (\sum p)(\sum q)}{n(\sum p^2) - (\sum p)^2} = \dfrac{3(9.8) - (9)(3.7)}{3(29) - 9^2} = \dfrac{-3.9}{6} \approx -0.7$

Intercept: $b = \dfrac{\sum q - m(\sum p)}{n} = \dfrac{3.7 - (-0.65)(9)}{3} = \dfrac{9.55}{3} \approx 3.2$

Thus, the regression line is

$$q = mp + b \approx -0.7p + 3.2.$$

When the selling price is \$350, $p = 3.5$, and so $q \approx -0.7(3.5) + 3.2 = 0.75$ billion, or 750 million smartphones.

19. Following is the table we use to compute the regression line:

x	y	xy	x^2
20	3	60	400
40	6	240	1,600
80	9	720	6,400
100	15	1,500	10,000
240	33	2,520	18,400

(The bottom row contains the column sums.)

$$n = 3 \text{ (number of data points)}$$

Slope: $\quad m = \dfrac{n(\sum xy) - (\sum x)(\sum y)}{n(\sum x^2) - (\sum x)^2} = \dfrac{4(2,520) - (240)(33)}{4(18,400) - 240^2} = \dfrac{2,160}{16,000} = 0.135$

Intercept: $b = \dfrac{\sum y - m(\sum x)}{n} = \dfrac{33 - (0.135)(240)}{4} = \dfrac{0.6}{4} = 0.15$

The regression model is therefore $y = mx + b = 0.135x + 0.15$.

$y(50) = 0.135(50) + 0.15 = 6.9$ million jobs

21. a. Data points (S, I): $(50, 0.6)$, $(60, 0.1)$, $(70, 0.3)$, $(80, 0.3)$

S	I	SI	S^2
50	0.6	30	2,500
60	0.1	6	3,600
70	0.3	21	4,900
80	0.3	24	6,400
260	1.3	81	17,400

(The bottom row contains the column sums.)

$$n = 4 \text{ (number of data points)}$$

Slope: $\quad m = \dfrac{n(\sum SI) - (\sum S)(\sum I)}{n(\sum S^2) - (\sum S)^2} = \dfrac{4(81) - (260)(1.3)}{4(17,400) - 260^2} = \dfrac{-14}{2,000} = -0.007$

Intercept: $b = \dfrac{\sum I - m(\sum S)}{n} = \dfrac{1.3 - (-0.007)(260)}{4} = \dfrac{3.12}{4} = 0.78$

Thus, the regression line is

$\quad I = mS + b = -0.007S + 0.78.$

Graph:

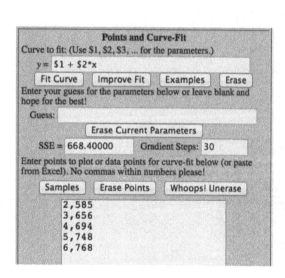

Independent variable is S and dependent variable is I.

b. The units of measurement of the slope are units of net income per unit of net sales: millions of dollars of net income per billion dollars of net sales. Thus, Amazon *lost* $0.007 billion ($7 million) in net income per additional billion dollars in net sales.

c. $I = -0.007S + 0.78$, and we are given $I = 0.5$. Substituting gives

$$0.5 = -0.007S + 0.78$$

Solving for S gives

$$S = \frac{-0.28}{-0.007} = 40$$

Thus, the company would need to earn $40 billion in net sales.

d. The graph shows a poor fit, so the linear model does not seem reasonable.

23. a. See the technology note accompanying Example 2 for the use of technology to obtain regression lines. The following result and plot were obtained using the Function Evaluator and Grapher on the Web site with the setup shown:

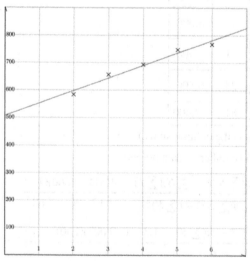

Regression equation: $L = 45.8n + 507$

b. The units of measurement of the slope are units of L per unit of n; that is, pages per edition; *Applied Calculus* is growing at a rate of 45.8 pages per edition.

25. a. Since production is a function of cultivated area, we take x as cultivated area, and y as production:

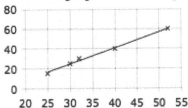

x	25	30	32	40	52
y	15	25	30	40	60

See the technology note accompanying Example 2 for the use of technology to obtain regression lines. We obtained the following regression line and plot in Excel. (coefficients rounded to two decimal places): $y = 1.62x - 23.87$

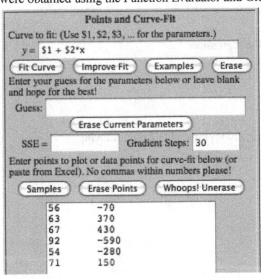

b. To interpret the slope $m = 1.62,$ recall that units of m are units of y per unit of x; that is, millions of tons of production of soybeans per million acres of cultivated land. Thus, production increases by 1.62 million tons of soybeans per million acres of cultivated land. More simply, each acre of cultivated land produces about 1.62 tons of soybeans.

27. a. y = Continental net income as a function of x = Price of oil. See the technology notes accompanying Example 2 and 3 for the use of technology to obtain regression lines and correlation coefficients. The following result and plot were obtained using the Function Evaluator and Grapher on the Web site with the setup shown:

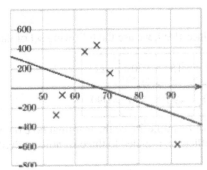

Regression equation: $y = -11.85x + 797.71$

Correlation coefficient: $r \approx -0.414$

b. As $|r| \approx 0.414$ is significantly less than 0.8, the values of x and y are not strongly correlated, so that Continental's net income does not appear correlated to the price of oil. **c.** The points in the graph are nowhere near the regression line, confirming the conclusion in (b).

29. a. Using x = Number of natural science doctorates and y = Number of engineering doctorates gives us the following table of values:

x	70	130	330	490	590	690
y	10	40	130	370	550	590

Using a technology method of Example 2 (see the marginal note on using technology), we obtain the following regression line and plot (coefficients rounded to three significant digits):

$y = x - 102$

59

Graph:

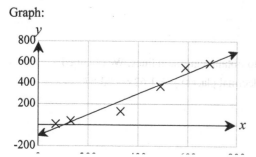

b. To interpret the slope, recall that units of the slope are units of y (engineering doctorates) per unit of x (natural science doctorates). Thus, $m \approx 1$ engineering doctorate per natural science doctorate, indicating that there is about one additional doctorate in engineering per additional doctorate in the natural sciences.

c. Using the technology method of Example 3, we can use technology to show the value of r^2:

$$r^2 \approx 0.9525$$

$$r = \sqrt{r^2} \approx \sqrt{0.9525} \approx 0.976$$

Since r is close to 1, the correlation between x and y is a strong one.

d. Yes; the graph suggests a linear relationship; the data points are close to the regression line and show no obvious pattern (such as a curve).

31. a. As t is time in years since 1990, we use the following set of data for the regression:

t	0	5	10	15	20	22
y	70	130	330	490	590	690

Using a technology method of Example 2 (see the marginal note on using technology), we obtain the following regression line and plot (coefficients rounded to three significant digits):

$$y = 28.9t + 37.0$$

Graph:

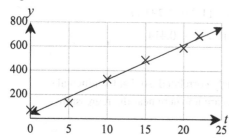

$r \approx \sqrt{0.9839} \approx 0.992$

b. Units of the slope are units of y (natural science doctorates) per unit of t (years); thus, doctorates per year. So, the number of natural science doctorates has been increasing at a rate of about 28.9 per year.

c. The slopes of successive pairs of points do not show an increasing nor decreasing trend as we go from left to right, so the number of natural science doctorates is increasing at a more-or-less constant rate.

d. Yes: If r had been equal to 1, then the points would lie exactly on the regression line, which would indicate that the number of doctorates is growing at a constant rate.

33. a. More-or-less constant rate; Exercise 29 suggests a roughly linear relationship between the number of natural science doctorates and the number of engineering doctorates, and Exercise 31 suggests that the number of natural science doctorates has been increasing at a more-or less constant rate. Therefore, the number of engineering doctorates is also increasing at a more-or-less constant rate.

b. No; $r = 1$ in Exercise 29 would indicate an exactly linear relationship between the number of natural science doctorates and the number of engineering doctorates, and so the conclusion would be the same.

c. No; $r = 1$ in Exercise 31 would indicate that the number of natural science doctorates has been increasing at a constant rate, and so the conclusion would be the same.

35. a. Using the method of Example 3, we obtain the following regression line and plot (coefficients rounded to two decimal places): $p = 0.13t + 0.22$; $r \approx 0.97$
Graph:

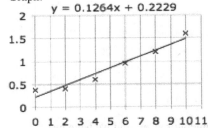

y = 0.1264x + 0.2229

b. The first and last points lie above the regression line, while the central points lie below it, suggesting a curve.
c. Here is a worksheet showing the computation of the residuals (based on Example 1 in the text):

◇	A	B	C	D
1	t	p (observed)	p (predicted)	Residual
2	0	0.38	=0.13*A2+0.22	=B2-C2
3	2	0.4		
4	4	0.6		
5	6	0.95		
6	8	1.2		
7	10	1.6		

◇	A	B	C	D
1	t	p (observed)	p (predicted)	Residual
2	0	0.38	0.22	0.16
3	2	0.4	0.48	-0.08
4	4	0.6	0.74	-0.14
5	6	0.95	1	-0.05
6	8	1.2	1.26	-0.06
7	10	1.6	1.52	0.08

Notice that the residuals are positive at first, become negative, and then become positive, confirming the impression from the graph.

Communication and reasoning exercises

37. The regression line is defined to be the line that gives the lowest sum-of-squares error, SSE. If we are given two points, (a, b) and (c, d) with $a \neq c$, then there is a line that passes through these two points, giving SSE = 0. Since 0 is the smallest value possible, this line must be the regression line.

39. If the points (x_1, y_1), (x_2, y_2), ..., (x_n, y_n) lie on a straight line, then the sum-of-squares error, SSE, for this line is zero. Since 0 is the smallest value possible, this line must be the regression line.

41. Calculation of the regression line:

x	y	xy	x^2
0	0	0	0
$-a$	a	$-a^2$	a^2
a	a	a^2	a^2
0	$2a$	0	$2a^2$

(The bottom row contains the column sums.)

$n = 3$ (number of data points)

Slope: $m = \dfrac{n(\sum xy) - (\sum x)(\sum y)}{n(\sum x^2) - (\sum x)^2} = \dfrac{3(0) - (0)(2a)}{3(2a^2) - 0^2} = 0$

Correlation coefficient $r = \dfrac{n(\sum xy) - (\sum x)(\sum y)}{\sqrt{n(\sum x^2) - (\sum x)^2} \cdot \sqrt{n(\sum y^2) - (\sum y)^2}}$ has the same numerator as m, and we have just

seen that this numerator is zero. Hence, $r = 0$.

43. No. The regression line through $(-1, 1)$, $(0, 0)$, and $(1, 1)$ passes through none of these points.

45. (Answers may vary.) The data in Exercise 35 give $r \approx 0.97$, yet the plotted points suggest a curve, not a straight line.

Chapter 1 Review

1. a. 1   **b.** −2   **c.** 0

d. $f(2) - f(-2) = 0 - 1 = -1$

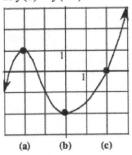

(a)   (b)   (c)

3. a. 1   **b.** 0   **c.** 0

d. $f(1) - f(-1) = 0 - 1 = -1$

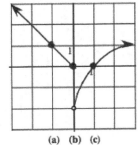

(a)   (b)   (c)

5. $y = -2x + 5$

y-intercept $= 5$, slope $= -2$

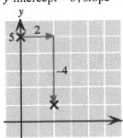
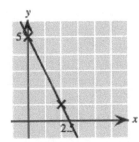

7. $y = \begin{cases} x/2 & \text{if} & -1 \le x \le 1 \\ x - 1 & \text{if} & 1 < x \le 3 \end{cases}$

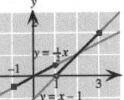
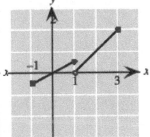

9. The graph of the function has a V-shape, indicating an absolute value function. Graph:

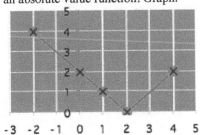

11. The graph of the function is a straight line, indicating a linear function. Graph:

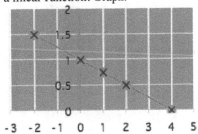

63

13. The parabolic shape of the graph indicates a quadratic model. Graph:

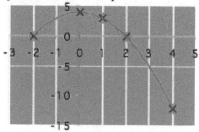

15. Through $(3, 2)$ with slope -3

Point: $(3, 2)$ **Slope:** $m = -3$ **Intercept:** $b = y_1 - mx_1 = 2 - (-3)(3) = 2 + 9 = 11$

Thus, the equation is $y = mx + b = -3x + 11$.

17. Through $(1, -3)$ and $(5, 2)$

Point: $(1, -3)$ **Slope:** $m = \dfrac{y_2 - y_1}{x_2 - x_1} = \dfrac{2 - (-3)}{5 - 1} = \dfrac{5}{4} = 1.25$

Intercept: $b = y_1 - mx_1 = -3 - (1.25)(1) = -4.25$

Thus, the equation is $y = mx + b = 1.25x - 4.25$.

19. Through $(1, 2)$ parallel to $x - 2y = 2$

Point: $(1, 2)$

Slope: Same as slope of $x - 2y = 2$. To find the slope, solve for y:

$$-2y = -x - 2$$

$$y = \frac{1}{2}x + 1, \text{ so that } m = \frac{1}{2}.$$

Intercept: $b = y_1 - mx_1 = 2 - \dfrac{1}{2}(1) = \dfrac{3}{2}$

Thus, the equation is $y = mx + b = \dfrac{1}{2}x + \dfrac{3}{2}$.

21. With slope 4 crossing $2x - 3y = 6$ at its x-intercept

We need the x-intercept of $2x - 3y = 6$. This is given by setting $y = 0$ and solving for x:

$$2x - 0 = 6$$

$$x = 3$$

Thus, the point is $(3, 0)$ because $y = 0$ on the x-axis.

Slope: $m = 4$

Intercept: $b = y_1 - mx_1 = 0 - 4(3) = -12$

Thus, the equation is

$$y = mx + b = 4x - 12$$

23.

$y = -x/2 + 1$:

x	Observed y	Predicted y	Residual^2
−1	1	1.5	0.25
1	2	0.5	2.25
2	0	0	0
		SSE:	2.5

$y = -x/4 + 1$:

x	Observed y	Predicted y	Residual^2
−1	1	1.25	0.0625
1	2	0.75	1.5625
2	0	0.5	0.25
		SSE:	1.875

The second line, $y = -x/4 + 1$, is a better fit.

25.

x	y	xy	x^2	y^2
−1	1	−1	1	1
1	2	2	1	4
2	0	0	4	0
2	3	1	6	5

(The bottom row contains the column sums.)

$n = 3$ (number of data points)

Slope: $m = \dfrac{n(\sum xy) - (\sum x)(\sum y)}{n(\sum x^2) - (\sum x)^2} = \dfrac{3(1) - (2)(3)}{3(6) - 2^2} = \dfrac{-3}{14} \approx -0.214$

Intercept: $b = \dfrac{\sum y - m(\sum x)}{n} = \dfrac{3 - (-0.214)(2)}{3} \approx 1.14$

Thus, the regression line is $y = mx + b = -0.214x + 1.14$.

$r = \dfrac{n(\sum xy) - (\sum x)(\sum y)}{\sqrt{n(\sum x^2) - (\sum x)^2} \cdot \sqrt{n(\sum y^2) - (\sum y)^2}} = \dfrac{3(1) - (2)(3)}{\sqrt{3(6) - 2^2}\sqrt{3(5) - 3^2}} \approx -0.33$

27. a. Graph:

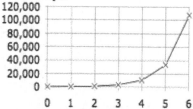

Since the data definitely suggests a curve, we rule out a linear function, leaving us with a choice of quadratic or exponential. Of the two, an exponential function would fit best, given the leveling off we see on the left; the graph of a quadratic function would not flatten out, but instead form a low point and begin rising again toward the left.**b.** The ratios (rounded to 1 decimal place) are:

$V(1)/V(0)$	$V(2)/V(1)$	$V(3)/V(2)$	$V(4)/V(3)$	$V(5)/V(4)$	$V(6)/V(5)$
$\dfrac{300}{100} = 3$	$\dfrac{1,000}{300} \approx 3.3$	$\dfrac{3,300}{1,000} = 3.3$	$\dfrac{10,500}{3,300} \approx 3.2$	$\dfrac{33,600}{10,500} \approx 3.2$	$\dfrac{107,400}{33,600} \approx 3.2$

They are close to 3.2.

c. The data suggest that website traffic is increasing by a factor of around 3.2 per year, so the prediction for next year (year 6) would be around $3.2 \times 107,400 \approx 343,700$ visits per day.

29. a. $c(x) = \begin{cases} 0.03x + 2 & \text{if } 0 \leq x \leq 50 \\ 0.05x + 1 & \text{if } x > 50 \end{cases}$,

Notice that x is *thousands* of visit per day, so 10,000 visits corresponds to $x = 10$, and the servers will crash an average of

$$c(10) = 0.03(10) + 2 = 2.3 \text{ times per day.}$$

(We used the first formula because 10 is in the interval [0, 50].) For 50,000 visitors,

$$c(50) = 0.03(50) + 2 = 3.5 \text{ crashes per day.}$$

(We again used the first formula because 50 is still in the interval [0, 50].) For 100,000 visitors,

$$c(100) = 0.05(100) + 1 = 6 \text{ crashes per day.}$$

(We used the second formula because 100 is in the interval $(50, +\infty)$.)

b. The coefficient 0.03 is the slope of the first formula, indicating that, for Web site traffic of up to 50,000 visits per day ($0 \leq x \leq 50$), the number of crashes is increasing by 0.03 per additional thousand visits.

c. To experience 8 crashes in a day, we desire $c(x) = 8$. If we try the first formula, we get

$$0.03x + 2 = 8$$

giving $x = (8 - 2)/0.03 = 200$, which is not in the domain of the first formula. So, we try the second formula:

$$0.05x + 1 = 8$$

$$0.05x = 7, \text{ so } x = \frac{7}{0.05} = 140,$$

which is in the domain of the second formula. Thus, we estimate that there were 140,000 visitors that day.

31.

t	1	2	3	4	5	6
$n(t)$	12.5	37.5	62.5	72.0	74.5	75.0

(a) Technology formulas:

    (A): `300/(4+100*5^(-t))`   (B): `13.3*t+8.0`

    (C): `-2.3*t^2+30.0*t-3.3`   (D): `7*3^(0.5*t)`

Here are the values for the four given models (rounded to 1 decimal place):

t	1	2	3	4	5	6
(A)	12.5	37.5	62.5	72.1	74.4	74.9
(B)	21.3	34.6	47.9	61.2	74.5	87.8
(C)	24.4	47.5	66.0	79.9	89.2	93.9
(D)	12.1	21.0	36.4	63.0	109.1	189.0

Model (A) gives an almost perfect fit, whereas the other models are not even close.

b. Looking at the table, we see the following behavior as t increases:

(A) Leveling off (B) Rising (C) Rising (begins to fall after 7 months, however) (D) Rising

33. a. Using $v(c) = -0.000005c^2 + 0.085c + 1,750,$ we get

$$v(5,000) = -0.000005(5,000)^2 + 0.085(5,000) + 1,750 = -125 + 425 + 1,750 = 2,050$$

$$v(6,000) = -0.000005(6,000)^2 + 0.085(6,000) + 1,750 = -180 + 510 + 1,750 = 2,080$$

Thus, increasing monthly advertising from $5,000 to $6,000 per month would result in $2,080 - 2,050 = 30$ more visits per day.

b. The following table shows the result of increasing expenditure by steps of $1,000:

Tech formula: `-0.000005*x^2+0.085*x+1750`

c	5,000	6,000	7,000	8,000	9,000	10,000
$v(c)$	2,050	2,080	2,100	2,110	2,110	2,100

The successive changes in the numbers of visits are:

$2,080 - 2,050 = 30; 2,100 - 2,080 = 20; 2,110 - 2,100 = 10; 2,110 - 2,110 = 0; 2,100 - 2,110 = -10,$

showing that the numbers of visits would increase at a slower and slower rate and then begin to decrease.

c. Here is a portion of the graph of v:

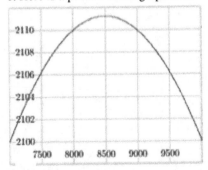

For $c = 8,500$ or larger, we see that Web site traffic is projected to decrease as advertising increases, and then drop toward zero. Thus, the model does not appear to give a reasonable prediction of traffic at expenditures larger than $8,500 per month.

35. a. Point: $(5,000, 2,050)$ **Slope:** $m = \dfrac{v_2 - v_1}{c_2 - c_1} = \dfrac{2,100 - 2,050}{6,000 - 5,000} = \dfrac{50}{1,000} = 0.05$

Intercept: $b = v_1 - mc_1 = 2,050 - (0.05)(5,000) = 1,800$

Thus, the equation is $v = mc + b = 0.05c + 1,800.$

b. A budget of $7,000 per month for banner ads corresponds to $v = 7,000.$

$$v(7,000) = 0.05(7,000) + 1,800 = 2,150 \text{ new visitors per day}$$

c. We are given $v = 2,500$ and want c.

$$2,500 = 0.05c + 1,800$$

$$0.05c = 2,500 - 1,800 = 700$$

Thus, $c = \dfrac{700}{0.05} = \$14,000$ per month.

37. Point: $(w, d) = (70, 74.5)$ **Slope:** $m = \dfrac{d_2 - d_1}{w_2 - w_1} = \dfrac{93.5 - 74.5}{90 - 70} = \dfrac{19}{20} = 0.95$

Intercept: $b = d_1 - mw_1 = 74.5 - (0.95)(70) = 8$

Thus, the equation is $d = mw + b = 0.95w + 8$.

OHagan dropped 90 m, so $d = 90$, and we want w.

$$90 = 0.95w + 8$$

$$0.95w = 90 - 8 = 82$$

Thus, $w = \dfrac{82}{.95} \approx 86$ kg.

39. a. Cost function: $C = mx + b$, where b = fixed cost = \$500 per week, and m = marginal cost = \$5.50 per album

Thus, the linear cost function is $C = 5.5x + 500$.

Revenue function: $R = mx + b$, where b = fixed revenue = 0, and m = marginal revenue = \$9.50 per album

Thus, the linear revenue function is $R = 9.5x$.

Profit function: $P = R - C$

$$P = 9.5x - (5.5x + 500)$$

$$= 4x - 500$$

b. For breakeven, $P = 0$

$$4x - 500 = 0$$

$$4x = 500$$

$$x = \dfrac{500}{4} = 125 \text{ albums per week}$$

To make a profit, the company should sell more than this number.

c. $R = 8.00x$

$$P = 8x - (5.5x + 500) = 2.5x - 500$$

For breakeven,

$$2.5x - 500 = 0, \text{ so } x = \dfrac{500}{2.5} = 200$$

To make a profit, the company should sell more than this number.

41. a. Demand: We are given two points: $(p, q) = (7, 500)$ and $(9.5, 300)$

Slope: $m = \dfrac{q_2 - q_1}{p_2 - p_1} = \dfrac{300 - 500}{9.5 - 7} = \dfrac{-200}{2.5} = -80$

Intercept: $b = q_1 - mp_1 = 500 - (-80)(7) = 500 + 560 = 1,060$

Thus, the demand equation is $q = mp + b = -80p + 1,060$.

b. When $p = \$12$, the demand is

$q = -80(12) + 1,060 = 100$ albums per week

c. From Exercise 39, the cost function is

$C = 5.5q + 500$ We use q for the monthly sales rather than x.

$\quad\; = 5.5(-80p + 1,060) + 500$ We want everything expressed in terms of p, so we used the demand equation.

$\quad\; = -440p + 5,830 + 500$

$C = -440p + 6,330$

To compute the profit in terms of price, we need the revenue as well:

$R = pq = p(-80p + 1,060) = -80p^2 + 1,060p$ **Profit:** $P = R - C$

$P = -80p^2 + 1,060p - (-440p + 6,330) = -80p^2 + 1,500p - 6,330$

Now we compare profits for the three prices:

$P(7.00) = -80(7)^2 + 1,500(7) - 6,330 = \$250$

$P(9.50) = -80(9.5)^2 + 1,500(9.5) - 6,330 = \$700$

$P(12) = -80(12)^2 + 1,500(12) - 6,330 = \$150$

Thus, charging \$9.50 will result in the largest weekly profit of \$700.

43. a. Calculation of the regression line:

x	y	xy	x^2
8	440	3,520	64
8.5	380	3,230	72.25
10	250	2,500	100
11.5	180	2,070	132.25
38	1,250	11,320	368.5

(The bottom row contains the column sums.)

$n = 4$ (number of data points)

Slope: $m = \dfrac{n(\sum xy) - (\sum x)(\sum y)}{n(\sum x^2) - (\sum x)^2} = \dfrac{4(11,320) - (38)(1,250)}{4(368.5) - 38^2} = \dfrac{-2,220}{30} = -74$

Intercept: $b = \dfrac{\sum y - m(\sum x)}{n} = \dfrac{1,250 - (-74)(38)}{4} = 1,015.5$

Thus, the regression line is $y = mx + b = -74x + 1,015.5$. Using the variable names p and q makes this equation

$q = -74p + 1,015.5$.

b. $q(10.50) = -74(10.50) + 1,015.5 = 238.5 \approx 239$ albums per week

Section 2.1

1. $PV = 2{,}000$, $r = 0.06$, $t = 1$

$INT = PVrt = 2{,}000(0.06)(1) = \$120$

$FV = PV + INT = 2{,}000 + 120 = \$2{,}120$

3. $PV = 4{,}000$, $i = 0.005$, $n = 8$

$INT = PVin = 4{,}000(0.005)(8) = \$160$

$FV = PV + INT = 4{,}000 + 160 = \$4{,}160$

5. $PV = 20{,}200$, $r = 0.05$, $t = \frac{1}{2}$

$INT = PVrt = 20{,}200(0.05)(\frac{1}{2}) = \$505$

$FV = PV + INT = 20{,}200 + 505 = \$20{,}705$

7. $PV = 10{,}000$, $r = 0.03$, $t = 10/12$

$INT = PVrt = 10{,}000(0.03)(10/12) = \$250$

$FV = PV + INT = 10{,}000 + 250 = \$10{,}250$

9. $PV = 12{,}000$, $i = 0.0005$, $n = 10$

$INT = PVin = 12{,}000(0.0005)(10) = \$60$

$FV = PV + INT = 12{,}000 + 60 = \$12{,}060$

11. $PV = FV/(1 + rt)$

$= 10{,}000/(1 + 0.02 \times 5) = \$9{,}090.91$

13. $PV = FV/(1 + rt)$

$= 1{,}000/(1 + 0.07 \times 0.5) = \$966.18$

15. $PV = FV/(1 + in)$

$= 15{,}000/(1 + 0.0003 \times 15) = \$14{,}932.80$

Applications

17. $FV = PV(1 + rt)$

$= 5{,}000(1 + 0.08 \times 0.5) = \$5{,}200$

19. $PV = FV/(1 + in)$

$= 1{,}000/(1 + 0.0002 \times 12) = \$997.61$

21. We are given the interest and asked to compute r.

$INT = 250$, $PV = 1{,}000$, $t = 5$

$INT = PVrt$

$250 = 1{,}000 \times r \times 5 = 5{,}000r$

$r = \dfrac{250}{5{,}000} = 0.05$, or 5%

23. Interest every six months: $INT = PVrt = (10{,}000)(0.034)(0.5) = \$170$

Total interest over the 10-year life of the bond: $INT = PVrt = (10{,}000)(0.034)(10) = \$3{,}400$

25. $PV = \dfrac{FV}{1 + rt} = \dfrac{8{,}840}{1 + (.0525)(2)} = \$8{,}000$

27. Goldman Sachs: $INT = PVrt = (5{,}000)(0.0615)(3) = \$922.50$

Wells Fargo: $INT = PVrt = (5{,}000)(0.035)(7) = \$1{,}225.00$

Wells Fargo would pay the most total interest, $\$1{,}225.00$.

29. We are given present and future values, and want to compute t.

$FV = 4{,}640$, $PV = 4{,}000$, $r = 0.08$

$FV = PV(1 + rt)$

70

$$4,640 = 4,000(1 + 0.08t)$$

$$1 + 0.08t = \frac{4,640}{4,000} = 1.160$$

$$0.08t = 0.160$$

$$t = \frac{0.160}{0.08} = 2 \text{ years}$$

31. $INT = PVin \Rightarrow 50 = 1,000 \times 4i = 4,000i;\ i = 50/4,000 = 0.0125,$ which is 1.25% weekly interest.

Annual rate $= 52 \times 0.0125 = 0.65,$ or 65%

33. You will pay $5,000 \times 0.09 \times 2 = \$900$ in interest on the loan. Adding the $100 fee, you pay the bank a total of $1,000 over the two years. To find the effective rate, we solve $1,000 = 5,000 \times 2r = 10,000r;\ r = 1,000/10,000 = 0.1,$ so the rate is 10%.

35. $5 = 69r/12;\ r = 12 \times 5/69 \approx 0.86957$ or 86.957%

37. Use $FV = PV(1 + in)$ with

$PV = 255.96,\ FV = 317.44,\ b = 6$ (months)

$$317.44 = 255.96(1 + 6i) = 255.96 + 1535.76i$$

$$i = (317.44 - 255.96)/1535.76 \approx 0.0400, \text{ or } 4.00\%$$

39. Selling in May would have given you a small loss while selling in any later month of the year would have gotten you a gain. Calculating the monthly returns as in the preceding exercises, we get the following figures:

Jun	Jul	Aug	Sep	Oct	Nov	Dec
4.24%	1.55%	2.56%	1.93%	4.10%	4.91%	4.32%

The largest monthly return would have been 4.91% if you had sold in November 2010.

41. No. Simple interest increase is linear. The graph is visibly not linear in that time period. Further, the slopes of the lines through the successive pairs of marked points are quite different.

43. 1950 population: $PV = 500,000;$ 2000 population: $FV = 2,800,000$

$$INT = FV - PV = 2,800,000 - 500,000 = 2,300,000$$

$$INT = PVrt$$

$$2,300,000 = 500,000r(50) = 25,000,000r$$

$$r = \frac{2,300,000}{25,000,000} \approx 0.092 \text{ or } 9.2\%$$

45. We are given: $PV = $ 1950 population $= 500,000;\ r = 0.092$ (from Exercise 43); $t = 60$ (years to 2010)

$$FV = PV(1 + rt) = 500,000(1 + 0.092 \times 60) = 3,260,000$$

47. $PV = $ 1950 population $= 500,000;\ r = 0.092$ (from Exercise 43). After t years, the population will be

$$FV = PV(1 + rt) = 500,000(1 + 0.092t) = 500,000 + 46,000t \text{ or } 500 + 46t \text{ thousand}$$

($t = $ time in years since 1950).

71

Graph:

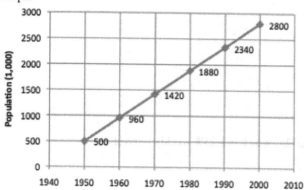

49. Actual discount $= 0.0025/2 = 0.00125$. Selling price $= 5{,}000 - (0.00125)(5{,}000) = \$4{,}993.75$.

$PV = \$4{,}993.75, \ FV = \$5{,}000, \ t = 0.5$

$FV = PV(1 + rt)$

$5{,}000 = 4{,}993.75(1 + 0.5r)$

$1 + 0.5r = 5{,}000/4{,}993.75$

$0.5r = 5{,}000/4{,}993.75 - 1$

$r = 2(5{,}000/4{,}993.75 - 1) \approx 0.002503, \text{ or } 0.2503\%$

51. To say that the discount rate is 3.705% means that its selling price (PV) is 3.705% lower than its maturity value (FV). To simplify the calculation, let us use a T-bill with a maturity value of $\$10{,}000$:

$PV = 10{,}000 - 10{,}000(0.03705/2) = 9814.75$

$10{,}000 = 9{,}814.75(1 + 0.5r) = 9{,}814.75 + 4{,}907.375r$

$r = (10{,}000 - 9{,}814.75)/4{,}907.375 = 0.03775, \text{ or } 3.775\%$

Communication and reasoning exercises

53. Graph (A) is the only possible choice, because the equation $FV = PV(1 + rt) = PV + PVrt$ gives the future value as a linear function of time.

55. $FV = PV(1 + in) = PV + PVin$. Since $FV = 1{,}000 + 0.5n$, we have

$PV = 1{,}000$

$PVin = 1{,}000in = 0.5n$

so

$i = \dfrac{0.5}{1{,}000} = 0.0005, \text{ or } 0.05\%$

57. $FV = PV(1 + rt)$;

$r = \text{ Annual rate} = 12 \times \text{Monthly rate} = 12i$;

$t = \text{ Number of years} = \dfrac{n}{12}$

so $\quad FV = PV(1 + rt) = PV\left(1 + (12i)\dfrac{n}{12}\right) = PV(1 + in)$

59. Wrong. In simple interest growth, the change each year is a fixed percentage of the *starting* value, and not the preceding year's value. (Also see the next exercise.)

61. Simple interest is always calculated on a constant amount, PV. If interest is paid into your account, then the amount on which interest is calculated does not remain constant.

1. We use $FV = PV(1 + i)^n$.

$PV = \$10,000,\ i = 0.002,\ n = 15$

$$FV = PV(1 + i)^n$$
$$= 10{,}000(1 + 0.002)^{15}$$
$$= 10{,}000(1.002)^{15} \approx \$10{,}304.24$$

Technology: 10000*(1+0.002)^15

3. We use $FV = PV(1 + i)^n$.

$PV = \$10,000,\ i = 0.002,\ n = 10 \times 12 = 120$

$$FV = PV(1 + i)^n$$
$$= 10{,}000(1 + 0.002)^{120}$$
$$= 10{,}000(1.002)^{120} \approx \$12{,}709.44$$

Technology: 10000*(1+0.002)^120

5. We use $FV = PV(1 + r/m)^{mt}$.

$PV = \$10,000,\ r = 0.03,\ m = 1,\ t = 10$

$$FV = PV(1 + r/m)^{mt}$$
$$= 10{,}000(1 + 0.03)^{10}$$
$$= 10{,}000(1.03)^{10} \approx \$13{,}439.16$$

Technology: 10000*(1+0.03)^10

7. We use $FV = PV(1 + i)^n$.

$PV = \$10,000,\ i = 0.025/4 = 0.00625,\ n = 4 \times 5 = 20$

$$FV = PV(1 + i)^n$$
$$= 10{,}000(1 + 0.00625)^{20}$$
$$= 10{,}000(1.00625)^{20} \approx \$11{,}327.08$$

Technology: 10000*(1+0.025/4)^(4*5)

9. We use $FV = PV(1 + r/m)^{mt}$.

$PV = \$10,000,\ r = 0.065,\ m = 365,\ t = 10$

$$FV = PV(1 + r/m)^{mt}$$
$$= 10{,}000\left(1 + \frac{0.065}{365}\right)^{365 \times 10}$$
$$= 10{,}000\left(1 + \frac{0.065}{365}\right)^{3{,}650} \approx \$19{,}154.30$$

Technology: 10000*
(1+0.065/365)^(365*10)

11. We use $PV = FV(1 + i)^{-n}$

$FV = \$1,000,\ n = 10,\ i = 0.05$

$$FV = FV(1 + i)^{-n}$$
$$= 1{,}000(1 + 0.05)^{-10}$$
$$= 1{,}000(1.05)^{-10} \approx \$613.91$$

Technology: 1000*(1+0.05)^(-10)

13. We use $PV = \dfrac{FV}{(1 + r/m)^{mt}}$

$FV = \$1,000,\ t = 5,\ r = 0.042,\ m = 52$

$$PV = \frac{FV}{(1 + r/m)^{mt}}$$
$$= \frac{1{,}000}{\left(1 + \frac{0.042}{52}\right)^{52 \times 5}}$$
$$= \frac{1{,}000}{\left(1 + \frac{0.042}{52}\right)^{260}} \approx \$810.65$$

Technology: 1000/(1+0.042/52)^(52*5)

15. We use $PV = FV(1 + i)^{-n}$

$FV = \$1,000,\ n = 4,\ i = -0.05$

$$PV = FV(1 + i)^{-n}$$
$$= 1{,}000(1 - 0.05)^{-4}$$
$$= 1{,}000(0.95^{-4}) \approx \$1{,}227.74$$

Technology: 1000*(1-0.05)^(-4)

17. $r_{\text{nom}} = 0.05$, $m = 4$

$$r_{\text{eff}} = (1 + r_{\text{nom}}/m)^m - 1$$

$$= \left(1 + \frac{0.05}{4}\right)^4 - 1$$

$$\approx 0.0509, \text{ or } 5.09\%$$

Technology: $(1+0.05/4)^4-1$

19. $r_{\text{nom}} = 0.10$, $m = 12$

$$r_{\text{eff}} = (1 + r_{\text{nom}}/m)^m - 1$$

$$= \left(1 + \frac{0.10}{12}\right)^{12} - 1$$

$$\approx 0.1047, \text{ or } 10.47\%$$

Technology: $(1+0.10/12)^{12}-1$

21. $r_{\text{nom}} = 0.10$, $m = 365 \times 24 = 8{,}760$

$$r_{\text{eff}} = (1 + r_{\text{nom}}/m)^m - 1$$

$$= \left(1 + \frac{0.10}{8{,}760}\right)^{8{,}760} - 1$$

$$\approx 0.1052, \text{ or } 10.52\%$$

Technology: $(1+0.10/8760)^{8760}-1$

Applications

23. $PV = \$1{,}000$, $i = 0.06/4 = 0.015$ $n = 4 \times 4 = 16$

$$FV = PV(1 + i)^n$$

$$= 1{,}000(1 + 0.015)^{16}$$

$$= 1{,}000(1.015)^{16} \approx \$1{,}268.99$$

The deposit will have grown by

$$\$1{,}268.99 - \$1{,}000 = \$268.99.$$

Technology: $1000*(1+0.06/4)^{\wedge}(16)$

25. $PV = \$3{,}000$, $i = -0.376$, $n = 3$

$$FV = PV(1 + i)^n$$

$$= 3{,}000(1 - 0.376)^3 \approx \$728.91$$

Technology: $3000*(1-0.376)^{\wedge}3$

27. $FV = \$5{,}000$, $i = 0.055$, $n = 10$

$$PV = \frac{FV}{(1 + i)^n}$$

$$= \frac{5{,}000}{(1 + 0.055)^{10}}$$

$$= \frac{5{,}000}{1.055^{10}} \approx \$2{,}927.15$$

Technology: $5000/(1+0.055)^{\wedge}10$

29. Gold: $PV = \$5{,}000$, $i = 0.10$, $n = 10$

$$FV = PV(1 + i)^n$$

$$= 5{,}000(1 + 0.10)^{10}$$

$$= 5{,}000(1.10)^{10} \approx \$12{,}968.71$$

CDs: $PV = \$5{,}000$, $i = 0.05/2 = 0.025$, $n = 2 \times 10 = 20$

$$FV = PV(1 + i)^n$$

$$= 5{,}000(1 + 0.025)^{20}$$

$$= 5{,}000(1.025)^{20} \approx \$8{,}193.08$$

Combined value after 10 years

$$= \$12{,}968.71 + \$8{,}193.08 = \$21{,}161.79$$

Technology:
$5000*(1+0.10)^{\wedge}10+5000*$
$(1+0.05/2)^{\wedge}(2*10)$

31. $PV = \$200,000$, $i = -0.02$, $n = 10$

$$FV = PV(1 + i)^n$$
$$= 200,000(1 - 0.02)^{10}$$
$$= 200,000(0.98)^{10} \approx \$163,414.56$$

Technology: `200000*(1-0.02)^10`

33. $FV = \$1,000,000$, $i = 0.06$, $n = 30$

$$PV = \frac{FV}{(1 + i)^n}$$
$$= \frac{1,000,000}{(1 + 0.06)^{30}}$$
$$= \frac{1,000,000}{1.06^{30}} \approx \$174,110$$

Technology: `1000000/(1+0.06)^30`

35. $FV = \$297.91$, $n = 6 \times 3 = 18$, $i = -0.05$

$$PV = \frac{FV}{(1 + i)^n}$$
$$= \frac{297.91}{(1 - 0.05)^{18}} = \frac{297.91}{0.95^{18}} \approx \$750.00$$

Technology: `297.91/(1-0.05)^18`

37. $FV = PV(1 + i)^n = 8,144.64(1 + 0.0512)^2 = \$9,000$

39. $PV = \dfrac{FV}{(1 + i)^n}$

$$= \frac{10,000}{(1 + 0.0297)^{10}} = \frac{10,000}{1.0297^{10}} \approx \$7,462.65$$

41. Total interest on $1 during the 8-year life of the bonds is

$$INT = FV - PV = (1 + 0.0441)^8 - 1 \approx \$0.41233$$

Thus, the monthly simple interest rate would be $\dfrac{0.41233}{8 \times 12} \approx 0.0043$, or 0.43%.

43. $FV = \$100,000$, $i = 0.04$, $n = 15$

$$PV = \frac{FV}{(1 + i)^n}$$
$$= \frac{100,000}{(1 + 0.04)^{15}}$$
$$= \frac{100,000}{1.04^{15}} \approx \$55,526.45 \text{ per year}$$

Technology: `100000/(1+0.04)^15`

45. $FV = \$30,000$, $n = 5$, $i = 0.02$

$$PV = \frac{FV}{(1 + i)^n}$$
$$= \frac{30,000}{(1 + 0.02)^5} = \frac{30,000}{1.02^5} \approx \$27,171.92$$

Technology: `30000/(1+0.02)^5`

47. $FV = \$200,000$, $n = 10$, $i = 0.06$

$$PV = \frac{FV}{(1+i)^n}$$

$$= \frac{200,000}{(1+0.06)^{10}} = \frac{200,000}{1.06^{10}} \approx \$111,678.96$$

Technology: 200000/(1+0.06)^10

49. Step 1: Calculate the future value of the investment:

$$PV = \$1,000, \ i = 0.05, \ n = 2$$

$$FV = PV(1+i)^n$$

$$= 1,000(1+0.05)^2$$

$$= 1,000(1.05)^2 \approx \$1,102.50$$

Technology: 1000*(1+0.05)^2

Step 2: Discount this value using inflation:

$$FV = \$1,102.50, \ i = 0.03, \ n = 2$$

$$PV = \frac{FV}{(1+i)^n}$$

$$= \frac{1,102.50}{(1+0.03)^2} = \frac{1,102.50}{1.03^2} \approx \$1,039.21$$

Technology: 1102.50/(1+0.03)^2

51. Compare the effective yields of the two investments:

First Investment:

$$r_{nom} = 0.12, \ m = 1$$

$$r_{eff} = r_{nom} = 0.12, \text{ or } 12\%$$

Second Investment:

$$r_{nom} = 0.119, \ m = 12$$

$$r_{eff} = (1 + r_{nom}/m)^m - 1$$

$$= \left(1 + \frac{0.119}{12}\right)^{12} - 1$$

$$\approx 0.1257, \text{ or } 12.57\%$$

The better investment is the second.

53. $PV = \$24$, $i = 0.063$, $n = 2015 - 1626 = 389$

$$FV = PV(1+i)^n$$

$$= 24(1+0.063)^{389}$$

$$= 24(1.063)^{389} \approx \$503,096 \text{ million}$$

This is more than the 2015 estimated market value of $362,524 million. Thus, the Lenape could have bought the island back in 2015.

Technology: 24*(1+0.063)^389

55. $PV = 100$ reals, $i = 0.099$, $n = 5$

$$FV = PV(1+i)^n$$

$$= 100(1+0.099)^5$$

$$= 100(1.099)^5 \approx 160 \text{ reals}$$

Technology: 100*(1+0.099)^5

57. $FV = 1,000$ bolivianos, $i = 0.043$, $n = 10$

$$PV = \frac{FV}{(1+i)^n}$$

$$= \frac{1,000}{(1+0.043)^{10}} = \frac{1,000}{1.043^{10}}$$

$$\approx 656 \text{ bolivianos}$$

Technology: 1000/(1+0.043)^10

59. Step 1: Calculate the future value of the investment:

$PV = 1,000$ bolivars, $r = 0.08$, $m = 2$, $t = 10$

$$FV = PV(1 + r/m)^{mt}$$

$$= 1,000\left(1 + \frac{0.08}{2}\right)^{2 \times 10}$$

$$= 1,000(1.04)^{20} \approx 2,191.12 \text{ bolivars}$$

Technology: `1000*(1+0.08/2)^20`

Step 2: Discount this amount using the inflation rate:

$FV = 2,191.12$, $i = 0.685$, $n = 10$

$$PV = \frac{FV}{(1 + i)^n}$$

$$= \frac{2,191.12}{(1 + 0.685)^{10}} = \frac{2,191.12}{1.685^{10}} \approx 12$$

bolivars

Technology: `2191.12/(1+0.685)^10`

61. We compare the future values of 1 unit of the currency for a 1-year period:

Mexico: $PV = 1$ peso, $i = 0.053$, $n = 1$

$$FV = PV(1 + i)^n$$

$$= 1(1 + 0.053)^1$$

$$= 1.053 \text{ pesos}$$

Now discount this using inflation:

$FV = 1.053$, $i = 0.025$, $n = 1$

$$PV = \frac{FV}{(1 + i)^n}$$

$$= \frac{1.053}{(1 + 0.025)^1} = \frac{1.053}{1.025} \approx 1.027 \text{ pesos}$$

Nicaragua: $PV = 1$ gold cordoba, $i = 0.06/2 = 0.03$, $n = 2$

$$FV = PV(1 + i)^n$$

$$= 1(1 + 0.03)^2$$

$$= 1.03^2 = 1.0609 \text{ gold cordobas}$$

Now discount this using inflation:

$FV = 1.0609$, $i = 0.03$, $n = 1$

$$PV = \frac{FV}{(1 + i)^n}$$

$$= \frac{1.0609}{(1 + 0.03)^1} = \frac{1.0609}{1.03}$$

$$= 1.03 \text{ gold cordobas}$$

The investment in Nicaragua is better.

63. $PV = 255.96$, $FV = 317.44$, $t = 6$ months $= 1/2$ year

$317.44 = 255.96(1 + r)^{1/2}$, so solving for r gives $r = (317.44/255.96)^2 - 1 \approx 0.5381$ or 53.81%

65. Selling in May would have given you a small loss while selling in any later month of the year would have gotten you a gain. Calculating the annual returns as in the preceding exercises, we get the following figures:

Jun	Jul	Aug	Sep	Oct	Nov	Dec
62.89%	19.93%	33.91%	24.78%	55.31%	65.99%	56.03%

The largest annual return would have been 65.99% if you had sold in November 2010.

67. No. Compound interest increase is exponential, and exponential curves either increase continually (in the case of appreciation) or decrease continually (in the case of depreciation). The graph of the stock price has both increases and decreases during the given period, so the curve cannot model compound interest change.

69. My investment: $PV = \$5,000$, $r = 0.054$, $m = 2$

$$FV = PV(1 + r/m)^{mt} = 5,000\left(1 + \frac{0.054}{2}\right)^{2t} = 5,000(1.027)^{2t}$$

Friend's investment: $PV = \$6,000$, $r = 0.048$, $m = 2$

$$FV = PV(1 + r/m)^{mt} = 6{,}000\left(1 + \frac{0.048}{2}\right)^{2t} = 6{,}000(1.024)^{2t}$$

Solution via logarithms:

$$5{,}000(1.027)^{2t} = 6{,}000(1.024)^{2t}$$

$$(1.027)^{2t} = 1.2(1.024)^{2t}$$

$$2t \log 1.027 = \log[1.2(1.024)^{2t}] = \log 1.2 + 2t \log 1.024$$

$$2t(\log 1.027 - \log 1.024) = \log 1.2$$

$$t = \frac{\log 1.2}{2(\log 1.027 - \log 1.024)} \approx 31$$

Solution via graphing:
Technology Formulas:

        Y₁ = 5000*1.027^(2*x)   Y₂ = 6000*1.024^(2*x)
Graph and zoomed-in view:

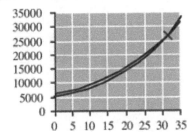 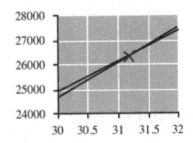

The graphs cross around $t \approx 31$ years. The value of the investment then is

$$5{,}000(1.027)^{2 \times 31} \approx \$26{,}100 \text{ (rounded to 3 significant digits)}$$

71. $PV = 40{,}000$, $i = 1.0$ (a 100% increase per period)

$$FV = PV(1 + i)^n = 40{,}000(1 + 1.0)^n = 40{,}000(2)^n$$

Logarithms:

$$40{,}000(2)^n = 1{,}000{,}000$$

$$2^n = 25$$

$$n \log 2 = \log 25$$

$$n = \frac{\log 25}{\log 2} \approx 4.65 \text{ months}$$

Technology:

        Y₁ = 40000*2^x   Y₂ = 1000000
Graph and zoomed-in view:

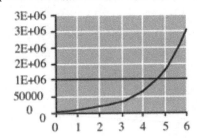 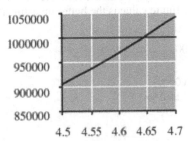

$n \approx 4.65$. Since n measures 6-month periods, this corresponds to $4.65/2 \approx 2.3$ years.

73. a. The amount you pay for the bond is its present value.

$FV = \$100,000, \; i = 0.15, \; t = 30$

$$PV = \frac{FV}{(1+i)^n} = \frac{100,000}{(1+0.15)^{30}} = \frac{100,000}{1.15^{30}} \approx \$1,510.31$$

b. Because the future value remains fixed at $100,000, the value of the bond at any time is its present value at that time, given the prevailing interest rate. Since it will pay $100,000 in 13 years' time, we have:

$FV = \$100,000, \; i = 0.0475, \; t = 13$

$$PV = \frac{FV}{(1+i)^n} = \frac{100,000}{(1+0.0.475)^{13}} = \frac{100,000}{1.0475^{13}} \approx \$54,701.29$$

c. By part (a) the bond cost you $1,510.31 and, by part (b), was worth $54,701.29 after 17 years.

$PV = \$1,510.31, \; FV = \$54,701.29$

$$FV = PV(1+i)^n$$

$$54,701.29 = 1,510.31(1+i)^{17}$$

$$\frac{54,701.29}{1,510.31} = (1+i)^{17}$$

$$1+i = \left(\frac{54,701.29}{1,510.31}\right)^{1/17}$$

$$i = \left(\frac{54,701.29}{1,510.31}\right)^{1/17} - 1 \approx 0.2351, \; \text{or } 23.51\%$$

Technology: `(54701.29/1510.31)^(1/17)-1`

Communication and reasoning exercises

75. The function $y = P(1 + r/m)^{mx}$ is not a linear function of x, but an exponential function. Thus, its graph is not a straight line.

77. Wrong. Its growth can be modeled by $0.01(1 + 0.10)^t = 0.01(1.10)^t$. This is an exponential function of t, not a linear one.

79. A compound interest investment behaves as though it were being compounded once a year at the effective rate. Thus, if two equal investments have the same effective rates, they grow at the same rate.

81. The effective rate exceeds the nominal rate when the interest is compounded more than once a year because then interest is being paid on interest accumulated during each year, resulting in a larger effective rate. Conversely, if the interest is compounded less often than once a year, the effective rate is less than the nominal rate.

83. Compare their future values in constant dollars. That is, compute their future values, and then discount each for inflation. The investment with the larger future value is the better investment.

85. $PV = 100, \; r = 0.10, \; m = 1, \; 10, \; 100, \; \ldots$

```
Y₁ = 100*(1.10)^x
Y₂ = 100*(1+ 0.10/10)^(10*x)
Y₃ = 100*(1+ 0.10/100)^(100*x)
...
```

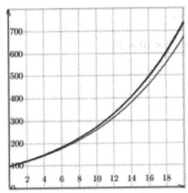

The graphs are approaching a particular curve (shown darker) as *m* gets larger, approximately the curve given by the largest value of *m*.

Section 2.3

1. $PMT = \$100$, $i =$ interest paid each period $= 0.05/12$, $n =$ total number of periods $= 12 \times 10 = 120$

$$FV = PMT\frac{(1+i)^n - 1}{i} = 100\frac{(1+0.05/12)^{120} - 1}{0.05/12} \approx \$15,528.23$$

Technology: `100*((1+0.05/12)^120-1)/(0.05/12)`

3. $PMT = \$1,000$, $i = 0.07/4$, $n = 4 \times 20 = 80$

$$FV = PMT\frac{(1+i)^n - 1}{i} = 1,000\frac{(1+0.07/4)^{80} - 1}{0.07/4} \approx \$171,793.82$$

Technology: `1000*((1+0.07/4)^80-1)/(0.07/4)`

5. $PV = \$5,000$, $PMT = \$100$, $i = 0.05/12$, $n = 12 \times 10 = 120$

We need to calculate the sum of $FV = PV(1+i)^n$ and $FV = PMT\dfrac{(1+i)^n - 1}{i}$.

$$FV = 5,000(1 + 0.05/12)^{120} + 100\frac{(1+0.05/12)^{120} - 1}{0.05/12} \approx \$23,763.28$$

Technology: `5000*(1+0.05/12)^120+100*((1+0.05/12)^120-1)/(0.05/12)`

7. $FV = \$10,000$, $i = 0.05/12$, $n = 12 \times 5 = 60$

$$PMT = FV\frac{i}{(1+i)^n - 1} = 10,000\frac{0.05/12}{(1+0.05/12)^{60} - 1} \approx \$147.05$$

Technology: `10000*0.05/12/((1+0.05/12)^60-1)`

9. $FV = \$75,000$, $i = 0.06/4$, $n = 4 \times 20 = 80$

$$PMT = FV\frac{i}{(1+i)^n - 1} = 75,000\frac{0.06/4}{(1+0.06/4)^{80} - 1} \approx \$491.12$$

Technology: `75000*0.06/4/((1+0.06/4)^80-1)`

11. We first account for the future value of the $10,000 already in the account: $PV = \$10,000$, $i = 0.05/12$, $n = 12 \times 5 = 60$

$$FV = PV(1+i)^n = 10,000(1 + 0.05/12)^{60}$$

We subtract this from $20,000 to get the future value of the payments, so:

$$PMT = FV\frac{i}{(1+i)^n - 1}$$

where

$$FV = \$20,000 - 10,000(1 + 0.05/12)^{60}$$

$$PMT = \frac{[20,000 - 10,000(1 + 0.05/12)^{60}](0.05/12)}{(1 + 0.05/12)^{60} - 1} \approx \$105.38$$

Technology: `(20000-10000*(1+0.05/12)^60)*0.05/12/((1+0.05/12)^60-1)`

13. $PMT = \$500$, $i = 0.03/12$, $n = 12 \times 20 = 240$

$$PV = PMT\frac{1 - (1+i)^{-n}}{i} = 500\frac{1 - (1+0.03/12)^{-240}}{0.03/12} \approx \$90,155.46$$

Technology: `500*(1-(1+0.03/12)^(-240))/(0.03/12)`

15. $PMT = \$1{,}500$, $i = 0.06/4$, $n = 4 \times 20 = 80$

$$PV = PMT \frac{1-(1+i)^{-n}}{i} = 1{,}500 \frac{1-(1+0.06/4)^{-80}}{0.06/4} \approx \$69{,}610.99$$

Technology: `1500*(1-(1+0.06/4)^(80))/(0.06/4)`

17. $FV = \$10{,}000$, $PMT = \$500$, $i = 0.03/12$, $n = 12 \times 20 = 240$

We need to fund both the future value and the payments, so the present value is the sum

$$PV = FV(1+i)^{-n} + PMT \frac{1-(1+i)^{-n}}{i}$$

$$= 10{,}000(1+0.03/12)^{-240} + 500 \frac{1-(1+0.03/12)^{-240}}{0.03/12} \approx \$95{,}647.68$$

Technology: `10000*(1+0.03/12)^(-240)+500*(1-(1+0.03/12)^(-240))/(0.03/12)`

19. $PV = \$100{,}000$, $i = 0.03/12$, $n = 12 \times 20 = 240$

$$PMT = PV \frac{i}{1-(1+i)^{-n}} = 100{,}000 \frac{0.03/12}{1-(1+0.03/12)^{-240}} \approx \$554.60$$

Technology: `100000*(0.03/12)/(1-(1+0.03/12)^(-240))`

21. $PV = \$75{,}000$, $i = 0.04/4 = 0.01$, $n = 4 \times 20 = 80$

$$PMT = PV \frac{i}{1-(1+i)^{-n}} = 75{,}000 \frac{0.01}{1-(1+0.01)^{-80}} \approx \$1{,}366.41$$

Technology: `75000*0.01/(1-(1+0.01)^(-80))`

23. $PV = \$100{,}000$, $FV = \$10{,}000$, $i = 0.03/12$, $n = 12 \times 20 = 240$

Part of the present value has to fund the future value of $10,000:

$$FV(1+i)^{-n} = 10{,}000(1+0.03/12)^{-240}$$

is the amount required; we subtract this from the present value and use

$$PV = 100{,}000 - 10{,}000(1+0.03/12)^{-240}$$

in the payment formula.

$$PMT = PV \frac{i}{1-(1+i)^{-n}} = \frac{(0.03/12)[100{,}000 - 10{,}000(1+0.03/12)^{-240}]}{1-(1+0.03/12)^{-240}} \approx \$524.14$$

Technology:
`(0.03/12)*(100000-10000*(1+0.03/12)^(-240))/(1-(1+0.03/12)^(-240))`

25. $PV = \$10{,}000$, $i = 0.09/12 = 0.0075$, $n = 4 \times 12 = 48$

$$PMT = PV \frac{i}{1-(1+i)^{-n}} = 10{,}000 \frac{0.0075}{1-(1+0.0075)^{-48}} \approx \$248.85$$

Technology: `10000*0.0075/(1-(1+0.0075)^(-48))`

27. $PV = \$100{,}000$, $i = 0.05/4$, $n = 4 \times 20 = 80$

$$PMT = PV \frac{i}{1-(1+i)^{-n}} = 100{,}000 \frac{0.05/4}{1-(1+0.05/4)^{-80}} \approx \$1{,}984.65$$

Technology: `100000*(0.05/4)/(1-(1+0.05/4)^(-80))`

29. $PV = 100,000$, $i = 0.043/12$, $n = 12 \times 30 = 360$

$$PMT = PV \frac{i}{1 - (1+i)^{-n}} = 100,000 \frac{0.043/12}{1 - (1+0.043/12)^{-360}} = \$494.87$$

Technology: `100000*(0.043/12)/(1-(1+0.043/12)^(-360))`

31. $PV = 1,000,000$, $i = 0.054/12$, $n = 12 \times 30 = 360$

$$PMT = PV \frac{i}{1 - (1+i)^{-n}} = 1,000,000 \frac{0.054/12}{1 - (1+0.054/12)^{-360}} = \$5,615.31$$

Technology: `1000000*(0.054/12)/(1-(1+0.054/12)^(-360))`

33. We calculated $PMT = \$494.87$ in Exercise 29.

$i = 0.043/12$, $n = 12 \times 30 = 360$, $k = 12 \times 10 = 120$, so $n - k = 240$

$$PV = PMT \frac{1 - (1+i)^{-(n-k)}}{i}$$

$$= 494.87 \frac{1 - (1+0.043/12)^{-240}}{0.043/12} = \$79,573.29$$

Technology: `494.87*(1-(1+0.043/12)^(-240))/(0.043/12)`

35. We calculated $PMT = \$5,615.31$ in Exercise 31.

$i = 0.054/12$, $n = 12 \times 30 = 360$, $k = 12 \times 5 = 60$, so $n - k = 300$

$$PV = PMT \frac{1 - (1+i)^{-(n-k)}}{i}$$

$$= 5,615.31 \frac{1 - (1+0.054/12)^{-300}}{0.054/12} = \$923,373.42$$

Technology: `5615.31*(1-(1+0.054/12)^(-300))/(0.054/12)`

37. First, we calculate the monthly payments:

$PV = 50,000$, $i = 0.085/12$, $n = 12 \times 200 = 2,400$

$$PMT = PV \frac{i}{1 - (1+i)^{-n}} = 50,000 \frac{0.085/12}{1 - (1+0.085/12)^{-2,400}} = \$354.17$$

Technology: `50000*(0.085/12)/(1-(1+0.085/12)^(-2400))`
Now calculate the outstanding balance based on the above payments:

$k = 12 \times 20 = 240$, so $n - k = 2,160$

$$PV = PMT \frac{1 - (1+i)^{-(n-k)}}{i}$$

$$= 354.17 \frac{1 - (1+0.085/12)^{-2,160}}{0.085/12} = \$50,000.46$$

Technology: `354.17*(1-(1+0.085/12)^(-2160))/(0.085/12)`
This is more than the original value of the loan. In a 200-year mortgage, the fraction of the initial payments going toward reducing the principal is so small that the rounding upward of the payment makes it appear that, for many years at the start of the mortgage, more is owed than the original amount borrowed.

39. The periodic payments are based on a 4.875% annual payment. For payments twice a year, this is
   $PMT = 1,000(0.04875/2) = \$24.375$

Since the bond yield is 4.880%, $i = 0.0488/2 = 0.0244$; $n = 2 \times 10 = 20$. The present value comes from the future value of $1,000 and the payments, which we treat like an annuity:

$$PV = FV(1+i)^{-n} + PMT\frac{1-(1+i)^{-n}}{i}$$

$$= 1{,}000(1+0.0244)^{-20} + 24.375\frac{1-(1+0.0244)^{-20}}{0.0244}$$

$$= 1{,}000(1.0244)^{-20} + 24.375\frac{1-1.0244^{-20}}{0.0244} \approx \$999.61$$

Technology: `1000*1.0244^(-20)+24.375*(1-1.0244^(-20))/0.0244`
Online Time Value of Money Utility:

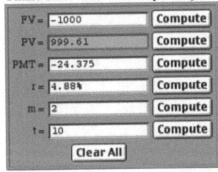

41. The periodic payments are based on a 3.625% annual payment. For payments twice a year, this is

$$PMT = 1{,}000(0.03625/2) = \$18.125$$

Since the bond yield is 3.705%, $i = 0.03705/2 = 0.018525$; $n = 2 \times 2 = 4$. The present value comes from the future value of $1,000 and the payments, which we treat like an annuity:

$$PV = FV(1+i)^{-n} + PMT\frac{1-(1+i)^{-n}}{i}$$

$$= 1{,}000(1+0.018525)^{-4} + 18.125\frac{1-(1+0.018525)^{-4}}{0.018525}$$

$$= 1{,}000(1.018525)^{-4} + 18.125\frac{1-1.018525^{-4}}{0.018525} \approx \$998.47$$

Technology: `1000*1.018525^(-4)+18.125*(1-1.018525^(-4))/0.018525`
Online Time Value of Money Utility:

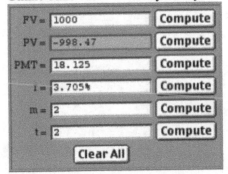

43. The periodic payments are based on a 5.5% annual payment. For payments twice a year, this is

$$PMT = 1{,}000(0.055/2) = \$27.5$$

Since the bond yield is 6.643%, $i = 0.06643/2 = 0.033215$; $n = 2 \times 10 = 20$. The present value comes from the future value of $1,000 and the payments, which we treat like an annuity:

$$PV = FV(1+i)^{-n} + PMT\frac{1-(1+i)^{-n}}{i}$$

$$= 1{,}000(1+0.033215)^{-20} + 27.5\frac{1-(1+0.033215)^{-20}}{0.033215}$$

$$= 1{,}000(1.033215)^{-20} + 27.5\frac{1-1.033215^{-20}}{0.033215} \approx \$917.45$$

Technology: `1000*1.033215^(-20)+27.5*(1-1.033215^(-20))/0.033215`

Online Time Value of Money Utility:

FV =	1000	Compute
PV =	-917.45	Compute
PMT =	27.5	Compute
r =	6.643%	Compute
m =	2	Compute
t =	10	Compute
	Clear All	

Applications

45. $PMT = \$400$, $i = 0.1483/12$, $n = 12 \times 20 = 240$

$$FV = PMT\frac{(1+i)^n - 1}{i} = 400\frac{(1+0.1483/12)^{240}-1}{0.1483/12} \approx \$584{,}686.94$$

Technology: `400*((1+0.1483/12)^240-1)/(0.1483/12)`

47. Current value: $PMT = \$500$, $i = 0.0277/12$, $n = 12 \times 15 = 180$

$$FV = PMT\frac{(1+i)^n - 1}{i} = 500\frac{(1+0.0277/12)^{180}-1}{0.0277/12} \approx \$111{,}422.99$$

Technology: `500*((1+0.0277/12)^180-1)/(0.0277/12)`

Value in 20 years: $PMT = \$500$, $i = 0.0267/12$, $n = 12 \times 20 = 240$

$$FV = PV(1+i)^n + PMT\frac{(1+i)^n - 1}{i}$$

$$= 111{,}422.99(1+0.0267/12)^{240} + 500\frac{(1+0.0267/12)^{240}-1}{0.0267/12}$$

$$\approx \$348{,}312.44$$

Technology: `111422.99*(1+0.0267/12)^240`
`+ 500*((1+0.0267/12)^240-1)/(0.0267/12)`

49. $FV = \$1{,}500{,}000$ $i = 0.0277/12$, $n = 12 \times 30 = 360$

$$PMT = FV\frac{i}{(1+i)^n - 1} = 1{,}500{,}000\frac{0.0277/12}{(1+0.0277/12)^{360}-1} \approx \$2{,}677.02$$

Technology: `1500000*0.0277/12/((1+0.0277/12)^360-1)`

51. Take x to be the amount you deposit each month into the stock fund. Then the deposit into the bond fund is four times that amount. The future value of the account is the sum of the future values of two investments: $\$x$ per month at

14.83% for $12 \times 25 = 300$ months, and $4x$ per month at 2.67% for 300 months:

$$FV = x\frac{(1+0.1483/12)^{300}-1}{0.1483/12} + 4x\frac{(1+0.0267/12)^{300}-1}{0.0267/12}$$

$$\approx 3{,}142.5166x + 1{,}704.1131x = 4{,}846.6297x$$

We need this amount to equal one million: $4{,}846.6297x = 1{,}000{,}000$, so $x = \dfrac{1{,}000{,}000}{4{,}846.6297} \approx 206.329$.

Thus, $206.33 should be invested in the stock fund each month.

The amount in the bond fund is four times that amount: $4 \times 206.33 = \$825.32$ per month.

53. Funding the $5,000 withdrawals:

$PMT = 5{,}000$, $i = 0.1483/12$, $n = 12 \times 10 = 120$

$$PV = PMT\frac{1-(1+i)^{-n}}{i} = 5{,}000\frac{1-(1+0.1483/12)^{-120}}{0.1483/12} \approx \$311{,}923.95$$

Technology: `5000*(1-(1 + 0.1483/12)^(-120))/(0.1483/12)`

To fund the lump sum of $30,000 after 10 years, we need the present value of $30,000 under compound interest:

$$PV = 30{,}000(1+0.1483/12)^{-120}$$

$$= \$6{,}870.84.$$

So, the total in the trust should be

$$\$311{,}923.95 + \$6{,}870.84 = \$318{,}794.79.$$

55. This is a two-stage process:
  (1) An accumulation stage to build the retirement fund
  (2) An amortization stage depleting the fund during retirement

Stage 1: Building the retirement fund. $PMT = \$1{,}200$, $i = 0.04/4 = 0.01$, $n = 4 \times 40 = 160$.

$$FV = PMT\frac{(1+i)^n - 1}{i}$$

$$= 1{,}200\frac{1.01^{160}-1}{0.01}$$

$$\approx \$469{,}659.16$$

Technology: `1200*(1.01^160-1)/0.01`

Stage 2: Depleting the fund. $PV = \$469{,}659.16$, $i = 0.04/4 = 0.01$, $n = 4 \times 25 = 100$.

$$PMT = PV\frac{i}{1-(1+i)^{-n}}$$

$$= \frac{469{,}659.16 \times 0.01}{1 - 1.01^{-100}}$$

$$\approx \$7{,}451.49$$

Technology: `469659.16*0.01/(1-1.01^(-100))`

57. This is a two-stage process:
  (1) An accumulation stage to build the retirement fund
  (2) An amortization stage depleting the fund during retirement

As in the text, we work backward, starting with Stage 2, where we have $PMT = \$5{,}000$, $i = 0.03/12 = 0.0025$, and

$n = 20 \times 12 = 240$, and we need to calculate the starting value PV.

$$PV = PMT \frac{1 - (1+i)^{-n}}{i}$$

$$= 5{,}000 \frac{1 - (1 + 0.0025)^{-240}}{0.0025}$$

$$= 5{,}000 \frac{1 - (1.0025)^{-240}}{0.0025}$$

$$\approx \$901{,}554.57$$

Technology: `5000*(1-1.0025^(-240))/0.0025`

Thus, in Stage 1, you need to accumulate $901,554.57 in the annuity: $FV = \$901{,}554.57$, $i = 0.03/12 = 0.0025$, $n = 40 \times 12 = 480$.

$$PMT = FV \frac{i}{(1+i)^n - 1}$$

$$= \frac{901{,}554.57 \times 0.0025}{1.0025^{480} - 1}$$

$$\approx \$973.54 \text{ per month}$$

Technology: `901554.57*0.0025/(1.0025^480-1)`

59. $FV = \$1{,}000{,}000$, $i = 0.048/12 = 0.004$, $n = 12 \times (87 - 30) = 684$

$$PMT = FV \frac{i}{(1+i)^n - 1} = 1{,}000{,}000 \frac{0.004}{1.004^{684} - 1} \approx \$278.92$$

Technology: `1000000*0.004/(1.004^684-1)`

61. Take x to be the current age of an insured person.

$FV = \$500{,}000$, $i = 0.048/12 = 0.004$;

$$PMT = FV \frac{i}{(1+i)^n - 1} = 500{,}000 \frac{0.004}{1.004^n - 1}$$

Males: $PMT = FV \dfrac{i}{(1+i)^n - 1} = 500{,}000 \dfrac{0.004}{1.004^{12(74-x)} - 1}$

Females: $PMT = FV \dfrac{i}{(1+i)^n - 1} = 500{,}000 \dfrac{0.004}{1.004^{12(77-x)} - 1}$

Technology: `500000*0.004/(1.004^(12*(74-x))-1)`
`500000*0.004/(1.004^(12*(77-x))-1)`

Result:

Age x	Male	Female
30	$276.62	$235.24
50	$927.11	$756.08
70	$9,469.40	$5,020.06

63. While in Mexico:

$FV = \$750{,}000$, $i = 0.048/12 = 0.004$, $n = 12 \times (73 - 22) = 612$

$$PMT = FV \frac{i}{(1+i)^n - 1} = 750{,}000 \frac{0.004}{1.004^{612} - 1} \approx \$285.47$$

When he moved to Canada eight years later, $n = 12 \times 8 = 96$, so the value of the policy was

$$FV = PMT \frac{(1+i)^n - 1}{i} = 285.47 \frac{1.004^{96} - 1}{0.004} \approx \$33,330.15.$$

To calculate the required payments in Canada, we use an annuity with a present value of \$33,330.15 and want the payments necessary to increase this amount to \$750,000 in $(80 - 30) = 50$ years ($n = 12 \times 50 = 600$), so we use the formula in the "Before we go on" discussion after after Example 1:

$$FV = PV(1+i)^n + PMT \frac{(1+i)^n - 1}{i} \text{ and solve for } PMT \text{ to obtain}$$

$$PMT = (FV - PV(1+i)^n) \frac{i}{(1+i)^n - 1}$$

$$= \left(750,000 - 33,330.15(1.004)^{600} \right) \frac{0.004}{1.004^{600} - 1}$$

$$\approx \$154.19. \text{So, the premiums decreased by } 285.47 - 154.19 = \$131.28.$$

65. We know the payments and need to calculate the present value:
$i = 0.0409/12$, $n = 12 \times 30 = 360$, $PMT = 600$

$$PMT = PV \frac{i}{1 - (1+i)^{-n}}$$

$$600 = PV \frac{0.0409/12}{1 - (1 + 0.0409/12)^{-360}}$$

$$\text{so } PV = 600 \frac{1 - (1 + 0.0409/12)^{-360}}{0.0409/12} \approx \$124,321.81.$$

This is the amount you can afford to finance. Including the down payment gives a total of
$$\$124,321.81 + 20,000 = \$144,321.81$$

67. Your hunch: Wait until December for the price to go down to \$140,000: $PV = \$140,000$, $i = 0.0409/12$, $n = 12 \times 30 = 360$

$$PMT = PV \frac{i}{1 - (1+i)^{-n}} = 140,000 \frac{0.0409/12}{1 - (1 + 0.0409/12)^{-360}} \approx \$675.67$$

Technology: `140000*(0.0409/12)/(1-(1+0.0409/12)^(-360))`

Broker's suggestion: Buy now for \$150,000: $PV = \$150,000$, $i = 0.0393/12$, $n = 12 \times 30 = 360$

$$PMT = PV \frac{i}{1 - (1+i)^{-n}} = 150,000 \frac{0.0393/12}{1 - (1 + 0.0393/12)^{-360}} \approx \$710.08$$

Technology: `150000*(0.0393/12)/(1-(1+0.0393/12)^(-360))`

Conclusion: Wait until December and pay $710.08 - 675.67 = \$34.41$ less per month.

69. We first calculate the payments:
$PV = \$150,000$, $i = 0.0393/12$, $n = 12 \times 30 = 360$

$$PMT = PV \frac{i}{1 - (1+i)^{-n}} = 150,000 \frac{0.0393/12}{1 - (1 + 0.0393/12)^{-360}} \approx \$710.08$$

Technology: `150000*(0.0393/12)/(1-(1+0.0393/12)^(-360))`
Now calculate the outstanding principal:
$k = 12 \times 15 = 180$, $n - k = 360 - 180 = 180$, $PMT = 710.08$

$$\text{Outstanding principal } = PMT \frac{1 - (1+i)^{-(n-k)}}{i}$$

$$= 710.08 \frac{1 - (1 + 0.0393/12)^{-180}}{0.0393/12} \approx \$96,454.02$$

Technology: 710.08*(1-(1+0.0393/12)^-180)/(0.0393/12)

71. We first calculate the payments on the original mortgage:

$PV = \$250,000$, $i = 0.0393/12$, $n = 12 \times 30 = 360$

$$PMT = PV \frac{i}{1 - (1 + i)^{-n}} = 250,000 \frac{0.0393/12}{1 - (1 + 0.0393/12)^{-360}} \approx \$1,183.47$$

Technology: 250000*(0.0393/12)/(1-(1+0.0393/12)^(-360))
Now calculate the outstanding principal:

$k = 12 \times 10 = 120$, $n - k = 360 - 120 = 240$, $PMT = 1,183.47$

$$\text{Outstanding principal } = PMT \frac{1 - (1 + i)^{-(n-k)}}{i}$$

$$= 1,183.47 \frac{1 - (1 + 0.0393/12)^{-240}}{0.0393/12} \approx \$196,492.38$$

Technology: 1183.47*(1-(1+0.0393/12)^-240)/(0.0393/12)
Now calculate the mortgage payments at the new rate on the the outstanding principal plus prepayment fee:

$PV = \$196,492.38 \times 1.04 \approx \$204,352.08$, $i = (0.0393/2)/12 = 0.0393/24$, $n = 12 \times 20 = 240$

$$PMT = PV \frac{i}{1 - (1 + i)^{-n}} = 204,352.08 \frac{0.0393/24}{1 - (1 + 0.0393/24)^{-240}} \approx \$1,030.40$$

Technology: 204352.08*(0.0393/24)/(1-(1+0.0393/24)^(-240))
Saving = $\$1,183.47 - 1,030.40 = \$153.07$

73. We first calculate the payments:

$PV = \$35,000$, $i = 0.0430/12$, $n = 12 \times 5 = 60$

$$PMT = PV \frac{i}{1 - (1 + i)^{-n}} = 35,000 \frac{0.0430/12}{1 - (1 + 0.0430/12)^{-60}} \approx \$649.33$$

Technology: 35000*(0.0430/12)/(1-(1+0.0430/12)^(-60))
Now calculate the amounts you could have financed in November and December:

November: $i = 0.0431/12$, $n = 60$, $PMT = 649.33$

$$PMT = PV \frac{i}{1 - (1 + i)^{-n}}$$

$$649.33 = PV \frac{0.0431/12}{1 - (1 + 0.0431/12)^{-60}}$$

so $PV = 649.33 \dfrac{1 - (1 + 0.0431/12)^{-60}}{0.0431/12} \approx \$34,991.58$.

December: $i = 0.0434/12$, $n = 60$, $PMT = 649.33$

$$PMT = PV \frac{i}{1 - (1 + i)^{-n}}$$

$$649.33 = PV \frac{0.0434/12}{1 - (1 + 0.0434/12)^{-60}}$$

so $PV = 649.33 \dfrac{1 - (1 + 0.0434/12)^{-60}}{0.0434/12} \approx \$34,965.95$.

75. Treat the card as a loan:

$PV = 5{,}000$, $i = 0.131/12$, $n = 12 \times 10 = 120$

$$PMT = PV \frac{i}{1 - (1 + i)^{-n}}$$

$$= 5{,}000 \frac{0.131/12}{1 - (1 + 0.131/12)^{-120}} \approx \$74.95.$$

Technology: `5000*(0.131/12)/(1-(1+0.131/12)^(-120))`

77. Solid Savings & Loan: $PV = \$10{,}000$, $i = 0.09/12 = 0.0075$, $n = 12 \times 4 = 48$.

$$PMT = PV \frac{i}{1 - (1 + i)^{-n}} = \frac{10{,}000 \times 0.0075}{1 - 1.0075^{-48}} \approx \$248.85$$

Technology: `10000*0.0075/(1-1.0075^(-48))`

Fifth Federal Bank & Trust: $PV = \$10{,}000$, $i = 0.07/12$, $n = 12 \times 3 = 36$.

$$PMT = PV \frac{i}{1 - (1 + i)^{-n}} = \frac{10{,}000 \times 0.07/12}{1 - (1 + 0.07/12)^{-36}} \approx \$308.77$$

Technology: `10000*(0.07/12)/(1-(1+0.07/12)^(-36))`

Answer: You should take the loan from Solid Savings & Loan: It will have payments of \$248.85 per month. The payments on the other loan would be more than \$300 per month.

79. We first calculate the original payments:

$PV = \$96{,}000$, $i = 0.0975/12$, $n = 12 \times 30 = 360$

$$PMT = PV \frac{i}{1 - (1 + i)^{-n}} = 96{,}000 \frac{0.0975/12}{1 - (1 + 0.0975/12)^{-360}} \approx \$824.79$$

Technology: `96000*(0.0975/12)/(1-(1+0.0975/12)^(-360))`

Now calculate the outstanding principal:

$k = 12 \times 4 = 48$, $n - k = 360 - 48 = 312$, $PMT = 824.79$

$$\text{Outstanding principal} = PMT \frac{1 - (1 + i)^{-(n-k)}}{i}$$

$$= 824.79 \frac{1 - (1 + 0.0975/12)^{-312}}{0.0975/12} \approx \$93{,}383.71$$

Technology: `824.79*(1-(1+0.0975/12)^(-312))/(0.0975/12)`

Total Interest Paid During the First Loan = Sum of payments − Reduction in principal

$$= 48 \times 824.79 - (96{,}000 - 93{,}383.71) = \$36{,}973.63$$

Now calculate the mortgage payments at the new rate on the outstanding principal:

$PV = \$93{,}383.71$, $i = 0.06875/12$, $n = 12 \times 30 = 360$

$$PMT = PV \frac{i}{1 - (1 + i)^{-n}} = 93{,}383.71 \frac{0.06875/12}{1 - (1 + 0.06875/12)^{-360}} \approx \$613.46$$

Technology: `93383.71*(0.06875/12)/(1-(1+0.06875/12)^(-360))`

Total Interest Paid During the New Loan = Sum of payments − Reduction in principal

$$= 360 \times 613.46 - 93{,}383.71 = \$127{,}461.89$$

Thus, the total interest paid over the duration of the two loans is $\$36{,}973.63 + 127{,}461.89 = \$164{,}435.52$.

Had the mortgage not been refinanced, the total interest would have been

Total Interest Paid = Sum of payments − Reduction in principal

$$= 360 \times 824.79 - 96{,}000 = \$20{,}0924.40. \text{Thus, the saving on interest is}$$

$$\$200{,}924.40 - 164{,}435.52 = \$36{,}488.88$$

81. We can construct an amortization table using the technique outlined in the Technology Guides. For example, using Excel, we might set it up as follows:

◇	A	B	C	D	E	F
1	Month	Outstanding Principal	Payment on Principal	Interest Payment		
2	0	$ 50,000.00			Rate	8%
3	1	=B2-C3	=F$4-D3	=DOLLAR(B2*F$2/12)	Years	15
4	2				Payment	=DOLLAR(-PMT(F2/12,F3*12,B2))
5	3					
6	4					
7	5					
8	6					

Adding the payments on principal (Column C) and interest payments (Column D) for each year will give the following table:

Year	Interest	Payment on Principal
1	$3,934.98	$1,798.98
2	$3,785.69	$1,948.27
3	$3,623.97	$2,109.99
4	$3,448.84	$2,285.12
5	$3,259.19	$2,474.77
6	$3,053.77	$2,680.19
7	$2,831.32	$2,902.64
8	$2,590.39	$3,143.57
9	$2,329.48	$3,404.48
10	$2,046.91	$3,687.05
11	$1,740.88	$3,993.08
12	$1,409.47	$4,324.49
13	$1,050.54	$4,683.42
14	$661.81	$5,072.15
15	$240.84	$5,491.80

83. The payments on the loan, ignoring the fee, are

$$PMT = 5,000 \frac{0.09/12}{1 - (1 + 0.09/12)^{-24}} = \$228.42.$$

Add to each payment $100/24 = \$4.17$ to get a new payment of $232.59. Now use technology to find the interest rate being charged on a 2-year $5,000 loan with this payment. For example, we can use the Online Time Value of Money Utility:

FV =	0	Compute
PV =	5000	Compute
PMT =	-232.59	Compute
r =	0.1081	Compute
m =	12	Compute
t =	2	Compute
	Clear all	Example

We see that the interest rate is 10.81%.

85. TI-83/84 Plus:

```
•N=153.5029583
 I%=4
 PV=0
 PMT=-500
 FV=100000
 P/Y=12
 C/Y=12
 PMT:END BEGIN
```

This gives $153.5/12 \approx 13$ years to retirement.

Online Time Value of Money Utility:

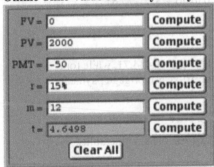

FV =	100000	Compute
PV =	0	Compute
PMT =	-500	Compute
r =	4%	Compute
m =	12	Compute
t =	12.7919	Compute
	Clear All	

87. TI-83/84 Plus:

```
•N=55.79763048
 I%=15
 PV=2000
 PMT=-50
 FV=0
 P/Y=12
 C/Y=12
 PMT:END BEGIN
```

This gives $55.798/12 \approx 4.5$ years to repay the debt.

Online Time Value of Money Utility:

FV =	0	Compute
PV =	2000	Compute
PMT =	-50	Compute
r =	15%	Compute
m =	12	Compute
t =	4.6498	Compute
	Clear All	

89. Graph the future value of both accounts using the formula for the future value of a sinking fund.

Your account: $i = 0.045/12 = 0.00375$, $PMT = \$100$.

$$FV = PMT \frac{(1+i)^n - 1}{i} = 100 \frac{1.00375^n - 1}{0.00375}$$

To graph this, use the technology formula `100*(1.00375^x-1)/0.00375`.

Lucinda's account: $i = 0.065/12$, $PMT = \$75$.

$$FV = PMT \frac{(1+i)^n - 1}{i} = 75 \frac{(1 + 0.065/12)^n - 1}{(0.065/12)}$$

To graph this, use the technology formula `75*((1+0.065/12)^x-1)/(0.065/12)`
Graph:

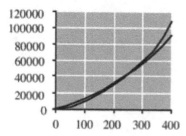

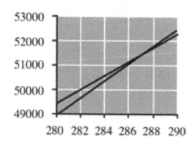

The graphs appear to cross around $n = 300$, so we zoom in there:

The graphs cross around $n = 287$, so the number of years is approximately $t = 287.5/12 \approx 24$ years.

Communication and reasoning exercises

91. He is wrong because his estimate ignores the interest that will be earned by your annuity—both while it is increasing and while it is decreasing. Your payments will be considerably smaller (depending on the interest earned).

93. Wrong; the split investment earns more. For instance, after ten years it earns $31,056.46 + $32,775.87 = $63,832.33, which is more than the $63,803.03 earned by the single investment.
(Mathematically, the future value does not depend linearly on the interest rate.)

95. He is not correct. For instance, the payments on a $100,000 10-year mortgage at 12% are $1,434.71, while for a 20-year mortgage at the same rate, they are $1,101.09, which is a lot more than half the 10-year mortgage payment.

97. $FV =$ maturity value $= \$1,000$, $PMT = 1,000(0.035/2) = \$17.50$, $PV =$ selling price $= \$994.69$,

$n = 2 \times 5 = 10$. Using technology, we compute the interest rate to be approximately 3.617%. (The online utility rounds this to two decimal places.)

TI-83/84 Plus:

Online Time Value of Money Utility
("Years" mode):

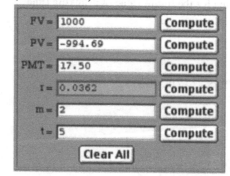

99. $PV = FV(1+i)^{-n} = PMT\dfrac{(1+i)^n - 1}{i}(1+i)^{-n} = PMT\dfrac{1-(1+i)^{-n}}{i}$

Chapter 2 Review

1. $FV = 6{,}000(1 + 0.0475 \times 5) = \$7{,}425.00$

3. $FV = 6{,}000(1 + 0.0475/12)^{60} = \$7{,}604.88$

5. $FV = 100 \dfrac{(1 + 0.0475/12)^{60} - 1}{0.0475/12} = \$6{,}757.41$

7. $PV = 6{,}000/(1 + 0.0475 \times 5) = \$4{,}848.48$

9. $PV = 6{,}000(1 + 0.0475/12)^{-60} = \$4{,}733.80$

11. $PV = 100 \dfrac{1 - (1 + 0.0475/12)^{-60}}{0.0475/12} = \$5{,}331.37$

13.

$PMT = 12{,}000 \dfrac{0.0475/12}{(1 + 0.0475/12)^{60} - 1} = \$177.58$

15. $PMT = 6{,}000 \dfrac{0.0475/12}{1 - (1 + 0.0475/12)^{-60}} = \$112.54$

17.

$PMT = 10{,}000 \dfrac{0.0475/12}{1 - (1 + 0.0475/12)^{-60}} = \$187.57$

19. $10{,}000 = 6{,}000(1 + 0.0475t) = 6{,}000 + 285t.\ t = (10{,}000 - 6{,}000)/285 = 14.0$ years.

21. $10{,}000 = 6{,}000(1 + 0.0475/12)^{12t}.$ To solve algebraically requires logarithms:

$$t = \frac{\log(10{,}000/6{,}000)}{12\log(1 + 0.0475/12)} \approx 10.8 \text{ years}$$

We could also find this using, for example, the TI-83/84 Plus TVM Solver.

23. $10{,}000 = 100 \dfrac{(1 + 0.0475/12)^{12t} - 1}{0.0475/12}.$ To solve algebraically requires logarithms:

$$t = \frac{\log[0.0475 \times 10{,}000/(100 \times 12) + 1]}{12\log(1 + 0.0475/12)} \approx 7.0 \text{ years}$$

We could also find this using, for example, the TI-83/84 Plus TVM Solver.

25. Each interest payment is $10{,}000 \times 0.06/2 = \$300$. For an annuity earning 7% and paying $300 every 6 months for 5 years, the present value is

$$PV = 300 \frac{1 - (1 + 0.07/2)^{-10}}{0.07/2} = \$2{,}494.98.$$

The present value of the $10,000 maturity value is

$$PV = 10{,}000(1 + 0.07/2)^{-10} = \$7{,}089.19.$$

The total price is $2,494.98 + 7,089.19 = \$9,584.17$.

27. 5.346% (using, for example, the TI-83/84 Plus TVM Solver)

29. $PV = 3.28,\ FV = 45.74,\ i = r/12,\ n = 9$ (months)

$45.74 = 3.28(1 + 92r/12) \approx 3.28 + 25.1467r$

$r \approx (45.74 - 3.28)/25.1467 \approx 1.6885 = 168.85\%$

31. The only dates on which she would have gotten an increase rather than a decrease were November 2009, February 2010, and August 2010. Calculating the annual returns as in the preceding exercises, we get the following figures:

Jan. 2007–Nov. 2009: 60.45%
Jan. 2007–Feb. 2010: 85.28%
Jan. 2007–Aug. 2010: 75.37%

The largest annual return was 85.28% if she sold in February 2010.

33. No. Simple interest increase is linear. We can compare slopes between successive points to see whether the slope remained roughly constant: From December 2002 to August 2004 the slope was $(16.31 - 3.28)/(20/12) = 7.818$, while from August 2004 to March 2005 the slope was $(33.95 - 16.31)/(7/12) = 30.24$. These slopes are quite different.

35. Use the compound interest formula: $FV = PV(1 + i)^n$, where $PV = \$150,000$, $i = 0.20$, $n = 1, 2, 3, 4, 5$.

2010: $FV = 150,000(1.20) = \$180,000$

2011: $FV = 150,000(1.20)^2 = \$216,000$

2012: $FV = 150,000(1.20)^3 = \$259,200$

2013: $FV = 150,000(1.20)^4 = \$311,040$

2014: $FV = 150,000(1.20)^5 = \$373,248$

Revenues first surpass $300,000 in 2013.

37. After the first day of trading, the value of each share will be $6. Thereafter, the shares appreciate in value by 8% per month for 6 months, and O'Hagan desires a future value of at least $500,000.

$FV = 500,000$, $i = 0.08$, $n = 6$

$PV = FV(1 + i)^{-n} = 500,000 \times 1.08^{-6} \approx \$315,084.81$

Therefore, since each share will be worth $6 after the first day, the number of shares they must sell is at least

$$\frac{315,084.81}{6} \approx 52,514.14.$$

Since they can offer only a whole number of shares, we must round this up to get the minimum desired future value. Thus, they should offer at least 52,515 shares.

39. $PV = 250,000$, $i = 0.095/12$, $n = 12 \times 10 = 120$

$$PMT = PV \frac{i}{1 - (1 + i)^{-n}} = \frac{250,000 \times 0.095/12}{1 - (1 + 0.095/12)^{-120}} \approx \$3,234.94$$

41. $i = 0.095/12$, $PMT = 3,000$, $n = 12 \times 10 = 120$

$$PV = PMT \frac{1 - (1 + i) - n}{i} = 3,000 \frac{1 - (1 + 0.095/12)^{-120}}{(0.095/12)} \approx \$231,844 \text{ (to the nearest dollar)}$$

43. $PV = 250,000$, $FV = 0$, $PMT = 3,000$, $n = 12 \times 10 = 120$, and we are seeking the interest rate.

Online Time Value of Money Utility:

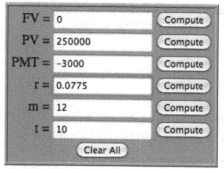

The interest rate would be 7.75%.

45. $PV = 50,000$, $PMT = 1,000 + 800$ (company contribution) $= 1,800$, $i = 0.073/12$, $n = 12 \times 10 = 120$.
Considering the contribution of the present \$50,000 as well as the payments, we get

$$FV = PV(1+i)^n + PMT\frac{(1+i)^n - 1}{i} = 50,000(1 + 0.073/12)^{120} + 1,800\frac{(1 + 0.073/12)^{120} - 1}{0.073/12} \approx \$420,275$$

(to the nearest dollar).
Technology: 50000*(1+0.073/12)^120+1800*((1+0.073/12)^120-1)/(0.073/12)

47. For the company's contribution, take $PMT = 800$, $i = 0.073/12$, $n = 12 \times 10 = 120$.

$$FV = PMT\frac{(1+i)^n - 1}{i} = 800\frac{(1 + 0.073/12)^{120} - 1}{0.073/12} \approx \$140,778$$

(to the nearest dollar).
Technology: 800*((1+0.073/12)^120-1)/(0.073/12)

49. We first take out the effect of the initial \$50,000: $PV = 50,000$, $i = 0.073/12$, $n = 12 \times 10 = 120$.

$$FV = PV(1+i)^n = 50,000(1 + 0.073/12)^{120}$$

Thus, the payments have to result in a future value of

$$FV = 500,000 - 50,000(1 + 0.073/12)^{120} \approx 396,475.163$$

We can now use the payment formula to determine the necessary payments:

$$PMT = FV\frac{i}{(1+i)^n - 1} \approx \frac{396,475.163 \times 0.073/12}{(1 + 0.073/12)^{120} - 1} \approx \$2,253.06$$

Since \$800 of this is contributed by the company, Callahan's payments should be
    \$2,253.06 − 800 = \$1,453.06.

51. First compute the amount Callahan needs at the start of retirement: $PMT = 5,000$, $i = 0.087/12$,
$n = 12 \times 30 = 360$

$$PV = PMT\frac{1 - (1+i)-n}{i} = 5,000\frac{1 - (1 + 0.087/12)^{-360}}{0.087/12} \approx \$638,461.93$$

Technology: 5000*(1-(1+0.087/12)^(-360))/(0.087/12)
In order to accumulate this amount, using the information from Exercise 45, we first discount the effect of the current
\$50,000 in the account: $PV = 50,000$, $FV = 638,461.93$, $i = 0.073/12$, $n = 12 \times 10 = 120$. The initial \$50,000 will
grow to

$$50,000(1 + 0.073/12)^{120}$$

so the payments need to result in a future value of only

$$638,461.93 - 50,000(1 + 0.073/12)^{120} \approx 534,937.093.$$

Now use the payment formula:

$$PMT = FV \frac{i}{(1+i)^n - 1} = \frac{534{,}937.093 \times 0.073/12}{(1 + 0.073/12)^{120} - 1} \approx \$3{,}039.90$$

Since $800 of this is contributed by the company, Callahan's payments should be

$3,039.90 − 800 = $2,239.90.

53. The bond will pay interest every 6 months amounting to

50,000(0.072/2) = $1,800.

For someone purchasing this bond after one year, there will be 9 years to maturity. Think of the bond as an investment that will pay the owner $1,800 every 6 months for 9 years, at which time it will pay $50,000. This is exactly the behavior of an annuity paired with an investment with future value $50,000

$$FV = 50{,}000, \quad PMT = 1{,}800, \quad i = 0.063/2 = 0.0315, \quad n = 2 \times 9 = 18$$

The present value has contributions both from the investment and the annuity:

$$PV = FV(1+i)^{-n} + PMT \frac{1 - (1+i)^{-n}}{i} = 50{,}000(1.0315)^{-18} + 1{,}800 \frac{1 - 1.0315^{-180}}{0.0315} \approx \$53{,}055.66$$

Technology: `50000*1.0315^(-18)+1800*(1-1.0315^(-18))/0.0315`

55. Here, $FV = 50{,}000$, $PV = 54{,}000$, $PMT = 1{,}800$, $n = 2 \times 8.5 = 17$, and we are seeking the interest rate.

Online Time Value of Money Utility:

FV =	50000	Compute
PV =	-54000	Compute
PMT =	1800	Compute
r =	0.0599	Compute
m =	2	Compute
t =	8.5	Compute
	Clear All	

The interest rate would have to be 5.99%.

Section 3.1

1. $2x - y = 1$

Solving for y gives $y = 2x - 1$, so the general solution is

$(x, y) = (x, 2x - 1)$; x arbitrary.

This is the solution parameterized by x. We get particular solutions by choosing values for x, such as

$x = -1 \Rightarrow (x, y) = (x, 2x - 1) = (x, 2(-1) - 1) = (-1, -3)$

$x = 0 \Rightarrow (x, y) = (x, 2x - 1) = (0, 2(0) - 1) = (0, -1)$

$x = 1 \Rightarrow (x, y) = (x, 2x - 1) = (1, 2(1) - 1) = (1, 1)$

For the solution parameterized by y, solve the original equation for x to get $x = \frac{1}{2}(y + 1)$, so the general solution

parameterized by y is

$(x, y) = \left(\frac{1}{2}(y + 1), y\right)$; y arbitrary

3. $3x + 4y = 2$

Solving for y gives $y = -\frac{3}{4}x + \frac{1}{2}$, so the general solution is

$(x, y) = \left(x, -\frac{3}{4}x + \frac{1}{2}\right)$; x arbitrary.

This is the solution parameterized by x. We get particular solutions by choosing values for x, such as

$x = -2 \Rightarrow (x, y) = \left(x, -\frac{3}{4}x + \frac{1}{2}\right) = \left(-2, -\frac{3}{4}(-2) + \frac{1}{2}\right) = (-3, 2)$

$x = 0 \Rightarrow (x, y) = \left(x, -\frac{3}{4}x + \frac{1}{2}\right) = \left(0, -\frac{3}{4}(0) + \frac{1}{2}\right) = \left(0, \frac{1}{2}\right)$

$x = 2 \Rightarrow (x, y) = \left(x, -\frac{3}{4}x + \frac{1}{2}\right) = \left(2, -\frac{3}{4}(2) + \frac{1}{2}\right) = (2, -1)$

For the solution parameterized by y, solve the original equation for x to get $x = \frac{1}{3}(-4y + 2)$, so the general solution

parameterized by y is

$(x, y) = \left(\frac{1}{3}(-4y + 2), y\right)$; y arbitrary.

5. $4x = -5$

We cannot solve for y, so the general solution cannot be parameterized by x.

For the solution parameterized by y, solve the original equation for x to get $x = -\frac{5}{4}$, so the general solution

parameterized by y is

$(x, y) = \left(-\frac{5}{4}, y\right)$; y arbitrary.

We get particular solutions by choosing values for y: $(-5/4, -1)$, $(-5/4, 0)$, $(-5/4, 1)$.

7. $x - y = 0 \qquad x + y = 4$

Adding gives

$$2x = 4 \Rightarrow x = 2.$$

Substituting $x = 2$ in the first equation:

$$2 - y = 0 \Rightarrow y = 2.$$

Solution: $(2, 2)$

Graph: $y = x;\ y = 4 - x$

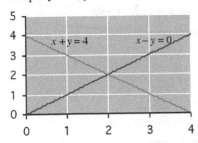

9. $x + y = 4 \qquad x - y = 2$

Adding gives

$$2x = 6 \Rightarrow x = 3.$$

Substituting $x = 3$ in the first equation:

$$3 + y = 4 \Rightarrow y = 4 - 3 = 1.$$

Solution: $(3, 1)$

Graph: $y = 4 - x;\ y = x - 2$

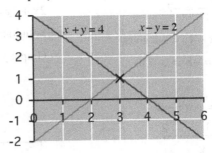

11. $3x - 2y = 6 \qquad 2x - 3y = -6$

Multiply the first equation by 2 and the second by -3:

$$6x - 4y = 12 \qquad -6x + 9y = 18.$$

Adding gives

$$5y = 30 \Rightarrow y = 6.$$

Substituting $y = 6$ in the first equation:

$$3x - 12 = 6 \Rightarrow 3x = 18 \Rightarrow x = 6.$$

Solution: $(6, 6)$

Graph: $y = \dfrac{3}{2}x - 3;\ y = \dfrac{2}{3}x + 2$

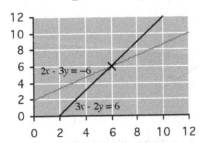

13. $0.5x + 0.1y = 0.7 \qquad 0.2x - 0.2y = 0.6$

Multiply both equations by 10:

$$5x + y = 7 \qquad 2x - 2y = 6.$$

Divide the second equation by 2:

$$5x + y = 7 \qquad x - y = 3.$$

Adding gives

$$6x = 10 \Rightarrow x = \frac{10}{6} = \frac{5}{3}.$$

Substituting $x = \dfrac{5}{3}$ in $x - y = 3$ gives

$$\frac{5}{3} - y = 3 \Rightarrow y = \frac{5}{3} - 3 = -\frac{4}{3}.$$

Solution: $\left(\dfrac{5}{3}, -\dfrac{4}{3} \right)$

Graph: $y = -5x + 7;\ y = x - 3$

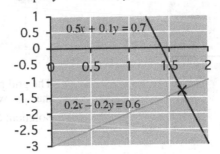

15. $\dfrac{x}{3} - \dfrac{y}{2} = 1$ $\dfrac{x}{4} + y = -2$

Multiply the first equation by 6 and the second by 4:

$2x - 3y = 6$ $x + 4y = -8.$

Multiply the second equation by -2:

$2x - 3y = 6$ $-2x - 8y = 16.$

Adding gives

$$-11y = 22 \Rightarrow y = \frac{22}{-11} = -2.$$

Substituting $y = -2$ into $x + 4y = -8$ gives

$x + 4(-2) = -8 \Rightarrow x - 8 = -8 \Rightarrow x = 0.$

Solution: $(0, -2)$

Graph: $y = (2/3)x - 2$; $y = -x/4 - 2$

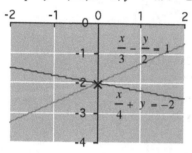

17. $2x + 3y = 1$ $-x - \dfrac{3y}{2} = -\dfrac{1}{2}$

Multiply the second equation by 2:

$2x + 3y = 1$ $-2x - 3y = -1.$

Adding gives $0 = 0$, indicating that the system is redundant: The graphs are the same, so there are infinitely many solutions. We obtain the solutions by solving either equation for y (or x).

$2x + 3y = 1 \Rightarrow 3y = -2x + 1 \Rightarrow y = (-2x + 1)/3$

Solution: $(x, (-2x + 1)/3)$; x arbitrary

Graph: $y = (-2x + 1)/3$

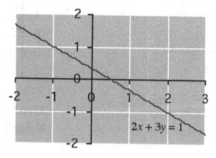

19. $2x + 3y = 2$ $-x - \dfrac{3y}{2} = -\dfrac{1}{2}$

Multiply the second equation by 2:

$2x + 3y = 2$ $-2x - 3y = -1.$

Adding gives $0 = 1$, indicating that the system is inconsistent. There is no solution; the lines are parallel.

Graph: $y = (-2x + 2)/3$; $y = (-2x + 1)/3$

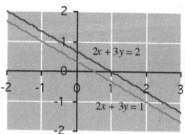

Parallel lines; no solution

21. $2x + 8y = 10$ $x + y = 5$

To graph these, solve for y:

$y = (-2x + 10)/8$ $y = -x + 5.$

Graph, with vertical and horizontal scales of 0.1:

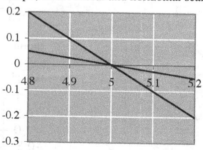

The grid point closest to the intersection of the lines gives the approximate solution (in this case the exact solution).

Solution: $(5, 0)$

Solutions Section 3.1

23. $3.1x - 4.5y = 6$ $4.5x + 1.1y = 0$

To graph these, solve for y:

$$y = (3.1x - 6)/4.5 \qquad y = -4.5x/1.1.$$

Graph, with vertical and horizontal scales of 0.1:

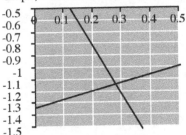

The grid point closest to the intersection of the lines gives the approximate solution.

Approximate solution: $(0.3, -1.1)$

25. $10.2x + 14y = 213$ $4.5x + 1.1y = 448$

To graph these, solve for y:

$$y = (-10.2x + 213)/14 \qquad y = (-4.5x + 448)/1.1.$$

Graph, with vertical and horizontal scales of 0.1:

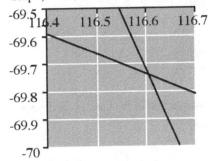

The grid point closest to the intersection of the lines gives the approximate solution.

Approximate solution: $(116.6, -69.7)$

27. Line through $(0, 1)$ and $(4.2, 2)$:

Point: $(0, 1)$ Slope: $m = \dfrac{y_2 - y_1}{x_2 - x_1} = \dfrac{2 - 1}{4.2 - 0} = \dfrac{1}{4.2}$ Intercept: $b = 1$ (Given)

Thus, the equation is $y = mx + b = \dfrac{1}{4.2} x + 1$.

Line through $(2.1, 3)$ and $(5.2, 0)$:

Point: $(5.2, 0)$ Slope: $m = \dfrac{y_2 - y_1}{x_2 - x_1} = \dfrac{0 - 3}{5.2 - 2.1} = -\dfrac{3}{3.1}$ Intercept: $b = y_1 - mx_1 = 0 - \left(-\dfrac{3}{3.1}\right)5.2 = \dfrac{15.6}{3.1}$

Thus, the equation is $y = mx + b = -\dfrac{3}{3.1} x + \dfrac{15.6}{3.1}$.

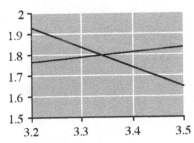

The grid point closest to the intersection of the lines gives the approximate solution.

Approximate solution: $(3.3, 1.8)$

29. Line through $(0, 0)$ and $(5.5, 3)$:

Point: $(0, 0)$ Slope: $m = \dfrac{y_2 - y_1}{x_2 - x_1} = \dfrac{3 - 0}{5.5 - 0} = \dfrac{3}{5.5}$ Intercept: 0 (Given)

Thus, the equation is $y = mx + b = \dfrac{3}{5.5} x$.

Line through $(5, 0)$ and $(0, 6)$:

Point: $(0, 6)$ Slope: $m = \dfrac{y_2 - y_1}{x_2 - x_1} = \dfrac{6 - 0}{0 - 5} = -\dfrac{6}{5}$ Intercept: 6 (Given)

Thus, the equation is $y = mx + b = -\dfrac{6}{5}x + 6$.

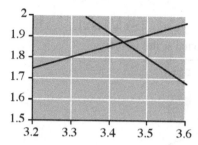

The grid point closest to the intersection of the lines gives the approximate solution.

Approximate solution: $(3.4, 1.9)$

31. Let $x =$ the number of soccer fans, and let $y =$ the number of football fans.

Reword the given statement using the phrases "the number of soccer fans," which is x, and "the number of football fans," which is y:

   The number of soccer fans is twice the number of football fans.
Translate the phrases into symbols:

   $x = 2y$.

In standard form, this becomes $x - 2y = 0$.

33. Let $x =$ the number of new clients, and let $y =$ the number of old clients.

Reword the given statement using the phrases "the number of new clients," which is x, and "the number of old clients," which is y:

   The number of new clients is 110% the number of old clients.
(No rewording necessary in this case.)
Translate the phrases into symbols:

   $x = \dfrac{110}{100}y = 1.10y$.

In standard form, this becomes $x - 1.10y = 0$.

35. Let $x =$ the number of gas giants, and let $y =$ the number of rocky planets.
We are given two pieces of information:
1. There are three times as many gas giants as rocky planets.
2. Their total is 12.
Thus, we expect to obtain two equations.

Reword the given information using the phrases "the number of gas giants," which is x, and "the number of rocky planets," which is y:

1. *The number of gas giants is three times the number of rocky planets.*
2. *The number of gas giants and the number of rocky planets add up to 12.*
Translate the phrases into symbols:

   1. $x = 3y$, or $x - 3y = 0$

   2. $x + y = 12$

37. Let $x =$ the number of ordinary shares, and let $y =$ the number of preferred shares.
We are given two pieces of information:
1. 20% of the total are preferred shares.
2. There are 15 more ordinary shares than preferred shares.

Reword the given information using the phrases "the number of ordinary shares," which is x, and "the number of preferred shares," which is y:

1. *The number of preferred shares is 20% of the total number (of ordinary and preferred shares).*
2. *The number of ordinary shares is 15 more than the number of preferred shares.*

Translate the phrases into symbols:

$$1. \ y = \frac{20}{100}(x+y) = \frac{1}{5}(x+y)$$

Simplify by multiplying both sides by 5:

$$5y = x + y \quad \Rightarrow \quad x = 4y, \text{ or } x - 4y = 0$$

$$2. \ x = 15 + y \text{ or } x - y = 15$$

Applications

39. Unknowns: x = the number of quarts of Creamy Vanilla; y = the number of quarts of Continental Mocha

Arrange the given information in a table with unknowns across the top:

	Vanilla (x)	Mocha (y)	Available
Eggs	2	1	500
Cream	3	3	900

We can now set up an equation for each of the items listed on the left:

Eggs: $\quad 2x + y = 500$

Cream: $\quad 3x + 3y = 900.$

Multiply the first equation by -1 and divide the second by 3:

$$-2x - y = -500 \qquad x + y = 300.$$

Adding gives

$$-x = -200 \Rightarrow x = 200 \text{ quarts of vanilla.}$$

Substituting $x = 200$ in the first equation gives

$$2(200) + y = 500 \Rightarrow 400 + y = 500 \Rightarrow y = 500 - 400 = 100 \text{ quarts of mocha.}$$

Solution: Make 200 quarts of vanilla and 100 quarts of mocha.

41. Unknowns: x = the number of servings of Mixed Cereal; y = the number of servings of Mango Tropical Fruit

Arrange the given information in a table with unknowns across the top:

	Cereal (x)	Mango (y)	Desired
Calories	60	80	200
Carbs.	11	21	43

We can now set up an equation for each of the items listed on the left:

Calories: $\quad 60x + 80y = 200$

Carbs: $\quad 11x + 21y = 43.$

Divide the first equation by 20:

$$3x + 4y = 10 \qquad 11x + 21y = 43.$$

Multiply the first equation by -11 and the second by 3:

$$-33x - 44y = -110 \qquad 33x + 63y = 129.$$

Adding gives

$$19y = 19 \Rightarrow y = 1 \text{ serving of Mango Tropical Fruit.}$$

Substituting $y = 1$ in the equation $3x + 4y = 10$ gives

$3x + 4(1) = 10 \Rightarrow 3x = 6 \Rightarrow x = 2$ servings of Mixed Cereal.

Solution: Use 1 serving of Mango Tropical Fruit and 2 servings of Mixed Cereal.

43. a. One half of the U.S. RDA for protein is 30 g.

Unknowns: x = the number of servings of Pork & Beans; y = the number of slices of white bread

Arrange the given information in a table with unknowns across the top:

	Pork & Beans (x)	Bread (y)	Desired
Protein	5	2	30
Carbs.	21	11	139

We can now set up an equation for each of the items listed on the left:

Protein: $\quad 5x + 2y = 30$

Carbs: $\qquad 21x + 11y = 139$.

Multiply the first equation by 11 and the second by -2:

$55x + 22y = 330 \qquad -42x - 22y = -278$.

Adding gives

$13x = 52 \Rightarrow x = \dfrac{52}{13} = 4$ servings of Pork & Beans.

Substituting $x = 4$ in the equation $5x + 2y = 30$ gives

$5(4) + 2y = 30 \Rightarrow 20 + 2y = 30 \Rightarrow 2y = 10 \Rightarrow y = 5$ slices of bread.

Solution: Prepare 4 servings of Pork & Beans and 5 slices of bread.

b. Decreasing the Carbohydrate total to 100 gives:

Protein: $\quad 5x + 2y = 30$

Carbs: $\qquad 21x + 11y = 100$.

Multiply the first equation by 11 and the second by -2:

$55x + 22y = 330 \qquad -42x - 22y = -200$.

Adding gives

$13x = 130 \Rightarrow x = \dfrac{130}{13} = 10$ servings of Pork & Beans.

Substituting $x = 10$ in the equation $5x + 2y = 30$ gives

$5(10) + 2y = 30 \Rightarrow 50 + 2y = 30 \Rightarrow 2y = 30 - 50 = -20 \Rightarrow y = -10$ slices of bread.

Since it is impossible to prepare a negative number of slices of bread, we conclude that it is not possible to prepare such a meal.

45. Unknowns: x = number of servings of Designer Whey; y = number of servings of Muscle Milk

Arrange the given information in a table with unknowns across the top:

	Designer Whey	Muscle Milk	Desired
Protein	18	32	280
Carbs	2	16	56
Cost	$0.50	$1.60	

We can now set up an equation for each nutrient listed on the left:

Protein: $\quad 18x + 32y = 280$

106

Carbs: $2x + 16y = 56$.

Multiply the second equation by -2 and add to the first:

$14x = 168 \Rightarrow x = 12$ servings of Designer Whey.

Substitute into the second equation to obtain

$24 + 16y = 56 \Rightarrow 16y = 32 \Rightarrow y = 2$ servings of Muscle Milk.

Cost $= 12(0.50) + 2(1.60) = \$9.20$.

Solution: Mix 12 servings of Designer Whey and 2 servings of Muscle Milk for a cost of \$9.20.

47. We first determine how many servings of each kind of supplement it contains:

Unknowns: $x =$ number of servings of Designer Whey; $y =$ number of servings of Pure Whey Protein Stack.

Arrange the given information in a table with unknowns across the top:

	Designer Whey	Pure Whey Stack	Total
Protein	18	24	540
Cost	0.50	0.60	14
Carbs	2	3	

We can now set up an equation for protein content and for cost:

Protein: $18x + 24y = 540$

Cost: $0.5x + 0.6y = 14$.

Divide the first equation by 6 and multiply the second by 10:

$3x + 4y = 90 \qquad 5x + 6y = 140$.

2 times the second minus 3 times the first gives $x = 10$ servings of Designer Whey.

Substituting in the first gives

$4y = 60 \Rightarrow y = 15$ Pure Whey Protein Stack.

From the table, Total amount of carbohydrates $= 10(2) + 15(3) = 65$ g.

49. Unknowns: $x =$ number of shares of TWTR; $y =$ number of shares of MSFT

Arrange the given information in a table with unknowns across the top:

	TWTR (x)	MSFT (y)	Total
Dec 1	40	48	22,400
Dec 31	36	45	20,700

We can now set up equations for the start and end of December:

Dec 1: $40x + 48y = 22,400$

Dec 31: $36x + 45y = 20,700$.

Divide the first equation by 8 and the second by 9:

$5x + 6y = 2,800 \qquad 4x + 5y = 2,300$.

Multiply the first by -4 and the second by 5:

$-20x - 24y = -11,200 \qquad 20x + 25y = 11,500$.

Adding gives

$y = 300$ shares of MSFT.

Substituting $y = 300$ into $5x + 6y = 2,800$ gives

$5x + 6(300) = 2,800 \Rightarrow 5x = 1,000 \Rightarrow x = 200$ shares of TWTR.

Solution: You bought 200 shares of TWTR and 300 shares of MSFT.

51. Unknowns: x = number of shares of TD; y = number of shares of CNA

The total investment was \$25,000 $\Rightarrow 45x + 40y = 25,000$.

You earned \$760 in dividends:

TD dividend = 4% of $45x$ invested $= 0.04(45x) = 1.8x$

CNA dividend = 2.5% of $40y$ invested $= 0.025(40y) = y$.

Thus, $1.8x + y = 760$.

We therefore have the following system:

$\qquad 45x + 40y = 25,000 \qquad 1.8x + y = 760$.

Divide the first equation by 5 and multiply the second equation by 5:

$\qquad 9x + 8y = 5,000 \qquad 9x + 5y = 3,800$.

Subtracting gives

$\qquad 3y = 1,200 \Rightarrow y = 400$ shares of CNA.

Substituting $y = 400$ in $9x + 8y = 5,000$ gives

$\qquad 9x + 8(400) = 5,000 \Rightarrow 9x + 3,200 = 5,000 \Rightarrow 9x = 1,800 \Rightarrow x = 200$ shares of TD.

Solution: You purchased 200 shares of TD and 400 shares of CNA.

53. Unknowns: x = number of members voting in favor; y = number of members voting against

We are given two pieces of information:

(1) Total number of votes is 435

$\qquad x + y = 435$

(2) 49 more members voted in favor than against. Rephrasing this gives:

$\qquad$ The number of members voting in favor exceeded the number of members voting against by 49.

$\qquad x - y = 49$

Thus we have two equations:

$\qquad x + y = 435 \qquad x - y = 49$.

Adding gives

$\qquad 2x = 484 \Rightarrow x = 242$.

Substituting $x = 242$ in the first equation gives

$\qquad 242 + y = 435 \Rightarrow y = 435 - 242 = 193$.

Solution: 242 voted in favor and 193 voted against.

55. Unknowns: x = number of soccer games won; y = number of football games won

We are given two pieces of information:

(1) The total number of games was 12 $\Rightarrow x + y = 12$

(2) Total number of points earned was 38 $\Rightarrow 2x + 4y = 38$

(Two points per soccer game and 4 per football game)

Thus we have two equations:

$\qquad x + y = 12 \qquad 2x + 4y = 38$.

Dividing the second by -2:

$\qquad x + y = 12 \Rightarrow \ -x - 2y = -19$.

Adding gives

$\qquad -y = -7 \Rightarrow y = 7$ football games.

Substituting $y = 7$ in the first equation gives

$\qquad x + 7 = 12 \Rightarrow x = 5$ soccer games.

Solution: Lombardi House won 5 soccer games and 7 football games.

57. Unknowns: x = number of brand X pens; y = number of brand Y pens

We are given two pieces of information:

(1) The total number of pens is $12 \Rightarrow x + y = 12$.

(2) The total amount spent was $42 \Rightarrow 4x + 2.8y = 42$

($4 per band X pen and $2.80 per brand Y pen).

Thus we have two equations:

$$x + y = 12 \qquad 4x + 2.8y = 42.$$

Multiply the second by 10:

$$x + y = 12 \Rightarrow 40x + 28y = 420.$$

Multiply the first by 10 and divide the second by -4:

$$10x + 10y = 120 \Rightarrow -10x - 7y = -105.$$

Adding,

$$3y = 15 \Rightarrow y = 5 \text{ brand } Y \text{ pens.}$$

Substituting this value in the first equation $x + y = 12$ gives

$$x + 5 = 12 \Rightarrow x = 7 \text{ brand } X \text{ pens.}$$

Solution: Elena purchased 7 brand X pens and 5 brand Y pens.

59. If we want demand to equal supply, we can simply set the two functions equal:

$$-60p + 150 = 80p - 60$$

$$140p = 210$$

$$p = \frac{210}{140} = \$1.50 \text{ per pet chia.}$$

(We do not need to solve for q.)

61. Demand: $D = 85 - 5P$ \qquad Supply: $S = 25 + 5P$

For the equilibrium, we can equate the supply and demand:

$$\text{Demand} = \text{Supply} \Rightarrow 85 - 5P = 25 + 5P \Rightarrow -10P = -60 \Rightarrow P = \$6 \text{ per widget.}$$

Substituting $P = \$6$ in the demand curve gives $D = 85 - 5(6) = 85 - 30 = 55$ widgets.

63. Demand: The given points are $(p, q) = (8, 15)$ and $(11, 3)$.

$$m = \frac{q_2 - q_1}{p_2 - p_1} = \frac{3 - 15}{11 - 8} = \frac{-12}{3} = -4$$

$$b = q_1 - mp_1 = 15 - (-4)(8) = 15 + 32 = 47$$

Thus, the demand equation is

$$q = mp + b = -4p + 47.$$

Supply: The given points are $(p, q) = (8, 3)$ and $(11, 15)$.

$$m = \frac{q_2 - q_1}{p_2 - p_1} = \frac{15 - 3}{11 - 8} = \frac{12}{3} = 4$$

$$b = q_1 - mp_1 = 3 - (4)(8) = 3 - 32 = -29$$

Thus, the supply equation is

$$q = mp + b = 4p - 29.$$

For the equilibrium price, we can equate the supply and demand:

$$\text{Supply} = \text{Demand} \Rightarrow -4p + 47 = 4p - 29 \Rightarrow -8p = -76 \Rightarrow p = \frac{76}{8} = \$9.50.$$

65. Unknowns: x = number of pairs of dirty socks; y = number of T-shirts

We are given two pieces of information:

(1) A total of 44 items were washed: $\Rightarrow x + y = 44$.

(2) There were three times as many pairs of dirty socks as T-shirts. Rephrasing this gives:

The number of pairs of dirty socks was three times the number of T-shirts, or

$$x = 3y.$$

Thus we have two equations:

$$x + y = 44 \qquad x - 3y = 0.$$

Subtracting (or multiplying the second by -1 and adding) gives

$$y = 44 \Rightarrow y = 11 \text{ T-shirts.}$$

Substituting $y = 11$ in the first equation gives

$$x + 11 = 44 \Rightarrow x = 44 - 11 = 33 \text{ pairs of dirty socks.}$$

Solution: Joe's roommate threw out 33 pairs of dirty socks and 11 T-shirts.

67. Unknowns: x = size of raise for each full-time employee; y = size of raise for each part-time employee

We are given two pieces of information:

(1) Total budget = \$6,000. There are 4 full-time employees each getting a raise of x and 2 part-time employees each getting a raise of y. Thus,

$$4x + 2y = 6,000.$$

(2) The raise received by each full-time employee is twice the raise that each of the part-time employees receives.

$$x = 2y$$

Thus we have two equations:

$$4x + 2y = 6,000 \qquad x - 2y = 0.$$

Adding gives

$$5x = 6,000 \Rightarrow x = \$1,200 \text{ per full-time employee.}$$

(We are not asked for the value of y.)

Communication and reasoning exercises

69. The three lines in a plane must intersect in a single point for there to be a unique solution. This can happen in two ways: (1) The three lines intersect in a single point, or (2) two of the lines are the same, and the third line intersects it in a single point.

71. The equilibrium price occurs at the point where the demand and supply lines cross. Even if two lines have negative slope, they will still intersect if the slopes differ. Therefore, there can be an equilibrium price.

73. You cannot round both of them up, since there will not be sufficient eggs and cream. Rounding both answers down will ensure that you will not run out of ingredients. It may be possible to round one answer down and the other up and still have sufficient eggs and cream, and this should be tried.

75. Since multiplying both sides of an equation by a nonzero number has no effect on its solutions, the graph (which represents the set of all solutions) is unchanged: (B).

77. The associated system has no solutions, and so the lines do not intersect. Thus, they must be parallel: (B).

79. Answers will vary.

81. Choosing two lines at random gives two random slopes (which are numbers). Since two randomly selected numbers are unlikely to be the same, it follows that two randomly chosen straight lines are very unlikely to be parallel (or the same line). Thus, the two lines are very likely to intersect in a point, giving a unique solution.

Section 3.2

1. $\begin{bmatrix} \boxed{1} & 1 & 4 \\ 1 & -1 & 2 \end{bmatrix} \begin{matrix} \\ R_2 - R_1 \end{matrix} \rightarrow \begin{bmatrix} 1 & 1 & 4 \\ 0 & -2 & -2 \end{bmatrix} (1/2)R_2 \rightarrow \begin{bmatrix} 1 & 1 & 4 \\ 0 & \boxed{-1} & -1 \end{bmatrix} \begin{matrix} R_1 + R_2 \\ \end{matrix} \rightarrow$

$\begin{bmatrix} 1 & 0 & 3 \\ 0 & -1 & -1 \end{bmatrix} \begin{matrix} \\ -R_2 \end{matrix} \rightarrow \begin{bmatrix} 1 & 0 & 3 \\ 0 & 1 & 1 \end{bmatrix}$

Translating back to equations gives the solution: $x = 3, y = 1$.

3. $\begin{bmatrix} \boxed{3} & -2 & 6 \\ 2 & -3 & -6 \end{bmatrix} 3R_2 - 2R_1 \rightarrow \begin{bmatrix} 3 & -2 & 6 \\ 0 & -5 & -30 \end{bmatrix} (1/5)R_2 \rightarrow \begin{bmatrix} 3 & -2 & 6 \\ 0 & \boxed{-1} & -6 \end{bmatrix} \begin{matrix} R_1 - 2R_2 \\ \end{matrix} \rightarrow$

$\begin{bmatrix} 3 & 0 & 18 \\ 0 & -1 & -6 \end{bmatrix} (1/3)R_1 \rightarrow \begin{bmatrix} 1 & 0 & 6 \\ 0 & -1 & -6 \end{bmatrix} \begin{matrix} \\ -R_2 \end{matrix} \rightarrow \begin{bmatrix} 1 & 0 & 6 \\ 0 & 1 & 6 \end{bmatrix}$

Translating back to equations gives the solution: $x = 6, y = 6$.

5. $\begin{bmatrix} 2 & 3 & 1 \\ -1 & -3/2 & -1/2 \end{bmatrix} 2R_2 \rightarrow \begin{bmatrix} \boxed{2} & 3 & 1 \\ -2 & -3 & -1 \end{bmatrix} \begin{matrix} \\ R_2 + R_1 \end{matrix} \rightarrow \begin{bmatrix} 2 & 3 & 1 \\ 0 & 0 & 0 \end{bmatrix} (1/2)R_1 \rightarrow$

$\begin{bmatrix} 1 & 3/2 & 1/2 \\ 0 & 0 & 0 \end{bmatrix}$

Translating back to equations gives: $x + \dfrac{3}{2} y = \dfrac{1}{2}$.

Solve for x: $x = \dfrac{1}{2} - \dfrac{3}{2} y$

General Solution: $x = \dfrac{1}{2} - \dfrac{3}{2} y$; y is arbitrary or, in coordinate form, $(\tfrac{1}{2}(1 - 3y), y)$; y arbitrary

7. $\begin{bmatrix} 2 & 3 & 2 \\ -1 & -3/2 & -1/2 \end{bmatrix} 2R_2 \rightarrow \begin{bmatrix} \boxed{2} & 3 & 2 \\ -2 & -3 & -1 \end{bmatrix} \begin{matrix} \\ R_2 + R_1 \end{matrix} \rightarrow \begin{bmatrix} 2 & 3 & 2 \\ 0 & 0 & 1 \end{bmatrix}$

Since Row 2 translates to the false statement $0 = 1$, there is no solution.

9. $\begin{bmatrix} \boxed{1} & 1 & 1 \\ 3 & -1 & 0 \\ 1 & -3 & -2 \end{bmatrix} \begin{matrix} \\ R_2 - 3R_1 \\ R_3 - R_1 \end{matrix} \rightarrow \begin{bmatrix} 1 & 1 & 1 \\ 0 & \boxed{-4} & -3 \\ 0 & -4 & -3 \end{bmatrix} \begin{matrix} 4R_1 + R_2 \\ \\ R_3 - R_2 \end{matrix} \rightarrow \begin{bmatrix} 4 & 0 & 1 \\ 0 & -4 & -3 \\ 0 & 0 & 0 \end{bmatrix} \begin{matrix} (1/4)R_1 \\ -(1/4)R_2 \\ \end{matrix} \rightarrow$

$\begin{bmatrix} 1 & 0 & 1/4 \\ 0 & 1 & 3/4 \\ 0 & 0 & 0 \end{bmatrix}$

Translating back to equations gives the solution: $x = \dfrac{1}{4}, y = \dfrac{3}{4}$.

11. $\begin{bmatrix} \boxed{1} & 1 & 0 \\ 3 & -1 & 1 \\ 1 & -1 & -1 \end{bmatrix} \begin{matrix} \\ R_2 - 3R_1 \\ R_3 - R_1 \end{matrix} \rightarrow \begin{bmatrix} 1 & 1 & 0 \\ 0 & \boxed{-4} & 1 \\ 0 & -2 & -1 \end{bmatrix} \begin{matrix} 4R_1 + R_2 \\ \\ 2R_3 - R_2 \end{matrix} \rightarrow \begin{bmatrix} 4 & 0 & 1 \\ 0 & -4 & 1 \\ 0 & 0 & -3 \end{bmatrix}$

Since Row 3 translates to the false statement $0 = -3$, there is no solution.

13. $\begin{bmatrix} 0.5 & 0.1 & 1.7 \\ 0.1 & -0.1 & 0.3 \\ 1 & 1 & 11/3 \end{bmatrix} \begin{matrix} 10R_1 \\ 10R_2 \\ 3R_3 \end{matrix} \rightarrow \begin{bmatrix} \boxed{5} & 1 & 17 \\ 1 & -1 & 3 \\ 3 & 3 & 11 \end{bmatrix} \begin{matrix} \\ 5R_2 - R_1 \\ 5R_3 - 3R_1 \end{matrix} \rightarrow \begin{bmatrix} 5 & 1 & 17 \\ 0 & -6 & -2 \\ 0 & 12 & 4 \end{bmatrix} \begin{matrix} \\ (1/2)R_2 \\ (1/4)R_3 \end{matrix} \rightarrow$

$$\begin{bmatrix} 5 & 1 & 17 \\ 0 & \boxed{-3} & -1 \\ 0 & 3 & 1 \end{bmatrix} \begin{matrix} 3R_1 + R_2 \\ \\ R_3 + R_2 \end{matrix} \rightarrow \begin{bmatrix} 15 & 0 & 50 \\ 0 & -3 & -1 \\ 0 & 0 & 0 \end{bmatrix} \begin{matrix} (1/15)R_1 \\ -(1/3)R_2 \end{matrix} \rightarrow \begin{bmatrix} 1 & 0 & 10/3 \\ 0 & 1 & 1/3 \\ 0 & 0 & 0 \end{bmatrix}$$

Translating back to equations gives the solution: $x = \dfrac{10}{3}, y = \dfrac{1}{3}$.

15. $\begin{bmatrix} \boxed{-1} & 2 & -1 & 0 \\ -1 & -1 & 2 & 0 \\ 2 & 0 & -1 & 4 \end{bmatrix} \begin{matrix} \\ R_2 - R_1 \\ R_3 + 2R_1 \end{matrix} \rightarrow \begin{bmatrix} -1 & 2 & -1 & 0 \\ 0 & -3 & 3 & 0 \\ 0 & 4 & -3 & 4 \end{bmatrix} \begin{matrix} \\ (1/3)R_2 \\ \end{matrix} \rightarrow$

$\begin{bmatrix} -1 & 2 & -1 & 0 \\ 0 & \boxed{-1} & 1 & 0 \\ 0 & 4 & -3 & 4 \end{bmatrix} \begin{matrix} R_1 + 2R_2 \\ \\ R_3 + 4R_2 \end{matrix} \rightarrow$

$\begin{bmatrix} -1 & 0 & 1 & 0 \\ 0 & -1 & 1 & 0 \\ 0 & 0 & \boxed{1} & 4 \end{bmatrix} \begin{matrix} R_1 - R_3 \\ R_2 - R_3 \\ \end{matrix} \rightarrow \begin{bmatrix} -1 & 0 & 0 & -4 \\ 0 & -1 & 0 & -4 \\ 0 & 0 & 1 & 4 \end{bmatrix} \begin{matrix} -R_1 \\ -R_2 \\ \end{matrix} \rightarrow \begin{bmatrix} 1 & 0 & 0 & 4 \\ 0 & 1 & 0 & 4 \\ 0 & 0 & 1 & 4 \end{bmatrix}$

Translating back to equations gives the solution: $x = 4, y = 4, z = 4$.

17. $\begin{bmatrix} 1 & 1 & 6 & -1 \\ 1/3 & -1/3 & 2/3 & 1 \\ 1/2 & 0 & 1 & 0 \end{bmatrix} \begin{matrix} \\ 3R_2 \\ 2R_3 \end{matrix} \rightarrow \begin{bmatrix} \boxed{1} & 1 & 6 & -1 \\ 1 & -1 & 2 & 3 \\ 1 & 0 & 2 & 0 \end{bmatrix} \begin{matrix} \\ R_2 - R_1 \\ R_3 - R_1 \end{matrix} \rightarrow \begin{bmatrix} 1 & 1 & 6 & -1 \\ 0 & -2 & -4 & 4 \\ 0 & -1 & -4 & 1 \end{bmatrix} \begin{matrix} \\ (1/2)R_2 \\ \end{matrix} \rightarrow$

$\begin{bmatrix} 1 & 1 & 6 & -1 \\ 0 & \boxed{-1} & -2 & 2 \\ 0 & -1 & -4 & 1 \end{bmatrix} \begin{matrix} R_1 + R_2 \\ \\ R_3 - R_2 \end{matrix} \rightarrow \begin{bmatrix} 1 & 0 & 4 & 1 \\ 0 & -1 & -2 & 2 \\ 0 & 0 & \boxed{-2} & -1 \end{bmatrix} \begin{matrix} R_1 + 2R_3 \\ R_2 - R_3 \\ \end{matrix} \rightarrow$

$\begin{bmatrix} 1 & 0 & 0 & -1 \\ 0 & -1 & 0 & 3 \\ 0 & 0 & -2 & -1 \end{bmatrix} \begin{matrix} \\ -R_2 \\ -(1/2)R_3 \end{matrix} \rightarrow$

$\begin{bmatrix} 1 & 0 & 0 & -1 \\ 0 & 1 & 0 & -3 \\ 0 & 0 & 1 & 1/2 \end{bmatrix}$

Translating back to equations gives the solution: $x = -1, y = -3, z = \dfrac{1}{2}$.

19. $\begin{bmatrix} -1/2 & 1 & -1/2 & 0 \\ -1/2 & -1/2 & 1 & 0 \\ 1 & -1/2 & -1/2 & 0 \end{bmatrix} \begin{matrix} 2R_1 \\ 2R_2 \\ 2R_3 \end{matrix} \rightarrow \begin{bmatrix} \boxed{-1} & 2 & -1 & 0 \\ -1 & -1 & 2 & 0 \\ 2 & -1 & -1 & 0 \end{bmatrix} \begin{matrix} \\ R_2 - R_1 \\ R_3 + 2R_1 \end{matrix} \rightarrow$

$\begin{bmatrix} -1 & 2 & -1 & 0 \\ 0 & -3 & 3 & 0 \\ 0 & 3 & -3 & 0 \end{bmatrix} \begin{matrix} \\ (1/3)R_2 \\ (1/3)R_3 \end{matrix} \rightarrow$

$\begin{bmatrix} -1 & 2 & -1 & 0 \\ 0 & \boxed{-1} & 1 & 0 \\ 0 & 1 & -1 & 0 \end{bmatrix} \begin{matrix} R_1 + 2R_2 \\ \\ R_3 + R_2 \end{matrix} \rightarrow \begin{bmatrix} -1 & 0 & 1 & 0 \\ 0 & -1 & 1 & 0 \\ 0 & 0 & 0 & 0 \end{bmatrix} \begin{matrix} -R_1 \\ -R_2 \\ \end{matrix} \rightarrow \begin{bmatrix} 1 & 0 & -1 & 0 \\ 0 & 1 & -1 & 0 \\ 0 & 0 & 0 & 0 \end{bmatrix}$

Translating back to equations gives: $x - z = 0, y - z = 0$.

Thus, the general solution is $x = z, y = z, z$ arbitrary, or (z, z, z); z arbitrary.

21. $\begin{bmatrix} 1 & 1 & 2 & -1 \\ 2 & 2 & 2 & 2 \\ 3/5 & 3/5 & 3/5 & 2/5 \end{bmatrix} 5R_3 \rightarrow \begin{bmatrix} 1 & 1 & 2 & -1 \\ 2 & 2 & 2 & 2 \\ 3 & 3 & 3 & 2 \end{bmatrix} (1/2)R_2 \rightarrow \begin{bmatrix} \boxed{1} & 1 & 2 & -1 \\ 1 & 1 & 1 & 1 \\ 3 & 3 & 3 & 2 \end{bmatrix} \begin{matrix} \\ R_2 - R_1 \\ R_3 - 3R_1 \end{matrix} \rightarrow$

$\begin{bmatrix} 1 & 1 & 2 & -1 \\ 0 & 0 & \boxed{-1} & 2 \\ 0 & 0 & -3 & 5 \end{bmatrix} \begin{matrix} R_1 + 2R_2 \\ \\ R_3 - 3R_2 \end{matrix} \rightarrow \begin{bmatrix} 1 & 1 & 0 & 3 \\ 0 & 0 & -1 & 2 \\ 0 & 0 & 0 & -1 \end{bmatrix}$

Since Row 3 translates to the false statement $0 = -1$, there is no solution.

23. $\begin{bmatrix} -0.5 & 0.5 & 0.5 & 1.5 \\ 4.2 & 2.1 & 2.1 & 0 \\ 0.2 & 0 & 0.2 & 0 \end{bmatrix} \begin{matrix} 2R_1 \\ 10R_2 \\ 5R_3 \end{matrix} \rightarrow \begin{bmatrix} -1 & 1 & 1 & 3 \\ 42 & 21 & 21 & 0 \\ 1 & 0 & 1 & 0 \end{bmatrix} (1/21)R_2 \rightarrow \begin{bmatrix} \boxed{-1} & 1 & 1 & 3 \\ 2 & 1 & 1 & 0 \\ 1 & 0 & 1 & 0 \end{bmatrix} \begin{matrix} \\ R_2 + 2R_1 \\ R_3 + R_1 \end{matrix}$

$\rightarrow$

$\begin{bmatrix} -1 & 1 & 1 & 3 \\ 0 & 3 & 3 & 6 \\ 0 & 1 & 2 & 3 \end{bmatrix} (1/3)R_2 \rightarrow \begin{bmatrix} -1 & 1 & 1 & 3 \\ 0 & \boxed{1} & 1 & 2 \\ 0 & 1 & 2 & 3 \end{bmatrix} \begin{matrix} R_1 - R_2 \\ \\ R_3 - R_2 \end{matrix} \rightarrow \begin{bmatrix} -1 & 0 & 0 & 1 \\ 0 & 1 & 1 & 2 \\ 0 & 0 & \boxed{1} & 1 \end{bmatrix} R_2 - R_3 \rightarrow$

$\begin{bmatrix} -1 & 0 & 0 & 1 \\ 0 & 1 & 0 & 1 \\ 0 & 0 & 1 & 1 \end{bmatrix} \begin{matrix} -R_1 \\ \\ \end{matrix} \rightarrow \begin{bmatrix} 1 & 0 & 0 & -1 \\ 0 & 1 & 0 & 1 \\ 0 & 0 & 1 & 1 \end{bmatrix}$

Translating back to equations gives the solution: $x = -1, y = 1, z = 1$.

25. $\begin{bmatrix} \boxed{2} & -1 & 1 & 4 \\ 3 & -1 & 1 & 5 \end{bmatrix} \begin{matrix} \\ 2R_2 - 3R_1 \end{matrix} \rightarrow \begin{bmatrix} 2 & -1 & 1 & 4 \\ 0 & \boxed{1} & -1 & -2 \end{bmatrix} \begin{matrix} R_1 + R_2 \\ \\ \end{matrix} \rightarrow \begin{bmatrix} 2 & 0 & 0 & 2 \\ 0 & 1 & -1 & -2 \end{bmatrix} (1/2)R_1 \rightarrow$

$\begin{bmatrix} 1 & 0 & 0 & 1 \\ 0 & 1 & -1 & -2 \end{bmatrix}$

Translating back to equations gives: $x = 1, y - z = -2$.

Thus, the general solution is $x = 1, \ y = z - 2, \ z$ arbitrary, or $(1, z - 2, z)$; z arbitrary.

27. $\begin{bmatrix} 0.75 & -0.75 & -1 & 4 \\ 1 & -1 & 4 & 0 \end{bmatrix} \begin{matrix} 4R_1 \\ \\ \end{matrix} \rightarrow \begin{bmatrix} \boxed{3} & -3 & -4 & 16 \\ 1 & -1 & 4 & 0 \end{bmatrix} \begin{matrix} \\ 3R_2 - R_1 \end{matrix} \rightarrow$

$\begin{bmatrix} 3 & -3 & -4 & 16 \\ 0 & 0 & 16 & -16 \end{bmatrix} \begin{matrix} \\ (1/16)R_2 \end{matrix} \rightarrow \begin{bmatrix} 3 & -3 & -4 & 16 \\ 0 & 0 & \boxed{1} & -1 \end{bmatrix} \begin{matrix} R_1 + 4R_2 \\ \\ \end{matrix} \rightarrow$

$\begin{bmatrix} 3 & -3 & 0 & 12 \\ 0 & 0 & 1 & -1 \end{bmatrix} \begin{matrix} (1/3)R_1 \\ \\ \end{matrix} \rightarrow \begin{bmatrix} 1 & -1 & 0 & 4 \\ 0 & 0 & 1 & -1 \end{bmatrix}$

Translating back to equations gives: $x - y = 4, z = -1$.

Thus, the general solution is $x = y + 4, \ y$ arbitrary, $z = -1$, or $(y + 4, y, -1)$; y arbitrary.

29. $\begin{bmatrix} 3 & 1 & -1 & 12 \end{bmatrix} (1/3)R_1 \rightarrow \begin{bmatrix} 1 & 1/3 & -1/3 & 4 \end{bmatrix}$

Translating back to equations gives: $x + \dfrac{1}{3}y - \dfrac{1}{3}z = 4$.

Thus, the general solution is $x = 4 - y/3 + z/3, \ y$ arbitrary, z arbitrary, or $(4 - y/3 + z/3, y, z)$; $y, \ z$ arbitrary.

31. $\begin{bmatrix} 1 & 1 & 2 & -1 \\ 2 & 2 & 2 & 2 \\ 0.75 & 0.75 & 1 & 0.25 \\ -1 & 0 & -2 & 21 \end{bmatrix} \begin{matrix} \\ \\ 4R_3 \\ \\ \end{matrix} \rightarrow \begin{bmatrix} 1 & 1 & 2 & -1 \\ 2 & 2 & 2 & 2 \\ 3 & 3 & 4 & 1 \\ -1 & 0 & -2 & 21 \end{bmatrix} \begin{matrix} \\ (1/2)R_2 \\ \\ \end{matrix} \rightarrow$

$$\begin{bmatrix} \boxed{1} & 1 & 2 & -1 \\ 1 & 1 & 1 & 1 \\ 3 & 3 & 4 & 1 \\ -1 & 0 & -2 & 21 \end{bmatrix} \begin{matrix} \\ R_2 - R_1 \\ R_3 - 3R_1 \\ R_4 + R_1 \end{matrix} \rightarrow$$

$$\begin{bmatrix} 1 & 1 & 2 & -1 \\ 0 & 0 & -1 & 2 \\ 0 & 0 & -2 & 4 \\ 0 & 1 & 0 & 20 \end{bmatrix} \begin{matrix} \\ \\ (1/2)R_3 \\ \\ \end{matrix} \rightarrow \begin{bmatrix} 1 & 1 & 2 & -1 \\ 0 & 0 & \boxed{-1} & 2 \\ 0 & 0 & -1 & 2 \\ 0 & 1 & 0 & 20 \end{bmatrix} \begin{matrix} R_1 + 2R_2 \\ \\ R_3 - R_2 \\ \\ \end{matrix} \rightarrow \begin{bmatrix} 1 & 1 & 0 & 3 \\ 0 & 0 & -1 & 2 \\ 0 & 0 & 0 & 0 \\ 0 & \boxed{1} & 0 & 20 \end{bmatrix} \begin{matrix} R_1 - R_4 \\ \\ \\ \end{matrix} \rightarrow$$

$$\begin{bmatrix} 1 & 0 & 0 & -17 \\ 0 & 0 & -1 & 2 \\ 0 & 0 & 0 & 0 \\ 0 & 1 & 0 & 20 \end{bmatrix} \begin{matrix} \\ -R_2 \\ \\ \end{matrix} \rightarrow \begin{bmatrix} 1 & 0 & 0 & -17 \\ 0 & 0 & 1 & -2 \\ 0 & 0 & 0 & 0 \\ 0 & 1 & 0 & 20 \end{bmatrix} \begin{matrix} \\ \\ R_4 \rightarrow R_2 \rightarrow R_3 \rightarrow R_4 \\ \\ \end{matrix} \rightarrow \begin{bmatrix} 1 & 0 & 0 & -17 \\ 0 & 1 & 0 & 20 \\ 0 & 0 & 1 & -2 \\ 0 & 0 & 0 & 0 \end{bmatrix}$$

Translating back to equations gives the solution: $x = -17, y = 20, z = -2$.

33. $\begin{bmatrix} \boxed{1} & 1 & 5 & 0 & 1 \\ 0 & 1 & 2 & 1 & 1 \\ 1 & 3 & 7 & 2 & 2 \\ 1 & 1 & 5 & 1 & 1 \end{bmatrix} \begin{matrix} \\ \\ R_3 - R_1 \\ R_4 - R_1 \end{matrix} \rightarrow \begin{bmatrix} 1 & 1 & 5 & 0 & 1 \\ 0 & \boxed{1} & 2 & 1 & 1 \\ 0 & 2 & 2 & 2 & 1 \\ 0 & 0 & 0 & 1 & 0 \end{bmatrix} \begin{matrix} R_1 - R_2 \\ \\ R_3 - 2R_2 \\ \end{matrix} \rightarrow$

$$\begin{bmatrix} 1 & 0 & 3 & -1 & 0 \\ 0 & 1 & 2 & 1 & 1 \\ 0 & 0 & -2 & 0 & -1 \\ 0 & 0 & 0 & 1 & 0 \end{bmatrix} \begin{matrix} 2R_1 + 3R_3 \\ R_2 + R_3 \\ \\ \end{matrix} \rightarrow \begin{bmatrix} 2 & 0 & 0 & -2 & -3 \\ 0 & 1 & 0 & 1 & 0 \\ 0 & 0 & -2 & 0 & -1 \\ 0 & 0 & 0 & \boxed{1} & 0 \end{bmatrix} \begin{matrix} R_1 + 2R_4 \\ R_2 - R_4 \\ \\ \end{matrix} \rightarrow$$

$$\begin{bmatrix} 2 & 0 & 0 & 0 & -3 \\ 0 & 1 & 0 & 0 & 0 \\ 0 & 0 & -2 & 0 & -1 \\ 0 & 0 & 0 & 1 & 0 \end{bmatrix} \begin{matrix} (1/2)R_1 \\ \\ -(1/2)R_3 \\ \end{matrix} \rightarrow \begin{bmatrix} 1 & 0 & 0 & 0 & -3/2 \\ 0 & 1 & 0 & 0 & 0 \\ 0 & 0 & 1 & 0 & 1/2 \\ 0 & 0 & 0 & 1 & 0 \end{bmatrix}$$

Translating back to equations gives the solution: $x = -\dfrac{3}{2}, y = 0, z = \dfrac{1}{2}, w = 0$.

35. $\begin{bmatrix} \boxed{1} & 1 & 5 & 0 & 1 \\ 0 & 1 & 2 & 1 & 1 \\ 1 & 1 & 5 & 1 & 1 \\ 1 & 2 & 7 & 2 & 2 \end{bmatrix} \begin{matrix} \\ \\ R_3 - R_1 \\ R_4 - R_1 \end{matrix} \rightarrow \begin{bmatrix} 1 & 1 & 5 & 0 & 1 \\ 0 & \boxed{1} & 2 & 1 & 1 \\ 0 & 0 & 0 & 1 & 0 \\ 0 & 1 & 2 & 2 & 1 \end{bmatrix} \begin{matrix} R_1 - R_2 \\ \\ \\ R_4 - R_2 \end{matrix} \rightarrow$

$$\begin{bmatrix} 1 & 0 & 3 & -1 & 0 \\ 0 & 1 & 2 & 1 & 1 \\ 0 & 0 & 0 & \boxed{1} & 0 \\ 0 & 0 & 0 & 1 & 0 \end{bmatrix} \begin{matrix} R_1 + R_3 \\ R_2 - R_3 \\ \\ R_4 - R_3 \end{matrix} \rightarrow \begin{bmatrix} 1 & 0 & 3 & 0 & 0 \\ 0 & 1 & 2 & 0 & 1 \\ 0 & 0 & 0 & 1 & 0 \\ 0 & 0 & 0 & 0 & 0 \end{bmatrix}$$

Translating back to equations gives: $x + 3z = 0, y + 2z = 1, w = 0$.

General solution: $x = -3z, \ y = 1 - 2z, \ z$ arbitrary, $w = 0$ or $(-3z, 1 - 2z, z, 0); z$ arbitrary

37. $\begin{bmatrix} \boxed{1} & -2 & 1 & -4 & 1 \\ 1 & 3 & 7 & 2 & 2 \\ 2 & 1 & 8 & -2 & 3 \end{bmatrix} \begin{matrix} \\ R_2 - R_1 \\ R_3 - 2R_1 \end{matrix} \rightarrow \begin{bmatrix} 1 & -2 & 1 & -4 & 1 \\ 0 & \boxed{5} & 6 & 6 & 1 \\ 0 & 5 & 6 & 6 & 1 \end{bmatrix} \begin{matrix} 5R_1 + 2R_2 \\ \\ R_3 - R_2 \end{matrix} \rightarrow$

$$\begin{bmatrix} 5 & 0 & 17 & -8 & 7 \\ 0 & 5 & 6 & 6 & 1 \\ 0 & 0 & 0 & 0 & 0 \end{bmatrix} \begin{matrix} (1/5)R_1 \\ (1/5)R_2 \\ \\ \end{matrix} \rightarrow \begin{bmatrix} 1 & 0 & 17/5 & -8/5 & 7/5 \\ 0 & 1 & 6/5 & 6/5 & 1/5 \\ 0 & 0 & 0 & 0 & 0 \end{bmatrix}$$

Translating back to equations gives: $x + \dfrac{17}{5}z - \dfrac{8}{5}w = \dfrac{7}{5}, y + \dfrac{6}{5}z + \dfrac{6}{5}w = \dfrac{1}{5}.$

General solution: $x = 7/5 - 17z/5 + 8w/5, \ y = 1/5 - 6z/5 - 6w/5, \ z, \ w$ arbitrary,

or $(7/5 - 17z/5 + 8w/5, 1/5 - 6z/5 - 6w/5, z, w); z, \ w$ arbitrary,

or $\left(\dfrac{1}{5}(7 - 17z + 8w), \dfrac{1}{5}(1 - 6z - 6w), z, w\right); z, \ w$ arbitrary

39. $\begin{bmatrix} 1 & 1 & 1 & 1 & 1 & 15 \\ 0 & 1 & -1 & 1 & -1 & -2 \\ 0 & 0 & 1 & 1 & 1 & 12 \\ 0 & 0 & 0 & 1 & -1 & -1 \\ 0 & 0 & 0 & 0 & 1 & 5 \end{bmatrix} \begin{matrix} R_1 - R_2 \\ \\ \\ \\ \\ \end{matrix} \rightarrow \begin{bmatrix} 1 & 0 & 2 & 0 & 2 & 17 \\ 0 & 1 & -1 & 1 & -1 & -2 \\ 0 & 0 & 1 & 1 & 1 & 12 \\ 0 & 0 & 0 & 1 & -1 & -1 \\ 0 & 0 & 0 & 0 & 1 & 5 \end{bmatrix} \begin{matrix} R_1 - 2R_3 \\ R_2 + R_3 \\ \\ \\ \\ \end{matrix} \rightarrow$

$\begin{bmatrix} 1 & 0 & 0 & -2 & 0 & -7 \\ 0 & 1 & 0 & 2 & 0 & 10 \\ 0 & 0 & 1 & 1 & 1 & 12 \\ 0 & 0 & 0 & 1 & -1 & -1 \\ 0 & 0 & 0 & 0 & 1 & 5 \end{bmatrix} \begin{matrix} R_1 + 2R_4 \\ R_2 - 2R_4 \\ R_3 - R_4 \\ \\ \\ \end{matrix} \rightarrow \begin{bmatrix} 1 & 0 & 0 & 0 & -2 & -9 \\ 0 & 1 & 0 & 0 & 2 & 12 \\ 0 & 0 & 1 & 0 & 2 & 13 \\ 0 & 0 & 0 & 1 & -1 & -1 \\ 0 & 0 & 0 & 0 & 1 & 5 \end{bmatrix} \begin{matrix} R_1 + 2R_5 \\ R_2 - 2R_5 \\ R_3 - 2R_5 \\ R_4 + R_5 \\ \\ \end{matrix} \rightarrow$

$\begin{bmatrix} 1 & 0 & 0 & 0 & 0 & 1 \\ 0 & 1 & 0 & 0 & 0 & 2 \\ 0 & 0 & 1 & 0 & 0 & 3 \\ 0 & 0 & 0 & 1 & 0 & 4 \\ 0 & 0 & 0 & 0 & 1 & 5 \end{bmatrix}$

Translating back to equations gives the solution: $x = 1, y = 2, z = 3, u = 4, v = 5.$

41. $\begin{bmatrix} 1 & -1 & 1 & -1 & 1 & 0 \\ 0 & 1 & -1 & 1 & -1 & -2 \\ 1 & 0 & 0 & 0 & -2 & -2 \\ 2 & -1 & 1 & -1 & -3 & -2 \\ 4 & -1 & 1 & -1 & -7 & -6 \end{bmatrix} \begin{matrix} \\ \\ R_3 - R_1 \\ R_4 - 2R_1 \\ R_5 - 4R_1 \end{matrix} \rightarrow \begin{bmatrix} 1 & -1 & 1 & -1 & 1 & 0 \\ 0 & 1 & -1 & 1 & -1 & -2 \\ 0 & 1 & -1 & 1 & -3 & -2 \\ 0 & 1 & -1 & 1 & -5 & -2 \\ 0 & 3 & -3 & 3 & -11 & -6 \end{bmatrix} \begin{matrix} R_1 + R_2 \\ \\ R_3 - R_2 \\ R_4 - R_2 \\ R_5 - 3R_2 \end{matrix} \rightarrow$

$\begin{bmatrix} 1 & 0 & 0 & 0 & 0 & -2 \\ 0 & 1 & -1 & 1 & -1 & -2 \\ 0 & 0 & 0 & 0 & -2 & 0 \\ 0 & 0 & 0 & 0 & -4 & 0 \\ 0 & 0 & 0 & 0 & -8 & 0 \end{bmatrix} \begin{matrix} \\ \\ (1/2)R_3 \\ (1/4)R_4 \\ (1/8)R_5 \end{matrix} \rightarrow \begin{bmatrix} 1 & 0 & 0 & 0 & 0 & -2 \\ 0 & 1 & -1 & 1 & -1 & -2 \\ 0 & 0 & 0 & 0 & -1 & 0 \\ 0 & 0 & 0 & 0 & -1 & 0 \\ 0 & 0 & 0 & 0 & -1 & 0 \end{bmatrix} \begin{matrix} \\ R_2 - R_3 \\ \\ R_4 - R_3 \\ R_5 - R_3 \end{matrix} \rightarrow$

$\begin{bmatrix} 1 & 0 & 0 & 0 & 0 & -2 \\ 0 & 1 & -1 & 1 & 0 & -2 \\ 0 & 0 & 0 & 0 & -1 & 0 \\ 0 & 0 & 0 & 0 & 0 & 0 \\ 0 & 0 & 0 & 0 & 0 & 0 \end{bmatrix} \begin{matrix} \\ \\ -R_3 \\ \\ \end{matrix} \rightarrow \begin{bmatrix} 1 & 0 & 0 & 0 & 0 & -2 \\ 0 & 1 & -1 & 1 & 0 & -2 \\ 0 & 0 & 0 & 0 & 1 & 0 \\ 0 & 0 & 0 & 0 & 0 & 0 \\ 0 & 0 & 0 & 0 & 0 & 0 \end{bmatrix}$

Translating back to equations gives: $x = -2, y - z + u = -2, v = 0.$

General solution: $x = -2, \ y = -2 + z - u, \ z$ arbitrary, u arbitrary, $v = 0$

or $(-2, -2 + z - u, z, u, 0); z, \ u$ arbitrary

43. $x + 2y - z + w = 30$

$2x - z + 2w = 30$

$x + 3y + 3z - 4w = 2$

$2x - 9y + w = 4$

Using technology:

Matrix #1

x	y	z	w	
1	2	-1	1	30
2	0	-1	2	30
1	3	3	-4	2
2	-9	0	1	4

Matrix #2

x	y	z	w	
1	2	-1	1	30
0	-4	1	0	-30
0	1	4	-5	-28
0	-13	2	-1	-56

Matrix #3

x	y	z	w	
2	0	-1	2	30
0	-4	1	0	-30
0	0	17	-20	-142
0	0	-5	-4	166

Matrix #4

x	y	z	w	
34	0	0	14	368
0	-68	0	20	-368
0	0	17	-20	-142
0	0	0	-168	2112

Matrix #5

x	y	z	w	
17	0	0	7	184
0	-17	0	5	-92
0	0	17	-20	-142
0	0	0	-7	88

Matrix #6

x	y	z	w	
17	0	0	0	272
0	-119	0	0	-204
0	0	119	0	-2754
0	0	0	-7	88

Matrix #7

x	y	z	w	
1	0	0	0	16
0	1	0	0	12/7
0	0	1	0	-162/7
0	0	0	1	-88/7

Solution: $(16, 12/7, -162/7, -88/7)$

45. $x + 2y + 3z + 4w + 5t = 6$

$2x + 3y + 4z + 5w + t = 5$

$3x + 4y + 5z + w + 2t = 4$

$4x + 5y + z + 2w + 3t = 3$

$5x + y + 2z + 3w + 4t = 2$

Using technology:

Matrix #1

x	y	z	w	t	
1	2	3	4	5	6
2	3	4	5	1	5
3	4	5	1	2	4
4	5	1	2	3	3
5	1	2	3	4	2

Matrix #2

x	y	z	w	t	
1	2	3	4	5	6
0	−1	−2	−3	−9	−7
0	−2	−4	−11	−13	−14
0	−3	−11	−14	−17	−21
0	−9	−13	−17	−21	−28

Matrix #3

x	y	z	w	t	
1	0	−1	−2	−13	−8
0	−1	−2	−3	−9	−7
0	0	0	−5	5	0
0	0	−5	−5	10	0
0	0	5	10	60	35

Matrix #4

x	y	z	w	t	
1	0	−1	−2	−13	−8
0	−1	−2	−3	−9	−7
0	0	0	−1	1	0
0	0	−1	−1	2	0
0	0	1	2	12	7

Matrix #5

x	y	z	w	t	
1	0	−1	0	−15	−8
0	−1	−2	0	−12	−7
0	0	0	−1	1	0
0	0	−1	0	1	0
0	0	1	0	14	7

Matrix #6

x	y	z	w	t	
1	0	0	0	−16	−8
0	−1	0	0	−14	−7
0	0	0	−1	1	0
0	0	−1	0	1	0
0	0	0	0	15	7

```
Matrix #7
x       y       z       w       t
15      0       0       0       0       -8
0       -15     0       0       0       -7
0       0       0       -15     0       -7
0       0       -15     0       0       -7
0       0       0       0       15      7
```

```
Matrix #8
x       y       z       w       t
1       0       0       0       0       -8/15
0       1       0       0       0       7/15
0       0       0       1       0       7/15
0       0       1       0       0       7/15
0       0       0       0       1       7/15
```

Solution: $(-8/15, 7/15, 7/15, 7/15, 7/15)$

47. $1.6x + 2.4y - 3.2z = 4.4$

$5.1x - 6.3y + 0.6z = -3.2$

$4.2x + 3.5y + 4.9z = 10.1$

We use the Excel Matrix Pivot Tool (on the Website):

x	y	z	
1.6	2.4	-3.2	4.4
5.1	-6.3	0.6	-3.2
4.2	3.5	4.9	10.1

→

x	y	z	
1	1.5	-2	2.75
0	-13.95	10.8	-17.225
0	-2.8	13.3	-1.45

→

x	y	z	
1	0	-0.8387097	0.89784946
0	1	-0.7741935	1.23476703
0	0	11.1322581	2.00734767

→

x	y	z	
1	0	0	1.049084
0	1	0	1.37436814
0	0	1	0.1803181

Solution (rounded to 1 decimal place): $(1.0, 1.4, 0.2)$

49. $-0.2x + 0.3y + 0.4z - t = 4.5$

$2.2x + 1.1y - 4.7z + 2t = 8.3$

$9.2y - 1.3t = 0$

$3.4x + 0.5z - 3.4t = 0.1$

We use the Excel Matrix Pivot Tool (on the Website):

x	y	z	t	
-0.2	0.3	0.4	-1	4.5
2.2	1.1	-4.7	2	8.3
0	9.2	0	-1.3	0
3.4	0	0.5	-3.4	0.1

→

x	y	z	t	
1	-1.5	-2	5	-22.5
0	4.4	-0.3	-9	57.8
0	9.2	0	-1.3	0
0	5.1	7.3	-20.4	76.6

→

x	y	z	t	
1	0	-2.1022727	1.93181818	-2.7954545
0	1	-0.0681818	-2.0454545	13.1363636
0	0	0.62727273	17.5181818	-120.85455
0	0	7.64772727	-9.9681818	9.60454545

→

x	y	z	t	
1	0	0	60.6431159	-407.83333
0	1	0	-0.1413043	1.7764E-15
0	0	1	27.9275362	-192.66667
0	0	0	-223.55036	1483.06667

→

x	y	z	t	
1	0	0	0	-5.5177974
0	1	0	0	-0.9374343
0	0	1	0	-7.3911984
0	0	0	1	-6.6341501

Solution (rounded to 1 decimal place): $(-5.5, -0.9, -7.4, -6.6)$

Communication and reasoning exercises

51. A pivot is an entry in a matrix that is selected to "clear a column"; that is, use the row operations of a certain type to obtain zeros everywhere above and below it. "Pivoting" is the procedure of clearing a column using a designated pivot.

53. $2R_1 + 5R_4$, or $6R_1 + 15R_4$ (which is less desirable, since it will produce a row in which every entry is divisible by 3)

55. It will include a row of zeros. (Subtracting the two rows produces a row of zeros.)

57. The claim is wrong. If there are more equations than unknowns, there can be a unique solution as well as row(s) of zeros in the reduced matrix, as in Example 6.

59. Since there are 5 columns, there are 4 unknowns. (The last column is for the right-hand sides.) Since there are 5 rows of which 3 are zero, that leaves 2 rows with pivots. Thus, there are 2 unknowns that are not parameters. The remaining 2 unknowns are arbitrary (parameters).

61. The number of pivots must equal the number of variables, since no variable will be used as a parameter.

63. A simple example is $x = 1; y - z = 1; x + y - z = 2$.

65. It has to be the zero solution (each unknown is equal to zero): Putting each unknown equal to zero causes each equation to be satisfied because the right-hand sides are zero. Thus, the zero solution is in fact a solution. Because the solution is unique, this solution is the *only* solution.

67. No: As was pointed out in Exercise 65, every homogeneous system has at least one solution (namely, the zero solution) and hence cannot be inconsistent.

Applications
Section 3.3

1. Unknowns: $x = $ the number of batches of vanilla, $y = $ the number of batches of mocha, $z = $ the number of batches of strawberryArrange the given information in a table with unknowns across the top:

	Vanilla (x)	Mocha (y)	Strawberry (z)	Avail.
Eggs	2	1	1	350
Milk	1	1	2	350
Cream	2	2	1	400

We can now set up an equation for each of the items listed on the left:

Eggs: $2x + y + z = 350$

Milk: $x + y + 2z = 350$

Cream: $2x + 2y + z = 400.$

$$\begin{bmatrix} \boxed{2} & 1 & 1 & 350 \\ 1 & 1 & 2 & 350 \\ 2 & 2 & 1 & 400 \end{bmatrix} \begin{matrix} \\ 2R_2 - R_1 \\ R_3 - R_1 \end{matrix} \rightarrow \begin{bmatrix} 2 & 1 & 1 & 350 \\ 0 & \boxed{1} & 3 & 350 \\ 0 & 1 & 0 & 50 \end{bmatrix} \begin{matrix} R_1 - R_2 \\ \\ R_3 - R_2 \end{matrix} \rightarrow \begin{bmatrix} 2 & 0 & -2 & 0 \\ 0 & 1 & 3 & 350 \\ 0 & 0 & -3 & -300 \end{bmatrix} \begin{matrix} (1/2)R_1 \\ \\ (1/3)R_3 \end{matrix} \rightarrow$$

$$\begin{bmatrix} 1 & 0 & -1 & 0 \\ 0 & 1 & 3 & 350 \\ 0 & 0 & \boxed{-1} & -100 \end{bmatrix} \begin{matrix} R_1 - R_3 \\ R_2 + 3R_3 \\ \end{matrix} \rightarrow \begin{bmatrix} 1 & 0 & 0 & 100 \\ 0 & 1 & 0 & 50 \\ 0 & 0 & -1 & -100 \end{bmatrix} \begin{matrix} \\ \\ -R_3 \end{matrix} \rightarrow \begin{bmatrix} 1 & 0 & 0 & 100 \\ 0 & 1 & 0 & 50 \\ 0 & 0 & 1 & 100 \end{bmatrix}$$

$x = 100, y = 50, z = 100$

Solution: Make 100 batches of vanilla, 50 batches of mocha, and 100 batches of strawberry.

3. Unknowns: $x = $ the number of sections of Finite Math, $y = $ the number of sections of Applied Calculus, $z = $ the number of sections of Computer Methods
We are given three pieces of information:

(1) There are a total of 6 sections: $x + y + z = 6.$

(2) The total number of students is 210: $40x + 40y + 10z = 210.$

(3) The total revenue is $260,000: $40,000x + 60,000y + 20,000z = 260,000$

or, working in thousands of dollars, $40x + 60y + 20z = 260.$

Thus, we have a system of 3 equations in 3 unknowns:

$x + y + z = 6$

$40x + 40y + 10z = 210$

$40x + 60y + 20z = 260.$

$$\begin{bmatrix} 1 & 1 & 1 & 6 \\ 40 & 40 & 10 & 210 \\ 40 & 60 & 20 & 260 \end{bmatrix} \begin{matrix} \\ (1/10)R_2 \\ (1/20)R_3 \end{matrix} \rightarrow \begin{bmatrix} \boxed{1} & 1 & 1 & 6 \\ 4 & 4 & 1 & 21 \\ 2 & 3 & 1 & 13 \end{bmatrix} \begin{matrix} \\ R_2 - 4R_1 \\ R_3 - 2R_1 \end{matrix} \rightarrow \begin{bmatrix} 1 & 1 & 1 & 6 \\ 0 & 0 & -3 & -3 \\ 0 & 1 & -1 & 1 \end{bmatrix} \begin{matrix} \\ (1/3)R_2 \\ \end{matrix} \rightarrow$$

$$\begin{bmatrix} 1 & 1 & 1 & 6 \\ 0 & 0 & \boxed{-1} & -1 \\ 0 & 1 & -1 & 1 \end{bmatrix} \begin{matrix} R_1 + R_2 \\ \\ R_3 - R_2 \end{matrix} \rightarrow \begin{bmatrix} 1 & 1 & 0 & 5 \\ 0 & 0 & -1 & -1 \\ 0 & \boxed{1} & 0 & 2 \end{bmatrix} \begin{matrix} R_1 - R_3 \\ \\ \end{matrix} \rightarrow \begin{bmatrix} 1 & 0 & 0 & 3 \\ 0 & 0 & -1 & -1 \\ 0 & 1 & 0 & 2 \end{bmatrix} \begin{matrix} \\ -R_2 \\ \end{matrix} \rightarrow$$

$$\begin{bmatrix} 1 & 0 & 0 & 3 \\ 0 & 0 & 1 & 1 \\ 0 & 1 & 0 & 2 \end{bmatrix} \qquad x = 3, y = 2, z = 1$$

Solution: Offer 3 sections of Finite Math, 2 sections of Applied Calculus, and 1 section of Computer Methods.

5. Unknowns: $x =$ revenue (in millions) earned from regional music, $y =$ revenue (in millions) earned from pop/rock music, $z =$ revenue (in millions) earned from tropical music

Total revenues were \$58 million: $\qquad x + y + z = 58$

Regional music brought in four times as much as tropical music. Reword this as follows: The revenue earned from regional music was four times the revenue earned from tropical music:

$\qquad x = 4z$, or $x - 4z = 0$.

Pop/rock music brought in \$10 million more than tropical music. Reword this as follows: The revenue earned from pop/rock music was \$10 million more than the revenue earned from tropical music:

$\qquad y = z + 10$, or $y - z = 10$.

We thus solve the system

$x + y + z = 58$

$x - 4z = 0$

$y - z = 10$.

$$\begin{bmatrix} \boxed{1} & 1 & 1 & 58 \\ 1 & 0 & -4 & 0 \\ 0 & 1 & -1 & 10 \end{bmatrix} \begin{matrix} \\ R_2 - R_1 \to \\ \\ \end{matrix} \begin{bmatrix} 1 & 1 & 1 & 58 \\ 0 & \boxed{-1} & -5 & -58 \\ 0 & 1 & -1 & 10 \end{bmatrix} \begin{matrix} R_1 + R_2 \\ \\ R_3 + R_2 \end{matrix} \to \begin{bmatrix} 1 & 0 & -4 & 0 \\ 0 & -1 & -5 & -58 \\ 0 & 0 & -6 & -48 \end{bmatrix} (1/6)R_3$$

$$\to \begin{bmatrix} 1 & 0 & -4 & 0 \\ 0 & -1 & -5 & -58 \\ 0 & 0 & \boxed{-1} & -8 \end{bmatrix} \begin{matrix} R_1 - 4R_3 \\ R_2 - 5R_3 \to \\ \\ \end{matrix} \begin{bmatrix} 1 & 0 & 0 & 32 \\ 0 & -1 & 0 & -18 \\ 0 & 0 & -1 & -8 \end{bmatrix} \begin{matrix} \\ -R_2 \to \\ -R_3 \end{matrix} \begin{bmatrix} 1 & 0 & 0 & 32 \\ 0 & 1 & 0 & 18 \\ 0 & 0 & 1 & 8 \end{bmatrix}$$

$x = 32, y = 18, z = 8$

Solution: Revenues were \$32 million for regional music, \$18 million for pop/rock music, and \$8 million for tropical music.

7. Unknowns: $x =$ the number of Airbus A330-300s, $y =$ the number of Boeing 767-300ERs, $z =$ the number of Boeing Dreamliner 787-9s

Passengers: $330x + 270y + 240z = 4,980$

Cost: $\qquad 250x + 200y + 250z = 4,300$

The number of Boeings is twice the number of Airbuses: $y + z = 2x$, or $2x - y - z = 0$.

Solving:

$$\begin{bmatrix} 2 & -1 & -1 & 0 \\ 330 & 270 & 240 & 4980 \\ 250 & 200 & 250 & 4300 \end{bmatrix} \begin{matrix} \\ (1/30)R_2 \to \\ (1/50)R_3 \end{matrix} \begin{bmatrix} \boxed{2} & -1 & -1 & 0 \\ 11 & 9 & 8 & 166 \\ 5 & 4 & 5 & 86 \end{bmatrix} \begin{matrix} \\ 2R_2 - 11R_1 \to \\ 2R_3 - 5R_1 \end{matrix}$$

$$\begin{bmatrix} 2 & -1 & -1 & 0 \\ 0 & \boxed{29} & 27 & 332 \\ 0 & 13 & 15 & 172 \end{bmatrix} \begin{matrix} 29R_1 + R_2 \\ \\ 29R_3 - 13R_2 \end{matrix} \to \begin{bmatrix} 58 & 0 & -2 & 332 \\ 0 & 29 & 27 & 332 \\ 0 & 0 & 84 & 672 \end{bmatrix} \begin{matrix} (1/2)R_1 \\ \\ (1/84)R_3 \end{matrix} \to$$

$$\begin{bmatrix} 29 & 0 & -1 & 166 \\ 0 & 29 & 27 & 332 \\ 0 & 0 & \boxed{1} & 8 \end{bmatrix} \begin{matrix} R_1 + R_3 \\ R_2 - 27R_3 \to \\ \\ \end{matrix} \begin{bmatrix} 29 & 0 & 0 & 174 \\ 0 & 29 & 0 & 116 \\ 0 & 0 & 1 & 8 \end{bmatrix} \begin{matrix} (1/29)R_1 \\ (1/29)R_2 \to \\ \\ \end{matrix}$$

$$\begin{bmatrix} 1 & 0 & 0 & 6 \\ 0 & 1 & 0 & 4 \\ 0 & 0 & 1 & 8 \end{bmatrix} \qquad x = 6, y = 4, z = 8$$

Solution: Order 6 Airbus A330-300s, 4 Boeing 767-300ERs, and 8 Dreamliners.

9. Unknowns: $x =$ the number of tons from CCC, $y =$ the number of tons from SSS, $z =$ the number of tons from BBF

Total order of cheese is 100 tons: $x + y + z = 100$.

Total cost = \$5,990: $\qquad\qquad 80x + 50y + 65z = 5,990$

Same amount from CCC and BBF: $x = z$, or $x - z = 0$

Solving:

$$\begin{bmatrix} 1 & 1 & 1 & 100 \\ 80 & 50 & 65 & 5990 \\ 1 & 0 & -1 & 0 \end{bmatrix} \begin{matrix} \\ (1/5)R_2 \\ \\ \end{matrix} \rightarrow \begin{bmatrix} \boxed{1} & 1 & 1 & 100 \\ 16 & 10 & 13 & 1198 \\ 1 & 0 & -1 & 0 \end{bmatrix} \begin{matrix} \\ R_2 - 16R_1 \rightarrow \\ R_3 - R_1 \end{matrix}$$

$$\begin{bmatrix} 1 & 1 & 1 & 100 \\ 0 & -6 & -3 & -402 \\ 0 & -1 & -2 & -100 \end{bmatrix} \begin{matrix} \\ (1/3)R_2 \\ \\ \end{matrix} \rightarrow \begin{bmatrix} 1 & 1 & 1 & 100 \\ 0 & \boxed{-2} & -1 & -134 \\ 0 & -1 & -2 & -100 \end{bmatrix} \begin{matrix} 2R_1 + R_2 \\ \\ 2R_3 - R_2 \end{matrix}$$

$$\begin{bmatrix} 2 & 0 & 1 & 66 \\ 0 & -2 & -1 & -134 \\ 0 & 0 & -3 & -66 \end{bmatrix} \begin{matrix} \\ \rightarrow \\ (1/3)R_3 \end{matrix} \begin{bmatrix} 2 & 0 & 1 & 66 \\ 0 & -2 & -1 & -134 \\ 0 & 0 & \boxed{-1} & -22 \end{bmatrix} \begin{matrix} R_1 + R_3 \\ R_2 - R_3 \rightarrow \\ \\ \end{matrix}$$

$$\begin{bmatrix} 2 & 0 & 0 & 44 \\ 0 & -2 & 0 & -112 \\ 0 & 0 & -1 & -22 \end{bmatrix} \begin{matrix} (1/2)R_1 \\ (1/2)R_2 \rightarrow \\ \\ \end{matrix} \begin{bmatrix} 1 & 0 & 0 & 22 \\ 0 & -1 & 0 & -56 \\ 0 & 0 & -1 & -22 \end{bmatrix} \begin{matrix} \\ -R_2 \rightarrow \\ -R_3 \end{matrix}$$

$$\begin{bmatrix} 1 & 0 & 0 & 22 \\ 0 & 1 & 0 & 56 \\ 0 & 0 & 1 & 22 \end{bmatrix} \qquad x = 22, y = 56, z = 22$$

Solution: The store ordered 22 tons from Cheesy Cream, 56 tons from Super Smooth & Sons, and 22 tons from Bagel's Best Friend.

11. Unknowns: $x =$ the number of evil sorcerers slain, $y =$ the number of trolls slain, $z =$ the number of orcs slain

Total number slain was 560: $\qquad x + y + z = 560$.

Total number of sword thrusts was 620: $\qquad 2x + 2y + z = 620$.

The number of trolls slain is five times the number of evil sorcerers slain: $\qquad y = 5x$, or $-5x + y = 0$.

Solving:

$$\begin{bmatrix} \boxed{1} & 1 & 1 & 560 \\ 2 & 2 & 1 & 620 \\ -5 & 1 & 0 & 0 \end{bmatrix} \begin{matrix} \\ R_2 - 2R_1 \rightarrow \\ R_3 + 5R_1 \end{matrix} \begin{bmatrix} 1 & 1 & 1 & 560 \\ 0 & 0 & \boxed{-1} & -500 \\ 0 & 6 & 5 & 2800 \end{bmatrix} \begin{matrix} R_1 + R_2 \\ \\ R_3 + 5R_2 \end{matrix} \rightarrow$$

$$\begin{bmatrix} 1 & 1 & 0 & 60 \\ 0 & 0 & -1 & -500 \\ 0 & 6 & 0 & 300 \end{bmatrix} \begin{matrix} \\ \rightarrow \\ (1/6)R_3 \end{matrix} \begin{bmatrix} 1 & 1 & 0 & 60 \\ 0 & 0 & -1 & -500 \\ 0 & \boxed{1} & 0 & 50 \end{bmatrix} \begin{matrix} R_1 - R_3 \\ \\ \rightarrow \end{matrix}$$

$$\begin{bmatrix} 1 & 0 & 0 & 10 \\ 0 & 0 & -1 & -500 \\ 0 & 1 & 0 & 50 \end{bmatrix} \begin{matrix} \\ -R_2 \rightarrow \\ \\ \end{matrix} \begin{bmatrix} 1 & 0 & 0 & 10 \\ 0 & 0 & 1 & 500 \\ 0 & 1 & 0 & 50 \end{bmatrix} \qquad x = 10, y = 50, z = 500$$

Solution: Halmar has slain 10 evil sorcerers, 50 trolls, and 500 orcs.

13. Unknowns: $x =$ amount of money donated to the MPBF, $y =$ amount of money donated to the SCN, $z =$ amount of money donated to the NY Jets

Given information:

(1) The society donated twice as much to the NY Jets as to the MPBF. Rephrase this as follows: The amount of money donated to the NY Jets was equal to twice the amount of money donated to the MPBF:

$z = 2x$, or $2x - z = 0$.

(2) The society donated equal amounts to the first two funds: $x = y$, or $x - y = 0$.

(3) Money donated back to the society: $x + 2y + 2z = 4,200$

Solving:

$$\begin{bmatrix} \boxed{2} & 0 & -1 & 0 \\ 1 & -1 & 0 & 0 \\ 1 & 2 & 2 & 4200 \end{bmatrix} \begin{matrix} \\ 2R_2 - R_1 \\ 2R_3 - R_1 \end{matrix} \rightarrow \begin{bmatrix} 2 & 0 & -1 & 0 \\ 0 & \boxed{-2} & 1 & 0 \\ 0 & 4 & 5 & 8400 \end{bmatrix} \begin{matrix} \\ \\ R_3 + 2R_2 \end{matrix} \rightarrow$$

$$\begin{bmatrix} 2 & 0 & -1 & 0 \\ 0 & -2 & 1 & 0 \\ 0 & 0 & 7 & 8400 \end{bmatrix} \begin{matrix} \\ \\ (1/7)R_3 \end{matrix} \rightarrow \begin{bmatrix} 2 & 0 & -1 & 0 \\ 0 & -2 & 1 & 0 \\ 0 & 0 & \boxed{1} & 1200 \end{bmatrix} \begin{matrix} R_1 + R_3 \\ R_2 - R_3 \\ \end{matrix} \rightarrow$$

$$\begin{bmatrix} 2 & 0 & 0 & 1200 \\ 0 & -2 & 0 & -1200 \\ 0 & 0 & 1 & 1200 \end{bmatrix} \begin{matrix} (1/2)R_1 \\ (1/2)R_2 \\ \end{matrix} \rightarrow \begin{bmatrix} 1 & 0 & 0 & 600 \\ 0 & -1 & 0 & -600 \\ 0 & 0 & 1 & 1200 \end{bmatrix} \begin{matrix} \\ -R_2 \\ \end{matrix} \rightarrow$$

$$\begin{bmatrix} 1 & 0 & 0 & 600 \\ 0 & 1 & 0 & 600 \\ 0 & 0 & 1 & 1200 \end{bmatrix}$$

Solution: It donated $600 to each of the MPBF and the SCN, and $1,200 to the Jets.

15. Unknowns: $x =$ the number of empty seats on United Continental, $y =$ the number of empty seats on American,

$z =$ the number of empty seats on Southwest

We are given the following information:

(1) United Continental, American, and Southwest flew a total of 210 empty seats: $x + y + z = 210$.

(2) The total cost of these seats was $89,760. (Note that the figures in the table are for one mile, but the trip was 3,000 miles):

$(3,000)0.149x + (3,000)0.146y + (3,000)0.124z = 89,760$. That is, $447x + 438y + 372z = 89,760$.

(3) United Continental had three times as many empty seats as American: $x = 3y$, or $x - 3y = 0$

We use the Pivot and Gauss-Jordan Tool on the Website:

1	1	1	210
447	438	372	89760
1	-3	0	0

$\rightarrow$

1	1	1	210
0	-3	-25	-1370
0	-4	-1	-210

$\rightarrow$

3	0	-22	-740
0	3	25	1370
0	0	1	50

$\rightarrow$

1	0	0	120
0	1	0	40
0	0	1	50

Solution: Continental: 120; American: 40; Southwest: 50

17. Unknowns: $x =$ amount invested in SHPIX, $y =$ amount invested in RYURX, $z =$ amount invested in RYCWX

The total investment was $9,000: $x + y + z = 9,000$.

You invested an equal amount in RYURX and RYCWX: $y = z$, or $y - z = 0$.

YTD loss from the first two funds was $260: $0.04x + 0.03y = 260$.

Solving:

$$\begin{bmatrix} 1 & 1 & 1 & 9000 \\ 0 & 1 & -1 & 0 \\ 0.04 & 0.03 & 0 & 260 \end{bmatrix} \begin{matrix} \\ \\ 100R_3 \end{matrix} \rightarrow \begin{bmatrix} \boxed{1} & 1 & 1 & 9000 \\ 0 & 1 & -1 & 0 \\ 4 & 3 & 0 & 26000 \end{bmatrix} \begin{matrix} \\ \\ R_3 - 4R_1 \end{matrix} \rightarrow$$

$$\begin{bmatrix} 1 & 1 & 1 & 9000 \\ 0 & \boxed{1} & -1 & 0 \\ 0 & -1 & -4 & -10000 \end{bmatrix} \begin{matrix} R_1 - R_2 \\ \\ R_3 + R_2 \end{matrix} \rightarrow \begin{bmatrix} 1 & 0 & 2 & 9000 \\ 0 & 1 & -1 & 0 \\ 0 & 0 & -5 & -10000 \end{bmatrix} \begin{matrix} \\ \\ (1/5)R_3 \end{matrix} \rightarrow$$

$$\begin{bmatrix} 1 & 0 & 2 & 9000 \\ 0 & 1 & -1 & 0 \\ 0 & 0 & \boxed{-1} & -2000 \end{bmatrix} \begin{matrix} R_1 + 2R_3 \\ R_2 - R_3 \\ \\ \end{matrix} \rightarrow \begin{bmatrix} 1 & 0 & 0 & 5000 \\ 0 & 1 & 0 & 2000 \\ 0 & 0 & -1 & -2000 \end{bmatrix} \begin{matrix} \\ \\ -R_3 \end{matrix} \rightarrow$$

$$\begin{bmatrix} 1 & 0 & 0 & 5000 \\ 0 & 1 & 0 & 2000 \\ 0 & 0 & 1 & 2000 \end{bmatrix} x = 5,000, \, y = 2,000, \, z = 2,000$$

Solution: You invested \$5,000 in SHPIX, \$2,000 in RYURX, \$2,000 in RYCWX.

19. Unknowns: $x =$ the number of shares of WSR, $y =$ the number of shares of HCC, $z =$ the number of shares of SNDK

The total investment was \$8,400.

Investment in WSR $= x$ shares @ \$16 $= 16x$

Investment in HCC $= y$ shares @ \$56 $= 56y$

Investment in SNDK $= z$ shares @ \$80 $= 80z$

Thus, $16x + 56y + 80z = 8,400$.

You expected to earn \$248 in dividends:

WSR dividend $= 7\%$ of $16x$ invested $= 0.07(16x) = 1.12x$

HCC dividend $= 2\%$ of $56y$ invested $= 0.02(56y) = 1.12y$

SNDK dividend $= 2\%$ of $80z$ invested $= 0.02(80z) = 1.6z$

Thus, $1.12x + 1.12y + 1.6z = 248$.

You purchased a total of 200 shares: $\quad x + y + z = 200$.

We therefore have the following system:

$x + y + z = 200$

$16x + 56y + 80z = 8,400$

$1.12x + 1.12y + 1.6z = 248$.

Solving:

$$\begin{bmatrix} 1 & 1 & 1 & 200 \\ 16 & 56 & 80 & 8400 \\ 1.12 & 1.12 & 1.6 & 248 \end{bmatrix} \begin{matrix} \\ \\ 25R_3 \end{matrix} \rightarrow \begin{bmatrix} 1 & 1 & 1 & 200 \\ 16 & 56 & 80 & 8400 \\ 28 & 28 & 40 & 6200 \end{bmatrix} \begin{matrix} \\ (1/8)R_2 \\ (1/4)R_3 \end{matrix} \rightarrow$$

$$\begin{bmatrix} \boxed{1} & 1 & 1 & 200 \\ 2 & 7 & 10 & 1050 \\ 7 & 7 & 10 & 1550 \end{bmatrix} \begin{matrix} \\ R_2 - 2R_1 \\ R_3 - 7R_1 \end{matrix} \rightarrow \begin{bmatrix} 1 & 1 & 1 & 200 \\ 0 & 5 & 8 & 650 \\ 0 & 0 & 3 & 150 \end{bmatrix} \begin{matrix} \\ \\ (1/3)R_3 \end{matrix} \rightarrow$$

$$\begin{bmatrix} 1 & 1 & 1 & 200 \\ 0 & \boxed{5} & 8 & 650 \\ 0 & 0 & 1 & 50 \end{bmatrix} \begin{matrix} 5R_1 - R_2 \\ \\ \end{matrix} \rightarrow \begin{bmatrix} 5 & 0 & -3 & 350 \\ 0 & 5 & 8 & 650 \\ 0 & 0 & \boxed{1} & 50 \end{bmatrix} \begin{matrix} R_1 + 3R_3 \\ R_2 - 8R_3 \\ \end{matrix} \rightarrow$$

$$\begin{bmatrix} 5 & 0 & 0 & 500 \\ 0 & 5 & 0 & 250 \\ 0 & 0 & 1 & 50 \end{bmatrix} \begin{matrix} (1/5)R_1 \\ (1/5)R_2 \\ \end{matrix} \rightarrow \begin{bmatrix} 1 & 0 & 0 & 100 \\ 0 & 1 & 0 & 50 \\ 0 & 0 & 1 & 50 \end{bmatrix}$$

Solution: You purchased 100 shares of WSR, 50 shares of HCC, and 50 shares of SNDK.

21. With $x, y, z,$ and u as indicated, the first piece of information we have is that $x + y + z + u = 284$.
The remaining equations must be written in standard form:

$$-3x + 3y + z - u = 6$$

$$x + y - z - u = 50$$

$$x - y + z - u = 42$$

The augmented matrix of the system is $\begin{bmatrix} 1 & 1 & 1 & 1 & 284 \\ -3 & 3 & 1 & -1 & 6 \\ 1 & 1 & -1 & -1 & 50 \\ 1 & -1 & 1 & -1 & 42 \end{bmatrix}$.

Using pivoting technology (for instance, press "reduce completely" in the Pivot and Guass-Jordan Tool on the Website), we obtain the solution $x = 88, y = 79, z = 75, u = 42$.
Microsoft: 88 million, Time Warner: 79 million, Yahoo: 75 million, Google: 42 million

23. Since the percentage market shares add up to 100, the third equation is $x + y + z + w = 100$.
If we rewrite the given equations in standard form, we get the second and third equations:

$$x - y - z = 1 \qquad z + 0.2w = 16.$$

Row reduction:

$\begin{bmatrix} 1 & 1 & 1 & 1 & 100 \\ 1 & -1 & -1 & 0 & 1 \\ 0 & 0 & 1 & 0.2 & 16 \end{bmatrix} 5R_3 \rightarrow \begin{bmatrix} \boxed{1} & 1 & 1 & 1 & 100 \\ 1 & -1 & -1 & 0 & 1 \\ 0 & 0 & 5 & 1 & 80 \end{bmatrix} R_2 - R_1 \rightarrow$

$\begin{bmatrix} 1 & 1 & 1 & 1 & 100 \\ 0 & \boxed{-2} & -2 & -1 & -99 \\ 0 & 0 & 5 & 1 & 80 \end{bmatrix} \begin{matrix} 2R_1 + R_2 \\ \\ \end{matrix} \rightarrow \begin{bmatrix} 2 & 0 & 0 & 1 & 101 \\ 0 & -2 & -2 & -1 & -99 \\ 0 & 0 & \boxed{5} & 1 & 80 \end{bmatrix} 5R_2 + 2R_3 \rightarrow$

$\begin{bmatrix} 2 & 0 & 0 & 1 & 101 \\ 0 & -10 & 0 & -3 & -335 \\ 0 & 0 & 5 & 1 & 80 \end{bmatrix} \begin{matrix} (1/2)R_1 \\ -(1/10)R_2 \\ (1/5)R_3 \end{matrix} \rightarrow \begin{bmatrix} 1 & 0 & 0 & 1/2 & 101/2 \\ 0 & 1 & 0 & 3/10 & 67/2 \\ 0 & 0 & 1 & 1/5 & 16 \end{bmatrix}$

Translating back to equations (and converting to decimals) gives

$$x + 0.5w = 50.5 \qquad y + 0.3w = 33.5 \qquad z + 0.2w = 16.$$

Solving for $x, y,$ and z in terms of w gives the general solution:

$$x = 50.5 - 0.5w$$

$$y = 33.5 - 0.3w$$

$$z = 16 - 0.2w$$

w arbitrary.

We now answer the question: Which of the three companies' market share is most impacted by the share held by other companies? Since w represents other companies," we look for which of $x, y,$ or z has the coefficient of w with the greatest absolute value—this is $x,$ representing State Farm. Thus, State Farm is most affected by other companies.

25. Unknowns: $x =$ the number of books sent from Brooklyn to Long Island, $y =$ the number of books sent from Brooklyn to Manhattan, $z =$ the number of books sent from Queens to Long Island, $w =$ the number of books sent from Queens to Manhattan
We represent the given information in a diagram:

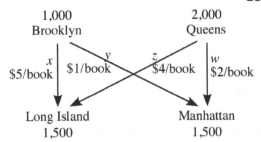

Note that, since a total of 3,000 books are ordered and there are a total of 3,000 in stock, both warehouses need to clear all their stocks.

Books from Brooklyn: $x + y = 1,000$ Books from Queens: $z + w = 2,000$

Books to Long Island: $x + z = 1,500$ Books to Manhattan Order: $y + w = 1,500$

a. Transportation budget:

$$5x + y + 4z + 2w = 9,000$$

We have five equations in four unknowns. Solving:

$$\begin{bmatrix} \boxed{1} & 0 & 1 & 0 & 1500 \\ 0 & 1 & 0 & 1 & 1500 \\ 1 & 1 & 0 & 0 & 1000 \\ 0 & 0 & 1 & 1 & 2000 \\ 5 & 1 & 4 & 2 & 9000 \end{bmatrix} \begin{array}{l} \\ \\ R_3 - R_1 \\ \\ R_5 - 5R_1 \end{array} \rightarrow \begin{bmatrix} 1 & 0 & 1 & 0 & 1500 \\ 0 & \boxed{1} & 0 & 1 & 1500 \\ 0 & 1 & -1 & 0 & -500 \\ 0 & 0 & 1 & 1 & 2000 \\ 0 & 1 & -1 & 2 & 1500 \end{bmatrix} \begin{array}{l} \\ \\ R_3 - R_2 \rightarrow \\ \\ R_5 - R_2 \end{array}$$

$$\begin{bmatrix} 1 & 0 & 1 & 0 & 1500 \\ 0 & 1 & 0 & 1 & 1500 \\ 0 & 0 & \boxed{-1} & -1 & -2000 \\ 0 & 0 & 1 & 1 & 2000 \\ 0 & 0 & -1 & 1 & 0 \end{bmatrix} \begin{array}{l} R_1 + R_3 \\ \\ \\ R_4 + R_3 \\ R_5 - R_3 \end{array} \rightarrow \begin{bmatrix} 1 & 0 & 0 & -1 & -500 \\ 0 & 1 & 0 & 1 & 1500 \\ 0 & 0 & -1 & -1 & -2000 \\ 0 & 0 & 0 & 0 & 0 \\ 0 & 0 & 0 & \boxed{2} & 2000 \end{bmatrix} \begin{array}{l} 2R_1 + R_5 \\ 2R_2 - R_5 \\ 2R_3 + R_5 \rightarrow \\ \\ \end{array}$$

$$\begin{bmatrix} 2 & 0 & 0 & 0 & 1000 \\ 0 & 2 & 0 & 0 & 1000 \\ 0 & 0 & -2 & 0 & -2000 \\ 0 & 0 & 0 & 0 & 0 \\ 0 & 0 & 0 & 2 & 2000 \end{bmatrix} \begin{array}{l} (1/2)R_1 \\ (1/2)R_2 \\ -(1/2)R_3 \rightarrow \\ \\ (1/2)R_5 \end{array} \begin{bmatrix} 1 & 0 & 0 & 0 & 500 \\ 0 & 1 & 0 & 0 & 500 \\ 0 & 0 & 1 & 0 & 1000 \\ 0 & 0 & 0 & 0 & 0 \\ 0 & 0 & 0 & 1 & 1000 \end{bmatrix} \text{Rearrange rows} \rightarrow$$

$$\begin{bmatrix} 1 & 0 & 0 & 0 & 500 \\ 0 & 1 & 0 & 0 & 500 \\ 0 & 0 & 1 & 0 & 1000 \\ 0 & 0 & 0 & 1 & 1000 \\ 0 & 0 & 0 & 0 & 0 \end{bmatrix}$$

Solution: Brooklyn to Long Island: 500 books; Brooklyn to Manhattan: 500 books; Queens to Long Island: 1,000 books; Queens to Manhattan: 1,000 books.

b. If we remove the equation that says that the total cost is $9,000, and solve, we get

$$\begin{bmatrix} \boxed{1} & 0 & 1 & 0 & 1500 \\ 0 & 1 & 0 & 1 & 1500 \\ 1 & 1 & 0 & 0 & 1000 \\ 0 & 0 & 1 & 1 & 2000 \end{bmatrix} \begin{array}{l} \\ \\ R_3 - R_1 \\ \\ \end{array} \rightarrow \begin{bmatrix} 1 & 0 & 1 & 0 & 1500 \\ 0 & \boxed{1} & 0 & 1 & 1500 \\ 0 & 1 & -1 & 0 & -500 \\ 0 & 0 & 1 & 1 & 2000 \end{bmatrix} \begin{array}{l} \\ \\ R_3 - R_2 \\ \\ \end{array} \rightarrow$$

$$\begin{bmatrix} 1 & 0 & 1 & 0 & 1500 \\ 0 & 1 & 0 & 1 & 1500 \\ 0 & 0 & \boxed{-1} & -1 & -2000 \\ 0 & 0 & 1 & 1 & 2000 \end{bmatrix} \begin{array}{l} R_1 + R_3 \\ \\ \\ R_4 + R_3 \end{array} \rightarrow \begin{bmatrix} 1 & 0 & 0 & -1 & -500 \\ 0 & 1 & 0 & 1 & 1500 \\ 0 & 0 & -1 & -1 & -2000 \\ 0 & 0 & 0 & 0 & 0 \end{bmatrix} \begin{array}{l} \\ \\ -R_3 \\ \\ \end{array} \rightarrow$$

$$\begin{bmatrix} 1 & 0 & 0 & -1 & -500 \\ 0 & 1 & 0 & 1 & 1500 \\ 0 & 0 & 1 & 1 & 2000 \\ 0 & 0 & 0 & 0 & 0 \end{bmatrix}$$

Translating back to equations gives: $x - w = -500$, $y + w = 1500$, $z + w = 2000$.

The general solution is $x = w - 500$, $y = 1,500 - w$, $z = 2,000 - w$.

The total cost is then $C = 5x + y + 4z + 2w = 5(w - 500) + 1,500 - w + 4(2,000 - w) + 2w = 7,000 + 2w$.

To decrease the cost, we should therefore make w as small as possible. We cannot set $w = 0$, as that would result in

$x = w - 500$ being negative; the smallest we can make w is 500. Using this value for w gives the following solution: Brooklyn to Long Island: 0 books; Brooklyn to Manhattan: 1,000 books; Queens to Long Island: 1,500 books; Queens to Manhattan: 500 books for a total cost of $7,000 + 2(500) = \$8,000$.

27. a. Unknowns: $x =$ the number of tourists from North America to Australia, $y =$ the number of tourists from

North America to South Africa, $z =$ the number of tourists from Europe to Australia, $w =$ the number of tourists from Europe to South Africa
We represent the given information in a diagram:

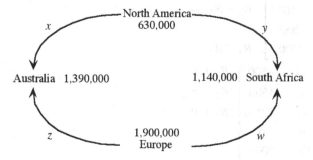

To avoid all the zeros, let us measure all these quantities in thousands.

North America: $x + y = 630$ Europe: $z + w = 1,900$

Australia: $x + z = 1,390$ South Africa: $y + w = 1,140$

$$\begin{bmatrix} \boxed{1} & 1 & 0 & 0 & 630 \\ 0 & 0 & 1 & 1 & 1900 \\ 1 & 0 & 1 & 0 & 1390 \\ 0 & 1 & 0 & 1 & 1140 \end{bmatrix} \begin{array}{c} \\ \\ R_3 - R_1 \\ \end{array} \rightarrow \begin{bmatrix} 1 & 1 & 0 & 0 & 630 \\ 0 & 0 & \boxed{1} & 1 & 1900 \\ 0 & -1 & 1 & 0 & 760 \\ 0 & 1 & 0 & 1 & 1140 \end{bmatrix} \begin{array}{c} \\ \\ R_3 - R_2 \\ \end{array} \rightarrow$$

$$\begin{bmatrix} 1 & 1 & 0 & 0 & 630 \\ 0 & 0 & 1 & 1 & 1900 \\ 0 & \boxed{-1} & 0 & -1 & -1140 \\ 0 & 1 & 0 & 1 & 1140 \end{bmatrix} \begin{array}{c} R_1 + R_3 \\ \\ \\ R_4 + R_3 \end{array} \rightarrow \begin{bmatrix} 1 & 0 & 0 & -1 & -510 \\ 0 & 0 & 1 & 1 & 1900 \\ 0 & -1 & 0 & -1 & -1140 \\ 0 & 0 & 0 & 0 & 0 \end{bmatrix} \begin{array}{c} \\ \\ -R_3 \\ \end{array} \rightarrow$$

$$\begin{bmatrix} 1 & 0 & 0 & -1 & -510 \\ 0 & 0 & 1 & 1 & 1900 \\ 0 & 1 & 0 & 1 & 1140 \\ 0 & 0 & 0 & 0 & 0 \end{bmatrix}$$

This system has infinitely many solutions, which is why the given information is not sufficient to determine the number of tourists from each region to each destination.

b. We are told that $x + y + z + y = 2,530$. However, this equation can be obtained by adding the first two equations in part (a). Thus, the new equation gives us no additional information, and so the associated system of linear equations will have the same infinite solution set as part (a).
c. The additional information is, rephrased:

The number of people from Europe to Australia = the number of people from Europe to South Africa: $z = w$, or

$z - w = 0$.

Adding this to our list of equations and solving:

$$\begin{bmatrix} \boxed{1} & 1 & 0 & 0 & 630 \\ 0 & 0 & 1 & 1 & 1900 \\ 1 & 0 & 1 & 0 & 1390 \\ 0 & 1 & 0 & 1 & 1140 \\ 0 & 0 & 1 & -1 & 0 \end{bmatrix} \begin{matrix} \\ \\ R_3 - R_1 \rightarrow \\ \\ \\ \end{matrix} \begin{bmatrix} 1 & 1 & 0 & 0 & 630 \\ 0 & 0 & \boxed{1} & 1 & 1900 \\ 0 & -1 & 1 & 0 & 760 \\ 0 & 1 & 0 & 1 & 1140 \\ 0 & 0 & 1 & -1 & 0 \end{bmatrix} \begin{matrix} \\ \\ R_3 - R_2 \rightarrow \\ \\ R_5 - R_2 \end{matrix}$$

$$\begin{bmatrix} 1 & 1 & 0 & 0 & 630 \\ 0 & 0 & 1 & 1 & 1900 \\ 0 & -1 & 0 & -1 & -1140 \\ 0 & 1 & 0 & 1 & 1140 \\ 0 & 0 & 0 & -2 & -1900 \end{bmatrix} \begin{matrix} \\ \\ \\ \\ (1/2)R_5 \end{matrix} \rightarrow \begin{bmatrix} 1 & 1 & 0 & 0 & 630 \\ 0 & 0 & 1 & 1 & 1900 \\ 0 & \boxed{-1} & 0 & -1 & -1140 \\ 0 & 1 & 0 & 1 & 1140 \\ 0 & 0 & 0 & -1 & -950 \end{bmatrix} \begin{matrix} R_1 + R_3 \\ \\ \\ R_4 + R_3 \\ \end{matrix} \rightarrow$$

$$\begin{bmatrix} 1 & 0 & 0 & -1 & -510 \\ 0 & 0 & 1 & 1 & 1900 \\ 0 & -1 & 0 & -1 & -1140 \\ 0 & 0 & 0 & 0 & 0 \\ 0 & 0 & 0 & \boxed{-1} & -950 \end{bmatrix} \begin{matrix} R_1 - R_5 \\ R_2 + R_5 \\ R_3 - R_5 \rightarrow \\ \\ \\ \end{matrix} \begin{bmatrix} 1 & 0 & 0 & 0 & 440 \\ 0 & 0 & 1 & 0 & 950 \\ 0 & -1 & 0 & 0 & -190 \\ 0 & 0 & 0 & 0 & 0 \\ 0 & 0 & 0 & -1 & -950 \end{bmatrix} \begin{matrix} \\ \\ -R_3 \rightarrow \\ \\ -R_5 \end{matrix}$$

$$\begin{bmatrix} 1 & 0 & 0 & 0 & 440 \\ 0 & 0 & 1 & 0 & 950 \\ 0 & 1 & 0 & 0 & 190 \\ 0 & 0 & 0 & 0 & 0 \\ 0 & 0 & 0 & 1 & 950 \end{bmatrix} \text{Rearrange rows} \rightarrow \begin{bmatrix} 1 & 0 & 0 & 0 & 440 \\ 0 & 1 & 0 & 0 & 190 \\ 0 & 0 & 1 & 0 & 950 \\ 0 & 0 & 0 & 1 & 950 \\ 0 & 0 & 0 & 0 & 0 \end{bmatrix}$$

Since this is a unique solution, we can determine the numbers from each country to each destination:
North America to Australia: 440,000, North America to South Africa: 190,000, Europe to Australia: 950,000, Europe to South Africa: 950,000.

29.

	Used alcohol	Alcohol-free	Totals
U.S.	x	y	14,000
Europe	z	w	95,000
Totals	63,550	45,450	

a. Using the row and column totals, we get four equations:

U.S.: $x + y = 14{,}000$ Europe: $z + w = 95{,}000$

Used Alcohol: $x + z = 63{,}550$ Alcohol-free: $y + w = 45{,}450$.

Row-reducing the augmented matrix gives:

$$\begin{bmatrix} \boxed{1} & 1 & 0 & 0 & 14000 \\ 0 & 0 & 1 & 1 & 95000 \\ 1 & 0 & 1 & 0 & 63550 \\ 0 & 1 & 0 & 1 & 45450 \end{bmatrix} \begin{matrix} \\ \\ R_3 - R_1 \\ \\ \end{matrix} \rightarrow \begin{bmatrix} 1 & 1 & 0 & 0 & 14000 \\ 0 & 0 & \boxed{1} & 1 & 95000 \\ 0 & -1 & 1 & 0 & 49550 \\ 0 & 1 & 0 & 1 & 45450 \end{bmatrix} \begin{matrix} \\ \\ R_3 - R_2 \\ \\ \end{matrix} \rightarrow$$

$$\begin{bmatrix} 1 & 1 & 0 & 0 & 14000 \\ 0 & 0 & 1 & 1 & 95000 \\ 0 & \boxed{-1} & 0 & -1 & -45450 \\ 0 & 1 & 0 & 1 & 45450 \end{bmatrix} \begin{matrix} R_1 + R_3 \\ \\ \\ R_4 + R_3 \end{matrix} \rightarrow \begin{bmatrix} 1 & 0 & 0 & -1 & -31450 \\ 0 & 0 & 1 & 1 & 95000 \\ 0 & -1 & 0 & -1 & -45450 \\ 0 & 0 & 0 & 0 & 0 \end{bmatrix} \begin{matrix} \\ \\ -R_3 \\ \end{matrix} \rightarrow$$

$$\begin{bmatrix} 1 & 0 & 0 & -1 & -31450 \\ 0 & 0 & 1 & 1 & 95000 \\ 0 & 1 & 0 & 1 & 45450 \\ 0 & 0 & 0 & 0 & 0 \end{bmatrix}$$

The row-reduced matrix shows that there are infinitely many solutions, and hence no unique solution. Thus, the given data are insufficient to obtain the missing data (x, y, z and w).

b. The number of US 10th graders who were alcohol-free was 50% more than the number who had used alcohol:

$y = x + 0.50x = 1.50x$, or $-1.50x + y = 0$.

If we include this additional equation, we get the system

$x + y = 14{,}000$

$z + w = 95{,}000$

$x + z = 63{,}550$

$y + w = 45{,}450$

$-1.50x + y = 0$.

$$\begin{bmatrix} 1 & 1 & 0 & 0 & 14000 \\ 0 & 0 & 1 & 1 & 95000 \\ 1 & 0 & 1 & 0 & 63550 \\ 0 & 1 & 0 & 1 & 45450 \\ -1.50 & 1 & 0 & 0 & 0 \end{bmatrix} 2R_5 \rightarrow \begin{bmatrix} \boxed{1} & 1 & 0 & 0 & 14000 \\ 0 & 0 & 1 & 1 & 95000 \\ 1 & 0 & 1 & 0 & 63550 \\ 0 & 1 & 0 & 1 & 45450 \\ -3 & 2 & 0 & 0 & 0 \end{bmatrix} \begin{matrix} \\ \\ R_3 - R_1 \\ \\ R_5 + 3R_1 \end{matrix} \rightarrow$$

$$\begin{bmatrix} 1 & 1 & 0 & 0 & 14000 \\ 0 & 0 & 1 & 1 & 95000 \\ 0 & -1 & 1 & 0 & 49550 \\ 0 & 1 & 0 & 1 & 45450 \\ 0 & 5 & 0 & 0 & 42000 \end{bmatrix} (1/5)R_5 \rightarrow \begin{bmatrix} 1 & 1 & 0 & 0 & 14000 \\ 0 & 0 & \boxed{1} & 1 & 95000 \\ 0 & -1 & 1 & 0 & 49550 \\ 0 & 1 & 0 & 1 & 45450 \\ 0 & 1 & 0 & 0 & 8400 \end{bmatrix} \begin{matrix} \\ \\ R_3 - R_2 \\ \\ \end{matrix} \rightarrow$$

$$\begin{bmatrix} 1 & 1 & 0 & 0 & 14000 \\ 0 & 0 & 1 & 1 & 95000 \\ 0 & \boxed{-1} & 0 & -1 & -45450 \\ 0 & 1 & 0 & 1 & 45450 \\ 0 & 1 & 0 & 0 & 8400 \end{bmatrix} \begin{matrix} R_1 + R_3 \\ \\ \\ R_4 + R_3 \\ R_5 + R_3 \end{matrix} \rightarrow \begin{bmatrix} 1 & 0 & 0 & -1 & -31450 \\ 0 & 0 & 1 & 1 & 95000 \\ 0 & -1 & 0 & -1 & -45450 \\ 0 & 0 & 0 & 0 & 0 \\ 0 & 0 & 0 & \boxed{-1} & -37050 \end{bmatrix} \begin{matrix} R_1 - R_5 \\ R_2 + R_5 \\ R_3 - R_5 \\ \\ \end{matrix} \rightarrow$$

$$\begin{bmatrix} 1 & 0 & 0 & 0 & 5600 \\ 0 & 0 & 1 & 0 & 57950 \\ 0 & -1 & 0 & 0 & -8400 \\ 0 & 0 & 0 & 0 & 0 \\ 0 & 0 & 0 & -1 & -37050 \end{bmatrix} \begin{matrix} \\ \\ -R_3 \\ \\ -R_5 \end{matrix} \rightarrow \begin{bmatrix} 1 & 0 & 0 & 0 & 5600 \\ 0 & 0 & 1 & 0 & 57950 \\ 0 & 1 & 0 & 0 & 8400 \\ 0 & 0 & 0 & 0 & 0 \\ 0 & 0 & 0 & 1 & 37050 \end{bmatrix} \text{Rearrange rows} \rightarrow \begin{bmatrix} 1 & 0 & 0 & 0 & 5600 \\ 0 & 1 & 0 & 0 & 8400 \\ 0 & 0 & 1 & 0 & 57950 \\ 0 & 0 & 0 & 1 & 37050 \\ 0 & 0 & 0 & 0 & 0 \end{bmatrix}$$

Thus, the missing data is: $x = 5{,}600, y = 8{,}400, z = 57{,}950, w = 37{,}050$.

31. $x =$ daily traffic flow along Eastward Blvd., $y =$ daily traffic flow along Northwest La., $z =$ daily traffic flow along Southwest La.

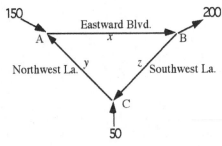

a. Intersection A: Traffic in = Traffic out: $150 + y = x$, or $x - y = 150$

   Intersection B: Traffic in = Traffic out: $x = 200 + z$, or $x - z = 200$

   Intersection C: Traffic in = Traffic out: $50 + z = y$, or $y - z = 50$

This gives us a system of three linear equations:

$$x - y = 150$$

$$x - z = 200$$

$$y - z = 50.$$

Solving:

$$\begin{bmatrix} \boxed{1} & -1 & 0 & 150 \\ 1 & 0 & -1 & 200 \\ 0 & 1 & -1 & 50 \end{bmatrix} \begin{array}{c} \\ R_2 - R_1 \\ \end{array} \rightarrow \begin{bmatrix} 1 & -1 & 0 & 150 \\ 0 & \boxed{1} & -1 & 50 \\ 0 & 1 & -1 & 50 \end{bmatrix} \begin{array}{c} R_1 + R_2 \\ \\ R_3 - R_2 \end{array} \rightarrow \begin{bmatrix} 1 & 0 & -1 & 200 \\ 0 & 1 & -1 & 50 \\ 0 & 0 & 0 & 0 \end{bmatrix}$$

Since there are infinitely many solutions, it is not possible to determine the daily flow of traffic along each of the three streets from the information given.

Translating the row-reduced matrix back into equations gives $x - z = 200$, or $x = 200 + z$, $y - z = 50$, or

$y = 50 + z$.

Thus, the general solution is:

$$x = 200 + z, \ y = 50 + z, \ z \geq 0 \text{ arbitrary,}$$

where z is traffic along Southwest Lane. Thus, it would suffice to know the traffic along Southwest to obtain the other traffic flows.

b. If we set $z = 60$, we obtain, from the general solution in part (a), $x = 200 + 60 = 260$ vehicles along Eastward Blvd., $y = 50 + 60 = 110$ vehicles along Northwest La., $z = 60$ vehicles along Southwest La.

c. From the general solution in part (a), the traffic flow along Northwest La. is given by $y = 50 + z$. Since $z \geq 0$, the value of y must be at least 50 vehicles per day.

33. $x =$ traffic on middle section of Bree, $y =$ traffic on middle section of Jeppe, $z =$ traffic on middle section of Simmons, $w =$ traffic on middle section of Harrison

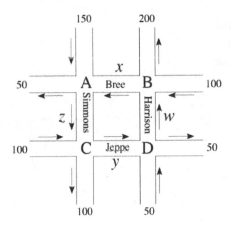

a. Intersection A: Traffic in = Traffic out: $150 + x = 50 + z$, or $x - z = -100$

Intersection B: Traffic in = Traffic out: $100 + w = 200 + x$, or $x - w = -100$

Intersection C: Traffic in = Traffic out: $100 + z = 100 + y$, or $y - z = 0$

Intersection D: Traffic in = Traffic out: $50 + y = 50 + w$, or $y - w = 0$

This gives us a system of three linear equations:

$x - z = -100$

$x - w = -100$

$y - z = 0$

$y - w = 0.$

Solving:

$$\begin{bmatrix} \boxed{1} & 0 & -1 & 0 & -100 \\ 1 & 0 & 0 & -1 & -100 \\ 0 & 1 & -1 & 0 & 0 \\ 0 & 1 & 0 & -1 & 0 \end{bmatrix} \begin{matrix} \\ R_2 - R_1 \\ \\ \\ \end{matrix} \rightarrow \begin{bmatrix} 1 & 0 & -1 & 0 & -100 \\ 0 & 0 & \boxed{1} & -1 & 0 \\ 0 & 1 & -1 & 0 & 0 \\ 0 & 1 & 0 & -1 & 0 \end{bmatrix} \begin{matrix} R_1 + R_2 \\ \\ R_3 + R_2 \\ \\ \end{matrix} \rightarrow$$

$$\begin{bmatrix} 1 & 0 & 0 & -1 & -100 \\ 0 & 0 & 1 & -1 & 0 \\ 0 & \boxed{1} & 0 & -1 & 0 \\ 0 & 1 & 0 & -1 & 0 \end{bmatrix} \begin{matrix} \\ \\ \\ R_4 - R_3 \end{matrix} \rightarrow \begin{bmatrix} 1 & 0 & 0 & -1 & -100 \\ 0 & 0 & 1 & -1 & 0 \\ 0 & 1 & 0 & -1 & 0 \\ 0 & 0 & 0 & 0 & 0 \end{bmatrix}$$

Translating back to equations gives: $x - w = -100$, $z - w = 0$, $y - w = 0$.

Thus, the general solution is: $x = w - 100, y = w, z = w, w$ arbitrary, where w is traffic along the middle section of Harrison.

b. Because $y = w$ has more than one possible value, it is not possible to determine the traffic flow along the middle section of Jeppe.

c. Given $x = 400$, the first equation of the general solution says that $400 = w - 100$, so $w = 500$. We need Simmons (z). But the third equation says $z = w$. So $w = 500$ cars down the middle section of Simmons.

d. If w is less than 100, x would become negative, by the first equation of the solution. Thus, $w \geq 100$.

e. Simmons is $z = w$, which can be as large as we like without making any of the variables negative (see the solution). Thus, there is no upper limit to the traffic on the middle section of Simmons. We can visualize this by imagining thousands of cars going around and around the center block without affecting the recorded numbers in the diagram.

35. a. Let us take the unknowns to be the net traffic flow going east on the three stretches of Broadway as shown in the figure. (If any of the unknowns is negative, it indicates a net positive flow in the opposite direction.)

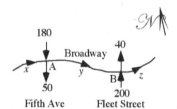

Intersection A: Traffic in = Traffic out:

$180 + x = 50 + y \Rightarrow x - y = -130$

Intersection B: Traffic in = Traffic out:

$200 + y = 40 + z \Rightarrow y - z = -160$

This gives us a system of two linear equations: $x - y = -130$ $y - z = -160$. Solving:

$$\begin{bmatrix} 1 & -1 & 0 & -130 \\ 0 & \boxed{1} & -1 & -160 \end{bmatrix} \begin{matrix} R_1 + R_2 \\ \\ \end{matrix} \rightarrow \begin{bmatrix} 1 & 0 & -1 & -290 \\ 0 & 1 & -1 & -160 \end{bmatrix}$$

Translating back to equations gives: $x - z = -290$, $y - z = -160$.

Thus, the general solution is: $x = z - 290, y = z - 160, z$ arbitrary.

As there are infinitely many solutions, we cannot determine the traffic flow along each stretch of Broadway. Knowing any one of $x, y,$ or z would enable us to solve for the other two unknowns uniquely.

b. The general solution from part (a) is $x = z - 290, y = z - 160, z$ arbitrary. East of Fleet Street, the traffic flow is z. If z is smaller than 160, the above solution shows that the values of x and y are negative, indicating a net flow to the west on those stretches.

37. We rewrite each given equation in standard form, using the given information that $M = 120$ billion dollars:

$$120 = C + D \Rightarrow C + D = 120$$

$$C = 0.2D \quad \Rightarrow C - 0.2D = 0$$

$$R = 0.1D \quad \Rightarrow R - 0.1D = 0$$

$$H = R + C \Rightarrow H - R - C = 0.$$

The matrices below are set up with the unknowns in the order: C, R, D, H.

$$\begin{bmatrix} 1 & 0 & 1 & 0 & 120 \\ 1 & 0 & -0.2 & 0 & 0 \\ 0 & 1 & -0.1 & 0 & 0 \\ -1 & -1 & 0 & 1 & 0 \end{bmatrix} \begin{matrix} \\ 5R_2 \\ 10R_3 \\ \end{matrix} \rightarrow \begin{bmatrix} \boxed{1} & 0 & 1 & 0 & 120 \\ 5 & 0 & -1 & 0 & 0 \\ 0 & 10 & -1 & 0 & 0 \\ -1 & -1 & 0 & 1 & 0 \end{bmatrix} \begin{matrix} \\ R_2 - 5R_1 \\ \\ R_4 + R_1 \end{matrix} \rightarrow$$

$$\begin{bmatrix} 1 & 0 & 1 & 0 & 120 \\ 0 & 0 & -6 & 0 & -600 \\ 0 & 10 & -1 & 0 & 0 \\ 0 & -1 & 1 & 1 & 120 \end{bmatrix} \begin{matrix} \\ (1/6)R_2 \\ \\ \\ \end{matrix} \rightarrow \begin{bmatrix} 1 & 0 & 1 & 0 & 120 \\ 0 & 0 & \boxed{-1} & 0 & -100 \\ 0 & 10 & -1 & 0 & 0 \\ 0 & -1 & 1 & 1 & 120 \end{bmatrix} \begin{matrix} R_1 + R_2 \\ \\ R_3 - R_2 \\ R_4 + R_2 \end{matrix} \rightarrow$$

$$\begin{bmatrix} 1 & 0 & 0 & 0 & 20 \\ 0 & 0 & -1 & 0 & -100 \\ 0 & 10 & 0 & 0 & 100 \\ 0 & -1 & 0 & 1 & 20 \end{bmatrix} \begin{matrix} \\ \\ (1/10)R_3 \\ \\ \end{matrix} \rightarrow \begin{bmatrix} 1 & 0 & 0 & 0 & 20 \\ 0 & 0 & -1 & 0 & -100 \\ 0 & \boxed{1} & 0 & 0 & 10 \\ 0 & -1 & 0 & 1 & 20 \end{bmatrix} \begin{matrix} \\ \\ \\ R_4 + R_3 \end{matrix} \rightarrow$$

$$\begin{bmatrix} 1 & 0 & 0 & 0 & 20 \\ 0 & 0 & -1 & 0 & -100 \\ 0 & 1 & 0 & 0 & 10 \\ 0 & 0 & 0 & 1 & 30 \end{bmatrix} \begin{matrix} \\ -R_2 \\ \\ \\ \end{matrix} \rightarrow \begin{bmatrix} 1 & 0 & 0 & 0 & 20 \\ 0 & 0 & 1 & 0 & 100 \\ 0 & 1 & 0 & 0 & 10 \\ 0 & 0 & 0 & 1 & 30 \end{bmatrix} \text{Rearrange rows} \rightarrow$$

$$\begin{bmatrix} 1 & 0 & 0 & 0 & 20 \\ 0 & 1 & 0 & 0 & 10 \\ 0 & 0 & 1 & 0 & 100 \\ 0 & 0 & 0 & 1 & 30 \end{bmatrix} C = 20, R = 10, D = 100, H = 30$$

We are asked for bank reserves R, which are therefore $10 billion.

39.

Adding the values along each beam gives $x + y + z = 3{,}050$

$x + z + 1{,}000 = 3{,}030$, or $x + z = 2{,}030$

$x + y + 1{,}000 = 3{,}020$, or $x + y = 2{,}020$.

$$\left[\begin{array}{cccc} \boxed{1} & 1 & 1 & 3050 \\ 1 & 0 & 1 & 2030 \\ 1 & 1 & 0 & 2020 \end{array}\right] \begin{array}{c} \\ R_2 - R_1 \\ R_3 - R_1 \end{array} \rightarrow \left[\begin{array}{cccc} 1 & 1 & 1 & 3050 \\ 0 & \boxed{-1} & 0 & -1020 \\ 0 & 0 & -1 & -1030 \end{array}\right] \begin{array}{c} R_1 + R_2 \\ \\ \end{array} \rightarrow$$

$$\left[\begin{array}{cccc} 1 & 0 & 1 & 2030 \\ 0 & -1 & 0 & -1020 \\ 0 & 0 & \boxed{-1} & -1030 \end{array}\right] \begin{array}{c} R_1 + R_3 \\ \\ \end{array} \rightarrow \left[\begin{array}{cccc} 1 & 0 & 0 & 1000 \\ 0 & -1 & 0 & -1020 \\ 0 & 0 & -1 & -1030 \end{array}\right] \begin{array}{c} \\ -R_2 \\ -R_3 \end{array} \rightarrow$$

$$\left[\begin{array}{cccc} 1 & 0 & 0 & 1000 \\ 0 & 1 & 0 & 1020 \\ 0 & 0 & 1 & 1030 \end{array}\right] x = 1{,}000, y = 1{,}020, z = 1{,}030$$

From the table, we find the corresponding components: $x =$ water, $y =$ gray matter, $z =$ tumor.

41.

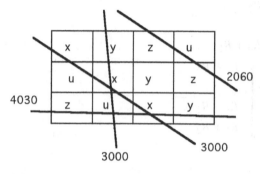

Adding the values along each beam gives

$x + y + z + u = 4{,}030$

$x + y + u = 3{,}000$

$3x + 2u = 3{,}000$

$2z + u = 2{,}060$.

Solving:

$$\left[\begin{array}{ccccc} \boxed{1} & 1 & 1 & 1 & 4030 \\ 1 & 1 & 0 & 1 & 3000 \\ 3 & 0 & 0 & 2 & 3000 \\ 0 & 0 & 2 & 1 & 2060 \end{array}\right] \begin{array}{c} \\ R_2 - R_1 \\ R_3 - 3R_1 \\ \end{array} \rightarrow \left[\begin{array}{ccccc} 1 & 1 & 1 & 1 & 4030 \\ 0 & 0 & \boxed{-1} & 0 & -1030 \\ 0 & -3 & -3 & -1 & -9090 \\ 0 & 0 & 2 & 1 & 2060 \end{array}\right] \begin{array}{c} R_1 + R_2 \\ \\ R_3 - 3R_2 \\ R_4 + 2R_2 \end{array} \rightarrow$$

$$\left[\begin{array}{ccccc} 1 & 1 & 0 & 1 & 3000 \\ 0 & 0 & -1 & 0 & -1030 \\ 0 & \boxed{-3} & 0 & -1 & -6000 \\ 0 & 0 & 0 & 1 & 0 \end{array}\right] \begin{array}{c} 3R_1 + R_3 \\ \\ \\ \end{array} \rightarrow \left[\begin{array}{ccccc} 3 & 0 & 0 & 2 & 3000 \\ 0 & 0 & -1 & 0 & -1030 \\ 0 & -3 & 0 & -1 & -6000 \\ 0 & 0 & 0 & \boxed{1} & 0 \end{array}\right] \begin{array}{c} R_1 - 2R_4 \\ \\ R_3 + R_4 \\ \end{array} \rightarrow$$

$$\left[\begin{array}{ccccc} 3 & 0 & 0 & 0 & 3000 \\ 0 & 0 & -1 & 0 & -1030 \\ 0 & -3 & 0 & 0 & -6000 \\ 0 & 0 & 0 & 1 & 0 \end{array}\right] \begin{array}{c} (1/3)R_1 \\ \\ (1/3)R_3 \\ \end{array} \rightarrow \left[\begin{array}{ccccc} 1 & 0 & 0 & 0 & 1000 \\ 0 & 0 & -1 & 0 & -1030 \\ 0 & -1 & 0 & 0 & -2000 \\ 0 & 0 & 0 & 1 & 0 \end{array}\right] \begin{array}{c} \\ -R_2 \\ -R_3 \\ \end{array} \rightarrow$$

$$\begin{bmatrix} 1 & 0 & 0 & 0 & 1000 \\ 0 & 0 & 1 & 0 & 1030 \\ 0 & 1 & 0 & 0 & 2000 \\ 0 & 0 & 0 & 1 & 0 \end{bmatrix} \text{Rearrange rows} \rightarrow \begin{bmatrix} 1 & 0 & 0 & 0 & 1000 \\ 0 & 1 & 0 & 0 & 2000 \\ 0 & 0 & 1 & 0 & 1030 \\ 0 & 0 & 0 & 1 & 0 \end{bmatrix}$$

$x = 1{,}000, y = 2{,}000, z = 1{,}030, u = 0$

From the table, we find the corresponding components: $x =$ water, $y =$ bone, $z =$ tumor, $u =$ air.

43. Since the three beams extending from left to right pass though the same four squares on the left, let us label those squares y. The vertical beam passes through an additional two squares, which we will label as z and u as shown:

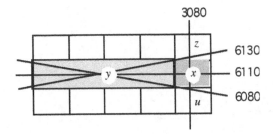

Adding the values along each beam gives

$x + z + u = 3{,}080$

$y + z = 6{,}130$

$x + y = 6{,}110$

$y + u = 6{,}080.$

Solving:

$$\begin{bmatrix} \boxed{1} & 0 & 1 & 1 & 3080 \\ 0 & 1 & 1 & 0 & 6130 \\ 1 & 1 & 0 & 0 & 6110 \\ 0 & 1 & 0 & 1 & 6080 \end{bmatrix} \begin{matrix} \\ \\ R_3 - R_1 \\ \\ \end{matrix} \rightarrow \begin{bmatrix} 1 & 0 & 1 & 1 & 3080 \\ 0 & \boxed{1} & 1 & 0 & 6130 \\ 0 & 1 & -1 & -1 & 3030 \\ 0 & 1 & 0 & 1 & 6080 \end{bmatrix} \begin{matrix} \\ \\ R_3 - R_2 \\ R_4 - R_2 \end{matrix} \rightarrow$$

$$\begin{bmatrix} 1 & 0 & 1 & 1 & 3080 \\ 0 & 1 & 1 & 0 & 6130 \\ 0 & 0 & \boxed{-2} & -1 & -3100 \\ 0 & 0 & -1 & 1 & -50 \end{bmatrix} \begin{matrix} 2R_1 + R_3 \\ 2R_2 + R_3 \\ \\ 2R_4 - R_3 \end{matrix} \rightarrow \begin{bmatrix} 2 & 0 & 0 & 1 & 3060 \\ 0 & 2 & 0 & -1 & 9160 \\ 0 & 0 & -2 & -1 & -3100 \\ 0 & 0 & 0 & 3 & 3000 \end{bmatrix} \begin{matrix} \\ \\ \\ (1/3)R_4 \end{matrix} \rightarrow$$

$$\begin{bmatrix} 2 & 0 & 0 & 1 & 3060 \\ 0 & 2 & 0 & -1 & 9160 \\ 0 & 0 & -2 & -1 & -3100 \\ 0 & 0 & 0 & \boxed{1} & 1000 \end{bmatrix} \begin{matrix} R_1 - R_4 \\ R_2 + R_4 \\ R_3 + R_4 \\ \\ \end{matrix} \rightarrow \begin{bmatrix} 2 & 0 & 0 & 0 & 2060 \\ 0 & 2 & 0 & 0 & 10160 \\ 0 & 0 & -2 & 0 & -2100 \\ 0 & 0 & 0 & 1 & 1000 \end{bmatrix} \begin{matrix} (1/2)R_1 \\ (1/2)R_2 \\ (1/2)R_3 \\ \\ \end{matrix} \rightarrow$$

$$\begin{bmatrix} 1 & 0 & 0 & 0 & 1030 \\ 0 & 1 & 0 & 0 & 5080 \\ 0 & 0 & -1 & 0 & -1050 \\ 0 & 0 & 0 & 1 & 1000 \end{bmatrix} \begin{matrix} \\ \\ -R_3 \\ \\ \end{matrix} \rightarrow \begin{bmatrix} 1 & 0 & 0 & 0 & 1030 \\ 0 & 1 & 0 & 0 & 5080 \\ 0 & 0 & 1 & 0 & 1050 \\ 0 & 0 & 0 & 1 & 1000 \end{bmatrix}$$

All we need is the value of x:1,030, which corresponds to tumor.

45. Unknowns: $x =$ the number of Democrats who voted in favor, $y =$ the number of Republicans who voted in favor, $z =$ the number of Others who voted in favor

Note: $333 - x$ Democrats voted against, $89 - y$ Republicans voted against, and $13 - z$ Others voted against.

Given information:

(1) There were 31 more votes in favor than against:

The total number of votes in favor $= x + y + z$

Total number of votes against $= (333 - x) + (89 - y) + (13 - z) = 435 - x - y - z$

Thus,

$\qquad x + y + z - (435 - x - y - z) = 31$, or $2x + 2y + 2z = 466$, or $x + y + z = 233$.

(2) 10 times as many Democrats voted for the bill as Republicans. Rephrasing this:

The number of Democrats voting for the bill was 10 times the number of Republicans voting for the bill:

$\qquad x = 10y$, or $x - 10y = 0$.

(3) 36 more non-Democrats voted against the bill than for it. Rephrasing this: The number of non-Democrats voting against the bill exceeded the number of non-Democrats voting for by 36:

$\qquad (89 - y) + (13 - z) - (y + z) = 36$, or $-2y - 2z = -66$, or $y + z = 33$.

Thus, we have three equations with three unknowns:

$\qquad x + y + z = 233$

$\qquad x - 10y = 0$

$\qquad y + z = 33.$

$$\begin{bmatrix} \boxed{1} & 1 & 1 & 233 \\ 1 & -10 & 0 & 0 \\ 0 & 1 & 1 & 33 \end{bmatrix} \begin{array}{l} \\ R_2 - R_1 \rightarrow \\ \\ \end{array} \begin{bmatrix} 1 & 1 & 1 & 233 \\ 0 & \boxed{-11} & -1 & -233 \\ 0 & 1 & 1 & 33 \end{bmatrix} \begin{array}{l} 11R_1 + R_2 \\ \\ 11R_3 + R_2 \end{array} \rightarrow$$

$$\begin{bmatrix} 11 & 0 & 10 & 2330 \\ 0 & -11 & -1 & -233 \\ 0 & 0 & 10 & 130 \end{bmatrix} \begin{array}{l} \\ \\ (1/10)R_3 \end{array} \rightarrow \begin{bmatrix} 11 & 0 & 10 & 2330 \\ 0 & -11 & -1 & -233 \\ 0 & 0 & \boxed{1} & 13 \end{bmatrix} \begin{array}{l} R_1 - 10R_3 \\ R_2 + R_3 \\ \end{array} \rightarrow$$

$$\begin{bmatrix} 11 & 0 & 0 & 2200 \\ 0 & -11 & 0 & -220 \\ 0 & 0 & 1 & 13 \end{bmatrix} \begin{array}{l} (1/11)R_1 \\ (1/11)R_2 \rightarrow \\ \end{array} \begin{bmatrix} 1 & 0 & 0 & 200 \\ 0 & -1 & 0 & -20 \\ 0 & 0 & 1 & 13 \end{bmatrix} \begin{array}{l} \\ -R_2 \rightarrow \\ \end{array}$$

$$\begin{bmatrix} 1 & 0 & 0 & 200 \\ 0 & 1 & 0 & 20 \\ 0 & 0 & 1 & 13 \end{bmatrix}$$

Solution: 200 Democrats, 20 Republicans, 13 of other parties voted for the bill.

47. $x =$ amount invested in company X, $y =$ amount invested in company Y, $z =$ amount invested in company Z,

$w =$ amount invested in company W

Investments totaled $65 million: $\qquad x + y + z + w = 65.$

Total return on investments was $8 million: $\qquad 0.15x - 0.20y + 0.20w = 8.$

Colossal invested twice as much in company X as in company Z. Rewording this: The amount invested in company X was twice the amount invested in company Z: $\qquad x = 2z$, or $x - 2z = 0.$

The amount invested in company W was 3 times the amount invested in company Z: $\qquad w = 3z$, or $-3z + w = 0.$

$$\begin{bmatrix} 1 & 1 & 1 & 1 & 65 \\ 0.15 & -0.20 & 0 & 0.20 & 8 \\ 1 & 0 & -2 & 0 & 0 \\ 0 & 0 & -3 & 1 & 0 \end{bmatrix} \begin{array}{l} \\ 20R_2 \\ \\ \\ \end{array} \rightarrow \begin{bmatrix} \boxed{1} & 1 & 1 & 1 & 65 \\ 3 & -4 & 0 & 4 & 160 \\ 1 & 0 & -2 & 0 & 0 \\ 0 & 0 & -3 & 1 & 0 \end{bmatrix} \begin{array}{l} \\ R_2 - 3R_1 \\ R_3 - R_1 \\ \end{array} \rightarrow$$

$$\begin{bmatrix} 1 & 1 & 1 & 1 & 65 \\ 0 & \boxed{-7} & -3 & 1 & -35 \\ 0 & -1 & -3 & -1 & -65 \\ 0 & 0 & -3 & 1 & 0 \end{bmatrix} \begin{array}{l} 7R_1 + R_2 \\ \\ 7R_3 - R_2 \\ \end{array} \rightarrow \begin{bmatrix} 7 & 0 & 4 & 8 & 420 \\ 0 & -7 & -3 & 1 & -35 \\ 0 & 0 & -18 & -8 & -420 \\ 0 & 0 & -3 & 1 & 0 \end{bmatrix} \begin{array}{l} \\ \\ (1/2)R_3 \\ \end{array} \rightarrow$$

$$\begin{bmatrix} 7 & 0 & 4 & 8 & 420 \\ 0 & -7 & -3 & 1 & -35 \\ 0 & 0 & \boxed{-9} & -4 & -210 \\ 0 & 0 & -3 & 1 & 0 \end{bmatrix} \begin{matrix} 9R_1 + 4R_3 \\ 3R_2 - R_3 \\ \\ 3R_4 - R_3 \end{matrix} \rightarrow \begin{bmatrix} 63 & 0 & 0 & 56 & 2940 \\ 0 & -21 & 0 & 7 & 105 \\ 0 & 0 & -9 & -4 & -210 \\ 0 & 0 & 0 & 7 & 210 \end{bmatrix} \begin{matrix} (1/7)R_1 \\ (1/7)R_2 \\ \\ (1/7)R_4 \end{matrix} \rightarrow$$

$$\begin{bmatrix} 9 & 0 & 0 & 8 & 420 \\ 0 & -3 & 0 & 1 & 15 \\ 0 & 0 & -9 & -4 & -210 \\ 0 & 0 & 0 & \boxed{1} & 30 \end{bmatrix} \begin{matrix} R_1 - 8R_4 \\ R_2 - R_4 \\ R_3 + 4R_4 \\ \\ \end{matrix} \rightarrow \begin{bmatrix} 9 & 0 & 0 & 0 & 180 \\ 0 & -3 & 0 & 0 & -15 \\ 0 & 0 & -9 & 0 & -90 \\ 0 & 0 & 0 & 1 & 30 \end{bmatrix} \begin{matrix} (1/9)R_1 \\ (1/3)R_2 \\ (1/9)R_3 \\ \\ \end{matrix} \rightarrow$$

$$\begin{bmatrix} 1 & 0 & 0 & 0 & 20 \\ 0 & -1 & 0 & 0 & -5 \\ 0 & 0 & -1 & 0 & -10 \\ 0 & 0 & 0 & 1 & 30 \end{bmatrix} \begin{matrix} \\ -R_2 \\ -R_3 \\ \\ \end{matrix} \rightarrow \begin{bmatrix} 1 & 0 & 0 & 0 & 20 \\ 0 & 1 & 0 & 0 & 5 \\ 0 & 0 & 1 & 0 & 10 \\ 0 & 0 & 0 & 1 & 30 \end{bmatrix}$$

$x = 20, y = 5, z = 10, w = 30$

As this is a unique solution, Smiley has sufficient information to piece together Colossal's investment portfolio: Colossal invested $20m in company X; $5m in company Y, $10m in company Z, and $30m in company W.

Communication and reasoning exercises

49. It is not realistic to expect to use exactly all of the ingredients. Solutions of the associated system may involve negative numbers or not exist. Only solutions with nonnegative values for all the unknowns correspond to being able to use up all of the ingredients.

51. The blend consists of 100 pounds of ingredient X. This says that $x = 100$, which is a linear equation.

53. The blend contains 30% ingredient Y by weight. Rephrasing: The weight of ingredient Y is 30% of the combined weights of X, Y, and Z: $y = 0.30(x + y + z) \Rightarrow y = 0.30x + 0.30y + 0.30z \Rightarrow 0.3x - 0.7y + 0.3z = 0$, which is a linear equation.

55. There is at least 30% ingredient Y by weight. Rephrasing: Tthe weight of ingredient Y is at least 30% of the combined weights of X, Y, and Z: $y \geq 0.30(x + y + z)$. This gives a linear inequality, not an equation.

57. Answers will vary.

Chapter 3 Review

1. To graph with technology, solve for y:

First line: $y = -x/2 + 2$

Second line: $y = 2x - 1$.

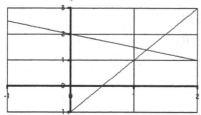

As the graphs intersect in a single point, there is a single (unique) solution.

3. To graph with technology, solve for y:

First line: $y = 2x/3$

Second line: $y = 2x/3$.

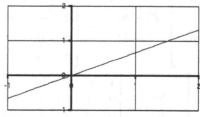

Infinitely many solutions

5. To graph with technology, solve for y:

First line: $y = -x + 1$

Second line: $y = -2x + 0.3$

Third line: $y = -3x/2 + 13/20$.

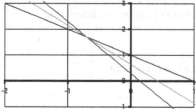

As the intersection of all three graphs is a single point, there is a single (unique) solution.

7.

$x + 2y = 4$

$2x - y = 1$

Multiply the second equation by 2:

$\quad x + 2y = 4$

$\quad 4x - 2y = 2$.

Adding gives $5x = 6 \Rightarrow x = \dfrac{6}{5}$.

Substituting $x = 6/5$ in the first equation gives $\dfrac{6}{5} + 2y = 4 \Rightarrow 2y = 4 - \dfrac{6}{5} = \dfrac{14}{5} \Rightarrow y = \dfrac{7}{5}$.

Solution: $x = \dfrac{6}{5}, y = \dfrac{7}{5}$

9.

$\frac{1}{2}x - \frac{3}{4}y = 0$

$6x - 9y = 0$

Multiply the first equation by 4 and divide the second by -3:

$\quad 2x - 3y = 0$

$-2x + 3y = 0.$

Adding gives $0 = 0$, so the system is redundant (the equations give the same lines).

To get the general solution, we solve for x: $2x = 3y \Rightarrow x = \dfrac{3y}{2}$.

General solution: $x = \dfrac{3y}{2}$, y arbitrary, or $\left(\dfrac{3y}{2}, y\right)$; y arbitrary

11.

$x + y = 1$

$2x + y = 0.3$

$3x + 2y = \dfrac{13}{10}$

$$\begin{bmatrix} 1 & 1 & 1 \\ 2 & 1 & 0.3 \\ 3 & 2 & 13/10 \end{bmatrix} \begin{matrix} \\ 10R_2 \\ 10R_3 \end{matrix} \rightarrow \begin{bmatrix} \boxed{1} & 1 & 1 \\ 20 & 10 & 3 \\ 30 & 20 & 13 \end{bmatrix} \begin{matrix} \\ R_2 - 20R_1 \\ R_3 - 30R_1 \end{matrix} \rightarrow \begin{bmatrix} 1 & 1 & 1 \\ 0 & \boxed{-10} & -17 \\ 0 & -10 & -17 \end{bmatrix} \begin{matrix} 10R_1 + R_2 \\ \\ R_3 - R_2 \end{matrix} \rightarrow$$

$$\begin{bmatrix} 10 & 0 & -7 \\ 0 & -10 & -17 \\ 0 & 0 & 0 \end{bmatrix} \begin{matrix} (1/10)R_1 \\ -(1/10)R_2 \\ \end{matrix} \rightarrow \begin{bmatrix} 1 & 0 & -7/10 \\ 0 & 1 & 17/10 \\ 0 & 0 & 0 \end{bmatrix}$$

Translating back to equations gives the solution: $x = -\dfrac{7}{10}, y = \dfrac{17}{10}.$

13.

$x + 2y = -3$

$x - z = 0$

$x + 3y - 2z = -2$

$$\begin{bmatrix} \boxed{1} & 2 & 0 & -3 \\ 1 & 0 & -1 & 0 \\ 1 & 3 & -2 & -2 \end{bmatrix} \begin{matrix} \\ R_2 - R_1 \\ R_3 - R_1 \end{matrix} \rightarrow \begin{bmatrix} 1 & 2 & 0 & -3 \\ 0 & \boxed{-2} & -1 & 3 \\ 0 & 1 & -2 & 1 \end{bmatrix} \begin{matrix} R_1 + R_2 \\ \\ 2R_3 + R_2 \end{matrix} \rightarrow \begin{bmatrix} 1 & 0 & -1 & 0 \\ 0 & -2 & -1 & 3 \\ 0 & 0 & -5 & 5 \end{bmatrix} \begin{matrix} \\ \\ (1/5)R_3 \end{matrix} \rightarrow$$

$$\begin{bmatrix} 1 & 0 & -1 & 0 \\ 0 & -2 & -1 & 3 \\ 0 & 0 & \boxed{-1} & 1 \end{bmatrix} \begin{matrix} R_1 - R_3 \\ R_2 - R_3 \\ \end{matrix} \rightarrow \begin{bmatrix} 1 & 0 & 0 & -1 \\ 0 & -2 & 0 & 2 \\ 0 & 0 & -1 & 1 \end{bmatrix} \begin{matrix} \\ (1/2)R_2 \\ \end{matrix} \rightarrow \begin{bmatrix} 1 & 0 & 0 & -1 \\ 0 & -1 & 0 & 1 \\ 0 & 0 & -1 & 1 \end{bmatrix} \begin{matrix} \\ -R_2 \\ -R_3 \end{matrix} \rightarrow$$

$$\begin{bmatrix} 1 & 0 & 0 & -1 \\ 0 & 1 & 0 & -1 \\ 0 & 0 & 1 & -1 \end{bmatrix}$$

Translating back to equations gives the solution: $x = -1, y = -1, z = -1.$

15.

$x - \dfrac{1}{2}y + z = 0$

$\dfrac{1}{2}x - \dfrac{1}{2}z = -1$

$\dfrac{3}{2}x - \dfrac{1}{2}y + \dfrac{1}{2}z = -1$

$$\begin{bmatrix} 1 & -1/2 & 1 & 0 \\ 1/2 & 0 & -1/2 & -1 \\ 3/2 & -1/2 & 1/2 & -1 \end{bmatrix} \begin{matrix} 2R_1 \\ 2R_2 \\ 2R_3 \end{matrix} \rightarrow \begin{bmatrix} \boxed{2} & -1 & 2 & 0 \\ 1 & 0 & -1 & -2 \\ 3 & -1 & 1 & -2 \end{bmatrix} \begin{matrix} \\ 2R_2 - R_1 \\ 2R_3 - 3R_1 \end{matrix} \rightarrow \begin{bmatrix} 2 & -1 & 2 & 0 \\ 0 & \boxed{1} & -4 & -4 \\ 0 & 1 & -4 & -4 \end{bmatrix} \begin{matrix} R_1 + R_2 \\ \\ R_3 - R_2 \end{matrix}$$

$$\rightarrow$$

$$\begin{bmatrix} 2 & 0 & -2 & -4 \\ 0 & 1 & -4 & -4 \\ 0 & 0 & 0 & 0 \end{bmatrix} \begin{matrix} (1/2)R_1 \\ \\ \end{matrix} \rightarrow \begin{bmatrix} 1 & 0 & -1 & -2 \\ 0 & 1 & -4 & -4 \\ 0 & 0 & 0 & 0 \end{bmatrix}$$

Translating back to equations gives: $x - z = -2$, $y - 4z = -4$.

General solution: $x = z - 2$, $y = 4z - 4 = 4(z - 1)$, z arbitrary, or $(z - 2, 4(z - 1), z)$; z arbitrary

17. Rewrite the given equations in standard form:

$$x - \frac{1}{2}y = 0$$

$$\frac{1}{2}x + \frac{1}{2}z = 2$$

$$-3x + y - z = 0.$$

$$\begin{bmatrix} 1 & -1/2 & 0 & 0 \\ 1/2 & 0 & 1/2 & 2 \\ -3 & 1 & -1 & 0 \end{bmatrix} \begin{matrix} 2R_1 \\ 2R_2 \\ \end{matrix} \rightarrow \begin{bmatrix} \boxed{2} & -1 & 0 & 0 \\ 1 & 0 & 1 & 4 \\ -3 & 1 & -1 & 0 \end{bmatrix} \begin{matrix} \\ 2R_2 - R_1 \\ 2R_3 + 3R_1 \end{matrix} \rightarrow \begin{bmatrix} 2 & -1 & 0 & 0 \\ 0 & \boxed{1} & 2 & 8 \\ 0 & -1 & -2 & 0 \end{bmatrix} \begin{matrix} R_1 + R_2 \\ \\ R_3 + R_2 \end{matrix} \rightarrow$$

$$\begin{bmatrix} 2 & 0 & 2 & 8 \\ 0 & 1 & 2 & 8 \\ 0 & 0 & 0 & 8 \end{bmatrix}$$

Since Row 3 translates to the false statement $0 = 8$, there is no solution.

19. $5F - 9C = 160$

We are given the additional information: $F = C$, or $F - C = 0$, giving us a second linear equation.

We can solve the resulting system of two equations in two unknowns by multiplying the second equation by -5:

$$5F - 9C = 160,$$

$$-5F + 5C = 0.$$

Adding gives $-4C = 160 \Rightarrow C = -40°$.

Since $C = F$, $F = -40°$ also.

21. $5F - 9C = 160$

We are told that the Fahrenheit temperature is 1.8 times the Celsius temperature: $F = 1.8C$, or $F - 1.8C = 0$, giving us the system

$$5F - 9C = 160$$

$$F - 1.8C = 0.$$

Multiplying the second equation by -5 gives

$$5F - 9C = 160 \qquad -5F + 9C = 0.$$

Adding gives the false statement $0 = 160$, showing that the given system is inconsistent (has no solution). Thus, it is not possible for the Fahrenheit temperature of an object to be 1.8 times its Celsius temperature.

23. The total population of the four cities is 10 million people: $x + y + z + w = 10$. This is a linear equation.

25. There are no people living in city D: $w = 0$. This is a linear equation.

27. City C has 30% more people than City B.

Rephrasing: The population of City C is 130% of the population of City B: $z = 1.30y$, or $-1.30y + z = 0$. This is a linear equation.

29. Unknowns: $x =$ the number of packages from Duffin House, $y =$ the number of packages from Higgins Press

Arrange the given information in a table:

	Duffin (x)	Higgins (y)	Desired totals
Horror	5	5	4,500
Romance	5	11	6,600

Horror: $5x + 5y = 4{,}500$

Romance: $5x + 11y = 6{,}600$

To solve, multiply the first equation by -1 and add, to obtain $6y = 2{,}100 \Rightarrow y = 350$.

The first equation can be divided by 5 and rewritten as $x + y = 900$. Substituting $y = 350$ gives us

$$x + 350 = 900 \Rightarrow x = 550.$$

Solution: Purchase 550 packages from Duffin House, 350 from Higgins Press.

31. Unknowns: $x =$ the number of packages from Duffin House, $y =$ the number of packages from Higgins Press

Cost: $50x + 150y = 90{,}000$

Also, you have promised to spend twice as much money for books from Duffin as from Higgins:

Amount spent on Duffin $= 2 \times$ amount spent on Higgins: $50x = 2(150)y$, or $50x - 300y = 0$.

Dividing each of these equations by 50 gives:

$$x + 3y = 1{,}800 \qquad x - 6y = 0.$$

Multiplying the first equation by 2 and adding gives: $3x = 3{,}600 \Rightarrow x = 1{,}200$.

Substituting $x = 1{,}200$ in the second equation gives $1{,}200 - 6y = 0 \Rightarrow 6y = 1{,}200 \Rightarrow y = 200$.

Solution: Purchase 1,200 packages from Duffin House, 200 from Higgins Press.

33. Demand: $q = -1{,}000p + 140{,}000$ \qquad Supply: $q = 2{,}000p + 20{,}000$

For the equilibrium price, we can equate the supply and demand:

$$\text{Demand} = \text{Supply} \Rightarrow -1{,}000p + 140{,}000 = 2{,}000p + 20{,}000 \Rightarrow -3{,}000p = -120{,}000$$

$$p = \frac{120{,}000}{3{,}000} = \$40 \text{ per book.}$$

35. Unknowns: $x =$ the number of baby sharks, $y =$ the number of piranhas, $z =$ the number of squids

Arrange the given data in a table with the unknowns across the top:

	Sharks (x)	Piranha (y)	Squid (z)	Total Consumed
Goldfish	1	1	1	21
Angelfish	2	0	1	21
Butterfly fish	2	3	0	35

Goldfish: $x + y + z = 21$

Angelfish: $2x + z = 21$

Butterfly fish: $2x + 3y = 35$

$$\begin{bmatrix} \boxed{1} & 1 & 1 & 21 \\ 2 & 0 & 1 & 21 \\ 2 & 3 & 0 & 35 \end{bmatrix} \begin{matrix} \\ R_2 - 2R_1 \\ R_3 - 2R_1 \end{matrix} \rightarrow \begin{bmatrix} 1 & 1 & 1 & 21 \\ 0 & \boxed{-2} & -1 & -21 \\ 0 & 1 & -2 & -7 \end{bmatrix} \begin{matrix} 2R_1 + R_2 \\ \\ 2R_3 + R_2 \end{matrix} \rightarrow \begin{bmatrix} 2 & 0 & 1 & 21 \\ 0 & -2 & -1 & -21 \\ 0 & 0 & -5 & -35 \end{bmatrix} \begin{matrix} \\ \\ (1/5)R_3 \end{matrix}$$

$$\rightarrow \begin{bmatrix} 2 & 0 & 1 & 21 \\ 0 & -2 & -1 & -21 \\ 0 & 0 & \boxed{-1} & -7 \end{bmatrix} \begin{matrix} R_1 + R_3 \\ R_2 - R_3 \\ \end{matrix} \rightarrow \begin{bmatrix} 2 & 0 & 0 & 14 \\ 0 & -2 & 0 & -14 \\ 0 & 0 & -1 & -7 \end{bmatrix} \begin{matrix} (1/2)R_1 \\ (1/2)R_2 \\ \end{matrix} \rightarrow \begin{bmatrix} 1 & 0 & 0 & 7 \\ 0 & -1 & 0 & -7 \\ 0 & 0 & -1 & -7 \end{bmatrix} \begin{matrix} \\ -R_2 \\ -R_3 \end{matrix}$$

$$\rightarrow \begin{bmatrix} 1 & 0 & 0 & 7 \\ 0 & 1 & 0 & 7 \\ 0 & 0 & 1 & 7 \end{bmatrix}$$

Solution: He has 7 of each type of carnivorous creature.

37. Unknowns: $x = $ the number of hits at OHaganBooks.com, $y = $ the number of hits at JungleBooks.com, $z = $ the number of hits at FarmerBooks.com

We are given the following information:

(1) Combined Web site traffic at the three sites is estimated at 10,000 hits per day: $x + y + z = 10,000.$

(2) The total number of orders is 1,500 per day: $0.10x + 0.20y + 0.20z = 1,500.$

(3) FarmerBooks.com gets as many book orders as the other two combined:

   $0.20z = 0.10x + 0.20y,$ or $0.10x + 0.20y - 0.20z = 0.$

Solving this system of three linear equations in two unknowns:

$$\begin{bmatrix} 1 & 1 & 1 & 10000 \\ 0.10 & 0.20 & 0.20 & 1500 \\ 0.10 & 0.20 & -0.20 & 0 \end{bmatrix} \begin{matrix} \\ 10R_2 \\ 10R_3 \end{matrix} \rightarrow \begin{bmatrix} \boxed{1} & 1 & 1 & 10000 \\ 1 & 2 & 2 & 15000 \\ 1 & 2 & -2 & 0 \end{bmatrix} \begin{matrix} \\ R_2 - R_1 \\ R_3 - R_1 \end{matrix} \rightarrow$$

$$\begin{bmatrix} 1 & 1 & 1 & 10000 \\ 0 & \boxed{1} & 1 & 5000 \\ 0 & 1 & -3 & -10000 \end{bmatrix} \begin{matrix} R_1 - R_2 \\ \\ R_3 - R_2 \end{matrix} \rightarrow \begin{bmatrix} 1 & 0 & 0 & 5000 \\ 0 & 1 & 1 & 5000 \\ 0 & 0 & -4 & -15000 \end{bmatrix} \begin{matrix} \\ \\ (1/4)R_3 \end{matrix} \rightarrow$$

$$\begin{bmatrix} 1 & 0 & 0 & 5000 \\ 0 & 1 & 1 & 5000 \\ 0 & 0 & \boxed{-1} & -3750 \end{bmatrix} \begin{matrix} \\ R_2 + R_3 \\ \end{matrix} \rightarrow \begin{bmatrix} 1 & 0 & 0 & 5000 \\ 0 & 1 & 0 & 1250 \\ 0 & 0 & -1 & -3750 \end{bmatrix} \begin{matrix} \\ \\ -R_3 \end{matrix} \rightarrow$$

$$\begin{bmatrix} 1 & 0 & 0 & 5000 \\ 0 & 1 & 0 & 1250 \\ 0 & 0 & 1 & 3750 \end{bmatrix}$$

Solution: 5,000 hits per day at OHaganBooks.com, 1,250 at JungleBooks.com, 3,750 at FarmerBooks.com

39. Unknowns: $x = $ the number of shares of HAL, $y = $ the number of shares of POM, $z = $ the number of shares of WELL

The total investment was $12,400:

Investment in HAL $= x$ shares @ $100 $= 100x$

Investment in POM $= y$ shares @ $20 $= 20y$

Investment in WELL $= z$ shares @ $25 $= 25z.$

Thus, $100x + 20y + 25z = 12,400$.

He earned $56 in dividends:

HAL dividend = 0.5% of $100x$ invested $= 0.005(100x) = 0.5x$

POM dividend = 1.5% of $20y$ invested $= 0.015(20y) = 0.3y$

WELL dividend = 0

Thus, $0.5x + 0.3y = 56$.

He purchased a total of 200 shares: $x + y + z = 200$.

We therefore have the following system:

$100x + 20y + 25z = 12,400$

$\quad 0.5x + 0.3y = 56$

$\quad\quad x + y + z = 200$

$$\begin{bmatrix} 100 & 20 & 25 & 12400 \\ 0.5 & 0.3 & 0 & 56 \\ 1 & 1 & 1 & 200 \end{bmatrix} \begin{matrix} \\ 10R_2 \\ \\ \end{matrix} \rightarrow \begin{bmatrix} 100 & 20 & 25 & 12400 \\ 5 & 3 & 0 & 560 \\ 1 & 1 & 1 & 200 \end{bmatrix} \begin{matrix} (1/5)R_1 \\ \\ \\ \end{matrix} \rightarrow$$

$$\begin{bmatrix} \boxed{20} & 4 & 5 & 2480 \\ 5 & 3 & 0 & 560 \\ 1 & 1 & 1 & 200 \end{bmatrix} \begin{matrix} \\ 4R_2 - R_1 \\ 20R_3 - R_1 \end{matrix} \rightarrow \begin{bmatrix} 20 & 4 & 5 & 2480 \\ 0 & \boxed{8} & -5 & -240 \\ 0 & 16 & 15 & 1520 \end{bmatrix} \begin{matrix} 2R_1 - R_2 \\ \\ R_3 - 2R_2 \end{matrix} \rightarrow$$

$$\begin{bmatrix} 40 & 0 & 15 & 5200 \\ 0 & 8 & -5 & -240 \\ 0 & 0 & 25 & 2000 \end{bmatrix} \begin{matrix} (1/5)R_1 \\ \\ (1/25)R_3 \end{matrix} \rightarrow \begin{bmatrix} 8 & 0 & 3 & 1040 \\ 0 & 8 & -5 & -240 \\ 0 & 0 & \boxed{1} & 80 \end{bmatrix} \begin{matrix} R_1 - 3R_3 \\ R_2 + 5R_3 \\ \end{matrix} \rightarrow$$

$$\begin{bmatrix} 8 & 0 & 0 & 800 \\ 0 & 8 & 0 & 160 \\ 0 & 0 & 1 & 80 \end{bmatrix} \begin{matrix} (1/8)R_1 \\ (1/8)R_2 \\ \end{matrix} \rightarrow \begin{bmatrix} 1 & 0 & 0 & 100 \\ 0 & 1 & 0 & 20 \\ 0 & 0 & 1 & 80 \end{bmatrix}$$

Solution: He purchased 100 shares of HAL, 20 shares of POM, and 80 shares of WELL.

41. Unknowns: $x =$ the number of credits of Liberal Arts, $y =$ the number of credits of Sciences, $z =$ the number of credits of Fine Arts, $w =$ the number of credits of Mathematics

Given information:

(1) The total number of credits is 124: $x + y + z + w = 124$.

(2) An equal number of Science and Fine Arts credits: $y = z$, or $y - z = 0$.

(3) Twice as many Mathematics credits as Science credits and Fine Arts credits combined. Rephrasing: The number of Mathematics credits is twice the sum of the numbers of Science and Fine Arts credits:

$\quad w = 2(y + z)$, or $2y + 2z - w = 0$.

(4) Liberal Arts credits exceed Mathematics credits by one third of the number of Fine Arts credits. Rephrasing: The number of Liberal Arts credits minus the number of Mathematics credits is one third of the number of Fine Arts credits:

$\quad x - w = \frac{1}{3}z$, or $x - \frac{1}{3}z - w = 0$

$$\begin{bmatrix} 1 & 1 & 1 & 1 & 124 \\ 0 & 1 & -1 & 0 & 0 \\ 0 & 2 & 2 & -1 & 0 \\ 1 & 0 & -1/3 & -1 & 0 \end{bmatrix} \begin{matrix} \\ \\ \\ 3R_4 \end{matrix} \rightarrow \begin{bmatrix} \boxed{1} & 1 & 1 & 1 & 124 \\ 0 & 1 & -1 & 0 & 0 \\ 0 & 2 & 2 & -1 & 0 \\ 3 & 0 & -1 & -3 & 0 \end{bmatrix} \begin{matrix} \\ \\ \\ R_4 - 3R_1 \end{matrix} \rightarrow$$

$$\begin{bmatrix} 1 & 1 & 1 & 1 & 124 \\ 0 & \boxed{1} & -1 & 0 & 0 \\ 0 & 2 & 2 & -1 & 0 \\ 0 & -3 & -4 & -6 & -372 \end{bmatrix} \begin{matrix} R_1 - R_2 \\ \\ R_3 - 2R_2 \\ R_4 + 3R_2 \end{matrix} \rightarrow \begin{bmatrix} 1 & 0 & 2 & 1 & 124 \\ 0 & 1 & -1 & 0 & 0 \\ 0 & 0 & \boxed{4} & -1 & 0 \\ 0 & 0 & -7 & -6 & -372 \end{bmatrix} \begin{matrix} 2R_1 - R_3 \\ 4R_2 + R_3 \\ \\ 4R_4 + 7R_3 \end{matrix} \rightarrow$$

$$\begin{bmatrix} 2 & 0 & 0 & 3 & 248 \\ 0 & 4 & 0 & -1 & 0 \\ 0 & 0 & 4 & -1 & 0 \\ 0 & 0 & 0 & -31 & -1488 \end{bmatrix} \begin{matrix} \\ \\ \\ (1/31)R_4 \end{matrix} \rightarrow \begin{bmatrix} 2 & 0 & 0 & 3 & 248 \\ 0 & 4 & 0 & -1 & 0 \\ 0 & 0 & 4 & -1 & 0 \\ 0 & 0 & 0 & \boxed{-1} & -48 \end{bmatrix} \begin{matrix} R_1+3R_4 \\ R_2-R_4 \\ R_3-R_4 \\ \\ \end{matrix} \rightarrow$$

$$\begin{bmatrix} 2 & 0 & 0 & 0 & 104 \\ 0 & 4 & 0 & 0 & 48 \\ 0 & 0 & 4 & 0 & 48 \\ 0 & 0 & 0 & -1 & -48 \end{bmatrix} \begin{matrix} (1/2)R_1 \\ (1/4)R_2 \\ (1/4)R_3 \\ \\ \end{matrix} \rightarrow \begin{bmatrix} 1 & 0 & 0 & 0 & 52 \\ 0 & 1 & 0 & 0 & 12 \\ 0 & 0 & 1 & 0 & 12 \\ 0 & 0 & 0 & -1 & -48 \end{bmatrix} \begin{matrix} \\ \\ \\ -R_4 \end{matrix} \rightarrow$$

$$\begin{bmatrix} 1 & 0 & 0 & 0 & 52 \\ 0 & 1 & 0 & 0 & 12 \\ 0 & 0 & 1 & 0 & 12 \\ 0 & 0 & 0 & 1 & 48 \end{bmatrix}$$

Solution: Billy-Sean is forced to take exactly the following combination: Liberal Arts: 52 credits, Sciences: 12 credits, Fine Arts: 12 credits, Mathematics: 48 credits.

43.

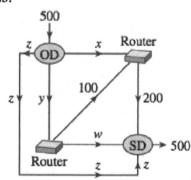

OD = Order department
SD = Shipping department

a. Order Department: Traffic in = Traffic out
$$500 = x + y + z$$
Top right router: Traffic in = Traffic out
$$100 + x = 200, \text{ or } x = 100$$
Bottom left router: Traffic in = Traffic out
$$y = 100 + w, \text{ or } y - w = 100$$
Shipping Department: Traffic in = Traffic out
$$z + w + 200 = 500, \text{ or } z + w = 300$$

$$\begin{bmatrix} \boxed{1} & 1 & 1 & 0 & 500 \\ 1 & 0 & 0 & 0 & 100 \\ 0 & 1 & 0 & -1 & 100 \\ 0 & 0 & 1 & 1 & 300 \end{bmatrix} \begin{matrix} \\ R_2-R_1 \\ \\ \\ \end{matrix} \rightarrow \begin{bmatrix} 1 & 1 & 1 & 0 & 500 \\ 0 & \boxed{-1} & -1 & 0 & -400 \\ 0 & 1 & 0 & -1 & 100 \\ 0 & 0 & 1 & 1 & 300 \end{bmatrix} \begin{matrix} R_1+R_2 \\ \\ R_3+R_2 \\ \\ \end{matrix} \rightarrow$$

$$\begin{bmatrix} 1 & 0 & 0 & 0 & 100 \\ 0 & -1 & -1 & 0 & -400 \\ 0 & 0 & \boxed{-1} & -1 & -300 \\ 0 & 0 & 1 & 1 & 300 \end{bmatrix} \begin{matrix} \\ R_2-R_3 \\ \\ R_4+R_3 \end{matrix} \rightarrow \begin{bmatrix} 1 & 0 & 0 & 0 & 100 \\ 0 & -1 & 0 & 1 & -100 \\ 0 & 0 & -1 & -1 & -300 \\ 0 & 0 & 0 & 0 & 0 \end{bmatrix} \begin{matrix} \\ -R_2 \\ -R_3 \\ \\ \end{matrix} \rightarrow$$

$$\begin{bmatrix} 1 & 0 & 0 & 0 & 100 \\ 0 & 1 & 0 & -1 & 100 \\ 0 & 0 & 1 & 1 & 300 \\ 0 & 0 & 0 & 0 & 0 \end{bmatrix}$$

Translating back to equations gives: $x = 100, y - w = 100, z + w = 300$.

General solution: $x = 100, y = 100 + w, z = 300 - w, w$ arbitrary

b. Because $y = 100 + w$ and w can be any number \geq 0 (w cannot be negative because it represents a number of book orders), the smallest possible value of y is 100 books per day.

c. The equation $z = 300 - w$ tells us that w cannot exceed 300 books per day, or else x would become negative.

d. If there is no traffic along z, then $z = 0$, giving:

$$x = 100 \qquad y = 100 + w \qquad 0 = 300 - w.$$

Thus, from the third equation, $w = 300$, giving us the particular solution

$$x = 100, y = 100 + 300 = 400, z = 0, w = 300.$$

e. If there is the same volume of traffic along y and z, then $y = z$, and so

$$100 + w = 300 - w \Rightarrow 2w = 200 \Rightarrow w = 100 \text{ books per day.}$$

45. Unknowns: $x =$ the number of packages from New York to Texas, $y =$ the number of packages from New York to California, $z =$ the number of packages from Illinois to Texas, $w =$ the number of packages from Illinois to California

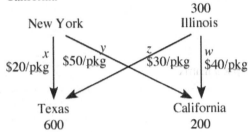

Books to Texas: $x + z = 600$

Books to California: $y + w = 200$

Books from Illinois: $z + w = 300$

Budget: $20x + 50y + 30z + 40w = 22{,}000$

$$\begin{bmatrix} 1 & 0 & 1 & 0 & 600 \\ 0 & 1 & 0 & 1 & 200 \\ 0 & 0 & 1 & 1 & 300 \\ 20 & 50 & 30 & 40 & 22000 \end{bmatrix} \begin{matrix} \\ \\ \\ (1/10)R_4 \end{matrix} \rightarrow \begin{bmatrix} \boxed{1} & 0 & 1 & 0 & 600 \\ 0 & 1 & 0 & 1 & 200 \\ 0 & 0 & 1 & 1 & 300 \\ 2 & 5 & 3 & 4 & 2200 \end{bmatrix} \begin{matrix} \\ \\ \\ R_4 - 2R_1 \end{matrix} \rightarrow$$

$$\begin{bmatrix} 1 & 0 & 1 & 0 & 600 \\ 0 & \boxed{1} & 0 & 1 & 200 \\ 0 & 0 & 1 & 1 & 300 \\ 0 & 5 & 1 & 4 & 1000 \end{bmatrix} \begin{matrix} \\ \\ \\ R_4 - 5R_2 \end{matrix} \rightarrow \begin{bmatrix} 1 & 0 & 1 & 0 & 600 \\ 0 & 1 & 0 & 1 & 200 \\ 0 & 0 & \boxed{1} & 1 & 300 \\ 0 & 0 & 1 & -1 & 0 \end{bmatrix} \begin{matrix} R_1 - R_3 \\ \\ \\ R_4 - R_3 \end{matrix} \rightarrow$$

$$\begin{bmatrix} 1 & 0 & 0 & -1 & 300 \\ 0 & 1 & 0 & 1 & 200 \\ 0 & 0 & 1 & 1 & 300 \\ 0 & 0 & 0 & -2 & -300 \end{bmatrix} \begin{matrix} \\ \\ \\ (1/2)R_4 \end{matrix} \rightarrow \begin{bmatrix} 1 & 0 & 0 & -1 & 300 \\ 0 & 1 & 0 & 1 & 200 \\ 0 & 0 & 1 & 1 & 300 \\ 0 & 0 & 0 & \boxed{-1} & -150 \end{bmatrix} \begin{matrix} R_1 - R_4 \\ R_2 + R_4 \\ R_3 + R_4 \\ \end{matrix} \rightarrow$$

$$\begin{bmatrix} 1 & 0 & 0 & 0 & 450 \\ 0 & 1 & 0 & 0 & 50 \\ 0 & 0 & 1 & 0 & 150 \\ 0 & 0 & 0 & -1 & -150 \end{bmatrix} \begin{matrix} \\ \\ \\ -R_4 \end{matrix} \rightarrow \begin{bmatrix} 1 & 0 & 0 & 0 & 450 \\ 0 & 1 & 0 & 0 & 50 \\ 0 & 0 & 1 & 0 & 150 \\ 0 & 0 & 0 & 1 & 150 \end{bmatrix}$$

Solution: New York to Texas: 450 packages, New York to California: 50 packages, Illinois to Texas: 150 packages, Illinois to California: 150 packages

Section 4.1

1. $A = \begin{bmatrix} 1 & 5 & 0 & \frac{1}{4} \end{bmatrix};$

A has 1 row and 4 columns. Therefore, it is a 1×4 matrix. a_{13} is the entry in row 1 and column 3, so $a_{13} = 0$.

3. $C = \begin{bmatrix} \frac{5}{2} \\ 1 \\ -2 \\ 8 \end{bmatrix}$ has 4 rows and 1 column. Therefore, it is a 4×1 matrix.

C_{11} is the entry in row 1 and column 1, so $C_{11} = \dfrac{5}{2}$.

5. $E = \begin{bmatrix} e_{11} & e_{12} & e_{13} & \cdots & e_{1q} \\ e_{21} & e_{22} & e_{23} & \cdots & e_{2q} \\ \vdots & \vdots & \vdots & \cdots & \vdots \\ e_{p1} & e_{p2} & e_{p3} & \cdots & e_{pq} \end{bmatrix}$ has p rows and q columns. Therefore, it is a $p \times q$ matrix.

E_{22} is the entry in row 2 and column 2, so $E_{22} = e_{22}$.

7. $B = \begin{bmatrix} 1 & 3 \\ 5 & -6 \end{bmatrix}$ has 2 rows and 2 columns. Therefore, it is a 2×2 matrix.

b_{12} is the entry in row 1 and column 2, so $b_{12} = 3$.

9. $D = \begin{bmatrix} d_1 & d_2 & \cdots & d_n \end{bmatrix}$ has 1 row and n columns. Therefore, it is a $1 \times n$ matrix.

D_{1r} is the entry in row 1 and column r, so $D_{1r} = d_r$.

11. $\begin{bmatrix} x+y & x+z \\ y+z & w \end{bmatrix} = \begin{bmatrix} 3 & 4 \\ 5 & 4 \end{bmatrix}$

Equating corresponding entries gives

$$x + y = 3$$
$$y + z = 5$$
$$z + w = 7$$
$$w = 4$$

Substitute $w = 4$ in the next-to-last equation to get $z = 3$. Then substitute into the second to get $y = 2$, then into the first to get $x = 1$.

Thus, $x = 1, y = 2, z = 3, w = 4$

13. We obtain $A + B$ by adding corresponding entries:

$$A + B = \begin{bmatrix} 0 & -1 \\ 1 & 0 \\ -1 & 2 \end{bmatrix} + \begin{bmatrix} 0.25 & -1 \\ 0 & 0.5 \\ -1 & 3 \end{bmatrix} = \begin{bmatrix} 0.25 & -2 \\ 1 & 0.5 \\ -2 & 5 \end{bmatrix}$$

15. To obtain $A + B - C$, add corresponding entries of A and B and subtract those of C:

$$A + B - C = \begin{bmatrix} 0 & -1 \\ 1 & 0 \\ -1 & 2 \end{bmatrix} + \begin{bmatrix} 0.25 & -1 \\ 0 & 0.5 \\ -1 & 3 \end{bmatrix} - \begin{bmatrix} 1 & -1 \\ 1 & 1 \\ -1 & -1 \end{bmatrix} = \begin{bmatrix} -0.75 & -1 \\ 0 & -0.5 \\ -1 & 6 \end{bmatrix}$$

17. $2A - C = 2\begin{bmatrix} 0 & -1 \\ 1 & 0 \\ -1 & 2 \end{bmatrix} - \begin{bmatrix} 1 & -1 \\ 1 & 1 \\ -1 & -1 \end{bmatrix} = \begin{bmatrix} -1 & -1 \\ 1 & -1 \\ -1 & 5 \end{bmatrix}$

19. The transpose, A^T, of A is obtained by writing the rows of A as columns:

$$A = \begin{bmatrix} 0 & -1 \\ 1 & 0 \\ -1 & 2 \end{bmatrix}; A^T = \begin{bmatrix} 0 & 1 & -1 \\ -1 & 0 & 2 \end{bmatrix}$$

Thus, $2A^T = 2\begin{bmatrix} 0 & 1 & -1 \\ -1 & 0 & 2 \end{bmatrix} = \begin{bmatrix} 0 & 2 & -2 \\ -2 & 0 & 4 \end{bmatrix}$

21. $A + B = \begin{bmatrix} 1 & -1 & 0 \\ 0 & 2 & -1 \end{bmatrix} + \begin{bmatrix} 3 & 0 & -1 \\ 5 & -1 & 1 \end{bmatrix} = \begin{bmatrix} 4 & -1 & -1 \\ 5 & 1 & 0 \end{bmatrix}$

23. $A - B + C = \begin{bmatrix} 1 & -1 & 0 \\ 0 & 2 & -1 \end{bmatrix} - \begin{bmatrix} 3 & 0 & -1 \\ 5 & -1 & 1 \end{bmatrix} + \begin{bmatrix} x & 1 & w \\ z & r & 4 \end{bmatrix} = \begin{bmatrix} -2+x & 0 & 1+w \\ -5+z & 3+r & 2 \end{bmatrix}$

25. $2A - B = 2\begin{bmatrix} 1 & -1 & 0 \\ 0 & 2 & -1 \end{bmatrix} - \begin{bmatrix} 3 & 0 & -1 \\ 5 & -1 & 1 \end{bmatrix} = \begin{bmatrix} -1 & -2 & 1 \\ -5 & 5 & -3 \end{bmatrix}$

27. $B = \begin{bmatrix} 3 & 0 & -1 \\ 5 & -1 & 1 \end{bmatrix} \Rightarrow B^T = \begin{bmatrix} 3 & 5 \\ 0 & -1 \\ -1 & 1 \end{bmatrix}$

Therefore, $3B^T = 3\begin{bmatrix} 3 & 5 \\ 0 & -1 \\ -1 & 1 \end{bmatrix} = \begin{bmatrix} 9 & 15 \\ 0 & -3 \\ -3 & 3 \end{bmatrix}$

29. TI-83/84 Plus format: `[A]-[C]`   Online Matrix Algebra Tool format: `A-C`

$$A - C = \begin{bmatrix} 1.5 & -2.35 & 5.6 \\ 44.2 & 0 & 12.2 \end{bmatrix} - \begin{bmatrix} 10 & 20 & 30 \\ -10 & -20 & -30 \end{bmatrix} = \begin{bmatrix} -8.5 & -22.35 & -24.4 \\ 54.2 & 20 & 42.2 \end{bmatrix}$$

31. TI-83/84 Plus format: `1.1[B]`   Online Matrix Algebra Tool format: `1.1*B`

$$1.1B = 1.1\begin{bmatrix} 1.4 & 7.8 \\ 5.4 & 0 \\ 5.6 & 6.6 \end{bmatrix} = \begin{bmatrix} 1.54 & 8.58 \\ 5.94 & 0 \\ 6.16 & 7.26 \end{bmatrix}$$

33. TI-83/84 Plus format: `[A]`^{T}`+4.2[B]`   Online Matrix Algebra Tool format: `A^T+4.2*B`

$$AT + 4.2B = \begin{bmatrix} 1.5 & 44.2 \\ -2.35 & 0 \\ 5.6 & 12.2 \end{bmatrix} + 4.2\begin{bmatrix} 1.4 & 7.8 \\ 5.4 & 0 \\ 5.6 & 6.6 \end{bmatrix} = \begin{bmatrix} 7.38 & 76.96 \\ 20.33 & 0 \\ 29.12 & 39.92 \end{bmatrix}$$

35. TI-83/84 Plus format: $(2.1[A]-2.3[C])^T$ Online Matrix Algebra Tool format: $(2.1*A-2.3*C)^T$

$$(2.1A - 2.3C)^T = \begin{bmatrix} -19.85 & 115.82 \\ -50.935 & 46 \\ -57.24 & 94.62 \end{bmatrix}$$

Applications

37. Regard each row of the given table as a matrix:

Sales in 2012 $= \begin{bmatrix} 17.0 & 135.8 & 65.7 \end{bmatrix}$

Change in 2013 $= \begin{bmatrix} 0.2 & 17.6 & 8.5 \end{bmatrix}$

Change in 2014 $= \begin{bmatrix} 2.4 & 39.2 & -10.8 \end{bmatrix}$

Sales in 2013 = Sales in 2012 + Change in 2013 $= \begin{bmatrix} 17.0 & 135.8 & 65.7 \end{bmatrix} + \begin{bmatrix} 0.2 & 17.6 & 8.5 \end{bmatrix} =$
$\begin{bmatrix} 17.2 & 153.4 & 74.2 \end{bmatrix}$

Sales in 2014 = Sales in 2013 + Change in 2014 $= \begin{bmatrix} 17.2 & 153.4 & 74.2 \end{bmatrix} + \begin{bmatrix} 2.4 & 39.2 & -10.8 \end{bmatrix} =$
$\begin{bmatrix} 19.6 & 192.6 & 63.4 \end{bmatrix}$

39. Write the given inventory table as a 2×3 matrix:

$$\text{Inventory} = \begin{bmatrix} 1,000 & 2,000 & 5,000 \\ 1,000 & 5,000 & 2,000 \end{bmatrix}$$

Write the given sales figures as a similar matrix:

$$\text{Sales} = \begin{bmatrix} 700 & 1,300 & 2,000 \\ 400 & 300 & 500 \end{bmatrix}$$

We then compute the remaining inventory by subtracting the sales:

Remaining Inventory = Inventory − Sales

$$= \begin{bmatrix} 1,000 & 2,000 & 5,000 \\ 1,000 & 5,000 & 2,000 \end{bmatrix} - \begin{bmatrix} 700 & 1,300 & 2,000 \\ 400 & 300 & 500 \end{bmatrix} = \begin{bmatrix} 300 & 700 & 3,000 \\ 600 & 4,700 & 1,500 \end{bmatrix}$$

41.

	A	B	C	D
1	Revenue			
2		2004	2005	2006
3	Full Boots	$10,000	$9,000	$11,000
4	Half Boots	$8,000	$7,200	$8,800
5	Sandals	$4,000	$5,000	$6,000
6				
7	Production Costs			
8		2004	2005	2006
9	Full Boots	$2,000	$1,800	$2,200
10	Half Boots	$2,400	$1,440	$1,760
11	Sandals	$1,200	$1,500	$2,000

Arrange the revenue and costs in two 3×3 matrices:

$$\text{Revenue} = \begin{bmatrix} 10,000 & 9,000 & 11,000 \\ 8,000 & 7,200 & 8,800 \\ 4,000 & 5,000 & 6,000 \end{bmatrix} \qquad \text{Cost} = \begin{bmatrix} 2,000 & 1,800 & 2,200 \\ 2,400 & 1,440 & 1,760 \\ 1,200 & 1,500 & 2,000 \end{bmatrix}$$

Profit = Revenue − Cost

$$= \begin{bmatrix} 10,000 & 9,000 & 11,000 \\ 8,000 & 7,200 & 8,800 \\ 4,000 & 5,000 & 6,000 \end{bmatrix} - \begin{bmatrix} 2,000 & 1,800 & 2,200 \\ 2,400 & 1,440 & 1,760 \\ 1,200 & 1,500 & 2,000 \end{bmatrix} = \begin{bmatrix} 8,000 & 7,200 & 8,800 \\ 5,600 & 5,760 & 7,040 \\ 2,800 & 3,500 & 4,000 \end{bmatrix}$$

43. Row vectors for the populations, in millions:

2000 distribution = A = [53.6 64.4 100.2 63.2]; 2010 distribution = B = [55.3 66.9 114.6 71.9]; Net

change 2000 to 2010 = $D = B - A$ = [1.7 2.5 14.4 8.7]

As all the net changes are positive, they represent net increases in the population.

Population in 2020 = $B + D$ = [55.3 66.9 114.6 71.9] + [1.7 2.5 14.4 8.7] = [57.0 69.4 129.0 80.6]

45. Regard each row in the table as a matrix.

Foreclosures in California = [55,900 51,900 54,100 56,200 59,400]

Foreclosures in Florida = [19,600 19,200 23,800 22,400 23,600]

Foreclosures in Texas = [8,800 9,100 9,300 10,600 10,100]

Total Foreclosures = Foreclosures in California + Foreclosures in Florida + Foreclosures in Texas

  = [55,900 51,900 54,100 56,200 59,400] + [19,600 19,200 23,800 22,400 23,600]

  + [8,800 9,100 9,300 10,600 10,100]

  = [84,300 80,200 87,200 89,200 93,100]

47. (Refer to the solution to Exercise 45.) The difference between the number of foreclosures in California and in Florida is given by

Difference = Foreclosures in California − Foreclosures in Florida

  = [55,900 51,900 54,100 56,200 59,400] − [19,600 19,200 23,800 22,400 23,600]

  = [36,300 32,700 30,300 33,800 35,800]

The difference was greatest in April.

49. First, organize the starting inventory as a 2×3 matrix:

	Proc	Mem	Tubes
Pom II	500	5,000	10,000
Pom Classic	200	2,000	20,000

Thus, Inventory = $\begin{bmatrix} 500 & 5,000 & 10,000 \\ 200 & 2,000 & 20,000 \end{bmatrix}$

The parts used in making one of each computer can also be arranged in a matrix:

Use = $\begin{bmatrix} 2 & 16 & 20 \\ 1 & 4 & 40 \end{bmatrix}$

a. After 2 months, the company has made 100 of each computer, so the inventory remaining is

Inventory Remaining = Inventory $-100 \cdot$ Use

$= \begin{bmatrix} 500 & 5,000 & 10,000 \\ 200 & 2,000 & 20,000 \end{bmatrix} - 100 \begin{bmatrix} 2 & 16 & 20 \\ 1 & 4 & 40 \end{bmatrix} = \begin{bmatrix} 300 & 3,400 & 8,000 \\ 100 & 1,600 & 16,000 \end{bmatrix}$

b. After n months, the company has made $50n$ of each computer. Therefore, the inventory remaining is

$\begin{bmatrix} 500 & 5,000 & 10,000 \\ 200 & 2,000 & 20,000 \end{bmatrix} - 50n \begin{bmatrix} 2 & 16 & 20 \\ 1 & 4 & 40 \end{bmatrix} = \begin{bmatrix} 500 - 100n & 5,000 - 800n & 10,000 - 1,000n \\ 200 - 50n & 2,000 - 200n & 20,000 - 2,000n \end{bmatrix}$

  The 1,1 entry is zero after $n = 5$ months   The 1,2 entry is zero after $n = 6.25$ months

  The 1,3 entry is zero after $n = 10$ months   The 2,1 entry is zero after $n = 4$ months

  The 2,2 entry is zero after $n = 10$ months   The 2,3 entry is zero after $n = 10$ months

Thus, after 4 months, the first entry becomes zero, meaning that the company will run out of Pom Classic processor chips after 4 months.

51. a. Arrange the 1998 tourism figures in a matrix: $A = \begin{bmatrix} 440 & 190 \\ 950 & 950 \\ 1790 & 200 \end{bmatrix}$

Then, arrange the predicted changes from 1998 to 2008 in a new matrix: $D = \begin{bmatrix} -20 & 40 \\ 50 & 50 \\ 0 & 100 \end{bmatrix}$

Tourism in 2008 = Tourism in 1998 + Change $= A + D = \begin{bmatrix} 420 & 230 \\ 1{,}000 & 1{,}000 \\ 1790 & 300 \end{bmatrix}$

b. The average of the tourism figures is obtained by taking the average of the entries in A and B. To obtain the average, add and divide by 2:   Average $= \frac{1}{2}(A + B)$

$A = $ 1998 figures $= \begin{bmatrix} 440 & 190 \\ 950 & 950 \\ 1{,}790 & 200 \end{bmatrix}$ $B = $ 2008 figures $= \begin{bmatrix} 420 & 230 \\ 1{,}000 & 1{,}000 \\ 1{,}790 & 300 \end{bmatrix}$ (From part (a))

Average $= \frac{1}{2}(A + B) = \frac{1}{2}\begin{bmatrix} 860 & 420 \\ 1{,}950 & 1{,}950 \\ 3{,}580 & 500 \end{bmatrix} = \begin{bmatrix} 430 & 210 \\ 975 & 975 \\ 1{,}790 & 250 \end{bmatrix}$

Communication and reasoning exercises

53. No; for two matrices to be equal, they must have the same dimensions.

55. $(A + B)_{ij} = A_{ij} + B_{ij}$

The left-hand side is the ijth entry of $A + B$. The right-hand side is the sum of the ijth entries of A and B. Thus, the equation tells us that the ij th entry of the sum $A + B$ is obtained by adding the ijth entries of A and B.

57. If A is any matrix, then $A_{11}, A_{22}, ..., A_{ii}, ...$ denote the entries going down the main diagonal (top left to bottom right). Thus, if $A_{ii} = 0$ for every i, then A would have zeros down the diagonal:

$A = \begin{bmatrix} 0 & \# & \# & \# & \# \\ \# & 0 & \# & \# & \# \\ \# & \# & 0 & \# & \# \\ \# & \# & \# & 0 & \# \\ \# & \# & \# & \# & 0 \end{bmatrix}$

(The symbols # indicate arbitrary numbers.)

59. The transpose of an $m \times n$ matrix is the $n \times m$ matrix obtained by writing its rows as columns. Thus, the entry originally in Row j and Column i winds up in Row i and Column j when a matrix is transposed. In other words, ijth entry of the transpose = jith entry of the original matrix: $(A^T)_{ij} = A_{ji}$

61. In a skew-symmetric matrix, the entry in the ij position is the negative of that in the ji position. In other words, each entry is the negative of its mirror image in the diagonal.

What about the diagonal entries? Each diagonal entry stays the same when the matrix is transposed. So, in a skew-symmetric matrix, each diagonal entry must equal its own negative. The only way this can happen is if the diagonal entries are zero.

Answers will vary.   **a.** $\begin{bmatrix} 0 & -4 \\ 4 & 0 \end{bmatrix}$ **b.** $\begin{bmatrix} 0 & -4 & 5 \\ 4 & 0 & 1 \\ -5 & -1 & 0 \end{bmatrix}$

63. The associativity of matrix addition is a consequence of the associativity of addition of numbers, since we add matrices by adding the corresponding entries (which are real numbers).

65. Answers will vary.

Section 4.2

1. $\begin{bmatrix} 1 & 3 & -1 \end{bmatrix} \begin{bmatrix} 9 \\ 1 \\ -1 \end{bmatrix} = \begin{bmatrix} 9+3+1 \end{bmatrix} = \begin{bmatrix} 13 \end{bmatrix}$

3. $\begin{bmatrix} -1 & \frac{1}{2} \end{bmatrix} \begin{bmatrix} -\frac{1}{3} \\ 1 \end{bmatrix} = \begin{bmatrix} \frac{1}{3} + \frac{1}{2} \end{bmatrix} = \begin{bmatrix} \frac{5}{6} \end{bmatrix}$

5. $\begin{bmatrix} 0 & -2 & 1 \end{bmatrix} \begin{bmatrix} x \\ y \\ z \end{bmatrix} = \begin{bmatrix} -2y + z \end{bmatrix}$

7. $\begin{bmatrix} 1 & 3 & 2 \end{bmatrix} \begin{bmatrix} 1 \\ -1 \end{bmatrix}$ is undefined. (The dimensions are 1×3 and 2×1; the 3 and the 2 do not match.)

9. $\begin{bmatrix} -1 & 1 \end{bmatrix} \begin{bmatrix} -3 & 1 & 4 & 3 \\ 0 & 1 & -2 & 1 \end{bmatrix} = \begin{bmatrix} (3+0) & (-1+1) & (-4-2) & (-3+1) \end{bmatrix} = \begin{bmatrix} 3 & 0 & -6 & -2 \end{bmatrix}$

11. $\begin{bmatrix} 1 & -1 & 2 & 3 \end{bmatrix} \begin{bmatrix} -1 & 2 & 0 \\ 2 & -1 & 0 \\ 0 & 5 & 2 \\ -1 & 8 & 1 \end{bmatrix} = \begin{bmatrix} (-1-2+0-3) & (2+1+10+24) & (0+0+4+3) \end{bmatrix} = \begin{bmatrix} -6 & 37 & 7 \end{bmatrix}$

13. $\begin{bmatrix} 1 & 0 & -1 \\ 1 & 1 & 2 \end{bmatrix} \begin{bmatrix} 0 & 1 & -1 \\ 1 & 0 & 1 \\ 4 & 8 & 0 \end{bmatrix} = \begin{bmatrix} (0+0-4) & (1+0-8) & (-1+0+0) \\ (0+1+8) & (1+0+16) & (-1+1+0) \end{bmatrix} = \begin{bmatrix} -4 & -7 & -1 \\ 9 & 17 & 0 \end{bmatrix}$

15. $\begin{bmatrix} 1 & 0 \\ 1 & -1 \end{bmatrix} \begin{bmatrix} 0 & 1 \\ 0 & 1 \end{bmatrix} = \begin{bmatrix} (0+0) & (1+0) \\ (0+0) & (1-1) \end{bmatrix} = \begin{bmatrix} 0 & 1 \\ 0 & 0 \end{bmatrix}$

17. $\begin{bmatrix} 0 & 1 \\ 0 & 1 \end{bmatrix} \begin{bmatrix} 1 & 0 \\ 1 & -1 \end{bmatrix} = \begin{bmatrix} (0+1) & (0-1) \\ (0+1) & (0-1) \end{bmatrix} = \begin{bmatrix} 1 & -1 \\ 1 & -1 \end{bmatrix}$

19. $\begin{bmatrix} 1 & -1 \\ 1 & -1 \end{bmatrix} \begin{bmatrix} 2 & 3 \\ 2 & 3 \end{bmatrix} = \begin{bmatrix} (2-2) & (3-3) \\ (2-2) & (3-3) \end{bmatrix} = \begin{bmatrix} 0 & 0 \\ 0 & 0 \end{bmatrix}$

21. $\begin{bmatrix} 1 & -1 \\ -1 & 1 \end{bmatrix} \begin{bmatrix} 2 & 3 \\ 2 & 3 \\ 1 & 1 \end{bmatrix}$ is undefined.

(The dimensions are 2×2 and 3×2; the 2 and the 3 do not match.)

23. $\begin{bmatrix} 1 & 0 & -1 \\ 2 & -2 & 1 \\ 0 & 0 & 1 \end{bmatrix} \begin{bmatrix} 1 & -1 & 4 \\ 1 & 1 & 0 \\ 0 & 4 & 1 \end{bmatrix} = \begin{bmatrix} (1+0+0) & (-1+0-4) & (4+0-1) \\ (2-2+0) & (-2-2+4) & (8+0+1) \\ (0+0+0) & (0+0+4) & (0+0+1) \end{bmatrix} = \begin{bmatrix} 1 & -5 & 3 \\ 0 & 0 & 9 \\ 0 & 4 & 1 \end{bmatrix}$

25.
$$\begin{bmatrix} 1 & 0 & 1 & 0 \\ -1 & 1 & 0 & 1 \\ -2 & 0 & 1 & 4 \\ 0 & -1 & 0 & 1 \end{bmatrix} \begin{bmatrix} 1 \\ -3 \\ 2 \\ 0 \end{bmatrix} = \begin{bmatrix} (1+0+2+0) \\ (-1-3+0+0) \\ (-2+0+2+0) \\ (0+3+0+0) \end{bmatrix} = \begin{bmatrix} 3 \\ -4 \\ 0 \\ 3 \end{bmatrix}$$

27. An on-line matrix algebra utility is available on the Website at

Student Web Site → Online Utilities → Matrix Algebra Tool

$$\begin{bmatrix} 1.1 & 2.3 & 3.4 & -1.2 \\ 3.4 & 4.4 & 2.3 & 1.1 \\ 2.3 & 0 & -2.2 & 1.1 \\ 1.2 & 1.3 & 1.1 & 1.1 \end{bmatrix} \begin{bmatrix} -2.1 & 0 & -3.3 \\ -3.4 & -4.8 & -4.2 \\ 3.4 & 5.6 & 1 \\ 1 & 2.2 & 9.8 \end{bmatrix} = \begin{bmatrix} 0.23 & 5.36 & -21.65 \\ -13.18 & -5.82 & -16.62 \\ -11.21 & -9.9 & 0.99 \\ -2.1 & 2.34 & 2.46 \end{bmatrix}$$

TI-83/84 Plus format: `[A]*[B]`   Online Matrix Algebra Tool format: `A*B`

29.

$$A = \begin{bmatrix} 0 & 1 & 1 & 1 \\ 0 & 0 & 1 & 1 \\ 0 & 0 & 0 & 1 \\ 0 & 0 & 0 & 0 \end{bmatrix}$$

$$A^2 = A \cdot A = \begin{bmatrix} 0 & 1 & 1 & 1 \\ 0 & 0 & 1 & 1 \\ 0 & 0 & 0 & 1 \\ 0 & 0 & 0 & 0 \end{bmatrix} \begin{bmatrix} 0 & 1 & 1 & 1 \\ 0 & 0 & 1 & 1 \\ 0 & 0 & 0 & 1 \\ 0 & 0 & 0 & 0 \end{bmatrix}$$

$$= \begin{bmatrix} (0+0+0+0) & (0+0+0+0) & (0+1+0+0) & (0+1+1+0) \\ (0+0+0+0) & (0+0+0+0) & (0+0+0+0) & (0+0+1+0) \\ (0+0+0+0) & (0+0+0+0) & (0+0+0+0) & (0+0+0+0) \\ (0+0+0+0) & (0+0+0+0) & (0+0+0+0) & (0+0+0+0) \\ (0+0+0+0) & (0+0+0+0) & (0+0+0+0) & (0+0+0+0) \end{bmatrix} = \begin{bmatrix} 0 & 0 & 1 & 2 \\ 0 & 0 & 0 & 1 \\ 0 & 0 & 0 & 0 \\ 0 & 0 & 0 & 0 \end{bmatrix}$$

$$A^3 = A \cdot A^2 = \begin{bmatrix} 0 & 1 & 1 & 1 \\ 0 & 0 & 1 & 1 \\ 0 & 0 & 0 & 1 \\ 0 & 0 & 0 & 0 \end{bmatrix} \begin{bmatrix} 0 & 0 & 1 & 2 \\ 0 & 0 & 0 & 1 \\ 0 & 0 & 0 & 0 \\ 0 & 0 & 0 & 0 \end{bmatrix}$$

$$= \begin{bmatrix} (0+0+0+0) & (0+0+0+0) & (0+0+0+0) & (0+1+0+0) \\ (0+0+0+0) & (0+0+0+0) & (0+0+0+0) & (0+0+0+0) \\ (0+0+0+0) & (0+0+0+0) & (0+0+0+0) & (0+0+0+0) \\ (0+0+0+0) & (0+0+0+0) & (0+0+0+0) & (0+0+0+0) \\ (0+0+0+0) & (0+0+0+0) & (0+0+0+0) & (0+0+0+0) \end{bmatrix} = \begin{bmatrix} 0 & 0 & 0 & 1 \\ 0 & 0 & 0 & 0 \\ 0 & 0 & 0 & 0 \\ 0 & 0 & 0 & 0 \end{bmatrix}$$

$$A^4 = A \cdot A^3 = \begin{bmatrix} 0 & 1 & 1 & 1 \\ 0 & 0 & 1 & 1 \\ 0 & 0 & 0 & 1 \\ 0 & 0 & 0 & 0 \end{bmatrix} \begin{bmatrix} 0 & 0 & 0 & 1 \\ 0 & 0 & 0 & 0 \\ 0 & 0 & 0 & 0 \\ 0 & 0 & 0 & 0 \end{bmatrix} = \begin{bmatrix} 0 & 0 & 0 & 0 \\ 0 & 0 & 0 & 0 \\ 0 & 0 & 0 & 0 \\ 0 & 0 & 0 & 0 \end{bmatrix}$$

Continuing to multiply by A continues to yield the zero matrix, so A^{100} = $\begin{bmatrix} 0 & 0 & 0 & 0 \\ 0 & 0 & 0 & 0 \\ 0 & 0 & 0 & 0 \\ 0 & 0 & 0 & 0 \end{bmatrix}$.

31. $AB = \begin{bmatrix} 0 & -1 & 0 & 1 \\ 10 & 0 & 1 & 0 \end{bmatrix} \begin{bmatrix} 0 & -1 \\ 1 & 1 \\ -1 & 3 \\ 5 & 0 \end{bmatrix} = \begin{bmatrix} (0-1+0+5) & (0-1+0+0) \\ (0+0-1+0) & (-10+0+3+0) \end{bmatrix} = \begin{bmatrix} 4 & -1 \\ -1 & -7 \end{bmatrix}$

TI-83/84 Plus format: `[A]*[B]`   Online Matrix Algebra Tool format: `A*B`

33. $A(B-C) = \begin{bmatrix} 0 & -1 & 0 & 1 \\ 10 & 0 & 1 & 0 \end{bmatrix} \left(\begin{bmatrix} 0 & -1 \\ 1 & 1 \\ -1 & 3 \\ 5 & 0 \end{bmatrix} - \begin{bmatrix} 1 & -1 \\ 1 & 1 \\ 1 & 1 \\ 1 & 1 \end{bmatrix} \right)$

$= \begin{bmatrix} 0 & -1 & 0 & 1 \\ 10 & 0 & 1 & 0 \end{bmatrix} \begin{bmatrix} -1 & 0 \\ 0 & 0 \\ -2 & 2 \\ 4 & -1 \end{bmatrix} = \begin{bmatrix} 4 & -1 \\ -12 & 2 \end{bmatrix}$

TI-83/84 Plus format: `[A]*([B]-[C])`   Online Matrix Algebra Tool format: `A*(B-C)`

35. $AB = \begin{bmatrix} 1 & -1 \\ 0 & 2 \\ 0 & -2 \end{bmatrix} \begin{bmatrix} 3 & 0 & -1 \\ 5 & -1 & 1 \end{bmatrix} = \begin{bmatrix} -2 & 1 & -2 \\ 10 & -2 & 2 \\ -10 & 2 & -2 \end{bmatrix}$

TI-83/84 Plus format: `[A]*[B]`   Online Matrix Algebra Tool format: `A*B`

37. $A(B+C) = \begin{bmatrix} 1 & -1 \\ 0 & 2 \\ 0 & -2 \end{bmatrix} \left(\begin{bmatrix} 3 & 0 & -1 \\ 5 & -1 & 1 \end{bmatrix} + \begin{bmatrix} x & 1 & w \\ z & r & 4 \end{bmatrix} \right)$

$= \begin{bmatrix} 1 & -1 \\ 0 & 2 \\ 0 & -2 \end{bmatrix} \begin{bmatrix} 3+x & 1 & -1+w \\ 5+z & -1+r & 5 \end{bmatrix} = \begin{bmatrix} -2+x-z & 2-r & -6+w \\ 10+2z & -2+2r & 10 \\ -10-2z & 2-2r & -10 \end{bmatrix}$

Online Matrix Algebra Tool format: `A*(B+C)`

39. a. $P^2 = P \cdot P = \begin{bmatrix} 0.2 & 0.8 \\ 0.2 & 0.8 \end{bmatrix} \begin{bmatrix} 0.2 & 0.8 \\ 0.2 & 0.8 \end{bmatrix} = \begin{bmatrix} (0.04+0.16) & (0.16+0.64) \\ (0.04+0.16) & (0.16+0.64) \end{bmatrix} = \begin{bmatrix} 0.2 & 0.8 \\ 0.2 & 0.8 \end{bmatrix}$

b. $P^4 = P^2 \cdot P^2 = \begin{bmatrix} 0.2 & 0.8 \\ 0.2 & 0.8 \end{bmatrix} \begin{bmatrix} 0.2 & 0.8 \\ 0.2 & 0.8 \end{bmatrix} = \begin{bmatrix} 0.2 & 0.8 \\ 0.2 & 0.8 \end{bmatrix}$ again.

c. $P^8 = P^4 \cdot P^4 = \begin{bmatrix} 0.2 & 0.8 \\ 0.2 & 0.8 \end{bmatrix} \begin{bmatrix} 0.2 & 0.8 \\ 0.2 & 0.8 \end{bmatrix} = \begin{bmatrix} 0.2 & 0.8 \\ 0.2 & 0.8 \end{bmatrix}$ again.

d. $P^{1,000} = P^8 \cdot P^8 \cdot P^8 \cdot ... \cdot P^8$ (125 times) $= \begin{bmatrix} 0.2 & 0.8 \\ 0.2 & 0.8 \end{bmatrix}$.

41. a. $P^2 = P \cdot P = \begin{bmatrix} 0.1 & 0.9 \\ 0 & 1 \end{bmatrix} \begin{bmatrix} 0.1 & 0.9 \\ 0 & 1 \end{bmatrix} = \begin{bmatrix} 0.01 & 0.99 \\ 0 & 1 \end{bmatrix}$

b. $P^4 = P^2 \cdot P^2 = \begin{bmatrix} 0.01 & 0.99 \\ 0 & 1 \end{bmatrix} \begin{bmatrix} 0.01 & 0.99 \\ 0 & 1 \end{bmatrix} = \begin{bmatrix} 0.0001 & 0.9999 \\ 0 & 1 \end{bmatrix}$

c. $P^8 = P^4 \cdot P^4 = \begin{bmatrix} 0.0001 & 0.9999 \\ 0 & 1 \end{bmatrix} \begin{bmatrix} 0.0001 & 0.9999 \\ 0 & 1 \end{bmatrix} \approx \begin{bmatrix} 0 & 1 \\ 0 & 1 \end{bmatrix}$ rounded to 4 decimal places.

d. $P^{1,000} = P^8 \cdot P^8 \cdot P^8 \cdot ... \cdot P^8$ (125 times) $\approx \begin{bmatrix} 0 & 1 \\ 0 & 1 \end{bmatrix}$

43. a. $P^2 = P \cdot P = \begin{bmatrix} 0.3 & 0.3 & 0.4 \\ 0.3 & 0.3 & 0.4 \\ 0.3 & 0.3 & 0.4 \end{bmatrix} \begin{bmatrix} 0.3 & 0.3 & 0.4 \\ 0.3 & 0.3 & 0.4 \\ 0.3 & 0.3 & 0.4 \end{bmatrix} = \begin{bmatrix} 0.3 & 0.3 & 0.4 \\ 0.3 & 0.3 & 0.4 \\ 0.3 & 0.3 & 0.4 \end{bmatrix} = P$

b. $P^4 = P^2 \cdot P^2 = P \cdot P = P$ again

c. $P^8 = P^4 \cdot P^4 = P \cdot P = P$ again

d. $P^{1,000} = P^8 \cdot P^8 \cdot P^8 \cdot \ldots \cdot P^8$ (125 times) $= P$ again

The rows of P are the same, and the entries in each row add up to 1. These calculations suggest that, if P is any square matrix with identical rows such that the entries in each row add up to 1, then $P \cdot P = P$.

45. $\begin{bmatrix} 2 & -1 & 4 \\ -4 & \frac{3}{4} & \frac{1}{3} \\ -3 & 0 & 0 \end{bmatrix} \begin{bmatrix} x \\ y \\ z \end{bmatrix} = \begin{bmatrix} 3 \\ -1 \\ 0 \end{bmatrix} \Rightarrow \begin{bmatrix} 2x - y + 4z \\ -4x + \frac{3}{4}y + \frac{1}{3}z \\ -3x \end{bmatrix} = \begin{bmatrix} 3 \\ -1 \\ 0 \end{bmatrix}$

Equating entries gives the following system of linear equations:

$2x - y + 4z = 3$

$-4x + \frac{3}{4}y + \frac{1}{3}z = -1$

$-3x = 0$

47. $\begin{bmatrix} 1 & -1 & 0 & 1 \\ 1 & 1 & 2 & 4 \end{bmatrix} \begin{bmatrix} x \\ y \\ z \\ w \end{bmatrix} = \begin{bmatrix} -1 \\ 2 \end{bmatrix} \Rightarrow \begin{bmatrix} x - y + w \\ x + y + 2z + 4w \end{bmatrix} = \begin{bmatrix} -1 \\ 2 \end{bmatrix}$

Equating entries gives the following system of linear equations:

$x - y + w = -1$

$x + y + 2z + 4w = 2$

49. $x - y = 4; \ 2x - y = 0$

The matrix form is $AX = B$, where A is the matrix of coefficients, X is the column matrix of unknowns, and B is the column matrix of right-hand sides.

$A = \begin{bmatrix} 1 & -1 \\ 2 & -1 \end{bmatrix}, X = \begin{bmatrix} x \\ y \end{bmatrix}, B = \begin{bmatrix} 4 \\ 0 \end{bmatrix}$

Thus, the matrix system is

$\begin{bmatrix} 1 & -1 \\ 2 & -1 \end{bmatrix} \begin{bmatrix} x \\ y \end{bmatrix} = \begin{bmatrix} 4 \\ 0 \end{bmatrix}$.

51. $x + y - z = 8; \ 2x + y + z = 4; \ \frac{3x}{4} + \frac{z}{2} = 1$

The matrix form is $AX = B$, where A is the matrix of coefficients, X is the column matrix of unknowns, and B is the column matrix of right-hand sides.

$A = \begin{bmatrix} 1 & 1 & -1 \\ 2 & 1 & 1 \\ \frac{3}{4} & 0 & \frac{1}{2} \end{bmatrix} X = \begin{bmatrix} x \\ y \\ z \end{bmatrix}, B = \begin{bmatrix} 8 \\ 4 \\ 1 \end{bmatrix}$

Thus, the matrix system is

$$\begin{bmatrix} 1 & 1 & -1 \\ 2 & 1 & 1 \\ \frac{3}{4} & 0 & \frac{1}{2} \end{bmatrix} \begin{bmatrix} x \\ y \\ z \end{bmatrix} = \begin{bmatrix} 8 \\ 4 \\ 1 \end{bmatrix}.$$

Applications

53. We have three prices and three quantities. Arrange the prices as a row matrix and quantities as a column matrix (so that we can multiply them):

$$\text{Price} = \begin{bmatrix} 15 & 10 & 12 \end{bmatrix} \qquad \text{Quantity} = \begin{bmatrix} 50 \\ 40 \\ 30 \end{bmatrix}$$

$$\text{Revenue} = \text{Price} \times \text{Quantity} = \begin{bmatrix} 15 & 10 & 12 \end{bmatrix} \begin{bmatrix} 50 \\ 40 \\ 30 \end{bmatrix} = \begin{bmatrix} 750 + 400 + 360 \end{bmatrix} = \begin{bmatrix} 1{,}510 \end{bmatrix}$$

55. We have three prices (stated as "costs") and three quantities. Arrange the prices as a row matrix and quantities as a column matrix (so that we can multiply them):

$$\text{Total Cost} = \text{Price} \times \text{Quantity} = \begin{bmatrix} 3.70 & 2.30 & 0.95 \end{bmatrix} \begin{bmatrix} 10 \\ 20 \\ 10 \end{bmatrix} = \begin{bmatrix} 92.5 \end{bmatrix}$$

Therefore, the total cost is $92.5 million.

57. We are asked to write the prices as a column matrix:

$$\text{Price:} \quad \begin{matrix} \text{Hard} \\ \text{Soft} \\ \text{Plastic} \end{matrix} \begin{bmatrix} 30 \\ 10 \\ 15 \end{bmatrix}$$

We are asked to find the revenue. Since the prices are given as a column matrix, we expect it to go on the right when we multiply rows by columns. So we write the formula for revenue with prices on the right:
    Revenue = Quantity × Price (Note the reversal of the usual order.)
This means that the quantities (hard, soft, plastic) should appear as rows in the quantity matrix:

$$\text{Quantity:} \begin{bmatrix} 700 & 1{,}300 & 2{,}000 \\ 400 & 300 & 500 \end{bmatrix}, \text{Price} = \begin{bmatrix} 30 \\ 10 \\ 15 \end{bmatrix}$$

$$\text{Revenue} = \text{Quantity} \times \text{Price} = \begin{bmatrix} 700 & 1{,}300 & 2{,}000 \\ 400 & 300 & 500 \end{bmatrix} \begin{bmatrix} 30 \\ 10 \\ 15 \end{bmatrix} = \begin{bmatrix} \$64{,}000 \\ \$22{,}500 \end{bmatrix}$$

59. We are given the mean income per person for each age group as well as the number of people.
Total income = Income per person × Number of females in 2020
Thus, we could write the income per person as a row matrix and the number of females in 2020 as a column matrix:
Income per person = $\begin{bmatrix} 14 & 42 & 48 & 29 \end{bmatrix}$ thousand dollars

$$\text{Number of females in 2020} = \begin{bmatrix} 23 \\ 43 \\ 43 \\ 31 \end{bmatrix} \text{million}$$

Total income = Income per person × Number of females in 2020 = $\begin{bmatrix} 14 & 42 & 48 & 29 \end{bmatrix} \begin{bmatrix} 23 \\ 43 \\ 43 \\ 31 \end{bmatrix}$

= \$5,091 thousand million, or \$5,091 billion

The total income, rounded to two significant digits, is \$5,100 billion (or \$5.1 trillion).

61. Take N to be the income per person; $N = \begin{bmatrix} 14 & 42 & 48 & 29 \end{bmatrix}$, take F and M to be, respectively, the female and male populations in 2020:

$$F = \begin{bmatrix} 23 \\ 43 \\ 43 \\ 31 \end{bmatrix} \qquad M = \begin{bmatrix} 24 \\ 43 \\ 41 \\ 20 \end{bmatrix}$$

Then the female income is NF and the male income is NM. The difference is therefore

$$D = NF - NM = N(F - M),$$

which is the single formula required. Computing,

$$N(F - M) = \begin{bmatrix} 14 & 42 & 48 & 29 \end{bmatrix} \left(\begin{bmatrix} 23 \\ 43 \\ 43 \\ 31 \end{bmatrix} - \begin{bmatrix} 24 \\ 43 \\ 41 \\ 20 \end{bmatrix} \right) = \begin{bmatrix} 14 & 42 & 48 & 29 \end{bmatrix} \begin{bmatrix} -1 \\ 0 \\ 2 \\ 11 \end{bmatrix}$$

= \$401 thousand million ≈ \$400 billion

63. $P = \begin{bmatrix} 3.2 & 14.9 \\ 26.3 & 20.1 \end{bmatrix}$

$\begin{bmatrix} -1 & 1 \end{bmatrix} P = \begin{bmatrix} -1 & 1 \end{bmatrix} \begin{bmatrix} 3.2 & 14.9 \\ 26.3 & 20.1 \end{bmatrix} = \begin{bmatrix} (-3.2 + 26.3) & (-14.9 + 20.1) \end{bmatrix} = \begin{bmatrix} 23.1 & 5.2 \end{bmatrix}$

To interpret this, look at what we are doing when we multiply the matrices: Multiplying $\begin{bmatrix} -1 & 1 \end{bmatrix}$ by the first column of P computes the per capita ice cream consumption in 1983 − per capita yogurt consumption in 1983, giving the amount, in pounds, by which per capita consumption of ice cream exceeded that of yogurt in 1983. Multiplying by the second column gives a similar calculation for 2013. Thus, $\begin{bmatrix} -1 & 1 \end{bmatrix} P$ represents the amount, in pounds, by which per capita consumption of ice cream exceeded that of yogurt in 2003 and 2013.

65. Take A to be the matrix given by the table:

$$A = \text{Total number of filings} = \begin{bmatrix} 54{,}100 & 56{,}200 & 59{,}400 \\ 23{,}800 & 22{,}400 & 23{,}600 \\ 9{,}300 & 10{,}600 & 10{,}100 \end{bmatrix}$$

The percentages for the three states handled by the firm can be represented by a row matrix:

Percentage handled by firm = $\begin{bmatrix} 0.10 & 0.05 & 0.20 \end{bmatrix}$

(We used a row so that the three percentages match the three states down each column of A.)

Number of foreclosures filings handled by firm = Percentage handled by firm × Total number

$$= \begin{bmatrix} 0.10 & 0.05 & 0.20 \end{bmatrix} \begin{bmatrix} 54{,}100 & 56{,}200 & 59{,}400 \\ 23{,}800 & 22{,}400 & 23{,}600 \\ 9{,}300 & 10{,}600 & 10{,}100 \end{bmatrix} = \begin{bmatrix} 8{,}460 & 8{,}860 & 9{,}140 \end{bmatrix}$$

67. $B = \begin{bmatrix} 1 & 1 & 0 \end{bmatrix}$, $A = \begin{bmatrix} 54{,}100 & 56{,}200 & 59{,}400 \\ 23{,}800 & 22{,}400 & 23{,}600 \\ 9{,}300 & 10{,}600 & 10{,}100 \end{bmatrix}$

When we multiply B by a column of A, we are adding the foreclosures in California and Florida for the corresponding month. Therefore, BA gives number of foreclosures in California and Florida combined in each of the months shown.

69. To compute the amount by which the combined foreclosures in California and Texas exceeded the foreclosures in Florida each month, we add the California and Texas entries and subtract the Florida entry in each column. This may be accomplished by multiplying each column by the row matrix $\begin{bmatrix} 1 & -1 & 1 \end{bmatrix}$:

$$\begin{bmatrix} 1 & -1 & 1 \end{bmatrix} \begin{bmatrix} 54{,}100 & 56{,}200 & 59{,}400 \\ 23{,}800 & 22{,}400 & 23{,}600 \\ 9{,}300 & 10{,}600 & 10{,}100 \end{bmatrix} = \begin{bmatrix} 39{,}600 & 44{,}400 & 45{,}900 \end{bmatrix}$$

This product gives the result for each of the three months. To add them up, we can multiply on the right by a column matrix whose entries are all 1:

$$\begin{bmatrix} 39{,}600 & 44{,}400 & 45{,}900 \end{bmatrix} \begin{bmatrix} 1 \\ 1 \\ 1 \end{bmatrix} = \begin{bmatrix} 129{,}900 \end{bmatrix}$$

Therefore, the matrix product we used was

$$\begin{bmatrix} 1 & -1 & 1 \end{bmatrix} \begin{bmatrix} 54{,}100 & 56{,}200 & 59{,}400 \\ 23{,}800 & 22{,}400 & 23{,}600 \\ 9{,}300 & 10{,}600 & 10{,}100 \end{bmatrix} \begin{bmatrix} 1 \\ 1 \\ 1 \end{bmatrix} = \begin{bmatrix} 129{,}900 \end{bmatrix}.$$

71. We are asked to organize the parts required data in a matrix, and the prices per part from each supplier in another.

Quantity of parts:

	Proc	Mem	Tubes	
Pom II	2	16	20	$= Q$
Pom Classic	1	4	40	

We multiply this by the prices per part matrix to obtain the total costs. Thus, we should write the price per parts matrix with the prices for each component as columns (to match the rows above):

Prices:

	Motorel	Intola	
Proc	100	150	
Mem	50	40	$= P$
Tubes	10	15	

$$\text{Total cost} = \text{Quantity} \times \text{Price} = QP = \begin{bmatrix} 2 & 16 & 20 \\ 1 & 4 & 40 \end{bmatrix} \begin{bmatrix} 100 & 150 \\ 50 & 40 \\ 10 & 15 \end{bmatrix} = \begin{bmatrix} 1{,}200 & 1{,}240 \\ 700 & 910 \end{bmatrix}$$

How are these costs organized? The clue is that the rows of the product QP correspond to the rows of the left-hand matrix, Q. Here Q has rows corresponding to Pom II and Classic. On the other hand, the *columns* of the product QP correspond to the columns of P, which are Motorel and Intola. Thus, the prices are organized as follows:

	Motorel	Intola
Pom II	$1,200	$1,240
Pom Classic	$700	$910

Solution: Motorel parts on the Pom II: $1,200; Intola parts on the Pom II: $1,240; Motorel parts on the Pom Classic: $700; Intola parts on the Pom Classic: $910

73. $A = \begin{bmatrix} 440 & 190 \\ 950 & 950 \\ 1{,}790 & 200 \end{bmatrix}$ $AB = \begin{bmatrix} 440 & 190 \\ 950 & 950 \\ 1{,}790 & 200 \end{bmatrix} \begin{bmatrix} 0.05 \\ 0.04 \end{bmatrix} = \begin{bmatrix} 29.6 \\ 85.5 \\ 97.5 \end{bmatrix}$

In computing this product, we are adding 5% of the number of people visiting Australia and 4% of the number visiting South Africa for each of the three region. These are exactly the percentages that decide to settle there. Thus, the entries of AB give the number of people from each of the three regions who settle in Australia or South Africa.

$$AC = \begin{bmatrix} 440 & 190 \\ 950 & 950 \\ 1{,}790 & 200 \end{bmatrix} \begin{bmatrix} 0.05 & 0 \\ 0 & 0.04 \end{bmatrix} = \begin{bmatrix} 22 & 7.6 \\ 47.5 & 38 \\ 89.5 & 8 \end{bmatrix}$$

Each row of the product is computed by taking 5% of all the tourists to Australia and 4% of those to South Africa separately. Thus, the entries of AC give the number of people from each of the three regions who settle in Australia, and the number that settle in South Africa.

75. Take P to be the matrix represented by the table:

$$P = \begin{bmatrix} 0.9923 & 0.0016 & 0.0042 & 0.0019 \\ 0.0018 & 0.9896 & 0.0047 & 0.0039 \\ 0.0056 & 0.0059 & 0.9827 & 0.0058 \\ 0.0024 & 0.0033 & 0.0044 & 0.9899 \end{bmatrix}$$

2008 Distribution: $\quad A = \begin{bmatrix} 54.1 & 65.7 & 112.9 & 70.3 \end{bmatrix}$

To compute the population distribution in 2009, compute $AP \approx \begin{bmatrix} 54.6 & 66.0 & 111.8 & 70.6 \end{bmatrix}$.

Communication and reasoning exercises

77. Answers will vary.

One example: $A = \begin{bmatrix} 1 & 2 \end{bmatrix}, B = \begin{bmatrix} 1 & 2 & 3 \\ 4 & 5 & 6 \end{bmatrix}$ $\quad$ Another example: $A = \begin{bmatrix} 1 \end{bmatrix}, B = \begin{bmatrix} 1 & 2 \end{bmatrix}$

79. We find that the addition and multiplication of 1×1 matrices is identical to the addition and multiplication of numbers.

81. The claim is correct. Every matrix equation represents the equality of two matrices. When two matrices are equal, each of their corresponding entries must be equal. Equating the corresponding entries gives a system of equations.

83. Here is a possible scenario: costs of items A, B, and C in 2013 $= \begin{bmatrix} 10 & 20 & 30 \end{bmatrix}$, Percentage increases in these costs in 2014 $= \begin{bmatrix} 0.5 & 0.1 & 0.20 \end{bmatrix}$, actual increases in costs $= \begin{bmatrix} 10 \times 0.5 & 20 \times 0.1 & 30 \times 0.20 \end{bmatrix}$.

85. It produces a matrix whose ij entry is the product of the ij entries of the two matrices.

Section 4.3

1. $A = \begin{bmatrix} 0 & 1 \\ 1 & 0 \end{bmatrix}$, $B = \begin{bmatrix} 0 & 1 \\ 1 & 0 \end{bmatrix} \Rightarrow AB = \begin{bmatrix} 0 & 1 \\ 1 & 0 \end{bmatrix}\begin{bmatrix} 0 & 1 \\ 1 & 0 \end{bmatrix} = \begin{bmatrix} 1 & 0 \\ 0 & 1 \end{bmatrix} = I$

Because A and B are square matrices with $AB = I$, A and B are inverses. (We do not have to check that $BA = I$ as well—see the note in the text after the definition of "inverse.")

3. $AB = \begin{bmatrix} 2 & 1 & 1 \\ 0 & 1 & 1 \\ 0 & 0 & 1 \end{bmatrix}\begin{bmatrix} \frac{1}{2} & -\frac{1}{2} & 0 \\ 0 & 1 & -1 \\ 0 & 0 & 1 \end{bmatrix} = \begin{bmatrix} 1 & 0 & 0 \\ 0 & 1 & 0 \\ 0 & 0 & 1 \end{bmatrix} = I$

Because A and B are square matrices with $AB = I$, A and B are inverses. (We do not have to check that $BA = I$ as well—see the note in the text after the definition of "inverse.")

5. $AB = \begin{bmatrix} a & 0 & 0 \\ 0 & b & 0 \\ 0 & 0 & 0 \end{bmatrix}\begin{bmatrix} a^{-1} & 0 & 0 \\ 0 & b^{-1} & 0 \\ 0 & 0 & 0 \end{bmatrix} = \begin{bmatrix} 1 & 0 & 0 \\ 0 & 1 & 0 \\ 0 & 0 & 0 \end{bmatrix} \neq I$

Because $AB \neq I$, A and B are not inverses.

7. To find the inverse of $\begin{bmatrix} 1 & 1 \\ 2 & 1 \end{bmatrix}$, we augment with the 2×2 identity matrix and row-reduce:

$\begin{bmatrix} \boxed{1} & 1 & 1 & 0 \\ 2 & 1 & 0 & 1 \end{bmatrix} \begin{matrix} \\ R_2 - 2R_1 \end{matrix} \rightarrow \begin{bmatrix} 1 & 1 & 1 & 0 \\ 0 & \boxed{-1} & -2 & 1 \end{bmatrix} \begin{matrix} R_1 + R_2 \\ \end{matrix} \rightarrow \begin{bmatrix} 1 & 0 & -1 & 1 \\ 0 & -1 & -2 & 1 \end{bmatrix} \begin{matrix} \\ -R_2 \end{matrix} \rightarrow$

$\begin{bmatrix} 1 & 0 & -1 & 1 \\ 0 & 1 & 2 & -1 \end{bmatrix}$.

The right-hand 2×2 block is the desired inverse: $\begin{bmatrix} 1 & 1 \\ 2 & 1 \end{bmatrix}^{-1} = \begin{bmatrix} -1 & 1 \\ 2 & -1 \end{bmatrix}$.

9. To find the inverse of $\begin{bmatrix} 0 & 1 \\ 1 & 0 \end{bmatrix}$, we augment with the 2×2 identity matrix and row-reduce:

$\begin{bmatrix} 0 & 1 & 1 & 0 \\ 1 & 0 & 0 & 1 \end{bmatrix} R_1 \leftrightarrow R_2 \rightarrow \begin{bmatrix} 1 & 0 & 0 & 1 \\ 0 & 1 & 1 & 0 \end{bmatrix}$.

The right-hand 2×2 block is the desired inverse: $\begin{bmatrix} 0 & 1 \\ 1 & 0 \end{bmatrix}^{-1} = \begin{bmatrix} 0 & 1 \\ 1 & 0 \end{bmatrix}$.

11. $\begin{bmatrix} \boxed{2} & 1 & 1 & 0 \\ 1 & 1 & 0 & 1 \end{bmatrix} \begin{matrix} \\ 2R_2 - R_1 \end{matrix} \rightarrow \begin{bmatrix} 2 & 1 & 1 & 0 \\ 0 & \boxed{1} & -1 & 2 \end{bmatrix} \begin{matrix} R_1 - R_2 \\ \end{matrix} \rightarrow \begin{bmatrix} 2 & 0 & 2 & -2 \\ 0 & 1 & -1 & 2 \end{bmatrix} \begin{matrix} (1/2)R_1 \\ \end{matrix} \rightarrow$

$\begin{bmatrix} 1 & 0 & 1 & -1 \\ 0 & 1 & -1 & 2 \end{bmatrix}$.

The right-hand 2×2 block is the desired inverse: $\begin{bmatrix} 2 & 1 \\ 1 & 1 \end{bmatrix}^{-1} = \begin{bmatrix} 1 & -1 \\ -1 & 2 \end{bmatrix}$.

13. $\begin{bmatrix} \boxed{2} & 1 & 1 & 0 \\ 4 & 2 & 0 & 1 \end{bmatrix} \begin{matrix} \\ R_2 - 2R_1 \end{matrix} \rightarrow \begin{bmatrix} 2 & 1 & 1 & 0 \\ 0 & 0 & -2 & 1 \end{bmatrix}$.

Since the left-hand 2×2 block has a row of zeros, we cannot reduce the matrix to obtain the 2×2 identity on the left. Therefore, the matrix is singular.

15. To find the inverse of a 3×3 matrix, we augment it with the 3×3 identity matrix and row-reduce:

$$\begin{bmatrix} 1 & 1 & 1 & 1 & 0 & 0 \\ 0 & \boxed{1} & 1 & 0 & 1 & 0 \\ 0 & 0 & 1 & 0 & 0 & 1 \end{bmatrix} \begin{matrix} R_1 - R_2 \\ \\ \end{matrix} \rightarrow \begin{bmatrix} 1 & 0 & 0 & 1 & -1 & 0 \\ 0 & 1 & 1 & 0 & 1 & 0 \\ 0 & 0 & \boxed{1} & 0 & 0 & 1 \end{bmatrix} \begin{matrix} \\ R_2 - R_3 \\ \\ \end{matrix} \rightarrow$$

$$\begin{bmatrix} 1 & 0 & 0 & 1 & -1 & 0 \\ 0 & 1 & 0 & 0 & 1 & -1 \\ 0 & 0 & 1 & 0 & 0 & 1 \end{bmatrix}.$$

The right-hand 3×3 block is the desired inverse: $\begin{bmatrix} 1 & 1 & 1 \\ 0 & 1 & 1 \\ 0 & 0 & 1 \end{bmatrix}^{-1} = \begin{bmatrix} 1 & -1 & 0 \\ 0 & 1 & -1 \\ 0 & 0 & 1 \end{bmatrix}.$

17. To find the inverse of a 3×3 matrix, we augment it with the 3×3 identity matrix and row-reduce:

$$\begin{bmatrix} \boxed{1} & 1 & 1 & 1 & 0 & 0 \\ 1 & 0 & 2 & 0 & 1 & 0 \\ 1 & -1 & 1 & 0 & 0 & 1 \end{bmatrix} \begin{matrix} \\ R_2 - R_1 \\ R_3 - R_1 \end{matrix} \rightarrow \begin{bmatrix} 1 & 1 & 1 & 1 & 0 & 0 \\ 0 & \boxed{-1} & 1 & -1 & 1 & 0 \\ 0 & -2 & 0 & -1 & 0 & 1 \end{bmatrix} \begin{matrix} R_1 + R_2 \\ \\ R_3 - 2R_2 \end{matrix} \rightarrow$$

$$\begin{bmatrix} 1 & 0 & 2 & 0 & 1 & 0 \\ 0 & -1 & 1 & -1 & 1 & 0 \\ 0 & 0 & \boxed{-2} & 1 & -2 & 1 \end{bmatrix} \begin{matrix} R_1 + R_3 \\ 2R_2 + R_3 \\ \end{matrix} \rightarrow \begin{bmatrix} 1 & 0 & 0 & 1 & -1 & 1 \\ 0 & -2 & 0 & -1 & 0 & 1 \\ 0 & 0 & -2 & 1 & -2 & 1 \end{bmatrix} \begin{matrix} \\ -(1/2)R_2 \\ -(1/2)R_3 \end{matrix} \rightarrow$$

$$\begin{bmatrix} 1 & 0 & 0 & 1 & -1 & 1 \\ 0 & 1 & 0 & 1/2 & 0 & -1/2 \\ 0 & 0 & 1 & -1/2 & 1 & -1/2 \end{bmatrix}.$$

The right-hand 3×3 block is the desired inverse: $\begin{bmatrix} 1 & 1 & 1 \\ 1 & 0 & 2 \\ 1 & -1 & 1 \end{bmatrix}^{-1} = \begin{bmatrix} 1 & -1 & 1 \\ 1/2 & 0 & -1/2 \\ -1/2 & 1 & -1/2 \end{bmatrix}.$

19.
$$\begin{bmatrix} \boxed{1} & 1 & 1 & 1 & 0 & 0 \\ 1 & -1 & 0 & 0 & 1 & 0 \\ 1 & 2 & 3 & 0 & 0 & 1 \end{bmatrix} \begin{matrix} \\ R_2 - R_1 \\ R_3 - R_1 \end{matrix} \rightarrow \begin{bmatrix} 1 & 1 & 1 & 1 & 0 & 0 \\ 0 & \boxed{-2} & -1 & -1 & 1 & 0 \\ 0 & 1 & 2 & -1 & 0 & 1 \end{bmatrix} \begin{matrix} 2R_1 + R_2 \\ \\ 2R_3 + R_2 \end{matrix} \rightarrow$$

$$\begin{bmatrix} 2 & 0 & 1 & 1 & 1 & 0 \\ 0 & -2 & -1 & -1 & 1 & 0 \\ 0 & 0 & \boxed{3} & -3 & 1 & 2 \end{bmatrix} \begin{matrix} 3R_1 - R_3 \\ 3R_2 + R_3 \\ \end{matrix} \rightarrow \begin{bmatrix} 6 & 0 & 0 & 6 & 2 & -2 \\ 0 & -6 & 0 & -6 & 4 & 2 \\ 0 & 0 & 3 & -3 & 1 & 2 \end{bmatrix} \begin{matrix} (1/2)R_1 \\ (1/2)R_2 \\ \end{matrix} \rightarrow$$

$$\begin{bmatrix} 3 & 0 & 0 & 3 & 1 & -1 \\ 0 & -3 & 0 & -3 & 2 & 1 \\ 0 & 0 & 3 & -3 & 1 & 2 \end{bmatrix} \begin{matrix} (1/3)R_1 \\ -(1/3)R_2 \\ (1/3)R_3 \end{matrix} \rightarrow \begin{bmatrix} 1 & 0 & 0 & 1 & 1/3 & -1/3 \\ 0 & 1 & 0 & 1 & -2/3 & -1/3 \\ 0 & 0 & 1 & -1 & 1/3 & 2/3 \end{bmatrix}.$$

The right-hand 3×3 block is the desired inverse: $\begin{bmatrix} 1 & 1 & 1 \\ 1 & -1 & 0 \\ 1 & 2 & 3 \end{bmatrix}^{-1} = \begin{bmatrix} 1 & 1/3 & -1/3 \\ 1 & -2/3 & -1/3 \\ -1 & 1/3 & 2/3 \end{bmatrix}.$

21.
$$\begin{bmatrix} \boxed{1} & 1 & 1 & 1 & 0 & 0 \\ 1 & 0 & 1 & 0 & 1 & 0 \\ 1 & -1 & 1 & 0 & 0 & 1 \end{bmatrix} \begin{matrix} \\ R_2 - R_1 \\ R_3 - R_1 \end{matrix} \rightarrow \begin{bmatrix} 1 & 1 & 1 & 1 & 0 & 0 \\ 0 & \boxed{-1} & 0 & -1 & 1 & 0 \\ 0 & -2 & 0 & -1 & 0 & 1 \end{bmatrix} \begin{matrix} R_1 + R_2 \\ \\ R_3 - 2R_2 \end{matrix} \rightarrow$$

$$\begin{bmatrix} 1 & 0 & 1 & 0 & 1 & 0 \\ 0 & -1 & 0 & -1 & 1 & 0 \\ 0 & 0 & 0 & 1 & -2 & 1 \end{bmatrix}.$$

Since the left-hand 3×3 block has a row of zeros, we cannot reduce the matrix to obtain the 3×3 identity on the left. Therefore, the matrix is singular.

23. To find the inverse of a 4×4 matrix, we augment it with the 4×4 identity matrix and row-reduce:

$$\left[\begin{array}{cccc|cccc} \boxed{1} & 0 & 1 & 0 & 1 & 0 & 0 & 0 \\ -1 & 1 & 0 & 1 & 0 & 1 & 0 & 0 \\ -1 & 0 & 0 & 1 & 0 & 0 & 1 & 0 \\ 0 & -1 & 0 & 1 & 0 & 0 & 0 & 1 \end{array}\right] \begin{array}{l} \\ R_2 + R_1 \\ R_3 + R_1 \\ \end{array} \rightarrow \left[\begin{array}{cccc|cccc} 1 & 0 & 1 & 0 & 1 & 0 & 0 & 0 \\ 0 & \boxed{1} & 1 & 1 & 1 & 1 & 0 & 0 \\ 0 & 0 & 1 & 1 & 1 & 0 & 1 & 0 \\ 0 & -1 & 0 & 1 & 0 & 0 & 0 & 1 \end{array}\right] \begin{array}{l} \\ \\ \\ R_4 + R_2 \end{array} \rightarrow$$

$$\left[\begin{array}{cccc|cccc} 1 & 0 & 1 & 0 & 1 & 0 & 0 & 0 \\ 0 & 1 & 1 & 1 & 1 & 1 & 0 & 0 \\ 0 & 0 & \boxed{1} & 1 & 1 & 0 & 1 & 0 \\ 0 & 0 & 1 & 2 & 1 & 1 & 0 & 1 \end{array}\right] \begin{array}{l} R_1 - R_3 \\ R_2 - R_3 \\ \\ R_4 - R_3 \end{array} \rightarrow \left[\begin{array}{cccc|cccc} 1 & 0 & 0 & -1 & 0 & 0 & -1 & 0 \\ 0 & 1 & 0 & 0 & 0 & 1 & -1 & 0 \\ 0 & 0 & 1 & 1 & 1 & 0 & 1 & 0 \\ 0 & 0 & 0 & \boxed{1} & 0 & 1 & -1 & 1 \end{array}\right] \begin{array}{l} R_1 + R_4 \\ \\ R_3 - R_4 \\ \end{array} \rightarrow$$

$$\left[\begin{array}{cccc|cccc} 1 & 0 & 0 & 0 & 0 & 1 & -2 & 1 \\ 0 & 1 & 0 & 0 & 0 & 1 & -1 & 0 \\ 0 & 0 & 1 & 0 & 1 & -1 & 2 & -1 \\ 0 & 0 & 0 & 1 & 0 & 1 & -1 & 1 \end{array}\right].$$

The right-hand 4×4 block is the desired inverse:

$$\left[\begin{array}{cccc} 1 & 0 & 1 & 0 \\ -1 & 1 & 0 & 1 \\ -1 & 0 & 0 & 1 \\ 0 & -1 & 0 & 1 \end{array}\right]^{-1} = \left[\begin{array}{cccc} 0 & 1 & -2 & 1 \\ 0 & 1 & -1 & 0 \\ 1 & -1 & 2 & -1 \\ 0 & 1 & -1 & 1 \end{array}\right].$$

25. To find the inverse of a 4×4 matrix, we augment it with the 4×4 identity matrix and row-reduce:

$$\left[\begin{array}{cccc|cccc} 1 & 2 & 3 & 4 & 1 & 0 & 0 & 0 \\ 0 & \boxed{1} & 2 & 3 & 0 & 1 & 0 & 0 \\ 0 & 0 & 1 & 2 & 0 & 0 & 1 & 0 \\ 0 & 0 & 0 & 1 & 0 & 0 & 0 & 1 \end{array}\right] \begin{array}{l} R_1 - 2R_2 \\ \\ \\ \end{array} \rightarrow \left[\begin{array}{cccc|cccc} 1 & 0 & -1 & -2 & 1 & -2 & 0 & 0 \\ 0 & 1 & 2 & 3 & 0 & 1 & 0 & 0 \\ 0 & 0 & \boxed{1} & 2 & 0 & 0 & 1 & 0 \\ 0 & 0 & 0 & 1 & 0 & 0 & 0 & 1 \end{array}\right] \begin{array}{l} R_1 + R_3 \\ R_2 - 2R_3 \\ \\ \end{array} \rightarrow$$

$$\left[\begin{array}{cccc|cccc} 1 & 0 & 0 & 0 & 1 & -2 & 1 & 0 \\ 0 & 1 & 0 & -1 & 0 & 1 & -2 & 0 \\ 0 & 0 & 1 & 2 & 0 & 0 & 1 & 0 \\ 0 & 0 & 0 & \boxed{1} & 0 & 0 & 0 & 1 \end{array}\right] \begin{array}{l} \\ R_2 + R_4 \\ R_3 - 2R_4 \\ \end{array} \rightarrow \left[\begin{array}{cccc|cccc} 1 & 0 & 0 & 0 & 1 & -2 & 1 & 0 \\ 0 & 1 & 0 & 0 & 0 & 1 & -2 & 1 \\ 0 & 0 & 1 & 0 & 0 & 0 & 1 & -2 \\ 0 & 0 & 0 & 1 & 0 & 0 & 0 & 1 \end{array}\right].$$

The right-hand 4×4 block is the desired inverse:

$$\left[\begin{array}{cccc} 1 & 2 & 3 & 4 \\ 0 & 1 & 2 & 3 \\ 0 & 0 & 1 & 2 \\ 0 & 0 & 0 & 1 \end{array}\right]^{-1} = \left[\begin{array}{cccc} 1 & -2 & 1 & 0 \\ 0 & 1 & -2 & 1 \\ 0 & 0 & 1 & -2 \\ 0 & 0 & 0 & 1 \end{array}\right].$$

27. $\left[\begin{array}{cc} a & b \\ c & d \end{array}\right] = \left[\begin{array}{cc} 1 & 1 \\ 1 & -1 \end{array}\right]$ $\det\left[\begin{array}{cc} 1 & 1 \\ 1 & -1 \end{array}\right] = ad - bc = (1)(-1) - (1)(1) = -2$

$\left[\begin{array}{cc} 1 & 1 \\ 1 & -1 \end{array}\right]^{-1} = \dfrac{1}{ad - bc}\left[\begin{array}{cc} d & -b \\ -c & a \end{array}\right] = \dfrac{1}{-2}\left[\begin{array}{cc} -1 & -1 \\ -1 & 1 \end{array}\right] = \left[\begin{array}{cc} 1/2 & 1/2 \\ 1/2 & -1/2 \end{array}\right]$

29. $\left[\begin{array}{cc} a & b \\ c & d \end{array}\right] = \left[\begin{array}{cc} 1 & 2 \\ 3 & 4 \end{array}\right]$ $\det\left[\begin{array}{cc} 1 & 2 \\ 3 & 4 \end{array}\right] = ad - bc = (1)(4) - (2)(3) = -2$

$$\begin{bmatrix} 1 & 2 \\ 3 & 4 \end{bmatrix}^{-1} = \frac{1}{ad-bc}\begin{bmatrix} d & -b \\ -c & a \end{bmatrix} = \frac{1}{-2}\begin{bmatrix} 4 & -2 \\ -3 & 1 \end{bmatrix} = \begin{bmatrix} -2 & 1 \\ 3/2 & -1/2 \end{bmatrix}$$

31. $\begin{bmatrix} a & b \\ c & d \end{bmatrix} = \begin{bmatrix} 1 & 0 \\ 0 & 1 \end{bmatrix}$ $\det\begin{bmatrix} 1/6 & -1/6 \\ 0 & 1/6 \end{bmatrix} = ad - bc = (1/6)(1/6) - (-1/6)(0) = 1/36$

$$\begin{bmatrix} 1/6 & -1/6 \\ 0 & 1/6 \end{bmatrix}^{-1} = \frac{1}{ad-bc}\begin{bmatrix} d & -b \\ -c & a \end{bmatrix} = 36\begin{bmatrix} 1/6 & 1/6 \\ 0 & 1/6 \end{bmatrix} = \begin{bmatrix} 6 & 6 \\ 0 & 6 \end{bmatrix}$$

33. $\begin{bmatrix} a & b \\ c & d \end{bmatrix} = \begin{bmatrix} 1 & 0 \\ 3/4 & 0 \end{bmatrix}$ $\det\begin{bmatrix} 1 & 0 \\ 3/4 & 0 \end{bmatrix} = ad - bc = (1)(0) - (0)(3/4) = 0$

Singular matrix

35. $\begin{bmatrix} 1.1 & 1.2 \\ 1.3 & -1 \end{bmatrix}^{-1} = \begin{bmatrix} 0.38 & 0.45 \\ 0.49 & -0.41 \end{bmatrix}$

TI-83/84 Plus format: $[A]^{-1}$ Spreadsheet: =MINVERSE(A1:B2) (Assuming the matrix is in cells A1–B2)
Online Matrix Algebra Tool format: A^(-1)   Student Website → Online Utilities → Matrix Algebra Tool

37. $\begin{bmatrix} 3.56 & 1.23 \\ -1.01 & 0 \end{bmatrix}^{-1} = \begin{bmatrix} 0.00 & -0.99 \\ 0.81 & 2.87 \end{bmatrix}$

TI-83/84 Plus format: $[A]^{-1}$ Spreadsheet: =MINVERSE(A1:B2) (Assuming the matrix is in cells A1–B2)
Online Matrix Algebra Tool format: A^(-1)   Student Website → Online Utilities → Matrix Algebra Tool

39. $\begin{bmatrix} 1.1 & 3.1 & 2.4 \\ 1.7 & 2.4 & 2.3 \\ 0.6 & -0.7 & -0.1 \end{bmatrix}$ is singular.

TI-83/84 Plus format: $[A]^{-1}$ Spreadsheet: =MINVERSE(A1:C3) (Assuming the matrix is in cells A1–C3)
Online Matrix Algebra Tool format: A^(-1)   Student Website → Online Utilities → Matrix Algebra Tool

41. $\begin{bmatrix} 0.01 & 0.32 & 0 & 0.04 \\ -0.01 & 0 & 0 & 0.34 \\ 0 & 0.32 & -0.23 & 0.23 \\ 0 & 0.41 & 0 & 0.01 \end{bmatrix}^{-1} = \begin{bmatrix} 91.35 & -8.65 & 0 & -71.30 \\ -0.07 & -0.07 & 0 & 2.49 \\ 2.60 & 2.60 & -4.35 & 1.37 \\ 2.69 & 2.69 & 0 & -2.10 \end{bmatrix}$

TI-83/84 Plus format: $[A]^{-1}$ Spreadsheet: =MINVERSE(A1:D4) (Assuming the matrix is in cells A1–D4)
Online Matrix Algebra Tool format: A^(-1)   Student Website → Online Utilities → Matrix Algebra Tool

43. $x + y = 4;$ $x - y = 1$

Matrix form: $\begin{bmatrix} 1 & 1 \\ 1 & -1 \end{bmatrix}\begin{bmatrix} x \\ y \end{bmatrix} = \begin{bmatrix} 4 \\ 1 \end{bmatrix}$

$AX = B \Rightarrow X = A^{-1}B$

$A^{-1} = \begin{bmatrix} 1 & 1 \\ 1 & -1 \end{bmatrix}^{-1} = \begin{bmatrix} 1/2 & 1/2 \\ 1/2 & -1/2 \end{bmatrix}$ See Exercise 27

Thus, $X = A^{-1}B = \begin{bmatrix} 1/2 & 1/2 \\ 1/2 & -1/2 \end{bmatrix}\begin{bmatrix} 4 \\ 1 \end{bmatrix} = \begin{bmatrix} 5/2 \\ 3/2 \end{bmatrix}.$

So, $(x, y) = (5/2, 3/2)$.

45. $\frac{x}{3} + \frac{y}{2} = 0;$ $\frac{x}{2} + y = -1$

Matrix form: $\begin{bmatrix} 1/3 & 1/2 \\ 1/2 & 1 \end{bmatrix}\begin{bmatrix} x \\ y \end{bmatrix} = \begin{bmatrix} 0 \\ -1 \end{bmatrix}$

$AX = B \Rightarrow X = A^{-1}B$

$A^{-1} = \begin{bmatrix} 1/3 & 1/2 \\ 1/2 & 1 \end{bmatrix}^{-1} = \begin{bmatrix} 12 & -6 \\ -6 & 4 \end{bmatrix}$

Thus, $X = A^{-1}B = \begin{bmatrix} 12 & -6 \\ -6 & 4 \end{bmatrix}\begin{bmatrix} 0 \\ -1 \end{bmatrix} = \begin{bmatrix} 6 \\ -4 \end{bmatrix}$.

So, $(x, y) = (6, -4)$.

47. $-x + 2y - z = 0$; $\quad -x - y + 2z = 0$; $\quad 2x - z = 6$

Matrix form: $\begin{bmatrix} -1 & 2 & -1 \\ -1 & -1 & 2 \\ 2 & 0 & -1 \end{bmatrix}\begin{bmatrix} x \\ y \\ z \end{bmatrix} = \begin{bmatrix} 0 \\ 0 \\ 6 \end{bmatrix}$

$AX = B \Rightarrow X = A^{-1}B$

$A^{-1} = \begin{bmatrix} -1 & 2 & -1 \\ -1 & -1 & 2 \\ 2 & 0 & -1 \end{bmatrix}^{-1} = \begin{bmatrix} 1/3 & 2/3 & 1 \\ 1 & 1 & 1 \\ 2/3 & 4/3 & 1 \end{bmatrix}$

Thus, $X = A^{-1}B = \begin{bmatrix} 1/3 & 2/3 & 1 \\ 1 & 1 & 1 \\ 2/3 & 4/3 & 1 \end{bmatrix}\begin{bmatrix} 0 \\ 0 \\ 6 \end{bmatrix} = \begin{bmatrix} 6 \\ 6 \\ 6 \end{bmatrix}$.

So, $(x, y, z) = (6, 6, 6)$.

49. The three systems of equations have the matrix form $AX = B$, where

$A = \begin{bmatrix} -1 & -4 & 2 \\ 1 & 2 & -1 \\ 1 & 1 & -1 \end{bmatrix}$; $A^{-1} = \begin{bmatrix} 1 & 2 & 0 \\ 0 & 1 & -1 \\ 1 & 3 & -2 \end{bmatrix}$.

a. $B = \begin{bmatrix} 4 \\ 3 \\ 8 \end{bmatrix}$ $\qquad X = A^{-1}B = \begin{bmatrix} 1 & 2 & 0 \\ 0 & 1 & -1 \\ 1 & 3 & -2 \end{bmatrix}\begin{bmatrix} 4 \\ 3 \\ 8 \end{bmatrix} = \begin{bmatrix} 10 \\ -5 \\ -3 \end{bmatrix}$

$(x, y, z) = (10, -5, -3)$

b. $B = \begin{bmatrix} 0 \\ 3 \\ 2 \end{bmatrix}$ $\qquad X = A^{-1}B = \begin{bmatrix} 1 & 2 & 0 \\ 0 & 1 & -1 \\ 1 & 3 & -2 \end{bmatrix}\begin{bmatrix} 0 \\ 3 \\ 2 \end{bmatrix} = \begin{bmatrix} 6 \\ 1 \\ 5 \end{bmatrix}$

$(x, y, z) = (6, 1, 5)$

c. $B = \begin{bmatrix} 0 \\ 0 \\ 0 \end{bmatrix}$ $\qquad X = A^{-1}B = \begin{bmatrix} 1 & 2 & 0 \\ 0 & 1 & -1 \\ 1 & 3 & -2 \end{bmatrix}\begin{bmatrix} 0 \\ 0 \\ 0 \end{bmatrix} = \begin{bmatrix} 0 \\ 0 \\ 0 \end{bmatrix}$

$(x, y, z) = (0, 0, 0)$

Applications

51. Unknowns: $x =$ the number of servings of Pork & Beans; $y =$ the number of slices of bread

a. Arrange the given information in a table with unknowns across the top:

	Pork & Beans (x)	Bread (y)	Desired
Protein	5	4	20
Carbs.	21	12	80

We can now set up an equation for each of the items listed on the left:

Protein: $5x + 4y = 20$

Carbs: $21x + 12y = 80$.

We put this system in matrix form:

$$\begin{bmatrix} 5 & 4 \\ 21 & 12 \end{bmatrix}\begin{bmatrix} x \\ y \end{bmatrix} = \begin{bmatrix} 20 \\ 80 \end{bmatrix}$$

$$AX = B \Rightarrow X = A^{-1}B$$

$$A^{-1} = \begin{bmatrix} 5 & 4 \\ 21 & 12 \end{bmatrix} = \begin{bmatrix} -1/2 & 1/6 \\ 7/8 & -5/24 \end{bmatrix}$$

Thus,

$$X = A^{-1}B = \begin{bmatrix} -1/2 & 1/6 \\ 7/8 & -5/24 \end{bmatrix}\begin{bmatrix} 20 \\ 80 \end{bmatrix} = \begin{bmatrix} 10/3 \\ 5/6 \end{bmatrix}$$

So, $(x, y) = (10/3, 5/6)$

Solution: Prepare 10/3 servings of beans, and 5/6 slices of bread.

b. We must solve the following system:

$5x + 4y = A$; $21x + 12y = B$

We put this system in matrix form:

$$\begin{bmatrix} 5 & 4 \\ 21 & 12 \end{bmatrix}\begin{bmatrix} x \\ y \end{bmatrix} = \begin{bmatrix} A \\ B \end{bmatrix}$$

Solving gives

$$\begin{bmatrix} x \\ y \end{bmatrix} = \begin{bmatrix} 5 & 4 \\ 21 & 12 \end{bmatrix}\begin{bmatrix} A \\ B \end{bmatrix} = \begin{bmatrix} -1/2 & 1/6 \\ 7/8 & -5/24 \end{bmatrix}\begin{bmatrix} A \\ B \end{bmatrix} = \begin{bmatrix} -A/2 + B/6 \\ 7A/8 - 5B/24 \end{bmatrix}.$$

Solution: Prepare $-A/2 + B/6$ servings of beans and $7A/8 - 5B/24$ slices of bread.

53. Unknowns: $x = $ the number of batches of vanilla; $y = $ the number of batches of mocha; $z = $ the number of batches of strawberry

Arrange the given information in a table with unknowns across the top:

	Vanilla (x)	Mocha (y)	Strawberry (z)
Eggs	2	1	1
Milk	1	1	2
Cream	2	2	1

a.   Eggs: $2x + y + z = 350$

Milk: $x + y + 2z = 350$

Cream: $2x + 2y + z = 400$

Matrix form:

$$\begin{bmatrix} 2 & 1 & 1 \\ 1 & 1 & 2 \\ 2 & 2 & 1 \end{bmatrix}\begin{bmatrix} x \\ y \\ z \end{bmatrix} = \begin{bmatrix} 350 \\ 350 \\ 400 \end{bmatrix}$$

$$AX = B \Rightarrow X = A^{-1}B$$

$$X = \begin{bmatrix} 2 & 1 & 1 \\ 1 & 1 & 2 \\ 2 & 2 & 1 \end{bmatrix}^{-1} \begin{bmatrix} 350 \\ 350 \\ 400 \end{bmatrix} = \begin{bmatrix} 1 & -1/3 & -1/3 \\ -1 & 0 & 1 \\ 0 & 2/3 & -1/3 \end{bmatrix} \begin{bmatrix} 350 \\ 350 \\ 400 \end{bmatrix} = \begin{bmatrix} 100 \\ 50 \\ 100 \end{bmatrix}$$

Solution: Use 100 batches of vanilla, 50 batches of mocha, and 100 batches of strawberry.

b. The requirements lead to the following matrix equation:

$$\begin{bmatrix} 2 & 1 & 1 \\ 1 & 1 & 2 \\ 2 & 2 & 1 \end{bmatrix} \begin{bmatrix} x \\ y \\ z \end{bmatrix} = \begin{bmatrix} 400 \\ 500 \\ 400 \end{bmatrix}$$

$$AX = B \Rightarrow X = A^{-1}B$$

$$X = \begin{bmatrix} 2 & 1 & 1 \\ 1 & 1 & 2 \\ 2 & 2 & 1 \end{bmatrix}^{-1} \begin{bmatrix} 400 \\ 500 \\ 400 \end{bmatrix} = \begin{bmatrix} 1 & -1/3 & -1/3 \\ -1 & 0 & 1 \\ 0 & 2/3 & -1/3 \end{bmatrix} \begin{bmatrix} 400 \\ 500 \\ 400 \end{bmatrix} = \begin{bmatrix} 100 \\ 0 \\ 200 \end{bmatrix}.$$

Solution: Use 100 batches of vanilla, no mocha, and 200 batches of strawberry.

c. The requirements lead to the following matrix equation:

$$\begin{bmatrix} 2 & 1 & 1 \\ 1 & 1 & 2 \\ 2 & 2 & 1 \end{bmatrix} \begin{bmatrix} x \\ y \\ z \end{bmatrix} = \begin{bmatrix} 400 \\ 500 \\ 400 \end{bmatrix}$$

$$AX = B \Rightarrow X = A^{-1}B$$

$$X = \begin{bmatrix} 2 & 1 & 1 \\ 1 & 1 & 2 \\ 2 & 2 & 1 \end{bmatrix}^{-1} \begin{bmatrix} A \\ B \\ C \end{bmatrix} = \begin{bmatrix} 1 & -1/3 & -1/3 \\ -1 & 0 & 1 \\ 0 & 2/3 & -1/3 \end{bmatrix} \begin{bmatrix} A \\ B \\ C \end{bmatrix} = \begin{bmatrix} A - B/3 - C/3 \\ -A + C \\ 2B/3 - C/3 \end{bmatrix}.$$

Solution: Use $A - B/3 - C/3$ batches of Vanilla, $-A + C$ batches of mocha, and $2B/3 - C/3$ batches of strawberry.

55. Unknowns: $x =$ amount invested in MYY, $y =$ amount invested in SH, $z =$ amount invested in REW

The total investment was \$9,000: $\qquad x + y + z = 9,000$

You invested an equal amount in SH and REW: $\qquad y = z$, or $y - z = 0$

Year-to-date (YTD) loss from the first two funds was \$400: $\qquad 0.06x + 0.05y = 400$
Resulting systems of equations:

$\qquad x + y + z = 9,000; \; y - z = 0; \; 0.06x + 0.05y = 400$

Matrix form: $\begin{bmatrix} 1 & 1 & 1 \\ 0 & 1 & -1 \\ 0.06 & 0.05 & 0 \end{bmatrix} \begin{bmatrix} x \\ y \\ z \end{bmatrix} = \begin{bmatrix} 9,000 \\ 0 \\ 400 \end{bmatrix}$

$$AX = B \Rightarrow X = A^{-1}B$$

$$X = \begin{bmatrix} 1 & 1 & 1 \\ 0 & 1 & -1 \\ 0.06 & 0.05 & 0 \end{bmatrix}^{-1} \begin{bmatrix} 9,000 \\ 0 \\ 400 \end{bmatrix} = \begin{bmatrix} -5/7 & -5/7 & 200/7 \\ 6/7 & 6/7 & -100/7 \\ 6/7 & -1/7 & -100/7 \end{bmatrix} \begin{bmatrix} 9,000 \\ 0 \\ 400 \end{bmatrix} = \begin{bmatrix} 5,000 \\ 2,000 \\ 2,000 \end{bmatrix}$$

Solution: You invested \$5,000 in MYY, \$2,000 in SH, \$2,000 in REW.

57. Unknowns: $x =$ the number of shares of WSR, $y =$ the number of shares of HCC, $z =$ the number of shares of SNDK
The total investment was \$8,400:

Investment in WSR $= x$ shares @ \$16 $= 16x$

Investment in HCC $= y$ shares @ \$56 $= 56y$

Investment in SNDK $= z$ shares @ \$80 $= 80z$

Thus, $16x + 56y + 80z = 8,400$.
You expected to earn \$248 in dividends:

WSR dividend $= 7\%$ of $16x$ invested $= 0.07(16x) = 1.12x$

HCC dividend $= 2\%$ of $56y$ invested $= 0.02(56y) = 1.12y$

SNDK dividend $= 2\%$ of $80z$ invested $= 0.02(80z) = 1.6z$

Thus, $1.12x + 1.12y + 1.6z = 248$.

You purchased a total of 200 shares: $\qquad x + y + z = 200$.

We therefore have the following system:

$\qquad x + y + z = 200;\ 16x + 56y + 80z = 8{,}400;\ 1.12x + 1.12y + 1.6z = 248$.

Matrix form:
$$\begin{bmatrix} 1 & 1 & 1 \\ 16 & 56 & 80 \\ 28/25 & 28/25 & 8/5 \end{bmatrix}\begin{bmatrix} x \\ y \\ z \end{bmatrix} = \begin{bmatrix} 200 \\ 8{,}400 \\ 248 \end{bmatrix}$$

$AX = B \Rightarrow X = A^{-1}B$

$$X = \begin{bmatrix} 1 & 1 & 1 \\ 16 & 56 & 80 \\ 28/25 & 28/25 & 8/5 \end{bmatrix}^{-1}\begin{bmatrix} 200 \\ 8{,}400 \\ 248 \end{bmatrix} = \begin{bmatrix} 0 & -1/40 & 5/4 \\ 10/3 & 1/40 & -10/3 \\ -7/3 & 0 & 25/12 \end{bmatrix}\begin{bmatrix} 200 \\ 8{,}400 \\ 248 \end{bmatrix} = \begin{bmatrix} 100 \\ 50 \\ 50 \end{bmatrix}$$

Solution: You purchased 100 shares of WSR, 50 shares of HCC, and 50 shares of SNDK.

59. Take P to be the matrix represented by the table:

$$P = \begin{bmatrix} 0.9923 & 0.0016 & 0.0042 & 0.0019 \\ 0.0018 & 0.9896 & 0.0047 & 0.0039 \\ 0.0056 & 0.0059 & 0.9827 & 0.0058 \\ 0.0024 & 0.0033 & 0.0044 & 0.9899 \end{bmatrix}$$

2009 Distribution: $\qquad A = \begin{bmatrix} 54.6 & 66.0 & 111.8 & 70.6 \end{bmatrix}$

To compute the population distribution in 2008, compute $A \cdot P^{-1} \approx \begin{bmatrix} 54.1 & 65.7 & 112.9 & 70.3 \end{bmatrix}$

(See the solution to Exercise 75 in the preceding section.)

TI-83/84 Plus format: $[A]*[B]^{-1}$

Spreadsheet: `=MMULT(A1:D1,MINVERSE(E1:H4))` (Assuming that A is in cells A1–D1 and P is in cells E1–H4)

Online Matrix Algebra Tool format: `A*P^(-1)` Student Website → Online Utilities → Matrix Algebra Tool

61. $R = \begin{bmatrix} \sqrt{1/2} & -\sqrt{1/2} \\ \sqrt{1/2} & \sqrt{1/2} \end{bmatrix} \approx \begin{bmatrix} 0.7071 & -0.7071 \\ 0.7071 & 0.7071 \end{bmatrix}$

a. The coordinates of a rotated point are given by

$$\begin{bmatrix} x' \\ y' \end{bmatrix} = R\begin{bmatrix} x \\ y \end{bmatrix} \approx \begin{bmatrix} 0.7071 & -0.7071 \\ 0.7071 & 0.7071 \end{bmatrix}\begin{bmatrix} 2 \\ 3 \end{bmatrix} = \begin{bmatrix} -0.7071 \\ 3.5355 \end{bmatrix}.$$

Thus, $(x', y') \approx (-0.7071, 3.5355)$.

b. Multiplication by R rotates points through $45°$. To rotate through $90°$, multiply by R again, obtaining

$$R\left[R\begin{bmatrix} x \\ y \end{bmatrix}\right] = R^2\begin{bmatrix} x \\ y \end{bmatrix}.$$

In other words, multiplying by R^2 will result in a counterclockwise rotation of $90°$.

To rotate by $135°$, multiply yet again by R. This amounts to multiplying the original column vector by R^3.

c. Let S be the matrix representing a clockwise rotation through $45°$. Since a rotation of $45°$ clockwise followed by a rotation of $45°$ counterclockwise results in no change:

$$R\left[S\begin{bmatrix} x \\ y \end{bmatrix}\right] = \begin{bmatrix} x \\ y \end{bmatrix} = I\begin{bmatrix} x \\ y \end{bmatrix}.$$

In other words, multiplication by RS should have the same effect as multiplication by the 2×2 identity matrix. This will happen if $S = R^{-1}$.

63. First, write the uncoded message as a string of numbers using A = 1, B = 2, C = 3, and so on:

"GO TO PLAN B" = $\begin{bmatrix} 7 & 15 & 0 & 20 & 15 & 0 & 16 & 12 & 1 & 14 & 0 & 2 \end{bmatrix}$.

First, arrange the numbers in a matrix with 2 rows:

Uncoded message = $\begin{bmatrix} 7 & 0 & 15 & 16 & 1 & 0 \\ 15 & 20 & 0 & 12 & 14 & 2 \end{bmatrix}$.

Then encode by multiplying by $A = \begin{bmatrix} 1 & 2 \\ 3 & 4 \end{bmatrix}$:

Coded message = $\begin{bmatrix} 1 & 2 \\ 3 & 4 \end{bmatrix}\begin{bmatrix} 7 & 0 & 15 & 16 & 1 & 0 \\ 15 & 20 & 0 & 12 & 14 & 2 \end{bmatrix} = \begin{bmatrix} 37 & 40 & 15 & 40 & 29 & 4 \\ 81 & 80 & 45 & 96 & 59 & 8 \end{bmatrix}$.

Arrange as a single row: $\begin{bmatrix} 37 & 81 & 40 & 80 & 15 & 45 & 40 & 96 & 29 & 59 & 4 & 8 \end{bmatrix}$.

65. Coded message = $\begin{bmatrix} 33 & 69 & 54 & 126 & 11 & 27 & 20 & 60 & 29 & 59 & 65 & 149 & 41 & 87 \end{bmatrix}$.

First, arrange the numbers in a matrix with two rows:

Coded message = $\begin{bmatrix} 33 & 54 & 11 & 20 & 29 & 65 & 41 \\ 69 & 126 & 27 & 60 & 59 & 149 & 87 \end{bmatrix}$.

To decode the message, multiply by A^{-1}:

$A^{-1} = \begin{bmatrix} 1 & 2 \\ 3 & 4 \end{bmatrix}^{-1} = \begin{bmatrix} -2 & 1 \\ 1.5 & -0.5 \end{bmatrix}$.

Decoded message:

$\begin{bmatrix} -2 & 1 \\ 1.5 & -0.5 \end{bmatrix}\begin{bmatrix} 33 & 54 & 11 & 20 & 29 & 65 & 41 \\ 69 & 126 & 27 & 60 & 59 & 149 & 87 \end{bmatrix} = \begin{bmatrix} 3 & 18 & 5 & 20 & 1 & 19 & 5 \\ 15 & 18 & 3 & 0 & 14 & 23 & 18 \end{bmatrix}$

Arrange as a single row: $\begin{bmatrix} 3 & 15 & 18 & 18 & 5 & 3 & 20 & 0 & 1 & 14 & 19 & 23 & 5 & 18 \end{bmatrix}$.

Translate to letters using A = 1, B = 2, C = 3, and so on:

Decoded message = "CORRECT ANSWER".

Communication and reasoning exercises

67. If $AB = I$ and $BA = I$, then A and B are inverse matrices. (See the definition of the inverse matrix in the text.) Thus, choice (A) is the correct choice.

69. The given matrix has two identical rows. If two rows of a matrix are the same, then row reducing it will lead to a row of zeros, and so it cannot be reduced to the identity. Thus, the inverse of the given matrix does not exist; the matrix is singular.

71. We check that the given matrix is the inverse of $\begin{bmatrix} a & b \\ c & d \end{bmatrix}$ by multiplying the two matrices:

$\left(\dfrac{1}{ad-bc}\begin{bmatrix} d & -b \\ -c & a \end{bmatrix}\right)\begin{bmatrix} a & b \\ c & d \end{bmatrix} = \dfrac{1}{ad-bc}\begin{bmatrix} ad-bc & db-bd \\ -ac+ca & -bc+ad \end{bmatrix} = \dfrac{1}{ad-bc}\begin{bmatrix} ad-bc & 0 \\ 0 & ad-bc \end{bmatrix} = \begin{bmatrix} 0 & 1 \\ 1 & 0 \end{bmatrix}$.

Since the product of the two matrices is the identity, they must be inverse matrices as claimed. In other words,

$\begin{bmatrix} a & b \\ c & d \end{bmatrix}^{-1} = \dfrac{1}{ad-bc}\begin{bmatrix} d & -b \\ -c & a \end{bmatrix}$ (provided $ad - bc \neq 0$).

73. When one or more of the d_i are zero; if that is the case, then the matrix $\begin{bmatrix} D & | & I \end{bmatrix}$ easily reduces to a matrix that has a row of zeros on the left-hand portion, so D is singular. Conversely, if none of the d_i are zero, then $\begin{bmatrix} D & | & I \end{bmatrix}$ easily reduces to a matrix of the form $\begin{bmatrix} I & | & E \end{bmatrix}$, showing that D is invertible.

75. To check that $B^{-1}A^{-1}$ is the inverse of AB, we multiply these two matrices and check that we obtain the identity matrix:

$$(AB)(B^{-1}A^{-1}) = A(BB^{-1})A^{-1} \qquad \text{Associative law}$$

$$= AIA^{-1} \qquad \text{Since } BB^{-1} = I$$

$$= AA^{-1} = I.$$

Since the product of the given matrices is the identity, they must be inverses as claimed.

77. If a square matrix A reduces to one with a row of zeros, then it cannot have an inverse. The reason is that, if A has an inverse, then every system of equations $AX = B$ has a unique solution, namely $X = A^{-1}B$. But if A reduces to a matrix with a row of zeros, then such a system has either infinitely many solutions or no solution at all.

Section 4.4

1. $e = RPC = \begin{bmatrix} 0 & 1 & 0 & 0 \end{bmatrix} \begin{bmatrix} 2 & 0 & -1 & 2 \\ -1 & 0 & 0 & -2 \\ -2 & 0 & 0 & 1 \\ 3 & 1 & -1 & 1 \end{bmatrix} \begin{bmatrix} 1 \\ 0 \\ 0 \\ 0 \end{bmatrix} = -1$

3. $e = RPC = \begin{bmatrix} 0.5 & 0.5 & 0 & 0 \end{bmatrix} \begin{bmatrix} 2 & 0 & -1 & 2 \\ -1 & 0 & 0 & -2 \\ -2 & 0 & 0 & 1 \\ 3 & 1 & -1 & 1 \end{bmatrix} \begin{bmatrix} 0 \\ 0 \\ 0.5 \\ 0.5 \end{bmatrix} = -0.25$

5. Since we are given a column matrix, we take our row strategy to be
$R = \begin{bmatrix} x & y & z & t \end{bmatrix}$.

$e = RPC = \begin{bmatrix} x & y & z & t \end{bmatrix} \begin{bmatrix} 0 & -1 & 5 \\ 2 & -2 & 4 \\ 0 & 3 & 0 \\ 1 & 0 & -5 \end{bmatrix} \begin{bmatrix} 0.25 \\ 0.75 \\ 0 \end{bmatrix} = \begin{bmatrix} x & y & z & t \end{bmatrix} \begin{bmatrix} -0.75 \\ -1 \\ 2.25 \\ 0.25 \end{bmatrix}$

$= -0.74x - y + 2.25z + 0.25t$.

The greatest coefficient is the coefficient of z, so we take $z = 1$ and $x = y = t = 0$. This gives the strategy
$R = \begin{bmatrix} 0 & 0 & 1 & 0 \end{bmatrix}$
and resulting payoff
$e = -0.74(0) - 0 + 2.25(1) + 0.25(0) = 2.25$.

7. Since we are given a row matrix, we take our column strategy to be
$C = \begin{bmatrix} x & y & z \end{bmatrix}^T$

$e = RPC = \begin{bmatrix} 1/2 & 0 & 1/4 & 1/4 \end{bmatrix} \begin{bmatrix} 0 & -1 & 5 \\ 2 & -2 & 4 \\ 0 & 3 & 0 \\ 1 & 0 & -5 \end{bmatrix} \begin{bmatrix} x \\ y \\ z \end{bmatrix} = \begin{bmatrix} 1/4 & 1/4 & 5/4 \end{bmatrix} \begin{bmatrix} x \\ y \\ z \end{bmatrix}$

$= (1/4)x + (1/4)y + (5/4)z$

The lowest coefficients are the coefficients of x and y, so we can either take $x = 1$ and $y = z = 0$, giving the strategy
$C = \begin{bmatrix} 1 & 0 & 0 \end{bmatrix}^T$ or we could take $y = 1$ and $x = z = 0$, giving the strategy $C = \begin{bmatrix} 0 & 1 & 0 \end{bmatrix}^T$ and resulting
expected payoff
$e = (1/4)(1) + (1/4)(0) + (5/4)(0) = 1/4$.

9. Checking the rows: Neither of rows a and b dominates the other: 1 is worse than 2, but 10 is better than -4.

Checking the columns: Column p dominates column q because each of the payoffs in column p is lower than or equal to the corresponding payoff in column q. We therefore eliminate column q to obtain

$$\begin{array}{c} \\ a \\ b \end{array} \begin{array}{cc} p & r \\ \begin{bmatrix} 1 & 10 \\ 2 & -4 \end{bmatrix} \end{array}$$

This matrix cannot be reduced further.

11. Checking the rows: Row 3 dominates both rows 1 and 2 because each of its payoffs is greater than or equal to the corresponding payoff in row 1 and 2, so we eliminate rows 1 and 2 to obtain

$$\begin{array}{c} \\ 3 \end{array} \begin{array}{ccc} a & b & c \\ \left[\begin{array}{ccc} 5 & 0 & -1 \end{array}\right] \end{array}$$

Checking the columns: Column c dominates columns a and b because -1 is less than 0 and 5. We therefore eliminate columns a and b to obtain

$$\begin{array}{c} \\ 3 \end{array} \begin{array}{c} c \\ \left[\begin{array}{c} -1 \end{array}\right] \end{array}$$

13. Checking the rows: Row q dominates rows p and s because each of its payoffs is greater than or equal to the corresponding payoff in row p and s, so we eliminate rows p and s to obtain

$$\begin{array}{c} \\ q \\ r \end{array} \begin{array}{ccc} a & b & c \\ \left[\begin{array}{ccc} 4 & 0 & 2 \\ 3 & -3 & 10 \end{array}\right] \end{array}$$

Checking the columns: Column b dominates the other two columns (each of its payoffs is less than or equal to the corresponding payoff in the other two columns). We therefore eliminate columns a and c to obtain

$$\begin{array}{c} \\ q \end{array} \begin{array}{c} b \\ \left[\begin{array}{c} 0 \end{array}\right] \end{array}$$

15. Circle the row minima and box the row maxima:

$$\begin{array}{c} \\ a \\ b \end{array} \begin{array}{cc} p & q \\ \left[\begin{array}{cc} ①& ① \\ 2 & -4 \end{array}\right] \end{array} \qquad \begin{array}{c} \\ a \\ b \end{array} \begin{array}{cc} p & q \\ \left[\begin{array}{cc} 1 & 1 \\ 2 & -4 \end{array}\right] \end{array}$$

Since the a, q-entry 1 was both circled and boxed, it is a saddle point, and the game is strictly determined with the row player's optimal strategy being a, the column player's optimal strategy being q, and the value = saddle point = 1.

17. Circle the row minima and box the row maxima:

$$\begin{array}{c} \\ a \\ b \end{array} \begin{array}{ccc} p & q & r \\ \left[\begin{array}{ccc} 2 & 0 & -2 \\ -1 & 3 & 0 \end{array}\right] \end{array} \qquad \begin{array}{c} \\ a \\ b \end{array} \begin{array}{ccc} p & q & r \\ \left[\begin{array}{ccc} 2 & 0 & -2 \\ -1 & 3 & 0 \end{array}\right] \end{array}$$

Since no entry is both circled and boxed, there are no saddle points, and so the game is not strictly determined.

19. Circle the row minima and box the row maxima:

$$\begin{array}{c} \\ P \\ Q \\ R \\ S \end{array} \begin{array}{ccc} a & b & c \\ \left[\begin{array}{ccc} 1 & -1 & -5 \\ 4 & -4 & 2 \\ 3 & -3 & -10 \\ 5 & -5 & -4 \end{array}\right] \end{array} \qquad \begin{array}{c} \\ P \\ Q \\ R \\ S \end{array} \begin{array}{ccc} a & b & c \\ \left[\begin{array}{ccc} 1 & -1 & -5 \\ 4 & -4 & 2 \\ 3 & -3 & -10 \\ 5 & -5 & -4 \end{array}\right] \end{array}$$

Since no entry is both circled and boxed, there are no saddle points, and so the game is not strictly determined.

21. a. Row strategy:

Take $R = \begin{bmatrix} x & 1-x \end{bmatrix}, C = \begin{bmatrix} 1 & 0 \end{bmatrix}^T$

$$e = RPC = \begin{bmatrix} x & 1-x \end{bmatrix} \begin{bmatrix} -1 & 2 \\ 0 & -1 \end{bmatrix} \begin{bmatrix} 1 \\ 0 \end{bmatrix} = -x.$$

Take $R = \begin{bmatrix} x & 1-x \end{bmatrix}, C = \begin{bmatrix} 0 & 1 \end{bmatrix}^T$

$$f = RPC = \begin{bmatrix} x & 1-x \end{bmatrix} \begin{bmatrix} -1 & 2 \\ 0 & -1 \end{bmatrix} \begin{bmatrix} 0 \\ 1 \end{bmatrix} = 2x - (1-x) = 3x - 1.$$

Graphs of e and f:

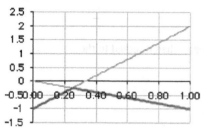

Lower (heavy) portion has its highest point at the intersection of the two lines. The x-coordinate of the intersection is given when $e = f$:

$$-x = 3x - 1 \Rightarrow 4x = 1 \Rightarrow x = 1/4.$$

So the optimal row strategy is

$$R = \begin{bmatrix} x & 1-x \end{bmatrix} = \begin{bmatrix} 1/4 & 3/4 \end{bmatrix}.$$

b. Column strategy:

Take $R = \begin{bmatrix} 1 & 0 \end{bmatrix}, C = \begin{bmatrix} x & 1-x \end{bmatrix}^T$

$$e = RPC = \begin{bmatrix} 1 & 0 \end{bmatrix} \begin{bmatrix} -1 & 2 \\ 0 & -1 \end{bmatrix} \begin{bmatrix} x \\ 1-x \end{bmatrix} = -x + 2(1-x) = -3x + 2$$

Take $R = \begin{bmatrix} 0 & 1 \end{bmatrix}, C = \begin{bmatrix} x & 1-x \end{bmatrix}^T$

$$e = RPC = \begin{bmatrix} 0 & 1 \end{bmatrix} \begin{bmatrix} -1 & 2 \\ 0 & -1 \end{bmatrix} \begin{bmatrix} x \\ 1-x \end{bmatrix} = x - 1.$$

Graphs of e and f:

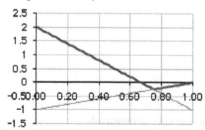

The upper (heavy) portion has its lowest point at the intersection of the two lines. The x-coordinate of the intersection is given when $e = f$:

$$-3x + 2 = x - 1 \Rightarrow 4x = 3 \Rightarrow x = 3/4.$$

So the optimal column strategy is

$$C = \begin{bmatrix} x & 1-x \end{bmatrix}^T = \begin{bmatrix} 3/4 & 1/4 \end{bmatrix}^T.$$

c. To compute the expected value of the game, compute the y-coordinate of the point on either graph with the given x-coordinate. For instance, using the second graph (part (b)),

$$e = x - 1 = 3/4 - 1 = -1/4.$$

Solution: $R = \begin{bmatrix} 1/4 & 3/4 \end{bmatrix}, C = \begin{bmatrix} 3/4 & 1/4 \end{bmatrix}^T, e = -1/4$

23. a. Row strategy:

Take $R = \begin{bmatrix} x & 1-x \end{bmatrix}, C = \begin{bmatrix} 1 & 0 \end{bmatrix}^T$

$$e = RPC = \begin{bmatrix} x & 1-x \end{bmatrix} \begin{bmatrix} -1 & -2 \\ -2 & 1 \end{bmatrix} \begin{bmatrix} 1 \\ 0 \end{bmatrix} = -x - 2(1-x) = x - 2.$$

Take $R = \begin{bmatrix} x & 1-x \end{bmatrix}, C = \begin{bmatrix} 0 & 1 \end{bmatrix}^T$

$$f = RPC = \begin{bmatrix} x & 1-x \end{bmatrix} \begin{bmatrix} -1 & -2 \\ -2 & 1 \end{bmatrix} \begin{bmatrix} 0 \\ 1 \end{bmatrix} = -2x + (1-x) = -3x + 1.$$

Graphs of e and f:

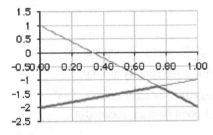

Lower (heavy) portion has its highest point at the intersection of the two lines. The x-coordinate of the intersection is given when $e = f$:

$$x - 2 = -3x + 1 \Rightarrow 4x = 3 \Rightarrow x = 3/4.$$

So the optimal row strategy is

$$R = \begin{bmatrix} x & 1-x \end{bmatrix} = \begin{bmatrix} 3/4 & 1/4 \end{bmatrix}.$$

b. Column strategy:

Take $R = \begin{bmatrix} 1 & 0 \end{bmatrix}, C = \begin{bmatrix} x & 1-x \end{bmatrix}^T$

$$e = RPC = \begin{bmatrix} 1 & 0 \end{bmatrix} \begin{bmatrix} -1 & -2 \\ -2 & 1 \end{bmatrix} \begin{bmatrix} x \\ 1-x \end{bmatrix} = -x - 2(1-x) = x - 2$$

Take $R = \begin{bmatrix} 0 & 1 \end{bmatrix}, C = \begin{bmatrix} x & 1-x \end{bmatrix}^T$

$$e = RPC = \begin{bmatrix} 0 & 1 \end{bmatrix} \begin{bmatrix} -1 & -2 \\ -2 & 1 \end{bmatrix} \begin{bmatrix} x \\ 1-x \end{bmatrix} = -2x + (1-x) = -3x + 1.$$

Note that e and f are the same as for part (a). Therefore, their graphs are the same, except that this time we are interested in the lowest point of the upper region:

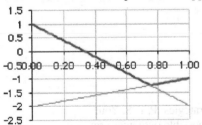

This is the same intersection point as in part (a), so $x = 3/4$ and the optimal column strategy is

$$C = \begin{bmatrix} x & 1-x \end{bmatrix}^T = \begin{bmatrix} 3/4 & 1/4 \end{bmatrix}^T.$$

c. To compute the expected value of the game, compute the y-coordinate of the point on either graph with the given x-coordinate. For instance, using the second graph (part (b)),

$$e = x - 2 = 3/4 - 2 = -5/4.$$

Solution: $R = \begin{bmatrix} 3/4 & 1/4 \end{bmatrix}, C = \begin{bmatrix} 3/4 & 1/4 \end{bmatrix}^T, e = -5/4$

Applications

25. The possible outcomes are:

> HH: You lose 1 point. Payoff: −1 HT: You win 1 point. Payoff: 1

> TH: You win 1 point. Payoff: 1 TT: You lose 1 point. Payoff: −1

Payoff matrix with you as the row player and your friend as the column player:

$$\begin{array}{c}\quad\quad\; H \quad\; T \\ \begin{array}{c} H \\ T \end{array}\left[\begin{array}{cc} -1 & 1 \\ 1 & -1 \end{array}\right]\end{array}$$

27. Take F = France; S = Sweden; N = Norway. Your strategies are to invade one of $F, S, N,$ and your opponent's strategies are to defend one of $F, S,$ or N. The possible outcomes in which you invade the same country as your opponent defends are $FF, SS, NN,$ and all of these have a payoff of −1 (you are defeated). Any other combination (you invade a country other than the one your opponent is defending) gives a payoff of 1 point (you are successful). Payoff matrix:

$$\begin{array}{c}\quad\quad\quad\quad\quad \textbf{Your Opponent Defends} \\ \quad\quad\quad\quad\quad F \quad\; S \quad\; N \\ \textbf{You Invade}\;\begin{array}{c} F \\ S \\ N \end{array}\left[\begin{array}{ccc} -1 & 1 & 1 \\ 1 & -1 & 1 \\ 1 & 1 & -1 \end{array}\right]\end{array}$$

29. Take B = Brakpan; N = Nigel; S = Springs. Your strategies are to locate in one of B, N, S, and your opponent's strategies are the same. The possible outcomes are:

BB, SS, or SS: You share the total business. No net gain or loss of sales. Payoff: 0

BN or NB: One of you locates at B, the other at N. These two cities provide the same potential market, so again there is no net gain or loss. Payoff: 0

SB or SN: You locate at S (total market: 1,000 burgers/day), your opponent locates at N or B (total market: 2,000 burgers/day). Your net loss is then 1,000 burgers/day. Payoff: −1,000

BS or NS: You locate at N or B (total market: 2,000 burgers/day), your opponent locates at S (total market: 1,000 bugers/day) Your net gain is then 1,000 burgers/day. Payoff: 1,000

Payoff matrix:

$$\begin{array}{c}\quad\quad\quad\quad\quad \textbf{Your Opponent} \\ \quad\quad\quad\quad B \quad\quad\quad N \quad\quad\quad S \\ \textbf{You}\;\begin{array}{c} B \\ N \\ S \end{array}\left[\begin{array}{ccc} 0 & 0 & 1{,}000 \\ 0 & 0 & 1{,}000 \\ -1{,}000 & -1{,}000 & 0 \end{array}\right]\end{array}$$

31. Let P = Pleasant Tap; T = Thunder Rumble; S = Strike the Gold, N = None. Your strategies: P, T, S. Your "opponent" is Nature, which decides the winner. Opponent's strategies: P, T, S, N. The possible outcomes are:

PP: You bet on P and P wins. Odds: 5–2. $2 wins you $5. Therefore $10 wins you 5 × $5 = $25.
  Payoff: 25

TT: You bet on T and T wins. Odds: 7–2. $2 wins you $7. Therefore $10 wins you 5 × $7 = $35.
  Payoff: 35

SS: You bet on S and S wins. Odds: 4–1. $1 wins you $4. Therefore $10 wins you 10 × $4 = $40.
  Payoff: 40

All other outcomes: You lose your $10 because your horse fails to win. Payoff: −10

Payoff matrix:

$$
\begin{array}{c}
\text{Winner}\\
\begin{array}{cccc}
P & T & S & N
\end{array}\\
\text{You Bet }
\begin{array}{c}
P\\
T\\
S
\end{array}
\left[\begin{array}{cccc}
25 & -10 & -10 & -10\\
-10 & 35 & -10 & -10\\
-10 & -10 & 40 & -10
\end{array}\right]
\end{array}
$$

33. $P = \begin{bmatrix} -60 & -40 \\ 30 & -50 \end{bmatrix}, R = \begin{bmatrix} .50 & .50 \end{bmatrix}, C = \begin{bmatrix} 0.20 \\ 0.80 \end{bmatrix}$

$e = RPC = \begin{bmatrix} .50 & .50 \end{bmatrix} \begin{bmatrix} -60 & -40 \\ 30 & -50 \end{bmatrix} \begin{bmatrix} 0.20 \\ 0.80 \end{bmatrix} = \begin{bmatrix} -15 & -45 \end{bmatrix} \begin{bmatrix} 0.20 \\ 0.80 \end{bmatrix} = -39$

You can expect to lose 39 customers.

35. $P = \begin{bmatrix} 200 & 150 & 140 \\ 130 & 220 & 130 \\ 110 & 110 & 220 \end{bmatrix}, R = \begin{bmatrix} x & y & z \end{bmatrix}, C = \begin{bmatrix} 0.2 \\ 0.5 \\ 0.3 \end{bmatrix}$

$e = RPC = \begin{bmatrix} x & y & z \end{bmatrix} \begin{bmatrix} 200 & 150 & 140 \\ 130 & 220 & 130 \\ 110 & 110 & 220 \end{bmatrix} \begin{bmatrix} 0.2 \\ 0.5 \\ 0.3 \end{bmatrix} = \begin{bmatrix} x & y & z \end{bmatrix} \begin{bmatrix} 157 \\ 175 \\ 143 \end{bmatrix} = 157x + 175y + 143z$

The largest coefficient is the coefficient of y, so we take $R = \begin{bmatrix} 0 & 1 & 0 \end{bmatrix}$ meaning that Option (II): Move to the suburbs (corresponding to y) is the best option.

37. a. $P = \begin{bmatrix} 90 & 70 & 70 \\ 40 & 90 & 40 \\ 60 & 40 & 90 \end{bmatrix}, R = \begin{bmatrix} 0.25 & 0.50 & 0.25 \end{bmatrix}, C = \begin{bmatrix} 0.25 \\ 0.50 \\ 0.25 \end{bmatrix}$

$e = RPC = \begin{bmatrix} 0.25 & 0.50 & 0.25 \end{bmatrix} \begin{bmatrix} 90 & 70 & 70 \\ 40 & 90 & 40 \\ 60 & 40 & 90 \end{bmatrix} \begin{bmatrix} 0.25 \\ 0.50 \\ 0.25 \end{bmatrix} = \begin{bmatrix} 0.25 & 0.50 & 0.25 \end{bmatrix} \begin{bmatrix} 75 \\ 65 \\ 57.5 \end{bmatrix} = 65.625$

You can expect to get about 66% on the test.

b. $P = \begin{bmatrix} 90 & 70 & 70 \\ 40 & 90 & 40 \\ 60 & 40 & 90 \end{bmatrix}, R = \begin{bmatrix} x & y & z \end{bmatrix}, C = \begin{bmatrix} 0.25 \\ 0.50 \\ 0.25 \end{bmatrix}$

$e = RPC = \begin{bmatrix} x & y & z \end{bmatrix} \begin{bmatrix} 90 & 70 & 70 \\ 40 & 90 & 40 \\ 60 & 40 & 90 \end{bmatrix} \begin{bmatrix} 0.25 \\ 0.50 \\ 0.25 \end{bmatrix} = \begin{bmatrix} x & y & z \end{bmatrix} \begin{bmatrix} 75 \\ 65 \\ 57.5 \end{bmatrix} = 75x + 65y + 57.5z$

The largest coefficient is the coefficient of x, so we take $R = \begin{bmatrix} 1 & 0 & 0 \end{bmatrix}$, meaning that game theory (corresponding to x) is the best option.

Your expected score is then $75(1) + 65(0) + 57.5(0) = 75\%$

c. $P = \begin{bmatrix} 90 & 70 & 70 \\ 40 & 90 & 40 \\ 60 & 40 & 90 \end{bmatrix}, R = \begin{bmatrix} 0.25 & 0.50 & 0.25 \end{bmatrix}, C = \begin{bmatrix} x \\ y \\ z \end{bmatrix}$

$e = RPC = \begin{bmatrix} 0.25 & 0.50 & 0.25 \end{bmatrix} \begin{bmatrix} 90 & 70 & 70 \\ 40 & 90 & 40 \\ 60 & 40 & 90 \end{bmatrix} \begin{bmatrix} x \\ y \\ z \end{bmatrix} = \begin{bmatrix} 57.5 & 72.5 & 60 \end{bmatrix} \begin{bmatrix} x \\ y \\ z \end{bmatrix} = 57.5x + 72.5y + 60z$

The lowest coefficient is the coefficient of x, so we take $C = \begin{bmatrix} 1 & 0 & 0 \end{bmatrix}^T$, meaning that game theory would be worst for you, and you could expect to earn $57.5(1) + 72.5(0) + 60(0) = 57.5\%$ for the test.

39. a. $P = \begin{bmatrix} -500{,}000 & -200{,}000 & 10{,}000 & 200{,}000 \\ -200{,}000 & 0 & 0 & 0 \\ -100{,}000 & 10{,}000 & -200{,}000 & -300{,}000 \end{bmatrix}, R = \begin{bmatrix} x & y & z \end{bmatrix}, C = \begin{bmatrix} 0.2 \\ 0.2 \\ 0.3 \\ 0.3 \end{bmatrix},$

(based on the past 10 years)

$e = RPC = \begin{bmatrix} x & y & z \end{bmatrix} \begin{bmatrix} -500{,}000 & -200{,}000 & 10{,}000 & 200{,}000 \\ -200{,}000 & 0 & 0 & 0 \\ -100{,}000 & 10{,}000 & -200{,}000 & -300{,}000 \end{bmatrix} \begin{bmatrix} 0.2 \\ 0.2 \\ 0.3 \\ 0.3 \end{bmatrix}$

$= \begin{bmatrix} x & y & z \end{bmatrix} \begin{bmatrix} -77{,}000 \\ -40{,}000 \\ -168{,}000 \end{bmatrix} = -77{,}000x - 40{,}000y - 168{,}000z$

The largest coefficient is the coefficient of y, so we take $R = \begin{bmatrix} 0 & 1 & 0 \end{bmatrix}$.

meaning that laying off 10 workers (corresponding to y) is the best option. The expected payoff is

$-77{,}000(0) - 40{,}000(1) - 168{,}000(0) = -40{,}000,$

corresponding to a cost of $40,000.

b. $P = \begin{bmatrix} -500{,}000 & -200{,}000 & 10{,}000 & 200{,}000 \\ -200{,}000 & 0 & 0 & 0 \\ -100{,}000 & 10{,}000 & -200{,}000 & -300{,}000 \end{bmatrix}, R = \begin{bmatrix} 0.50 & 0 & 0.50 \end{bmatrix}, C = \begin{bmatrix} x \\ y \\ z \\ t \end{bmatrix},$

$e = RPC = \begin{bmatrix} 0.50 & 0 & 0.50 \end{bmatrix} \begin{bmatrix} -500{,}000 & -200{,}000 & 10{,}000 & 200{,}000 \\ -200{,}000 & 0 & 0 & 0 \\ -100{,}000 & 10{,}000 & -200{,}000 & -300{,}000 \end{bmatrix} \begin{bmatrix} x \\ y \\ z \\ t \end{bmatrix}$

$= \begin{bmatrix} -300{,}000 & -95{,}000 & -95{,}000 & -50{,}000 \end{bmatrix} \begin{bmatrix} x \\ y \\ z \\ t \end{bmatrix}$

$= -300{,}000x - 95{,}000y - 95{,}000z - 50{,}000t$

The lowest coefficient is the coefficient of x, so we take $C = \begin{bmatrix} 1 & 0 & 0 & 0 \end{bmatrix}^T$, meaning that 0 inches of snow would be worst for him, costing him $300,000.

c. The strategy from part (a) is $R = \begin{bmatrix} 0 & 1 & 0 \end{bmatrix}$; laying off 10 workers. If he does so, the worst that could happen is 0 inches of snow (which would cost him $20,000). Since he feels that the Gods of Chaos are planning on 0 inches of snow, his best option would be to lay off 15 workers (according to the payoff matrix) and cut his losses to $100,000.

41. a. The payoff matrix is $P = \begin{bmatrix} 15 & 60 & 80 \\ 15 & 60 & 60 \\ 10 & 20 & 40 \end{bmatrix}$.

Checking the rows: Row 1 dominates each of the other rows, so we eliminate rows 2 and 3, leaving

$\begin{bmatrix} 15 & 60 & 80 \end{bmatrix}$.

Checking the columns: Column 1 dominates each of the other columns, so we eliminate columns 2 and 3, leaving the 1×1 game

$\begin{bmatrix} 15 \end{bmatrix}$,

which corresponds to the first CE strategy (charge $1,000) and the first GCS strategy (charge $900). So, CE should

charge $1,000 and GCS should charge $900. Since the payoff is 15, a 15% gain in market share for CE results.
b. Look at the original (unreduced) payoff matrix. CE is aware that GCS is planning to charge $900. For the best market share, CE should charge either $1,000 or $1,200 because either will result in the best market share (15% gain) under the circumstances. Thus, in terms of market share, the added information has no effect. However, in terms of revenue, the better of the two options would be to charge the larger price: $1,200 (the more CE can charge for the same market, the better!).

43. Take CCC as the row player and MMA as the column player. Thus, CCC's strategies are Pablo, Sal and Edison, while MMA's strategies are Carlos, Marcus, and Noto. To set up the payoff matrix, enter 1 for every combination in which the CCC wrestler beats the MMA wrestler, −1 for every combination in which the MMA wrestler beats the CCC wrestler, and 0 in all the evenly matched combinations. The resulting payoff matrix is

		MMA	
	Carlos	Marcus	Noto
Pablo	1	1	0
CCC Sal	0	−1	0
Edison	0	−1	−1

Comparing rows, we see that Row 1 dominates the other two rows, so we eliminate Rows 2 and 3, leaving

	Carlos	Marcus	Noto
Pablo	1	1	0

Comparing columns, we see that Column 3 dominates the other two columns, so we eliminate Columns 1 and 2, leaving us with

	Noto
Pablo	0

Thus the game is reduced to Pablo vs. Noto. Since the payoff is 0, the game is evenly matched.

45. We first reduce by dominance (following the "FAQ" in the textbook):
Comparing the rows, we find that neither row dominates the other.
Comparing the columns, we see than Column 1 dominates Column 2. So we eliminate Column 2, leaving

	Northern Route
Northern Route	2
Southern Route	1

Comparing the rows, we now see that Row 1 dominates Row 2, so we eliminate Row 2, leaving us with

	Northern Route
Northern Route	2

Thus, both commanders should use the northern route, resulting in an estimate of two days' bombing time.

47. Take C = Confess and N = Do not confess. If we take the (negative) payoffs as the amount of time Slim faces behind bars, we get:

		Joe	
		C	N
Slim	C	−2	0
	N	−10	−5

As Row 1 dominates Row 2, Slim's optimal strategy is to confess.

49. a. Let F represent a visit to Florida and O a visit to Ohio. The outcomes are:

FF and OO: Both candidates visit the same state, so Romney still has a 24% chance of winning, and so the payoff is 24 in both cases.

FO: Romney visits Florida, giving him a $60 + 10 = 70\%$ chance in that state, and Obama visits Ohio, reducing

Romney's chances there to $40 - 10 = 30\%$. Thus, the probability of Romney winning both states is $0.70 \times 0.30 = 0.21$

OF: Romney visits Ohio, giving him a $40 + 10 = 50\%$ chance in that state, and Obama visits Florida, reducing

Romney's chances there to $60 - 10 = 50\%$. Thus, the probability of Romney winning both states is $0.50 \times 0.50 = 0.25$
This gives the following payoff matrix:

Obama

$$\text{Romney} \begin{array}{c} F \\ O \end{array} \begin{bmatrix} 24 & 21 \\ 25 & 24 \end{bmatrix} \quad \begin{array}{c} F \quad O \end{array}$$

b. In the payoff matrix in part (a), Row 2 dominates Row 1 (meaning that Romney should visit Ohio). In that case):

$$O \begin{bmatrix} 25 & 24 \end{bmatrix} \quad \begin{array}{c} F \quad O \end{array}$$

Column 2 now dominates Column 1, leaving us with the following solution

$$O \begin{bmatrix} 24 \end{bmatrix} \quad \begin{array}{c} O \end{array}$$

meaning that both candidates should visit Ohio, leaving Romney with a 24% chance of winning the election.

51. The payoff matrix (payoffs in thousands of dollars) is

Spliish

$$\text{Softex} \begin{array}{c} \text{WISH} \\ \text{WASH} \end{array} \begin{bmatrix} -20 & 100 \\ 0 & -20 \end{bmatrix} \quad \begin{array}{c} \text{WISH} \quad \text{WASH} \end{array}$$

We are asked to provide the optimal row strategy.
Take $R = \begin{bmatrix} x & 1-x \end{bmatrix}, C = \begin{bmatrix} 1 & 0 \end{bmatrix}^T$

$$e = RPC = \begin{bmatrix} x & 1-x \end{bmatrix} \begin{bmatrix} -20 & 100 \\ 0 & -20 \end{bmatrix} \begin{bmatrix} 1 \\ 0 \end{bmatrix} = -20x + 0(1-x) = -20x$$

Take $R = \begin{bmatrix} x & 1-x \end{bmatrix}, C = \begin{bmatrix} 0 & 1 \end{bmatrix}^T$

$$f = RPC = \begin{bmatrix} x & 1-x \end{bmatrix} \begin{bmatrix} -20 & 100 \\ 0 & -20 \end{bmatrix} \begin{bmatrix} 0 \\ 1 \end{bmatrix} = 100x - 20(1-x) = 120x - 20$$

Graphs of e and f:

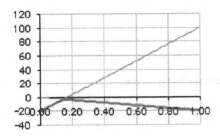

Lower (heavy) portion has its highest point at the intersection of the two lines. The x-coordinate of the intersection is given when $e = f$:

$$-20x = 120x - 20 \Rightarrow 140x = 20 \Rightarrow x = 1/7.$$

So the optimal row strategy is

$$R = \begin{bmatrix} x & 1-x \end{bmatrix} = \begin{bmatrix} 1/7 & 6/7 \end{bmatrix}.$$

This solution corresponds to allocating 1/7 of its advertising budget to WISH and the rest (6/7) to WASH. The value of the game is then

$$e = -20x = -20/7 \approx -2.86.$$

So Softex can expect to lose approximately $2,860.

Communication and reasoning exercises

53. Like a saddle point in a payoff matrix, the center of a saddle is a low point (minimum height) in one direction and a high point (maximum) in a perpendicular direction.

55. Although there is a saddle point in the (2,4) position, you would be wrong to use saddle points (based on the minimax criterion) to reach the conclusion that row strategy 2 is best. One reason is that the entries in the matrix do not represent payoffs, since high numbers of employees in an area do not necessarily represent benefit to the row player. Another reason for this is that there is no opponent deciding what your job will be in such a way as to force you into the least populated job.

57. If you strictly alternate the two strategies, the column player will know which pure strategy you will play on each move and can choose a pure strategy accordingly. For example, consider the game

$$\begin{array}{cc} & \begin{array}{cc} a & b \end{array} \\ \begin{array}{c} A \\ B \end{array} & \begin{bmatrix} 1 & 0 \\ 0 & 1 \end{bmatrix}. \end{array}$$

By the analysis of Example 3 (or the symmetry of the game), the best strategy for the row player is $\begin{bmatrix} 0.5 & 0.5 \end{bmatrix}$, and the best strategy for the column player is $\begin{bmatrix} 0.5 & 0.5 \end{bmatrix}^T$. This gives an expected value of 0.5 for the game. However, suppose that the row player alternates A and B strictly and that the column player catches on to this. Then, whenever the row player plays A the column player will play b, and whenever the row player plays B the column player will play a. This gives a payoff of 0 each time, worse for the row player than the expected value of 0.5.

Section 4.5

1. $A = \begin{bmatrix} 0.2 & 0.05 \\ 0.8 & 0.01 \end{bmatrix}$; Sector 1 = Paper; Sector 2 = Wood

a. Wood → Paper = Sector 2 → Sector 1 ; The corresponding entry of the technology matrix is $a_{21} = 0.8$.

b. Paper → Paper = Sector 1 → Sector 1 ; The corresponding entry of the technology matrix is $a_{11} = 0.2$.

c. Paper → Wood = Sector 1 → Sector 2
The corresponding entry of the technology matrix is $a_{12} = 0.05$.

3. Sector 1 = Television news, Sector 2 = Radio news

a_{11} = units of Television news needed to produce one unit of Television news = 0.2

a_{12} = units of Television news needed to produce one unit of Radio news = 0.1

a_{21} = units of Radio news needed to produce one unit of Television news = 0.5

a_{22} = units of Radio news needed to produce one unit of Radio news = 0

Thus, $A = \begin{bmatrix} 0.2 & 0.1 \\ 0.5 & 0 \end{bmatrix}$.

5. $X = (I - A)^{-1}D = \left(\begin{bmatrix} 1 & 0 \\ 0 & 1 \end{bmatrix} - \begin{bmatrix} 0.5 & 0.4 \\ 0 & 0.5 \end{bmatrix} \right)^{-1} \begin{bmatrix} 10{,}000 \\ 20{,}000 \end{bmatrix} = \begin{bmatrix} 0.5 & -0.4 \\ 0 & 0.5 \end{bmatrix}^{-1} \begin{bmatrix} 10{,}000 \\ 20{,}000 \end{bmatrix}$

$= \begin{bmatrix} 2 & 1.6 \\ 0 & 2 \end{bmatrix} \begin{bmatrix} 10{,}000 \\ 20{,}000 \end{bmatrix} = \begin{bmatrix} 52{,}000 \\ 40{,}000 \end{bmatrix}$

7. $X = (I - A)^{-1}D = \left(\begin{bmatrix} 1 & 0 \\ 0 & 1 \end{bmatrix} - \begin{bmatrix} 0.1 & 0.4 \\ 0.2 & 0.5 \end{bmatrix} \right)^{-1} \begin{bmatrix} 25{,}000 \\ 15{,}000 \end{bmatrix} = \begin{bmatrix} 0.9 & -0.4 \\ -0.2 & 0.5 \end{bmatrix}^{-1} \begin{bmatrix} 25{,}000 \\ 15{,}000 \end{bmatrix}$

$= \begin{bmatrix} 50/37 & 40/37 \\ 20/37 & 90/37 \end{bmatrix} \begin{bmatrix} 25{,}000 \\ 15{,}000 \end{bmatrix} = \begin{bmatrix} 50{,}000 \\ 50{,}000 \end{bmatrix}$

9. $X = (I - A)^{-1}D = \left(\begin{bmatrix} 1 & 0 & 0 \\ 0 & 1 & 0 \\ 0 & 0 & 1 \end{bmatrix} - \begin{bmatrix} 0.5 & 0.1 & 0 \\ 0 & 0.5 & 0.1 \\ 0 & 0 & 0.5 \end{bmatrix} \right)^{-1} \begin{bmatrix} 1{,}000 \\ 1{,}000 \\ 2{,}000 \end{bmatrix}$

$= \begin{bmatrix} 0.5 & -0.1 & 0 \\ 0 & 0.5 & -0.1 \\ 0 & 0 & 0.5 \end{bmatrix}^{-1} \begin{bmatrix} 1{,}000 \\ 1{,}000 \\ 2{,}000 \end{bmatrix} = \begin{bmatrix} 2 & 0.4 & 0.08 \\ 0 & 2 & 0.4 \\ 0 & 0 & 2 \end{bmatrix} \begin{bmatrix} 1{,}000 \\ 1{,}000 \\ 2{,}000 \end{bmatrix} = \begin{bmatrix} 2{,}560 \\ 2{,}800 \\ 4{,}000 \end{bmatrix}$

11. $X = (I - A)^{-1}D = \left(\begin{bmatrix} 1 & 0 & 0 \\ 0 & 1 & 0 \\ 0 & 0 & 1 \end{bmatrix} - \begin{bmatrix} 0.2 & 0.2 & 0 \\ 0.2 & 0.4 & 0.2 \\ 0 & 0.2 & 0.2 \end{bmatrix}\right)^{-1} \begin{bmatrix} 16{,}000 \\ 8{,}000 \\ 8{,}000 \end{bmatrix}$

$= \begin{bmatrix} 0.8 & -0.2 & 0 \\ -0.2 & 0.6 & -0.2 \\ 0 & -0.2 & 0.8 \end{bmatrix}^{-1} \begin{bmatrix} 16{,}000 \\ 8{,}000 \\ 8{,}000 \end{bmatrix} = \begin{bmatrix} 11/8 & 1/2 & 1/8 \\ 1/2 & 2 & 1/2 \\ 1/8 & 1/2 & 11/8 \end{bmatrix} \begin{bmatrix} 16{,}000 \\ 8{,}000 \\ 8{,}000 \end{bmatrix} = \begin{bmatrix} 27{,}000 \\ 28{,}000 \\ 17{,}000 \end{bmatrix}$

13. Change in Production $= (I - A)^{-1} \times$ Change in Demand

$= \left(\begin{bmatrix} 1 & 0 \\ 0 & 1 \end{bmatrix} - \begin{bmatrix} 0.1 & 0.4 \\ 0.2 & 0.5 \end{bmatrix}\right)^{-1} \begin{bmatrix} 50 \\ 30 \end{bmatrix} = \begin{bmatrix} 0.9 & -0.4 \\ -0.2 & 0.5 \end{bmatrix}^{-1} \begin{bmatrix} 50 \\ 30 \end{bmatrix}$

$= \begin{bmatrix} 50/37 & 40/37 \\ 20/37 & 90/37 \end{bmatrix} \begin{bmatrix} 50 \\ 30 \end{bmatrix} = \begin{bmatrix} 100 \\ 100 \end{bmatrix}$

15. Change in Production $= (I - A)^{-1} \times$ Change in Demand

$= \begin{bmatrix} 1.5 & 0.1 & 0 \\ 0.2 & 1.2 & 0.1 \\ 0.1 & 0.7 & 1.6 \end{bmatrix} \begin{bmatrix} 1 \\ 0 \\ 0 \end{bmatrix} = \begin{bmatrix} 1.5 \\ 0.2 \\ 0.1 \end{bmatrix}$

Note that this is the first column of $(I - A)^{-1}$.

In general, the ith column of $(I - A)^{-1}$ gives the change in production necessary to meet an increase in external demand of one unit for the product of Sector i.

17. Given table:

		To		
		A	**B**	**C**
	A	1,000	2,000	3,000
From	**B**	0	4,000	0
	C	0	1,000	3,000
	Total Output	5,000	5,000	6,000

We obtain the technology matrix from the input-output table by dividing each column by its total:

$A = \begin{bmatrix} 1000/5000 & 2000/5000 & 3000/6000 \\ 0 & 4000/5000 & 0 \\ 0 & 1000/5000 & 3000/6000 \end{bmatrix} = \begin{bmatrix} 0.2 & 0.4 & 0.5 \\ 0 & 0.8 & 0 \\ 0 & 0.2 & 0.5 \end{bmatrix}$.

Applications

19. First obtain the technology matrix from the input-output table by dividing each column by its total:

$A = \begin{bmatrix} 10{,}000/50{,}000 & 20{,}000/40{,}000 \\ 5000/50{,}000 & 0 \end{bmatrix} = \begin{bmatrix} 0.2 & 0.5 \\ 0.1 & 0 \end{bmatrix}$

Production $= (I - A)^{-1} \times$ Demand

$= \left(\begin{bmatrix} 1 & 0 \\ 0 & 1 \end{bmatrix} - \begin{bmatrix} 0.2 & 0.5 \\ 0.1 & 0 \end{bmatrix}\right)^{-1} \begin{bmatrix} 45{,}000 \\ 30{,}000 \end{bmatrix} = \begin{bmatrix} 0.8 & -.5 \\ -.1 & 1 \end{bmatrix}^{-1} \begin{bmatrix} 45{,}000 \\ 30{,}000 \end{bmatrix}$

$$= \begin{bmatrix} 4/3 & 2/3 \\ 2/15 & 16/15 \end{bmatrix} \begin{bmatrix} 45,000 \\ 30,000 \end{bmatrix} = \begin{bmatrix} 80,000 \\ 30,000 \end{bmatrix}.$$

Solution: The Main DR had to produce $80,000 worth of food, while Bits & Bytes had to produce $38,000$ worth of food.

21. First obtain the technology matrix from the input-output table by dividing each column by its total:

$$A = \begin{bmatrix} 6000/90,000 & 500/140,000 \\ 24,000/90,000 & 30,000/140,000 \end{bmatrix} = \begin{bmatrix} 1/15 & 1/280 \\ 4/15 & 3/14 \end{bmatrix}$$

Production $= (I - A)^{-1} \times$ Demand

$$= \left(\begin{bmatrix} 1 & 0 \\ 0 & 1 \end{bmatrix} - \begin{bmatrix} 1/15 & 1/280 \\ 4/15 & 3/14 \end{bmatrix} \right)^{-1} \begin{bmatrix} 80,000 \\ 90,000 \end{bmatrix} = \begin{bmatrix} 14/15 & -1/280 \\ -4/15 & 11/14 \end{bmatrix}^{-1} \begin{bmatrix} 80,000 \\ 90,000 \end{bmatrix}$$

$$\approx \begin{bmatrix} 1.07282 & 0.00488 \\ 0.36411 & 1.27438 \end{bmatrix} \begin{bmatrix} 80,000 \\ 90,000 \end{bmatrix} \approx \begin{bmatrix} 86,265 \\ 143,823 \end{bmatrix}.$$

Solution: Equipment Sector production approximately $86,000 million, Components Sector production approximately $140,000 million.

23. $(I - A)^{-1} = \begin{bmatrix} 1.228 & 0.182 \\ 0.006 & 1.1676 \end{bmatrix}$

Sector 1 = Textiles, Sector 2 = Clothing & footwear

a. The missing term is the number of units of Sector 2 needed to produce one additional unit of Sector 1. This quantity is given by the 2,1-entry of $(I - A)^{-1}$, or 0.006.

b. The missing terms refer to the 1,2-entry of $(I - A)^{-1}$, which gives the number of units of Sector 1 needed to produce one additional unit of Sector 2—in other words, the additional dollars worth of <u>textiles</u> that must be produced to meet a one-dollar increase in the demand for <u>clothing and footwear</u>.

25. As the technology matrix is obtained from the input-output table by dividing each column by the production total for that column, we perform this process in reverse to obtain the input-output table; that is, multiply each entry in the technology matrix by the production total for that column. Thus, the input-output matrix is

		To		
		Primary	**Seconday**	**Tertiary**
	Primary	53.1	330	0
From	**Seconday**	82.6	2,530	768
	Tertiary	41.3	1,320	1,440
	Total output	590	11,000	9,600

(Entries are in billions of pesos.)

27. $A = \begin{bmatrix} 0.09 & 0.03 & 0.00 \\ 0.14 & 0.23 & 0.08 \\ 0.07 & 0.12 & 0.15 \end{bmatrix}$, $X = \begin{bmatrix} 590 \\ 11,000 \\ 9,600 \end{bmatrix}$

Amount available for external use $= D = X - AX = \begin{bmatrix} 590 \\ 11,000 \\ 9,600 \end{bmatrix} - \begin{bmatrix} 0.09 & 0.03 & 0.00 \\ 0.14 & 0.23 & 0.08 \\ 0.07 & 0.12 & 0.15 \end{bmatrix} \begin{bmatrix} 590 \\ 11,000 \\ 9,600 \end{bmatrix}$

$$= \begin{bmatrix} 590 \\ 11,000 \\ 9,600 \end{bmatrix} - \begin{bmatrix} 383.1 \\ 3380.6 \\ 2801.3 \end{bmatrix} = \begin{bmatrix} 206.9 \\ 7,619.4 \\ 6,798.7 \end{bmatrix} \approx \begin{bmatrix} 210 \\ 7,600 \\ 6,800 \end{bmatrix}$$

210 billion pesos of raw materials, 7,600 billion pesos of manufactured goods, and 6,800 billion pesos of services.

29. To determine how each sector would need to react to an increase (or decrease) in demand, we compute

$$(I - A)^{-1} = \begin{bmatrix} 0.91 & -0.03 & 0 \\ -0.14 & 0.77 & -0.08 \\ -0.07 & -0.12 & 0.85 \end{bmatrix}^{-1} \approx \begin{bmatrix} 1.106 & 0.044 & 0.004 \\ 0.214 & 1.326 & 0.125 \\ 0.121 & 0.191 & 1.194 \end{bmatrix}.$$

TI-83/84 Plus Format: `(Identity(3)-[A])`$^{-1}$ Matrix Algebra Tool: `(I-A)^-1`

We are given external demand increases of 2,000, 0, and −1,000 billion pesos in the sectors, so we multiply $(I - A)^{-1}$ by the column matrix

$$D = \begin{bmatrix} 2,000 \\ 0 \\ -1,000 \end{bmatrix} \text{ and obtain } (I - A)^{-1}D \approx \begin{bmatrix} 2,208 \\ 303 \\ -952 \end{bmatrix}.$$

Thus, production in the primary sector would rise by around 2,208 billion pesos, production in the secondary sector would rise by around 303 billion pesos, and prodution in the tertiary sector would drop by 952 billion pesos.

31. To determine how each sector would need to react to an increase in demand, we compute $(I - A)^{-1}D$, where D is the increase in demand. First obtain the technology matrix from the input-output table by dividing each column by its total:

$$A \approx \begin{bmatrix} 0.12206 & 0.00007 & 0.00744 & 0.01801 \\ 0.27250 & 0.03553 & 0.00000 & 0.09993 \\ 0.00000 & 0.00000 & 0.02998 & 0.00128 \\ 0.05545 & 0.09082 & 0.20503 & 0.00455 \end{bmatrix}.$$

Then compute the matrix $(I - A)^{-1}$.

TI-83/84 Plus Format: `(Identity(4)-[A])`$^{-1}$ Matrix Algebra Tool: `(I-A)^-1`

$$(I - A)^{-1} = \begin{bmatrix} 1.14099 & 0.00205 & 0.01317 & 0.02087 \\ 0.33210 & 1.04734 & 0.02605 & 0.11118 \\ 0.00012 & 0.00013 & 1.03119 & 0.00135 \\ 0.09388 & 0.09569 & 0.21550 & 1.01615 \end{bmatrix}$$

We are given external demand increases of $1,000 million in each sector, so we multiply $(I - A)^{-1}$ by the column matrix

$$D = \begin{bmatrix} 1,000 \\ 1,000 \\ 1,000 \\ 1,000 \end{bmatrix} \text{ and obtain } (I - A)^{-1}D \approx \begin{bmatrix} 1,177 \\ 1,517 \\ 1,033 \\ 1,421 \end{bmatrix}.$$

The entries of this matrix show the amounts, in millions of dollars, by which each sector needs to increase production to meet the additional demand. (Round answers to four significant digits.)

33. a. To determine how each sector would need to react to an increase in demand, we compute $(I - A)^{-1}D$, where D is the increase in demand. First obtain the technology matrix from the input-output table by dividing each column by its total:

$$A = \begin{bmatrix} 0.07216 & 0.00540 & 0.47754 & 0.21242 \\ 0.00165 & 0.01006 & 0.00244 & 0.02584 \\ 0.00503 & 0.00627 & 0.12763 & 0.03026 \\ 0.03324 & 0.03223 & 0.01189 & 0.14393 \end{bmatrix}.$$

Then compute the matrix $(I - A)^{-1}$.

TI-83/84 Plus Format: `(Identity(4)-[A])`$^{-1}$ Matrix Algebra Tool: `(I-A)^-1`

$$(I - A)^{-1} = \begin{bmatrix} 1.09155 & 0.01929 & 0.60157 & 0.29270 \\ 0.00295 & 1.01123 & 0.00488 & 0.03143 \\ 0.00779 & 0.00873 & 1.15118 & 0.04289 \\ 0.04260 & 0.03894 & 0.03953 & 1.18127 \end{bmatrix}$$

We are given external demand increases of $100 in agriculture and 0 in all others, so we multiply $(I - A)^{-1}$ by the column matrix

$$D = \begin{bmatrix} 100 \\ 0 \\ 0 \\ 0 \end{bmatrix} \text{ and obtain } X = (I - A)^{-1}D = \begin{bmatrix} 109.155 \\ 0.295 \\ 0.779 \\ 4.260 \end{bmatrix}$$

The additional production required from the meat and milk sector (Sector 3) is the (1,3)-entry: $0.78 (rounded to the nearest 1¢).

b. The diagonal entries in $(I - A)^{-1}$ show the additional production required from that sector to meet a $1 increase for the product of that sector. Since the largest diagonal entry is the 4,4-entry: 1.18127, we conclude that Sector 4 (other food products) requires the most of its own product in order to meet a $1 increase in external demand for that product.

Communication and reasoning exercises

35. It would mean that all of the sectors require neither their own product nor the product of any other sector.

37. The sum of the entries in a row of an input-output table gives the total internal demand for that sector's products. If that total was equal to the total output for that sector, it would mean that all of the output of that sector was used internally in the economy. Thus, none of the output was available for export and no importing was necessary.

39. If an entry in the matrix $(I - A)^{-1}$ is zero, then an increase in demand for one sector (the column sector) has no effect on the production of another sector (the row sector).

41. The off-diagonal entries in $(I - A)^{-1}$ show the additional production required by each sector to meet a one-unit increase for the product of some other sector. Usually, to produce one unit of one sector requires less than one unit of input from another. We would expect then that an increase in demand of one unit for one sector would require a smaller increase in production in another sector.

Chapter 4 Review

1. The sum of two matrices is defined only when they have the same dimensions. Because A is 2×3 and B is 2×2, their dimensions differ, so the sum $A + B$ is undefined.

3. $A^T = \begin{bmatrix} 1 & 4 \\ 2 & 5 \\ 3 & 6 \end{bmatrix}$, so $2A^T + C = 2\begin{bmatrix} 1 & 4 \\ 2 & 5 \\ 3 & 6 \end{bmatrix} + \begin{bmatrix} -1 & 0 \\ 1 & 1 \\ 0 & 1 \end{bmatrix} = \begin{bmatrix} 1 & 8 \\ 5 & 11 \\ 6 & 13 \end{bmatrix}$

5. $A^T B = \begin{bmatrix} 1 & 4 \\ 2 & 5 \\ 3 & 6 \end{bmatrix}\begin{bmatrix} 1 & -1 \\ 0 & 1 \end{bmatrix} = \begin{bmatrix} (1+0) & (-1+4) \\ (2+0) & (-2+5) \\ (3+0) & (-3+6) \end{bmatrix} = \begin{bmatrix} 1 & 3 \\ 2 & 3 \\ 3 & 3 \end{bmatrix}$

7. $B^2 = \begin{bmatrix} 1 & -1 \\ 0 & 1 \end{bmatrix}\begin{bmatrix} 1 & -1 \\ 0 & 1 \end{bmatrix} = \begin{bmatrix} (1+0) & (-1-1) \\ (0+0) & (0+1) \end{bmatrix} = \begin{bmatrix} 1 & -2 \\ 0 & 1 \end{bmatrix}$

9. $AC + B = \begin{bmatrix} 1 & 2 & 3 \\ 4 & 5 & 6 \end{bmatrix}\begin{bmatrix} -1 & 0 \\ 1 & 1 \\ 0 & 1 \end{bmatrix} + \begin{bmatrix} 1 & -1 \\ 0 & 1 \end{bmatrix} = \begin{bmatrix} 1 & 5 \\ 1 & 11 \end{bmatrix} + \begin{bmatrix} 1 & -1 \\ 0 & 1 \end{bmatrix} = \begin{bmatrix} 2 & 4 \\ 1 & 12 \end{bmatrix}$

11. To find the inverse of $\begin{bmatrix} 1 & -1 \\ 0 & 1 \end{bmatrix}$, we augment with the 2×2 identity matrix and row-reduce:

$\begin{bmatrix} 1 & -1 & 1 & 0 \\ 0 & 1 & 0 & 1 \end{bmatrix} \begin{matrix} R_1 + R_2 \\ \\ \end{matrix} \rightarrow \begin{bmatrix} 1 & 0 & 1 & 1 \\ 0 & 1 & 0 & 1 \end{bmatrix}$.

The right-hand 2×2 block is the desired inverse: $\begin{bmatrix} 1 & -1 \\ 0 & 1 \end{bmatrix}^{-1} = \begin{bmatrix} 1 & 1 \\ 0 & 1 \end{bmatrix}$.

13. To find the inverse of a 3×3 matrix, we augment it with the 3×3 identity matrix and row-reduce:

$\begin{bmatrix} 1 & 2 & 3 & 1 & 0 & 0 \\ 0 & 4 & 1 & 0 & 1 & 0 \\ 0 & 0 & 1 & 0 & 0 & 1 \end{bmatrix} \begin{matrix} 2R_1 - R_2 \\ \\ \\ \end{matrix} \rightarrow \begin{bmatrix} 2 & 0 & 5 & 2 & -1 & 0 \\ 0 & 4 & 1 & 0 & 1 & 0 \\ 0 & 0 & 1 & 0 & 0 & 1 \end{bmatrix} \begin{matrix} R_1 - 5R_3 \\ R_2 - R_3 \\ \\ \end{matrix} \rightarrow$

$\begin{bmatrix} 2 & 0 & 0 & 2 & -1 & -5 \\ 0 & 4 & 0 & 0 & 1 & -1 \\ 0 & 0 & 1 & 0 & 0 & 1 \end{bmatrix} \begin{matrix} (1/2)R_1 \\ (1/4)R_2 \\ \\ \end{matrix} \rightarrow \begin{bmatrix} 1 & 0 & 0 & 1 & -1/2 & -5/2 \\ 0 & 1 & 0 & 0 & 1/4 & -1/4 \\ 0 & 0 & 1 & 0 & 0 & 1 \end{bmatrix}$.

The right-hand 3×3 block is the desired inverse:

$\begin{bmatrix} 1 & 2 & 3 \\ 0 & 4 & 1 \\ 0 & 0 & 1 \end{bmatrix}^{-1} = \begin{bmatrix} 1 & -1/2 & -5/2 \\ 0 & 1/4 & -1/4 \\ 0 & 0 & 1 \end{bmatrix}$.

15. To find the inverse of a 4×4 matrix, we augment it with the 4×4 identity matrix and row-reduce:

$\begin{bmatrix} 1 & 2 & 3 & 4 & 1 & 0 & 0 & 0 \\ 2 & 3 & 3 & 3 & 0 & 1 & 0 & 0 \\ 0 & 1 & 2 & 3 & 0 & 0 & 1 & 0 \\ 0 & 0 & 1 & 2 & 0 & 0 & 0 & 1 \end{bmatrix} \begin{matrix} \\ R_2 - 2R_1 \\ \\ \\ \end{matrix} \rightarrow \begin{bmatrix} 1 & 2 & 3 & 4 & 1 & 0 & 0 & 0 \\ 0 & -1 & -3 & -5 & -2 & 1 & 0 & 0 \\ 0 & 1 & 2 & 3 & 0 & 0 & 1 & 0 \\ 0 & 0 & 1 & 2 & 0 & 0 & 0 & 1 \end{bmatrix} \begin{matrix} R_1 + 2R_2 \\ \\ R_3 + R_2 \\ \\ \end{matrix} \rightarrow$

$$\begin{bmatrix} 1 & 0 & -3 & -6 & -3 & 2 & 0 & 0 \\ 0 & -1 & -3 & -5 & -2 & 1 & 0 & 0 \\ 0 & 0 & \boxed{-1} & -2 & -2 & 1 & 1 & 0 \\ 0 & 0 & 1 & 2 & 0 & 0 & 0 & 1 \end{bmatrix} \begin{matrix} R_1 - 3R_3 \\ R_2 - 3R_3 \\ \\ R_4 + R_3 \end{matrix} \rightarrow \begin{bmatrix} 1 & 0 & 0 & 0 & 3 & -1 & -3 & 0 \\ 0 & -1 & 0 & 1 & 4 & -2 & -3 & 0 \\ 0 & 0 & -1 & -2 & -2 & 1 & 1 & 0 \\ 0 & 0 & 0 & 0 & -2 & 1 & 1 & 1 \end{bmatrix}.$$

Since the left-hand 4×4 block has a row of zeros, we cannot reduce the matrix to obtain the 4×4 identity on the left. Therefore, the matrix is singular.

17. $x + 2y = 0$

$3x + 4y = 2$

Matrix form:

$$\begin{bmatrix} 1 & 2 \\ 3 & 4 \end{bmatrix} \begin{bmatrix} x \\ y \end{bmatrix} = \begin{bmatrix} 0 \\ 2 \end{bmatrix}$$

Solving gives

$$\begin{bmatrix} x \\ y \end{bmatrix} = \begin{bmatrix} 1 & 2 \\ 3 & 4 \end{bmatrix}^{-1} \begin{bmatrix} 0 \\ 2 \end{bmatrix} = \begin{bmatrix} -2 & 1 \\ 3/2 & -1/2 \end{bmatrix} \begin{bmatrix} 0 \\ 2 \end{bmatrix} = \begin{bmatrix} 2 \\ -1 \end{bmatrix}.$$

Solution: $(x, y) = (2, -1)$

19. $x + y + z = 2$

$x + 2y + z = 3$

$x + y + 2z = 1$

Matrix form:

$$\begin{bmatrix} 1 & 1 & 1 \\ 1 & 2 & 1 \\ 1 & 1 & 2 \end{bmatrix} \begin{bmatrix} x \\ y \\ z \end{bmatrix} = \begin{bmatrix} 2 \\ 3 \\ 1 \end{bmatrix}$$

Solving gives

$$\begin{bmatrix} x \\ y \\ z \end{bmatrix} = \begin{bmatrix} 1 & 1 & 1 \\ 1 & 2 & 1 \\ 1 & 1 & 2 \end{bmatrix}^{-1} \begin{bmatrix} 2 \\ 3 \\ 1 \end{bmatrix} = \begin{bmatrix} 3 & -1 & -1 \\ -1 & 1 & 0 \\ -1 & 0 & 1 \end{bmatrix} \begin{bmatrix} 2 \\ 3 \\ 1 \end{bmatrix} = \begin{bmatrix} 2 \\ 1 \\ -1 \end{bmatrix}.$$

Solution: $(x, y, z) = (2, 1, -1)$

21. Reduce by dominance: Row 1 dominates row 2, so eliminate row 2: $\begin{bmatrix} 2 & 1 & 3 & 2 \\ 2 & 0 & 1 & 3 \end{bmatrix}$.

Now column 2 dominates all the other columns, so eliminate all but column 2: $\begin{bmatrix} 1 \\ 0 \end{bmatrix}$.

Row 1 dominates row 2, so we eliminate row 2, leaving just the single entry 1. This means that the optimal strategies are pure strategies, corresponding to the first row and second column: $R = \begin{bmatrix} 1 & 0 & 0 \end{bmatrix}$, $C = \begin{bmatrix} 0 & 1 & 0 & 0 \end{bmatrix}^T$ The corresponding expected value is the entry in the first row, second column, $e = 1$.

23. We begin by reducing by dominance.

The second row dominates the first, so eliminate the first: $\begin{bmatrix} -1 & 3 & 0 \\ 3 & 3 & -1 \end{bmatrix}$.

Now, the first column dominates the second, so eliminate the second: $\begin{bmatrix} -1 & 0 \\ 3 & -1 \end{bmatrix}$.

We can't reduce any further, so now we look for the players' optimal mixed strategies. We begin with the row player. Take $R = \begin{bmatrix} x & 1-x \end{bmatrix}$ and $C = \begin{bmatrix} 1 & 0 \end{bmatrix}^T$. $e = RPC = -4x + 3$. If we take $R = \begin{bmatrix} x & 1-x \end{bmatrix}$ and $C = \begin{bmatrix} 0 & 1 \end{bmatrix}^T$, we

186

get $f = RPC = x - 1$. Here are the graphs of e and f:

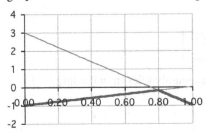

The lower (heavy) portion has its highest point at the intersection of the two lines:

$$-4x + 3 = x - 1 \quad \Rightarrow \quad -5x = -4 \quad \Rightarrow \quad x = 4/5 = 0.8$$

So the optimal row strategy is $R = \begin{bmatrix} 0.8 & 0.2 \end{bmatrix}$ for the 2×2 game, which corresponds to $\begin{bmatrix} 0 & 0.8 & 0.2 \end{bmatrix}$ for the original game.

For the column player we take $R = \begin{bmatrix} 1 & 0 \end{bmatrix}$ and $C = \begin{bmatrix} x & 1-x \end{bmatrix}^T$, getting $e = RPC = -x$. Taking $R = \begin{bmatrix} 0 & 1 \end{bmatrix}$ and $C = \begin{bmatrix} x & 1-x \end{bmatrix}^T$ we get $f = RPC = 4x - 1$. Here are the graphs of e and f:

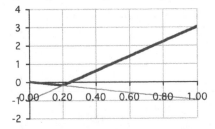

The upper (heavy) portion has its lowest point at the intersection of the two lines:

$$-x = 4x - 1 \quad \Rightarrow \quad -5x = -1 \quad \Rightarrow \quad x = 1/5 = 0.2.$$

So, the optimal column strategy is $C = \begin{bmatrix} 0.2 & 0.8 \end{bmatrix}^T$ for the 2×2 game, which corresponds to $\begin{bmatrix} 0.2 & 0 & 0.8 \end{bmatrix}^T$ for the original game.

The expected value is the second coordinate of either intersection point above, or the product RPC with the optimal strategies. In any case, it is $e = -0.2$.

25. $(I - A)^{-1} = \begin{bmatrix} 0.7 & -0.1 \\ 0 & 0.7 \end{bmatrix}^{-1} = \begin{bmatrix} 10/7 & 10/49 \\ 0 & 10/7 \end{bmatrix}$, and so $X = (I - A)^{-1}D = \begin{bmatrix} 1,100 \\ 700 \end{bmatrix}$

27. $(I - A)^{-1} = \begin{bmatrix} 0.8 & -0.2 & -0.2 \\ 0 & 0.8 & -0.2 \\ 0 & 0 & 0.8 \end{bmatrix}^{-1} = \begin{bmatrix} 5/4 & 5/16 & 25/64 \\ 0 & 5/4 & 5/16 \\ 0 & 0 & 5/4 \end{bmatrix}$, and so $X = (I - A)^{-1}D = \begin{bmatrix} 48,125 \\ 22,500 \\ 10,000 \end{bmatrix}$

29. Inventory $= \begin{bmatrix} 2,500 & 4,000 & 3,000 \\ 1,500 & 3,000 & 1,000 \end{bmatrix}$, Sales $= \begin{bmatrix} 300 & 500 & 100 \\ 100 & 600 & 200 \end{bmatrix}$, Purchases $= \begin{bmatrix} 400 & 400 & 300 \\ 200 & 400 & 300 \end{bmatrix}$

To obtain the inventory in the warehouses at the end of June, we subtract the June sales from the June 1 stock (inventory) and add the new books purchased:

Inventory − Sales + Purchases

$$= \begin{bmatrix} 2,500 & 4,000 & 3,000 \\ 1,500 & 3,000 & 1,000 \end{bmatrix} - \begin{bmatrix} 300 & 500 & 100 \\ 100 & 600 & 200 \end{bmatrix} + \begin{bmatrix} 400 & 400 & 300 \\ 200 & 400 & 300 \end{bmatrix} = \begin{bmatrix} 2,600 & 3,900 & 3,200 \\ 1,600 & 2,800 & 1,100 \end{bmatrix}.$$

31. Inventory on July 1 = $\begin{bmatrix} 2{,}600 & 3{,}900 & 3{,}200 \\ 1{,}600 & 2{,}800 & 1{,}100 \end{bmatrix}$ from Exercise 29

Sales each month = $\begin{bmatrix} 280 & 550 & 100 \\ 50 & 500 & 120 \end{bmatrix}$ Purchases each month = $\begin{bmatrix} 400 & 400 & 300 \\ 200 & 400 & 300 \end{bmatrix}$

To obtain the inventory in the warehouses x months after July 1, we subtract x times the monthly sales from the July 1 stock and add x times the new books purchased:

Inventory $-x$Sales $+x$Purchases

$$= \begin{bmatrix} 2{,}600 & 3{,}900 & 3{,}200 \\ 1{,}600 & 2{,}800 & 1{,}100 \end{bmatrix} -x \begin{bmatrix} 280 & 550 & 100 \\ 50 & 500 & 120 \end{bmatrix} +x \begin{bmatrix} 400 & 400 & 300 \\ 200 & 400 & 300 \end{bmatrix}$$

$$= \begin{bmatrix} 2{,}600 & 3{,}900 & 3{,}200 \\ 1{,}600 & 2{,}800 & 1{,}100 \end{bmatrix} +x \begin{bmatrix} 120 & -150 & 200 \\ 150 & -100 & 180 \end{bmatrix};$$

Nevada Sci Fi inventory is given by the (2,2)-entry: $2{,}800 + x(-500 + 400) = 2{,}950 - 100x$, which is zero when $x = 28$ months from July 1, or December 1 of next year.

33. Projected July sales figures are $\begin{bmatrix} 280 & 550 & 100 \\ 50 & 500 & 120 \end{bmatrix}$.

Revenue = Quantity × Price = $\begin{bmatrix} 280 & 550 & 100 \\ 50 & 500 & 120 \end{bmatrix} \begin{bmatrix} 5 \\ 6 \\ 5.5 \end{bmatrix} = \begin{bmatrix} 5{,}250 \\ 3{,}910 \end{bmatrix}$ Texas Nevada

35. Unknowns: $x =$ Number of shares purchased on July 1, $y =$ Number of shares purchased on August 1, $z =$ Number of shares purchased on September 1
Given information:

Total of 5,000 shares purchased: $x + y + z = 5{,}000$

Total of \$50,000 invested: $20x + 10y + 5z = 50{,}000$

Total dividends of \$300 on shares held as of August 15: $0.10(x + y) = 300$ $\Rightarrow$ $0.10x + 0.10y = 300$

We have three equations in three unknowns:

$$x + y + z = 5{,}000; \quad 20x + 10y + 5z = 50{,}000; \quad 0.10x + 0.10y = 300.$$

Matrix form: $\begin{bmatrix} 1 & 1 & 1 \\ 20 & 10 & 5 \\ 0.1 & 0.1 & 0 \end{bmatrix} \begin{bmatrix} x \\ y \\ z \end{bmatrix} = \begin{bmatrix} 5{,}000 \\ 50{,}000 \\ 300 \end{bmatrix}$

Solving gives

$$\begin{bmatrix} x \\ y \\ z \end{bmatrix} = \begin{bmatrix} 1 & 1 & 1 \\ 20 & 10 & 5 \\ 0.1 & 0.1 & 0 \end{bmatrix}^{-1} \begin{bmatrix} 5{,}000 \\ 50{,}000 \\ 300 \end{bmatrix} = \begin{bmatrix} -.5 & 0.1 & -5 \\ 0.5 & -.1 & 15 \\ 1 & 0 & -10 \end{bmatrix} \begin{bmatrix} 5{,}000 \\ 50{,}000 \\ 300 \end{bmatrix} = \begin{bmatrix} 1{,}000 \\ 2{,}000 \\ 2{,}000 \end{bmatrix}.$$

Solution: The company made the following investments: July 1: 1,000 shares, August 1: 2,000 shares, September 1: 2,000 shares

37. We first need to solve #35:

Unknowns: $x =$ Number of shares purchased on July 1, $y =$ Number of shares purchased on August 1. $z =$ Number of shares purchased on September 1
Given information:

Total of 5,000 shares purchased: $x + y + z = 5{,}000$

Total of \$50,000 invested: $20x + 10y + 5z = 50{,}000$

Total dividends of \$300 on shares held as of August 15: $0.10(x + y) = 300 \quad \Rightarrow \quad 0.10x + 0.10y = 300$

We have three equations in three unknowns:

$$x + y + z = 5{,}000; 20x + 10y + 5z = 50{,}000; 0.10x + 0.10y = 300.$$

Matrix form: $\begin{bmatrix} 1 & 1 & 1 \\ 20 & 10 & 5 \\ 0.1 & 0.1 & 0 \end{bmatrix} \begin{bmatrix} x \\ y \\ z \end{bmatrix} = \begin{bmatrix} 5{,}000 \\ 50{,}000 \\ 300 \end{bmatrix}$

Solving gives

$$\begin{bmatrix} x \\ y \\ z \end{bmatrix} = \begin{bmatrix} 1 & 1 & 1 \\ 20 & 10 & 5 \\ 0.1 & 0.1 & 0 \end{bmatrix}^{-1} \begin{bmatrix} 5{,}000 \\ 50{,}000 \\ 300 \end{bmatrix} = \begin{bmatrix} -.5 & 0.1 & -5 \\ 0.5 & -.1 & 15 \\ 1 & 0 & -10 \end{bmatrix} \begin{bmatrix} 5{,}000 \\ 50{,}000 \\ 300 \end{bmatrix} = \begin{bmatrix} 1{,}000 \\ 2{,}000 \\ 2{,}000 \end{bmatrix}.$$

To answer the current question, let us write the number of shares purchased (calculated in Exercise 35) as a row matrix:

Shares Purchased $= \begin{bmatrix} 1{,}000 & 2{,}000 & 2{,}000 \end{bmatrix}$.

Then we can calculate the loss by computing the total purchase cost and subtracting the total proceeds (dividends plus selling price):

Loss = Number of shares × (Purchase price − Dividends − Selling price)

$$= \begin{bmatrix} 1{,}000 & 2{,}000 & 2{,}000 \end{bmatrix} \left(\begin{bmatrix} 20 \\ 10 \\ 5 \end{bmatrix} - \begin{bmatrix} 0.10 \\ 0.10 \\ 0 \end{bmatrix} - \begin{bmatrix} 3 \\ 1 \\ 1 \end{bmatrix} \right) = \begin{bmatrix} 42{,}700 \end{bmatrix}.$$

39. July 1 customers $= \begin{bmatrix} 2{,}000 & 4{,}000 & 4{,}000 \end{bmatrix}$

Customers at the end of July $= \begin{bmatrix} 2{,}000 & 4{,}000 & 4{,}000 \end{bmatrix} \begin{bmatrix} 0.8 & 0.1 & 0.1 \\ 0.4 & 0.6 & 0 \\ 0.2 & 0 & 0.8 \end{bmatrix} = \begin{bmatrix} 4{,}000 & 2{,}600 & 3{,}400 \end{bmatrix}$

41. The matrix shows that no JungleBooks.com customers switched directly to FarmerBooks.com, so the only way to get to FarmerBooks.com is via OHaganBooks.com.

43. $P = \begin{bmatrix} 0 & -60 & -40 \\ 30 & 20 & 10 \\ 20 & 0 & 15 \end{bmatrix}, R = \begin{bmatrix} x & y & z \end{bmatrix}, C = \begin{bmatrix} 0.4 \\ 0.2 \\ 0.4 \end{bmatrix},$

$$e = RPC = \begin{bmatrix} x & y & z \end{bmatrix} \begin{bmatrix} 0 & -60 & -40 \\ 30 & 20 & 10 \\ 20 & 0 & 15 \end{bmatrix} \begin{bmatrix} 0.4 \\ 0.2 \\ 0.4 \end{bmatrix} = \begin{bmatrix} x & y & z \end{bmatrix} \begin{bmatrix} -28 \\ 20 \\ 14 \end{bmatrix} = -28 + 20y + 14z$$

The largest coefficient is the coefficient of y, so we take $R = \begin{bmatrix} 0 & 1 & 0 \end{bmatrix}$,

meaning that you should go with the "3 for 1" promotion. The resulting effect on your customer base is then

$$e = -28(0) + 20(1) + 14(0) = 20,$$

so you will gain 20,000 customers from JungleBooks.com.

45. Since JungleBooks.com now knows that OHaganBooks.com is assuming it will use the mixed column strategy of Exercise 43, it also knows that OHaganBooks must logically respond by going with the "3 for 1" promotion to counter this. Therefore, its best response will be to use its third strategy "3 for 2" and thus cut its losses to 10,000 customers. But, having seen the e-mail, O'Hagan knows this as well, and so its logical move will be to go with the *Finite Math* promo and thereby gain 15,000 customers!

47. As neither company is now certain about the strategy of the other, the fundamental principle of game theory comes into effect and so we solve the game to find OHaganBooks.com's optimal minimax strategy. You start by reducing by dominance, which leads to throwing out the "no promotion" options for both players, leaving the following 2×2

game:

$$P = \begin{bmatrix} 20 & 10 \\ 0 & 15 \end{bmatrix}.$$

We need only find your optimal strategy for the row player. Take $R = \begin{bmatrix} x & 1-x \end{bmatrix}$ and $C = \begin{bmatrix} 1 & 0 \end{bmatrix}^T$.

$e = RPC = 20x$. If we take $R = \begin{bmatrix} x & 1-x \end{bmatrix}$ and $C = \begin{bmatrix} 0 & 1 \end{bmatrix}^T T$, we get $f = RPC = -5x + 15$. Here are the graphs of e and f:

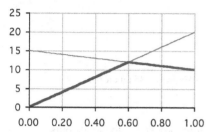

The lower (heavy) portion has its highest point at the intersection of the two lines:

$$20x = -5x + 15 \quad \Rightarrow \quad 25x = 15 \quad \Rightarrow \quad x = 3/5 = 0.6.$$

So the optimal row strategy is $R = \begin{bmatrix} 0.6 & 0.4 \end{bmatrix}$ for the 2×2 game, which corresponds to $\begin{bmatrix} 0 & 0.6 & 0.4 \end{bmatrix}$ for the original game. The expected value is the height of the point of intersection, $e = 20(0.6) = 12$, corresponding to a gain of 12,000 customers.

49. We obtain the technology matrix A by dividing each entry in the input-output table by the total in the last row:

$$A = \begin{bmatrix} 0.1 & 0.5 \\ 0.01 & 0.05 \end{bmatrix}$$

51. First, compute

$$(I - A)^{-1} = \left(\begin{bmatrix} 1 & 0 \\ 0 & 1 \end{bmatrix} - \begin{bmatrix} 0.1 & 0.5 \\ 0.01 & 0.05 \end{bmatrix} \right)^{-1} = \begin{bmatrix} 0.9 & -.5 \\ -0.01 & 0.95 \end{bmatrix}^{-1} = \begin{bmatrix} 19/17 & 10/17 \\ 1/85 & 18/17 \end{bmatrix}.$$

We are told that the total (external) demand for Bruno Mills' products is $\text{Demand} = \begin{bmatrix} 170 \\ 1,700 \end{bmatrix}$.

$$\text{Production} = (I - A)^{-1} \times \text{Demand} = \begin{bmatrix} 19/17 & 10/17 \\ 1/85 & 18/17 \end{bmatrix} \begin{bmatrix} 170 \\ 1,700 \end{bmatrix} = \begin{bmatrix} 1,190 \\ 1,802 \end{bmatrix}$$

Thus, $1,190 worth of paper and $1,802 worth of books must be produced.

Section 5.1

1. $2x + y \leq 10$

First sketch the graph of $2x + y = 10$ (graph on the left).

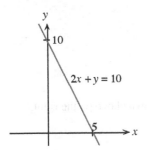
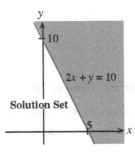

Choose $(0, 0)$ as a test point: $2(0) + (0) \leq 10.$ ✓

Since $(0, 0)$ is in the solution set, we block out the region on the other side of the line as shown above on the right. Since the solution set is not completely enclosed, it is unbounded.

3. $-x - 2y \leq 8$

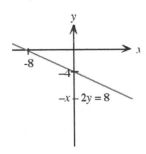
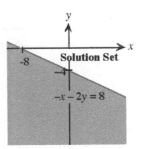

Choose $(0, 0)$ as a test point: $-0 - 2(0) \leq 8.$ ✓

Since $(0, 0)$ is in the solution set, we block out the region on the other side of the line as shown above on the right. Since the solution set is not completely enclosed, it is unbounded.

5. $3x + 2y \geq 5$

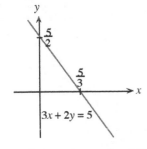
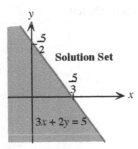

Choose $(0, 0)$ as a test point: $3(0) + 2(0) \geq 5.$ ✗

Since $(0, 0)$ is not in the solution set, we block out the region on the same side of the line as shown above on the right. Since the solution set is not completely enclosed, it is unbounded.

7. $x \leq 3y$

To sketch $x = 3y$, solve for y to obtain $y = \frac{1}{3}x$. This is a line of slope $\frac{1}{3}$ passing through the origin.

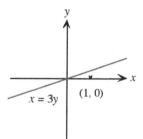

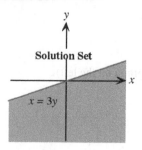

As $(0, 0)$ is on the line, we choose another test point—say, $(1, 0)$: $1 \leq 3(0)$. ✗

Since $(1, 0)$ is not in the solution set, we block out the region on the same side of the line as shown above on the right. Since the solution set is not completely enclosed, it is unbounded.

9. $\dfrac{3x}{4} - \dfrac{y}{4} \leq 1$

To sketch the associated line, replace the inequality by equality and solve for y:

$$\frac{3x}{4} - \frac{y}{4} = 1 \Rightarrow \frac{y}{4} = \frac{3x}{4} - 1 \Rightarrow y = 3x - 4.$$

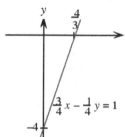

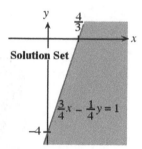

Choose $(0, 0)$ as a test point: $\dfrac{3(0)}{4} - \dfrac{0}{4} \leq 1$. ✓

Since $(0, 0)$ is in the solution set, we block out the region on the other side of the line as shown above on the right. Since the solution set is not completely enclosed, it is unbounded.

11. $x \geq -5$

The line $x = -5$ is a vertical line as shown on the left:

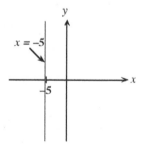

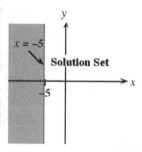

Choose $(0, 0)$ as a test point: $0 \geq -5$. ✓

Since $(0, 0)$ is in the solution set, we block out the region on the other side of the line as shown above on the right. Since the solution set is not completely enclosed, it is unbounded.

13. $4x - y \leq 8$ $x + 2y \leq 2$

The two associated lines are shown below on the left:

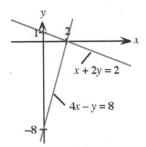 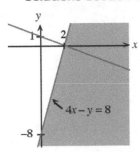 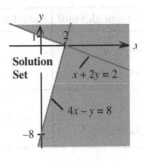

Choose $(0,0)$ as a test point for the region $4x - y \le 8$: $\qquad 4(0) - 0 \le 8.$ ✓

Therefore, we shade to the right of the line $4x - y = 8$ as shown above center.

Choose $(0,0)$ as a test point for the region $x + 2y \le 2$: $\qquad 0 + 2(0) \le 2.$ ✓

Therefore, we shade above the line $x + 2y = 2$ as shown above on the right.

The white region shown is the solution set, which, being not entirely enclosed, is unbounded.
For the corner point, we solve the system

$$4x - y = 8 \qquad x + 2y = 2.$$

Multiplying the first equation by 2 and adding gives $9x = 18$, so $x = 2$.

Substituting $x = 2$ in the first equation gives $8 - y = 8$, so $y = 0$.

Therefore, the corner point is $(2, 0)$.

15. $3x + 2y \ge 6 \qquad 3x - 2y \le 6 \qquad x \ge 0$
The three associated lines are shown below on the left.

Choose $(0,0)$ as a test point for the region $3x + 2y \ge 6$: $\qquad 3(0) + 2(0) \ge 6.$ ✗

Therefore, we shade to the left of the line $3x + 2y = 6$ as shown in the second figure below.

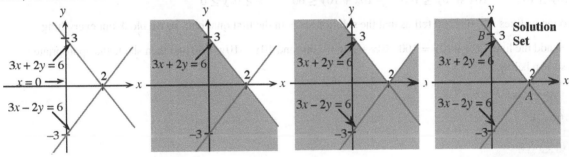

Now choose $(0,0)$ as a test point for the region $3x - 2y \le 6$: $\qquad 3(0) - 2(0) \le 6.$ ✓

Therefore, we shade below the line $3x - 2y = 6$ as shown in the third figure above.

Now choose $(1,0)$ as a test point for the region $x \ge 0$: $\qquad 1 \ge 0.$ ✓

Therefore, we shade to the left of $x = 0$ as shown above on the right, leaving the solution set as the unshaded area.

Again, the solution set is unbounded, since it is no entirely enclosed. The corner points are shown in the following table:

Corner Point	Lines through Point	Coordinates
A	$3x + 2y = 6$ $3x - 2y = 6$	$(2, 0)$
B	$x = 0$ $3x + 2y = 6$	$(0, 3)$

17. $x + y \geq 5$ $x \leq 10$ $y \leq 8$ $x \geq 0, y \geq 0$

The last two inequalities $x \geq 0, y \geq 0$ tell us that the solution set is in the first quadrant, so we block out everything else, as shown on the left below:

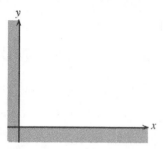
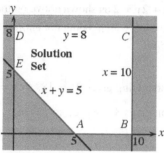

We then add the lines $x + y = 5, x = 10,$ and $y = 8$ and then shade the appropriate regions, as shown on the right above. The solution set is bounded, since it is entirely enclosed.

Corner points: We can easily read these off the graph: A:$(5, 0)$, B:$(10, 0)$, C:$(10, 8)$, D:$(0, 8)$, E:$(0, 5)$.

19. $20x + 10y \leq 100$ $10x + 20y \leq 100$ $10x + 10y \leq 60$ $x \geq 0, y \geq 0$

The last two inequalities $x \geq 0, y \geq 0$ tell us that the solution set is in the first quadrant, so we block out everything

else. We then add the lines $20x + 10y = 100, \ 10x + 20y = 100,$ and $10x + 10y = 60$ and then shade the appropriate regions, as shown below.

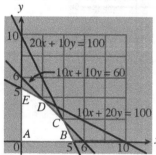

The solution set is the bounded white region. The corner points are shown in the following table:

Corner Point	Lines through Point	Coordinates
A	$x = 0, y = 0$	$(0, 0)$
B	$y = 0$ $20x + 10y = 100$	$(5, 0)$
C	$20x + 10y = 100$ $10x + 10y = 60$	$(4, 2)$
D	$10x + 10y = 60$ $10x + 20y = 100$	$(2, 4)$
E	$x = 0$ $10x + 20y = 100 (0, 5)$	

21. $20x + 10y \geq 100$ $10x + 20y \geq 100$ $10x + 10y \geq 80$ $x \geq 0, y \geq 0$

Proceeding as before, we obtain the unbounded solution set shown below:

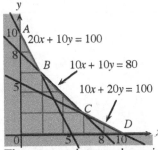

The corner points are shown in the following table:

Corner Point	Lines through Point	Coordinates
A	$x = 0$ $20x + 10y = 100$	$(0, 10)$
B	$20x + 10y = 100$ $10x + 10y = 80$	$(2, 6)$
C	$10x + 10y = 80$ $10x + 20y = 100$	$(6, 2)$
D	$10x + 20y = 100$ $y = 0 (10, 0)$	

23. $-3x + 2y \leq 5$ $3x - 2y \leq 6$ $x \leq 2y$ $x \geq 0, y \geq 0$

Solution set (unbounded):

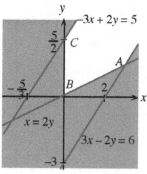

Corner points: We can read two of the corner points, $B:(0,0)$ and $C:(0, \frac{5}{2})$, directly from the graph. The remaining corner point is A, at the intersection of the lines $x = 2y$ and $3x - 2y = 6$. Solving this system of two equations gives the point A as $A:(3, \frac{3}{2})$.

25. $2x - y \geq 0 \qquad x - 3y \leq 0 \qquad x \geq 0, y \geq 0$

To sketch the lines, solve for y in each case:

$$2x - y = 0 \Rightarrow y = 2x \qquad x - 3y = 0 \Rightarrow y = \frac{1}{3}x.$$

Solution set (unbounded):

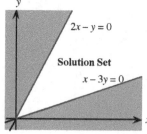

The only corner point is the origin: $(0,0)$.

27. To draw the region $2.1x - 4.3y \geq 9.7$ using technology, solve the associated equation for y:

$$4.3y = 2.1x - 9.7 \Rightarrow y = \frac{2.1}{4.3}x - \frac{8.7}{4.3}.$$

Solution set (unbounded):

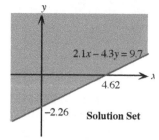

29. To draw the region $-0.2x + 0.7y \geq 3.3$; $1.1x + 3.4y \geq 0$ using technology, solve the associated equations for y:

$$y = \frac{0.2}{0.7}x + \frac{3.3}{0.7} = \frac{2}{7}x + \frac{33}{7} \qquad y = -\frac{1.1}{3.4}x = -\frac{11}{34}x.$$

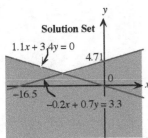

To obtain the coordinates of (the only) corner point, zoom in to it until you can read off the coordinates to two decimal places: $(-7.74, 2.50)$.

31. To draw the region $4.1x - 4.3y \le 4.4$; $7.5x - 4.4y \le 5.7$; $4.3x + 8.5y \le 10$ using technology, solve the associated equations for y:

$$y = \frac{4.1}{4.3}x - \frac{4.4}{4.3} \qquad y = \frac{7.5}{4.4}x - \frac{5.7}{4.4} \qquad y = -\frac{4.3}{8.5}x + \frac{10}{4.3}.$$

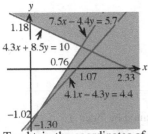

To obtain the coordinates of the corner points, zoom in to it until you can read off the coordinates to two decimal places: A:$(0.36, -0.68)$, B:$(1.12, 0.61)$.

Applications

33. Unknowns: $x =$ Number of quarts of Creamy Vanilla, $y =$ Number of quarts of Continental Mocha

Arrange the given information in a table with unknowns across the top:

	Vanilla (x)	Mocha (y)	Available
Eggs	2	1	500
Cream	3	3	900

We can now set up an inequality for each of the items listed on the left:

Eggs: $2x + y \le 500$

Cream: $3x + 3y \le 900$ or $x + y \le 300$

Since the factory cannot manufacture negative amounts, we also have $x \ge 0, y \ge 0$.

The solution and corner points are shown below:

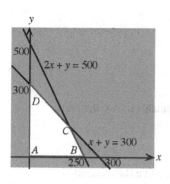

Corner Point	Lines through Point	Coordinates
A	$x = 0, y = 0$	$(0, 0)$
B	$y = 0$ $2x + y = 500$	$(250, 0)$
C	$2x + y = 500$ $x + y = 300$	$(200, 100)$
D	$x + y = 300$ $x = 0$	$(0, 300)$

35. Unknowns: $x =$ Number of ounces of chicken, $y =$ Number of ounces of grain

Arrange the given information in a table with unknowns across the top:

	Chicken (x)	Grain (y)	Required
Protein	10	2	200
Fat	5	2	150

We can now set up an inequality for each of the items listed on the left (note that "at least" is represented by "\geq"):

Protein: $10x + 2y \geq 200$, or $5x + y \geq 100$

Fat: $5x + 2y \geq 150$

The amounts of ingredients cannot be negative: $x \geq 0, y \geq 0$.

The solution set is the unshaded region shown below:

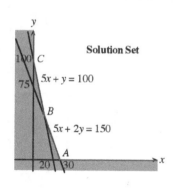

Corner Point	Lines through Point	Coordinates
A	$y = 0$ $5x + 2y = 150$	$(30, 0)$
B	$5x + 2y = 150$ $5x + y = 100$	$(10, 50)$
C	$5x + y = 100$ $x = 0$	$(0, 100)$

37. Unknowns: $x =$ Number of servings of Mixed Cereal for Baby, $y =$ Number of servings of Mango Tropical Fruit Dessert

Arrange the given information in a table with unknowns across the top:

	Cereal (x)	Mango (y)	Required
Calories	60	80	140
Carbs.	11	21	32

We can now set up an inequality for each of the items listed on the left (note that "at least" is represented by "\geq"):

Calories: $60x + 80y \geq 140$ or $3x + 4y \geq 7$

Carbs: $11x + 21y \geq 32$

The values of the unknowns cannot be negative: $x \geq 0, y \geq 0$.

The solution set is the unshaded region shown below:

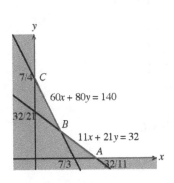

Corner Point	Lines through Point	Coordinates
A	$y = 0$ $11x + 21y = 32$	$(32/11, 0)$
B	$11x + 21y = 32$ $3x + 4y = 7$	$(1, 1)$
C	$3x + 4y = 7$ $x = 0$	$(0, 7/4)$

39. Unknowns: $x = $ Number of dollars in MHI, $y = $ Number of dollars in BKK

Total invested is up to $80,000: $x + y \leq 80,000$.

Interest earned is at least $4,200: $0.06x + 0.05y \geq 4,200$ or $6x + 5y \geq 420,000$.

The values of the unknowns cannot be negative: $x \geq 0, y \geq 0$.

The solution set is the unshaded region shown below:

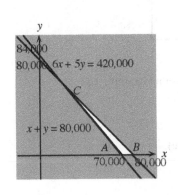

Corner Point	Lines through Point	Coordinates
A	$y = 0$ $6x + 5y = 420,000$	$(70,000, 0)$
B	$y = 0$ $x + y = 80,000$	$(80,000, 0)$
C	$x + y = 80,000$ $6x + 5y = 420,000$	$(20,000, 60,000)$

41. Unknowns: $x = $ Number of shares of TD; $y = $ Number of shares of CNA

You have up to $25,000 to invest: $45x + 40y \leq 25,000$.

You wish to earn at least $760 in dividends:

TD dividend $= 4\%$ of $45x$ invested $= 0.04(45x) = 1.8x$

CNA dividend $= 2.5\%$ of $40y$ invested $= 0.025(40y) = y$.

Thus, $1.8x + y \geq 760$.

The values of the unknowns cannot be negative: $x \geq 0, y \geq 0$.

The solution set is the unshaded region shown below:

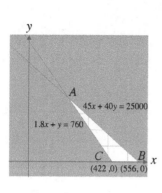

Corner Point	Lines through Point	Coordinates (rounded)
A	$45x + 40y = 25,000$ $1.8x + y = 760$	$(200, 400)$
B	$45x + 40y = 25,000$ $y = 0$	$(556, 0)$
C	$1.8x + y = 760$ $y = 0$	$(422, 0)$

43. Unknowns: $x =$ Number of full-page ads in *Sports Illustrated*, $y =$ Number of full-page ads in *GQ*

Readership: $0.65x + 0.15y \geq 3$

At least 3 full-page ads in each magazine: $x \geq 3, y \geq 3$

The values of the unknowns cannot be negative: $x \geq 0, y \geq 0$.

The solution set is the unshaded region shown below:

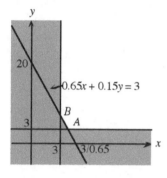

Corner Point	Lines through Point	Coordinates (rounded)
A	$y = 3$ $0.65x + 0.15y = 3$	$(4, 3)$
B	$x = 3$ $0.65x + 0.15y = 3$	$(3, 7)$

45. Many of the systems of inequalities in the earlier exercises have unbounded solution sets. Another example is:
$x \geq 0, y \geq 0, x + y \geq 1$.

47. The given triangle is the region enclosed by the lines $x = 0, y = 0,$ and $x + 2y = 2$ (see figure).

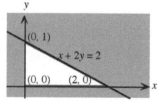

Thus, the region can be described as the solution set of the system $x \geq 0, y \geq 0, x + 2y \leq 2$.

49. (Answers may vary.) One limitation is that the method is suitable only for situations with two unknown quantities. Another limitation is accuracy, which is never perfect when graphing.

51. There should be at least 3 more grams of ingredient A than ingredient B. Rephrasing this statement gives:
The number of grams of ingredient A exceeds the grams of ingredient B by 3: $x - y \geq 3$ — Choice (C).

53. There should be at least 3 parts (by weight) of ingredient A to 2 parts of ingredient B. That is, $3/2 = 1.5$ parts of ingredient A to 1 part of ingredient B. Rephrasing this statement gives: The number of grams of ingredient A is 1.5

times the number of grams of ingredient B:

$x \geq 1.5y \Rightarrow 2x \geq 3y \Rightarrow 2x - 3y \geq 0$ — Choice (B).

55. There are no feasible solutions; that is, it is impossible to satisfy all the constraints.

57. Answers will vary.

Section 5.2

1. Maximize $p = x + y$ subject to $x + 2y \leq 9$, $2x + y \leq 9$, $x \geq 0$, $y \geq 0$.

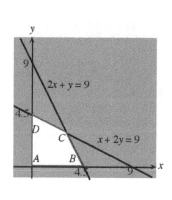

Corner Point	Lines through Point	Coordinates	$p = x + y$
A	$x = 0$, $y = 0$	$(0, 0)$	0
B	$y = 0$ $2x + y = 9$	$(4.5, 0)$	4.5
C	$2x + y = 9$ $x + 2y = 9$	$(3, 3)$	6
D	$x + 2y = 9$ $x = 0$	$(0, 4.5)$	4.5

Maximum value occurs at point C: $p = 6$ for $x = 3$, $y = 3$.

3. Minimize $c = x + y$ subject to $x + 2y \geq 6$, $2x + y \geq 6$, $x \geq 0$, $y \geq 0$.

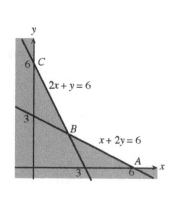

Corner Point	Lines through Point	Coordinates	$c = x + y$
A	$y = 0$ $x + 2y = 6$	$(6, 0)$	6
B	$x + 2y = 6$ $2x + y = 6$	$(2, 2)$	4
C	$2x + y = 6$ $x = 0$	$(0, 6)$	6

Although the feasible region is unbounded, there is no need to add a bounding rectangle since, by the FAQ at the end of the section in the text:

If you are minimizing $c = ax + by$ with a and b nonnegative, $x \geq 0$, and $y \geq 0$, then optimal solutions always exist.

Minimum value occurs at point B: $c = 4$, $x = 2$, $y = 2$.

5. Maximize $p = 3x + y$ subject to $3x - 7y \leq 0$, $7x - 3y \geq 0$, $x + y \leq 10$, $x \geq 0$, $y \geq 0$.

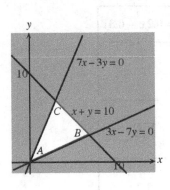

Corner Point	Lines through Point	Coordinates	$p = 3x + y$
A	$x = 0$, $y = 0$	$(0, 0)$	0
B	$3x - 7y = 0$ $x + y = 10$	$(7, 3)$	24
C	$7x - 3y = 0$ $x + y = 10$	$(3, 7)$	16

Maximum value occurs at point B: $p = 24$, $x = 7$, $y = 3$.

7. Maximize $p = 3x + 2y$ subject to $0.2x + 0.1y \leq 1$, $0.15x + 0.3y \leq 1.5$, $10x + 10y \leq 60$, $x \geq 0$, $y \geq 0$.

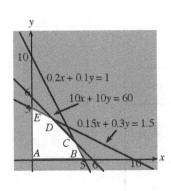

Corner Point	Lines through Point	Coordinates	$p = 3x + 2y$
A	$x = 0$, $y = 0$	$(0, 0)$	0
B	$y = 0$ $0.2x + 0.1y = 1$	$(5, 0)$	15
C	$0.2x + 0.1y = 1$ $10x + 10y = 60$	$(4, 2)$	16
D	$10x + 10y = 60$ $0.15x + 0.3y = 1.5$	$(2, 4)$	14
E	$0.15x + 0.3y = 1.5$ $x = 0$	$(0, 5)$	10

Maximum value occurs at point C: $p = 16$, $x = 4$, $y = 2$.

9. Minimize $c = 0.2x + 0.3y$ subject to $0.2x + 0.1y \geq 1$, $0.15x + 0.3y \geq 1.5$, $10x + 10y \geq 80$, $x \geq 0$, $y \geq 0$.

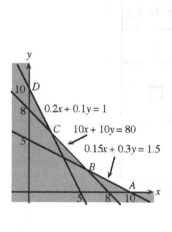

Corner Point	Lines through Point	Coordinates	$c = 0.2x + 0.3y$
A	$y = 0$ $0.15x + 0.3y = 1.5$	$(10, 0)$	2
B	$0.15x + 0.3y = 1.5$ $10x + 10y = 80$	$(6, 2)$	1.8
C	$10x + 10y = 80$ $0.2x + 0.1y = 1$	$(2, 6)$	2.2
D	$0.2x + 0.1y = 1$ $x = 0$	$(0, 10)$	3

Although the feasible region is unbounded, there is no need to add a bounding rectangle as the coefficients of c are nonnegative.

Maximum value occurs at point B: $c = 1.8$, $x = 6$, $y = 2$.

11. Maximize and minimize $p = x + 2y$ subject to $x + y \geq 2$, $x + y \leq 10$, $x - y \leq 2$, $x - y \geq -2$.

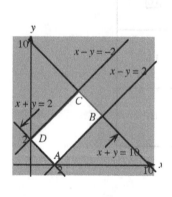

Corner Point	Lines through Point	Coordinates	$p = x + 2y$
A	$x + y = 2$ $x - y = 2$	$(2, 0)$	2 (Min)
B	$x - y = 2$ $x + y = 10$	$(6, 4)$	14
C	$x + y = 10$ $x - y = -2$	$(4, 6)$	16 (Max)
D	$x - y = -2$ $x + y = 2$	$(0, 2)$	4

Maximum value occurs at point C: $p = 16$, $x = 4$, $y = 6$; minimum value occurs at point A: $p = 2$, $x = 2$, $y = 0$.

13. Maximize $p = 2x - y$ subject to $x + 2y \geq 6$, $x \leq 8$, $x \geq 0$, $y \geq 0$.

The feasible region is unbounded, and the problem does not satisfy any of the conditions in the FAQ at the end of the text for this section, so we need to add a bounding rectangle, adding two new corner points to the feasible region as shown:

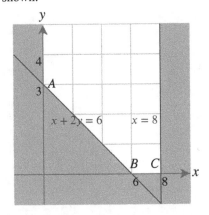 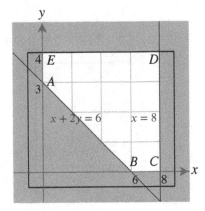

Corner Point	Lines through Point	Coordinates	$p = 2x - y$
A	$x + 2y = 6$ $x = 0$	$(0, 3)$	-3
B	$x + 2y = 6$ $y = 0$	$(6, 0)$	12
C	$x = 8$, $y = 0$	$(8, 0)$	16 (Max)
D	$x = 8$, $y = 4$	$(8, 4)$	12
E	$x = 0$, $y = 4$	$(0, 4)$	-4

The maximum occurs at the point C, which is a corner point of the original feasible region, so the LP problem has an optimal solution: $p = 16$, $x = 8$, $y = 0$.

15. Maximize $p = 2x + 3y$ subject to $0.1x + 0.2y \geq 1$, $2x + y \geq 10$, $x \geq 0$, $y \geq 0$.

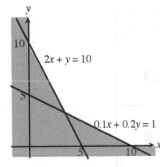

This is an unbounded region, and we wish to maximize $p = 2x + 3y$. According to the FAQ at the end of the text for this section, if the feasible region is unbounded:

If you are maximizing $p = ax + by$ with a and both positive, then there is no optimal solution.

Thus, there is no optimal solution.

17. Minimize $c = x - 3y$ subject to $3x + y \geq 5$, $2x - y \geq 0$, $x - 3y \leq 0$, $x \geq 0$, $y \geq 0$.

The feasible region is unbounded, and the problem does not satisfy any of the conditions in the FAQ at the end of the text for this section, so we need to add a bounding rectangle, adding two new corner points to the feasible region as shown:

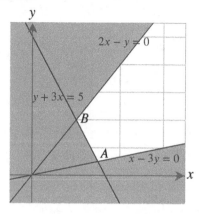 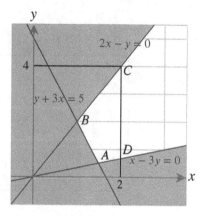

Corner Point	Lines through Point	Coordinates	$c = x - 3y$
A	$3x + y = 5$ $x - 3y = 0$	$(3/2, 1/2)$	0
B	$3x + y = 5$ $2x - y = 0$	$(1, 2)$	-5
C	$x = 2$ $2x - y = 0$	$(2, 4)$	-10 (Min)
D	$x = 2$ $x - 3y = 0$	$(2, 2/3)$	0

The minimum occurs only at the point C, which is not a corner point of the original feasible region, so the LP problem has no optimal solution. (The value of the objective function is unbounded.)

19. Minimize $c = 2x + 4y$ subject to $0.1x + 0.1y \geq 1$, $x + 2y \geq 14$, $x \geq 0$, $y \geq 0$.

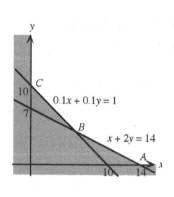

Corner Point	Lines through Point	Coordinates	$c = 2x + 4y$
A	$y = 0$ $x + 2y = 14$	$(14, 0)$	$\boxed{28}$
B	$x + 2y = 14$ $0.1x + 0.1y = 1$	$(6, 4)$	$\boxed{28}$
C	$0.1x + 0.1y = 1$ $x = 0$	$(0, 10)$	40

Although the feasible region is unbounded, there is no need to add a bounding rectangle since the coefficients of c are nonnegative.

Minimum value occurs at points A and B: $c = 28$; $(x, y) = (14, 0)$ and $(6, 4)$, and the line connecting them.

21. Minimize $c = 3x - 3y$ subject to $\dfrac{x}{4} \leq y$, $y \leq \dfrac{2x}{3}$, $x + y \geq 5$, $x + 2y \leq 10$, $x \geq 0$, $y \geq 0$.

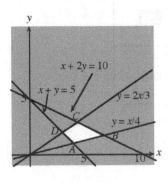

Corner Point	Lines through Point	Coordinates	$c = 3x - 3y$
A	$y = x/4$ $x + y = 5$	$(4, 1)$	9
B	$y = x/4$ $x + 2y = 10$	$(20/3, 5/3)$	15
C	$x + 2y = 10$ $y = 2x/3$	$(30/7, 20/7)$	30/7
D	$y = 2x/3$ $x + y = 5$	$(3, 2)$	3

Minimum value occurs at point D: $c = 3$, $x = 3$, $y = 2$.

23. Maximize $p = x + y$ subject to $x + 2y \geq 10$, $2x + 2y \leq 10$, $2x + y \geq 10$, $x \geq 0$, $y \geq 0$.

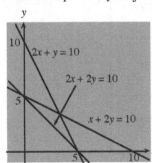

Feasible region is empty—no solutions.

Applications

25. (To set up the inequalities, see the solution to #33 in the preceding section.)

Unknowns: $x = $ # quarts of Creamy Vanilla, $y = $ # quarts of Continental Mocha

Maximize $p = 3x + 2y$ subject to $2x + y \leq 500$, $3x + 3y \leq 900$ (or $x + y \leq 300$), $x \geq 0$, $y \geq 0$.

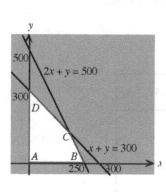

Corner Point	Lines through Point	Coordinates	$p = 3x + 2y$
A	$x = 0,\ y = 0$	$(0, 0)$	0
B	$y = 0$ $2x + y = 500$	$(250, 0)$	750
C	$2x + y = 500$ $x + y = 300$	$(200, 100)$	800
D	$x + y = 300$ $x = 0$	$(0, 300)$	600

Maximum value occurs at the point C: $p = 800$; $x = 200$, $y = 100$.

Solution: You should make 200 quarts of Creamy Vanilla and 100 quarts of Continental Mocha.

27. (To set up the inequalities, see the solution to #35 in the preceding section.)

Unknowns: $x =$ # ounces of chicken, $y =$ # ounces of grain

Minimize $c = 10x + y$ subject to $10x + 2y \geq 200$, $5x + 2y \geq 150$, $x \geq 0$, $y \geq 0$.

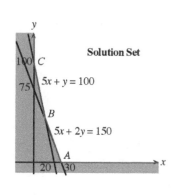

Corner Point	Lines through Point	Coordinates	$c = 10x + y$
A	$y = 0$ $5x + 2y = 150$	$(30, 0)$	300
B	$5x + 2y = 150$ $5x + y = 100$	$(10, 50)$	150
C	$5x + y = 100$ $x = 0$	$(0, 100)$	100

Although the feasible region is unbounded, there is no need to add a bounding rectangle as the coefficients of c are nonnegative.

Minimum value occurs at the point C: $c = 800$; $x = 0$, $y = 100$.

Solution: Ruff, Inc., should use 100 oz of grain and no chicken.

29. (To set up the inequalities, see the solution to #37 in the preceding section.)

Unknowns: $x =$ # servings of Mixed Cereal $y =$ # servings of Mango Tropical Fruit

Minimize $c = 30x + 50y$ subject to $11x + 21y \geq 32$, $3x + 4y \geq 7$, $x \geq 0$, $y \geq 0$.

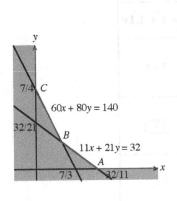

Corner Point	Lines through Point	Coordinates	$c = 30x + 50y$
A	$y = 0$ $11x + 21y = 32$	$(32/11, 0)$	87.3
B	$11x + 21y = 32$ $3x + 4y = 7$	$(1, 1)$	$\boxed{80}$
C	$3x + 4y = 7$ $x = 0$	$(0, 7/4)$	87.5

Although the feasible region is unbounded, there is no need to add a bounding rectangle as the coefficients of c are nonnegative.

Minimum value occurs at the point B: $c = 80$; $x = 1$, $y = 1$.

Solution: Feed your child 1 serving of cereal and 1 serving of dessert.

31. Unknowns: $x = $ # compact fluorescent light bulbs, $y = $ # square ft of insulation

Maximize $p = 2x + 0.2y$ subject to $4x + y \leq 1{,}200$, $x \leq 60$, $y \leq 1{,}100$, $x \geq 0$, $y \geq 0$.

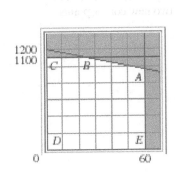

Corner Point	Lines through Point	Coordinates	$p = 2x + 0.2y$
A	$4x + y = 1{,}200$ $x = 60$	$(60, 960)$	$\boxed{312}$
B	$y = 1{,}100$ $4x + y = 1{,}200$	$(25, 1{,}100)$	270
C	$y = 1{,}100$, $x = 0$	$(0, 1{,}100)$	220
D	$y = 0\ x = 0$	$(0, 0)$	0
E	$x = 60$, $y = 0$	$(60, 0)$	120

Maximum value occurs at the point A: $p = 312$; $x = 60$, $y = 960$.

Solution: Purchase 60 compact fluorescent light bulbs and 960 square feet of insulation for a total saving of $312 per year in energy costs.

33. Unknowns: $x = $ # servings of Xtend, $y = $ # servings of Gainz

Minimize $c = x + 1.1y$ subject to $2y \geq 4$, $2.5x + 3y \geq 36$, $7x + 6y \geq 84$, $x \geq 0$, $y \geq 0$.

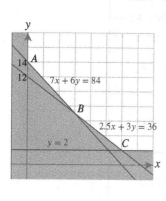

Corner Point	Lines through Point	Coordinates	$c = x + 1.1y$
A	$7x + 6y = 84$ $x = 0$	$(0, 14)$	15.4
B	$2.5x + 3y = 36$ $7x + 6y = 84$	$(6, 7)$	$\boxed{13.7}$
C	$2.5x + 3y = 36$ $y = 2$	$(12, 2)$	14.2

Although the feasible region is unbounded, there is no need to add a bounding rectangle as the coefficients of c are nonnegative.

Minimum value occurs at the point B: $c = 13.7$; $x = 6$, $y = 7$.

Solution: Mix 6 servings of Xtend and 7 servings of Gainz for a cost of $13.70.

35. Unknowns: $x = \#$ servings of Gainz, $y = \#$ servings of Strongevity.

Constraints: $2x + 2.5y \geq 40$, $3x + y \geq 38$, $6x \leq 90$, $x \geq 0$, $y \geq 0$.

In part (a) you are maximizing $p = x - y$ and in part (b) you are maximizing $p = y - x$.

The feasible region is unbounded, and the LP problems in both parts (a) and (b) do not satisfy any of the conditions in the FAQ at the end of the text for this section, so we need to add a bounding rectangle, adding two new corner points to the feasible region as shown:

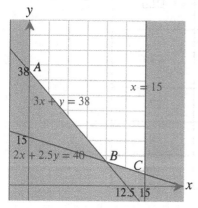
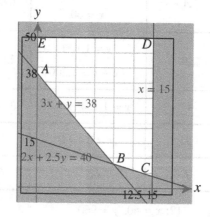

Corner Point	Lines through Point	Coordinates	a. $p = x - y$	b. $p = y - x$
A	$3x + y = 38$ $x = 0$	$(0, 38)$	-38	38
B	$2x + 2.5y = 40$ $3x + y = 38$	$(10, 8)$	2	-2
C	$2x + 2.5y = 40$ $x = 15$	$(15, 4)$	$\boxed{11}$	-11
D	$x = 15, \ y = 50$	$(15, 50)$	-35	35
E	$x = 0, \ y = 50$	$(0, 50)$	-50	$\boxed{50}$

a. The maximum occurs at the point C, which is a corner point of the original feasible region, so the LP problem has an optimal solution: $p = 11$, $x = 15$, $y = 4$. Use 15 servings of Gainz and 4 servings of Strongevity.

b. The maximum occurs only at the point E, which is not a corner point of the original feasible region, so the LP problem has no optimal solution. (The value of the objective function is unbounded.)

How do we understand this result in terms of the supplements? Note that we can make $p = y - x$ as large as we like by increasing y, which means going vertically upwards in the graph on the left, without leaving the feasible region.

Increasing y corresponds to increasing the amount of Strongevity. So, you could use as much Strongevity as you like without violating your trainer's specifications (Strongevity contains no BCAAs).

37. Unknowns: $x = $ # Dracula salamis, $y = $ # Frankenstein sausages

Maximize $p = x + 3y$ subject to $x + 2y \le 1,000$, $3x + 2y \le 2,400$, $y \le 2x$, $x \ge 0$, $y \ge 0$.

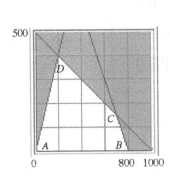

Corner Point	Lines through Point	Coordinates	$p = x + 3y$
A	$x = 0, \ y = 0$	$(0, 0)$	0
B	$3x + 2y = 2,400$ $y = 0$	$(800, 0)$	800
C	$x + 2y = 1,000$ $3x + 2y = 2,400$	$(700, 150)$	$1,150$
D	$x + 2y = 1,000$ $y - 2x = 0$	$(200, 400)$	$\boxed{1,400}$

Maximum value occurs at the point D: $p = 1,400$; $x = 200$, $y = 400$.

Solution: You should make 200 Dracula Salamis and 400 Frankenstein Sausages, for a profit of $1,400.

39. Unknowns: $x = $ # spots on *The Big Bang Theory*, $y = $ # spots on *American Dad*

Maximize $E = 1.8x + 1.5y$, subject to $x + y \ge 30$, $3x + y \le 120$, $-x + y \le 0$, $x \ge 0$, $y \ge 0$.

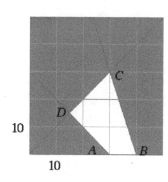

Corner Point	Lines through Point	Coordinates	$E = 1.8x + 1.5y$
A	$x + y = 30$ $y = 0$	$(30, 0)$	54
B	$3x + y = 120$ $y = 0$	$(40, 0)$	72
C	$3x + y = 120$ $-x + y = 0$	$(30, 30)$	99
D	$x + y = 30$ $-x + y = 0$	$(15, 15)$	49.5

Maximum value occurs at the point C: $E = 99$; $x = 30$, $y = 30$.

Solution: Purchase 30 spots on *The Big Bang Theory* and 30 spots on *American Dad*.

41. Unknowns: $x = $ # OCR shares, $y = $ # RCKY shares

Minimize $p = 28x + 6y$ subject to $90x + 20y \le 10,000$ or $9x + 2y \le 1,000$,

$(0.01)(90x) + (0.02)(20y) \ge 120$, or $0.9x + 0.4y \ge 120$ or $9x + 4y \ge 1,200$,

$x \ge 0$, $y \ge 0$.

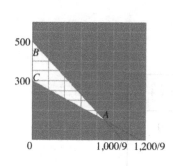

Corner Point	Lines through Point	Coordinates	$p = 28x + 6y$
A	$9x + 2y = 1,000$ $9x + 4y = 1,200$	$(88.89, 100)$	$3,088.89$
B	$9x + 2y = 1,000$ $x = 0$	$(0, 500)$	$3,000$
C	$9x + 4y = 1,200$ $x = 0$	$(0, 300)$	$1,800$

Minimum value occurs at the point C: $p = 1,800$; $x = 0$, $y = 300$

Solution: Purchase no shares of OCR and 300 shares of RCKY.

43. Unknowns: $x = $ # TD shares; $y = $ # CNA shares

Minimize $c = 3x + 2y$ subject to $45x + 40y \le 25,000$, $0.04(45x) + 0.025(40y) \ge 760$ or $1.8x + y \ge 760$, $x \ge 0$, $y \ge 0$.

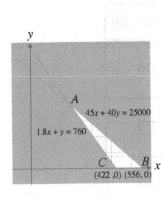

Corner Point	Lines through Point	Coordinates (rounded)	$c = 3x + 2y$
A	$45x + 40y = 25{,}000$ $1.8x + y = 760$	$(200, 400)$	$1{,}400$
B	$45x + 40y = 25{,}000$ $y = 0$	$(555.6, 0)$	$1{,}666.67$
C	$1.8x + y = 760$ $y = 0$	$(422.2, 0)$	$\boxed{1{,}266.67}$

Minimum value occurs at the point C: $c \approx 1{,}266.67$; $x \approx 422.2$, $y = 0$.

Solution: Purchase 422.2 shares of TD and no shares of CNA.

45. Unknowns: $x = $ # hours spent in battle instruction per week, $y = $ # hours spent per week in diplomacy instruction

Maximize $p = 50x + 40y$ subject to $x + y \le 50$, $x \ge 2y$, $y \ge 10$, $10x + 5y \ge 400$, $x \ge 0$, $y \ge 0$.

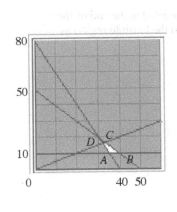

Corner Point	Lines through Point	Coordinates	$p = 50x + 40y$
A	$y = 10$ $10x + 5y = 400$	$(35, 10)$	$2{,}150$
B	$x + y = 50$ $y = 10$	$(40, 10)$	$\boxed{2{,}400}$
C	$x + y = 50$ $x - 2y = 0$	$(100/3, 50/3)$	$7{,}000/3$
D	$x - 2y = 0$ $10x + 5y = 400$	$(32, 16)$	$2{,}240$

Maximum value occurs at the point B: $p = 2{,}400$; $x = 40$, $y = 10$.

Solution: He should instruct in diplomacy for 10 hours per week and in battle for 40 hours per week, giving a weekly profit of 2,400 ducats.

47. Unknowns: $x = $ # sleep spells, $y = $ # shock spells

a. Minimize $c = 50x + 20y$ subject to $3x + y \ge 24$, $2x + 4y \ge 26$, $x - y \ge 0$, $x - 3y \le 3$, $x \ge 0$, $y \ge 0$.

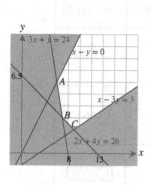

Corner Point	Lines through Point	Coordinates	$c = 50x + 20y$
A	$3x + y = 24$ $x - y = 0$	$(6, 6)$	420
B	$3x + y = 24$ $2x + 4y = 26$	$(7, 3)$	410
C	$2x + 4y = 26$ $x - 3y = 3$	$(9, 2)$	490

Although the feasible region is unbounded, there is no need to add a bounding rectangle since, by the FAQ at the end of the section in the text:

If you are minimizing $c = ax + by$ with a and b nonnegative, $x \geq 0$, and $y \geq 0$, then optimal solutions always exist.

Minimum value occurs at point B: $c = 410$, $x = 7$, $y = 3$. Use 7 sleep spells and 3 shock spells.

b. and **c.** Minimize (b) $c = 40x - 10y$ or (c) $c = 10x - 40y$ subject to $3x + y \geq 24$, $2x + 4y \geq 26$, $x - y \geq 0$,

$x - 3y \leq 3$, $x \geq 0$, $y \geq 0$.

The feasible region is unbounded, and the problem does not satisfy any of the conditions in the FAQ at the end of the text for this section, so we need to add a bounding rectangle, adding three new corner points to the feasible region as shown:

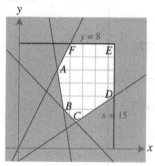

Corner Point	Lines through Point	Coordinates	$c = 40x - 10y$	$c = 10x - 40y$
A	$3x + y = 24$ $x - y = 0$	$(6, 6)$	$\boxed{180}$	-180
B	$3x + y = 24$ $2x + 4y = 26$	$(7, 3)$	250	-50
C	$2x + 4y = 26$ $x - 3y = 3$	$(9, 2)$	340	10
D	$x = 15$ $x - 3y = 3$	$(15, 4)$	560	-10
E	$x = 15, \ y = 8$	$(15, 8)$	520	-170
F	$x - y = 0$ $y = 8$	$(8, 8)$	240	$\boxed{-240}$

(b) The minimum occurs at the point B, which is a corner point of the original feasible region, so the LP problem has an optimal solution: $c = 180$, $x = 6$, $y = 6$: Use 6 sleep spells and 6 shock spells.

(c) The minimum occurs at the point F, which is not a corner point of the original feasible region, so the LP problem has no optimal solution. Net expenditure of aural energy can be an arbitrarily large negative number.

49. Unknowns: $x = $ # hours for new customers, $y = $ # hours for old customers

Maximize $p = 10x + 30y$ subject to $10x + 30y \geq 1{,}200$, $x + y \leq 160$, $x \geq 100$, $x \geq 0$, $y \geq 0$.

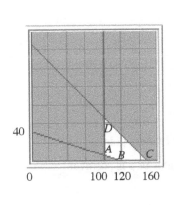

Corner Point	Lines through Point	Coordinates	$p = 10x + 30y$
A	$10x + 30y = 1{,}200$ $x = 100$	$(100, 20/3)$	$1{,}200$
B	$10x + 30y = 1{,}200$ $y = 0$	$(120, 0)$	$1{,}200$
C	$x + y = 160$ $y = 0$	$(160, 0)$	$1{,}600$
D	$x + y = 160$ $x = 100$	$(100, 60)$	$\boxed{2{,}800}$

Maximum value occurs at the point D: $p = 2{,}800$; $x = 100$, $y = 60$.

Solution: Allocate 100 hours per week for new customers and 60 hours per week for old customers.

Communication and reasoning exercises

51. By the Fundamental Theorem of Linear Programming, linear programming problems with bounded, nonempty feasible regions always have optimal solutions (Choice (A)).

53. Every point along the line connecting them is also an optimal solution.

55. a. Minimizing $c = 4x + y$ satisfies the first condition in the FAQ at the end of the text for this section, so optimal solutions exist, and it is not necessary to add a bounding rectangle.

b. Maximizing $p = 2x$ does not satisfy any of the conditions in the FAQ, so we need to add a bounding rectangle to check if optimal solutions exist.

c. Maximizing $p = 4x - y$ does not satisfy any of the conditions in the FAQ, so we need to add a bounding rectangle to check if optimal solutions exist.

d. Maximizing $p = 2x + y$ satisfies the second condition in the FAQ, so no optimal solution exists, and it is not necessary to add a bounding rectangle.

57. Here are two simple examples:

(1) (Empty feasible region) Maximize $p = x + y$ subject to $x + y \leq 10$; $x + y \geq 11$, $x \geq 0$, $y \geq 0$.

(2) (Unbounded feasible region and no optimal solutions) Minimize $c = x - y$ subject to $x + y \geq 10$, $x \geq 0$, $y \geq 0$.

59. Answers may vary.

61. A simple example is the following: Maximize profit $p = 2x + y$ subject to $x \geq 0$, $y \geq 0$. Then p can be made as large as we like by choosing large values of x and/or y. Thus, there is no optimal solution to the problem.

63. Mathematically, this means that there are infinitely many possible solutions: one for each point along the line joining the two corner points in question. In practice, select those points with integer solutions (since x and y must be whole numbers in this problem) that are in the feasible region and close to this line, and choose the one that gives the largest profit.

65. The corner points are shown in the following figure:

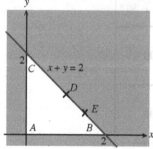

The corner points have coordinates A: $(0,0)$, B: $(2,0)$, C: $(0,2)$, and so $p = xy$ is zero at each of these points.

However, at infinitely many points along the diagonal such as D: $(1,1)$ and E: $(1.5, 0.5)$, the value of p is positive.

(For example, at D, $p = (1)(1) = 1$, and at E, $p = (1.5)(0.5) = 0.75$.)

Section 5.3

1. Introduce slack variables and rewrite the constraints and objective function in standard form, and set up the first tableau:

$x + 2y + s = 6$
$-x + y + t = 4$
$x + y + u = 4$
$-2x - y + p = 0$

	x	y	s	t	u	p	
s	1	2	1	0	0	0	6
t	-1	1	0	1	0	0	4
u	1	1	0	0	1	0	4
p	-2	-1	0	0	0	1	0

The most negative entry in the bottom row is the -2 in the x-column, so we use this column as the pivot column. The test ratios are: s: 6/1, u: 4/1. The smallest test ratio is u: 4/1. Thus we pivot on the 1 in the u-row.

	x	y	s	t	u	p		
s	1	2	1	0	0	0	6	$R_1 - R_3$
t	-1	1	0	1	0	0	4	$R_2 + R_3$
u	1	1	0	0	1	0	4	
p	-2	-1	0	0	0	1	0	$R_4 + 2R_3$

	x	y	s	t	u	p	
s	0	1	1	0	-1	0	2
t	0	2	0	1	1	0	8
x	1	1	0	0	1	0	4
p	0	1	0	0	2	1	8

As there are no more negative numbers in the bottom row, we are done, and read off the solution:
Optimal solution:
$p = 8/1 = 8$; $x = 4/1 = 4, y = 0$.

3. Introduce slack variables and rewrite the constraints and objective function in standard form, and set up the first tableau:

$5x - 5y + s = 20$
$2x - 10y + t = 40$
$-x + y + p = 0$

	x	y	s	t	p	
s	5	-5	1	0	0	20
t	2	-10	0	1	0	40
p	-1	1	0	0	1	0

The most negative entry in the bottom row is the -1 in the x-column, so we use this column as the pivot column. The test ratios are: s: 20/5, t: 40/2. The smallest test ratio is s: 20/5. Thus we pivot on the 5 in the s-row.

	x	y	s	t	p		
s	5	−5	1	0	0	20	
t	2	−10	0	1	0	40	$5R_2 - 2R_1$
p	−1	1	0	0	1	0	$5R_3 + R_1$

	x	y	s	t	p	
x	5	−5	1	0	0	20
t	0	−40	−2	5	0	160
p	0	0	1	0	5	20

As there are no more negative numbers in the bottom row, we are done, and read off the solution:
Optimal solution:
$p = 20/5 = 4$; $x = 20/5 = 4$, $y = 0$.

5. Introduce slack variables and rewrite the constraints and objective function in standard form, and set up the first tableau:
$5x + 5z + s = 100$
$5y - 5z + t = 50$
$5x - 5y + u = 50$
$-5x + 4y - 3z + p = 0$

	x	y	z	s	t	u	p	
s	5	0	5	1	0	0	0	100
t	0	5	−5	0	1	0	0	50
u	5	−5	0	0	0	1	0	50
p	−5	4	−3	0	0	0	1	0

The most negative entry in the bottom row is the −5 in the x-column, so we use this column as the pivot column. The test ratios are: s: 100/5, u: 50/5. The smallest test ratio is u: 50/5. Thus we pivot on the 5 in the u-row.

	x	y	z	s	t	u	p		
s	5	0	5	1	0	0	0	100	$R_1 - R_3$
t	0	5	−5	0	1	0	0	50	
u	5	−5	0	0	0	1	0	50	
p	−5	4	−3	0	0	0	1	0	$R_4 + R_3$

	x	y	z	s	t	u	p	
s	0	5	5	1	0	−1	0	50
t	0	5	−5	0	1	0	0	50
x	5	−5	0	0	0	1	0	50
p	0	−1	−3	0	0	1	1	50

The most negative entry in the bottom row is the −3 in the z-column, so we use this column as the pivot column. The only positive entry in this column is the 5 in the s-row, so we pivot on that entry.

	x	y	z	s	t	u	p	
s	0	5	5	1	0	−1	0	50
t	0	5	−5	0	1	0	0	50
x	5	−5	0	0	0	1	0	50
p	0	−1	−3	0	0	1	1	50

$R_2 + R_1$

$5R_4 + 3R_1$

	x	y	z	s	t	u	p	
z	0	5	5	1	0	−1	0	50
t	0	10	0	1	1	−1	0	100
x	5	−5	0	0	0	1	0	50
p	0	10	0	3	0	2	5	400

As there are no more negative numbers in the bottom row, we are done, and read off the solution:
Optimal solution:
$p = 400/5 = 80; \ x = 50/5 = 10, y = 0, z = 50/5 = 10.$

7. Introduce slack variables and rewrite the constraints and objective function in standard form, and set up the first tableau:
$x + y − z + s = 3$
$x + 2y + z + t = 8$
$x + y + u = 5$
$−7x − 5y − 6z + p = 0$

	x	y	z	s	t	u	p	
s	1	1	−1	1	0	0	0	3
t	1	2	1	0	1	0	0	8
u	1	1	0	0	0	1	0	5
p	−7	−5	−6	0	0	0	1	0

The most negative entry in the bottom row is the −7 in the x-column, so we use this column as the pivot column. The test ratios are: s: 3/1, t: 8/1, u: 5/1. The smallest test ratio is s: 3/1. Thus we pivot on the 1 in the s-row.

	x	y	z	s	t	u	p	
s	1	1	−1	1	0	0	0	3
t	1	2	1	0	1	0	0	8
u	1	1	0	0	0	1	0	5
p	−7	−5	−6	0	0	0	1	0

$R_2 − R_1$

$R_3 − R_1$

$R_4 + 7R_1$

	x	y	z	s	t	u	p	
x	1	1	−1	1	0	0	0	3
t	0	1	2	−1	1	0	0	5
u	0	0	1	−1	0	1	0	2
p	0	2	−13	7	0	0	1	21

The most negative entry in the bottom row is the −13 in the z-column, so we use this column as the pivot column. The test ratios are: t: 5/2, u: 2/1. The smallest test ratio is u: 2/1. Thus we pivot on the 1 in the u-row.

	x	y	z	s	t	u	p		
x	1	1	−1	1	0	0	0	3	$R_1 + R_3$
t	0	1	2	−1	1	0	0	5	$R_2 - 2R_3$
u	0	0	[1]	−1	0	1	0	2	
p	0	2	−13	7	0	0	1	21	$R_4 + 13R_3$

	x	y	z	s	t	u	p	
x	1	1	0	0	0	1	0	5
t	0	1	0	1	1	−2	0	1
z	0	0	1	−1	0	1	0	2
p	0	2	0	−6	0	13	1	47

The most negative entry in the bottom row is the −6 in the s-column, so we use this column as the pivot column. The only positive entry in this column is the 1 in the t-row, so we pivot on that entry.

	x	y	z	s	t	u	p		
x	1	1	0	0	0	1	0	5	
t	0	1	0	[1]	1	−2	0	1	
z	0	0	1	−1	0	1	0	2	$R_3 + R_2$
p	0	2	0	−6	0	13	1	47	$R_4 + 6R_2$

	x	y	z	s	t	u	p	
x	1	1	0	0	0	1	0	5
s	0	1	0	1	1	−2	0	1
z	0	1	1	0	1	−1	0	3
p	0	8	0	0	6	1	1	53

As there are no more negative numbers in the bottom row, we are done, and read off the solution:
Optimal solution:
$p = 53/1 = 53$; $x = 5/1 = 5$, $y = 0$, $z = 3/1 = 3$.

9. Introduce slack variables and rewrite the constraints and objective function in standard form, and set up the first tableau:
$$5x_1 - x_2 + x_3 + s = 1,500$$
$$2x_1 + 2x_2 + x_3 + t = 2,500$$
$$4x_1 + 2x_2 + x_3 + u = 2,000$$
$$-3x_1 - 7x_2 - 8x_3 + z = 0$$

	x_1	x_2	x_3	s	t	u	z	
s	5	−1	1	1	0	0	0	1500
t	2	2	1	0	1	0	0	2500
u	4	2	1	0	0	1	0	2000
z	−3	−7	−8	0	0	0	1	0

The most negative entry in the bottom row is the –8 in the x_3-column, so we use this column as the pivot column. The test ratios are: s: 1500/1, t: 2500/1, u: 2000/1. The smallest test ratio is s: 1500/1. Thus we pivot on the 1 in the s-row.

	x_1	x_2	x_3	s	t	u	z		
s	5	−1	1	1	0	0	0	1500	
t	2	2	1	0	1	0	0	2500	$R_2 - R_1$
u	4	2	1	0	0	1	0	2000	$R_3 - R_1$
z	−3	−7	−8	0	0	0	1	0	$R_4 + 8R_1$

	x_1	x_2	x_3	s	t	u	z	
x_3	5	−1	1	1	0	0	0	1500
t	−3	3	0	−1	1	0	0	1000
u	−1	3	0	−1	0	1	0	500
z	37	−15	0	8	0	0	1	12000

The most negative entry in the bottom row is the –15 in the x_2-column, so we use this column as the pivot column. The test ratios are: t: 1000/3, u: 500/3. The smallest test ratio is u: 500/3. Thus we pivot on the 3 in the u-row.

	x_1	x_2	x_3	s	t	u	z		
x_3	5	−1	1	1	0	0	0	1500	$3R_1 + R_3$
t	−3	3	0	−1	1	0	0	1000	$R_2 - R_3$
u	−1	3	0	−1	0	1	0	500	
z	37	−15	0	8	0	0	1	12000	$R_4 + 5R_3$

	x_1	x_2	x_3	s	t	u	z	
x_3	14	0	3	2	0	1	0	5000
t	−2	0	0	0	1	−1	0	500
x_2	−1	3	0	−1	0	1	0	500
z	32	0	0	3	0	5	1	14500

As there are no more negative numbers in the bottom row, we are done, and read off the solution:
Optimal solution:
$z = 14,500/1 = 14,500$; $x_1 = 0, x_2 = 500/3, x_3 = 5,000/3$.

11. Introduce slack variables and rewrite the constraints and objective function in standard form, and set up the first tableau:
$x + y + z + s = 3$
$y + z + w + t = 4$
$x + z + w + u = 5$
$x + y + w + v = 6$
$-x - y - z - w + p = 0$

	x	y	z	w	s	t	u	v	p	
s	1	1	1	0	1	0	0	0	0	3
t	0	1	1	1	0	1	0	0	0	4
u	1	0	1	1	0	0	1	0	0	5
v	1	1	0	1	0	0	0	1	0	6
p	−1	−1	−1	−1	0	0	0	0	1	0

The most negative entry in the bottom row is the −1 in the x-column, so we use this column as the pivot column. The test ratios are: s: 3/1, u: 5/1, v: 6/1. The smallest test ratio is s: 3/1. Thus we pivot on the 1 in the s-row.

	x	y	z	w	s	t	u	v	p		
s	1	1	1	0	1	0	0	0	0	3	
t	0	1	1	1	0	1	0	0	0	4	
u	1	0	1	1	0	0	1	0	0	5	$R_3 - R_1$
v	1	1	0	1	0	0	0	1	0	6	$R_4 - R_1$
p	−1	−1	−1	−1	0	0	0	0	1	0	$R_5 + R_1$

	x	y	z	w	s	t	u	v	p	
x	1	1	1	0	1	0	0	0	0	3
t	0	1	1	1	0	1	0	0	0	4
u	0	−1	0	1	−1	0	1	0	0	2
v	0	0	−1	1	−1	0	0	1	0	3
p	0	0	0	−1	1	0	0	0	1	3

The most negative entry in the bottom row is the −1 in the w-column, so we use this column as the pivot column. The test ratios are: t: 4/1, u: 2/1, v: 3/1. The smallest test ratio is u: 2/1. Thus we pivot on the 1 in the u-row.

	x	y	z	w	s	t	u	v	p		
x	1	1	1	0	1	0	0	0	0	3	
t	0	1	1	1	0	1	0	0	0	4	$R_2 - R_3$
u	0	−1	0	1	−1	0	1	0	0	2	
v	0	0	−1	1	−1	0	0	1	0	3	$R_4 - R_3$
p	0	0	0	−1	1	0	0	0	1	3	$R_5 + R_3$

	x	y	z	w	s	t	u	v	p	
x	1	1	1	0	1	0	0	0	0	3
t	0	2	1	0	1	1	−1	0	0	2
w	0	−1	0	1	−1	0	1	0	0	2
v	0	1	−1	0	0	0	−1	1	0	1
p	0	−1	0	0	0	0	1	0	1	5

The most negative entry in the bottom row is the −1 in the y-column, so we use this column as the pivot column. The test ratios are: x: 3/1, t: 2/2, v: 1/1. The smallest test ratio is t: 2/2. Thus we pivot on the 2 in the t-row.

	x	y	z	w	s	t	u	v	p		
x	1	1	1	0	1	0	0	0	0	3	$2R_1 - R_2$
t	0	2	1	0	1	1	−1	0	0	2	
w	0	−1	0	1	−1	0	1	0	0	2	$2R_3 + R_2$
v	0	1	−1	0	0	0	−1	1	0	1	$2R_4 - R_2$
p	0	−1	0	0	0	0	1	0	1	5	$2R_5 + R_2$

	x	y	z	w	s	t	u	v	p	
x	2	0	1	0	1	−1	1	0	0	4
y	0	2	1	0	1	1	−1	0	0	2
w	0	0	1	2	−1	1	1	0	0	6
v	0	0	−3	0	−1	−1	−1	2	0	0
p	0	0	1	0	1	1	1	0	2	12

As there are no more negative numbers in the bottom row, we are done, and read off the solution:
Optimal solution:
$p = 12/2 = 6$; $x = 4/2 = 2, y = 2/2 = 1, z = 0, w = 6/2 = 3$.

13. Introduce slack variables and rewrite the constraints and objective function in standard form, and set up the first tableau:
$x + y + s = 1$
$y + z + t = 2$
$z + w + r = 3$
$w + v + q = 4$
$-x - y - z - w - v + p = 0$

	x	y	z	w	v	s	t	r	q	p	
s	1	1	0	0	0	1	0	0	0	0	1
t	0	1	1	0	0	0	1	0	0	0	2
r	0	0	1	1	0	0	0	1	0	0	3
q	0	0	0	1	1	0	0	0	1	0	4
p	−1	−1	−1	−1	−1	0	0	0	0	1	0

The most negative entry in the bottom row is the −1 in the x-column, so we use this column as the pivot column. The only positive entry in this column is the 1 in the s-row, so we pivot on that entry.

	x	y	z	w	v	s	t	r	q	p		
s	1	1	0	0	0	1	0	0	0	0	1	
t	0	1	1	0	0	0	1	0	0	0	2	
r	0	0	1	1	0	0	0	1	0	0	3	
q	0	0	0	1	1	0	0	0	1	0	4	
p	−1	−1	−1	−1	−1	0	0	0	0	1	0	$R_5 + R_1$

	x	y	z	w	v	s	t	r	q	p	
x	1	1	0	0	0	1	0	0	0	0	1
t	0	1	1	0	0	0	1	0	0	0	2
r	0	0	1	1	0	0	0	1	0	0	3
q	0	0	0	1	1	0	0	0	1	0	4
p	0	0	−1	−1	−1	1	0	0	0	1	1

The most negative entry in the bottom row is the −1 in the z-column, so we use this column as the pivot column. The test ratios are: t: 2/1, r: 3/1. The smallest test ratio is t: 2/1. Thus we pivot on the 1 in the t-row.

	x	y	z	w	v	s	t	r	q	p		
x	1	1	0	0	0	1	0	0	0	0	1	
t	0	1	1	0	0	0	1	0	0	0	2	
r	0	0	1	1	0	0	0	1	0	0	3	$R_3 - R_2$
q	0	0	0	1	1	0	0	0	1	0	4	
p	0	0	−1	−1	−1	1	0	0	0	1	1	$R_5 + R_2$

	x	y	z	w	v	s	t	r	q	p	
x	1	1	0	0	0	1	0	0	0	0	1
z	0	1	1	0	0	0	1	0	0	0	2
r	0	−1	0	1	0	0	−1	1	0	0	1
q	0	0	0	1	1	0	0	0	1	0	4
p	0	1	0	−1	−1	1	1	0	0	1	3

The most negative entry in the bottom row is the −1 in the w-column, so we use this column as the pivot column. The test ratios are: r: 1/1, q: 4/1. The smallest test ratio is r: 1/1. Thus we pivot on the 1 in the r-row.

	x	y	z	w	v	s	t	r	q	p		
x	1	1	0	0	0	1	0	0	0	0	1	
z	0	1	1	0	0	0	1	0	0	0	2	
r	0	−1	0	1	0	0	−1	1	0	0	1	
q	0	0	0	1	1	0	0	0	1	0	4	$R_4 - R_3$
p	0	1	0	−1	−1	1	1	0	0	1	3	$R_5 + R_3$

	x	y	z	w	v	s	t	r	q	p	
x	1	1	0	0	0	1	0	0	0	0	1
z	0	1	1	0	0	0	1	0	0	0	2
w	0	−1	0	1	0	0	−1	1	0	0	1
q	0	1	0	0	1	0	1	−1	1	0	3
p	0	0	0	0	−1	1	0	1	0	1	4

The most negative entry in the bottom row is the −1 in the v-column, so we use this column as the pivot column. The only positive entry in this column is the 1 in the q-row, so we pivot on that entry.

	x	y	z	w	v	s	t	r	q	p	
x	1	1	0	0	0	1	0	0	0	0	1
z	0	1	1	0	0	0	1	0	0	0	2
w	0	−1	0	1	0	0	−1	1	0	0	1
q	0	1	0	0	$\boxed{1}$	0	1	−1	1	0	3
p	0	0	0	0	−1	1	0	1	0	1	4

$R_5 + R_4$

	x	y	z	w	v	s	t	r	q	p	
x	1	1	0	0	0	1	0	0	0	0	1
z	0	1	1	0	0	0	1	0	0	0	2
w	0	−1	0	1	0	0	−1	1	0	0	1
v	0	1	0	0	1	0	1	−1	1	0	3
p	0	1	0	0	0	1	1	0	1	1	7

As there are no more negative numbers in the bottom row, we are done, and read off the solution:
Optimal solution:
$p = 7/1 = 7$; $x = 1/1 = 1, y = 0, z = 2/1 = 2, w = 1/1 = 1, v = 3/1 = 3$.
Another solution is $x = 1$, $y = 0$, $z = 2$, $w = 0$, $v = 4$.

15. You can use the online Pivot and Gauss-Jordan Tool in decimal mode to do the pivoting, or use the online Simplex Method Tool. When entering problems in the Simplex Method Tool there is no need to enter the inequalities $x \geq 0$, $y \geq 0$, etc.

	x	y	z	s	t	u	p	
s	0.1	1	−2.2	1	0	0	0	4.5
t	2.1	1	1	0	1	0	0	8
u	1	$\boxed{2.2}$	0	0	0	1	0	5
p	−2.5	−4.2	−2	0	0	0	1	0

	x	y	z	s	t	u	p	
s	−0.35	0	−2.2	1	0	−0.45	0	2.23
t	1.65	0	$\boxed{1}$	0	1	−0.45	0	5.73
y	0.45	1	0	0	0	0.45	0	2.27
p	−0.59	0	−2	0	0	1.91	1	9.55

	x	y	z	s	t	u	p	
s	3.27	0	0	1	2.2	−1.45	0	14.83
z	1.65	0	1	0	1	−0.45	0	5.73
y	0.45	1	0	0	0	0.45	0	2.27
p	2.7	0	0	0	2	1	1	21

Optimal solution:
$p = 21$; $x = 0, y = 2.27, z = 5.73$.

17. You can use the online Pivot and Gauss-Jordan Tool in decimal mode to do the pivoting, or use the online Simplex Method Tool. When entering problems in the Simplex Method Tool there is no need to enter the inequalities $x \geq 0$, $y \geq 0$, etc.

	x	y	z	w	s	t	u	v	p	
s	1	2	3	0	1	0	0	0	0	3
t	0	1	1	2.2	0	1	0	0	0	4
u	1	0	1	2.2	0	0	1	0	0	5
v	1	1	0	2.2	0	0	0	1	0	6
p	−1	−2	−3	−1	0	0	0	0	1	0

	x	y	z	w	s	t	u	v	p	
z	0.33	0.67	1	0	0.33	0	0	0	0	1
t	−0.33	0.33	0	2.2	−0.33	1	0	0	0	3
u	0.67	−0.67	0	2.2	−0.33	0	1	0	0	4
v	1	1	0	2.2	0	0	0	1	0	6
p	0	0	0	−1	1	0	0	0	1	3

	x	y	z	w	s	t	u	v	p	
z	0.33	0.67	1	0	0.33	0	0	0	0	1
w	−0.15	0.15	0	1	−0.15	0.45	0	0	0	1.36
u	1	−1	0	0	0	−1	1	0	0	1
v	1.33	0.67	0	0	0.33	−1	0	1	0	3
p	−0.15	0.15	0	0	0.85	0.45	0	0	1	4.36

	x	y	z	w	s	t	u	v	p	
z	0	1	1	0	0.33	0.33	−0.33	0	0	0.67
w	0	0	0	1	−0.15	0.3	0.15	0	0	1.52
x	1	−1	0	0	0	−1	1	0	0	1
v	0	2	0	0	0.33	0.33	−1.33	1	0	1.67
p	0	0	0	0	0.85	0.3	0.15	0	1	4.52

Optimal solution:
$p = 4.52$; $x = 1$, $y = 0$, $z = 0.67$, $w = 1.52$.
Another solution is $x = 1.67$, $y = 0.67$, $z = 0$, $w = 1.52$.

19. You can use the online Pivot and Gauss-Jordan Tool in decimal mode to do the pivoting, or use the online Simplex Method Tool. When entering problems in the Simplex Method Tool there is no need to enter the inequalities $x \geq 0$, $y \geq 0$, etc.

	x	y	z	w	v	s	t	r	q	p	
s	1	1	0	0	0	1	0	0	0	0	1.1
t	0	1	1	0	0	0	1	0	0	0	2.2
r	0	0	1	1	0	0	0	1	0	0	3.3
q	0	0	0	1	1	0	0	0	1	0	4.4
p	−1	1	−1	1	−1	0	0	0	0	1	0

	x	y	z	w	v	s	t	r	q	p	
x	1	1	0	0	0	1	0	0	0	0	1.1
t	0	1	1	0	0	0	1	0	0	0	2.2
r	0	0	1	1	0	0	0	1	0	0	3.3
q	0	0	0	1	1	0	0	0	1	0	4.4
p	0	2	−1	1	−1	1	0	0	0	1	1.1

	x	y	z	w	v	s	t	r	q	p	
x	1	1	0	0	0	1	0	0	0	0	1.1
z	0	1	1	0	0	0	1	0	0	0	2.2
r	0	−1	0	1	0	0	−1	1	0	0	1.1
q	0	0	0	1	1	0	0	0	1	0	4.4
p	0	3	0	1	−1	1	1	0	0	1	3.3

	x	y	z	w	v	s	t	r	q	p	
x	1	1	0	0	0	1	0	0	0	0	1.1
z	0	1	1	0	0	0	1	0	0	0	2.2
r	0	−1	0	1	0	0	−1	1	0	0	1.1
v	0	0	0	1	1	0	0	0	1	0	4.4
p	0	3	0	2	0	1	1	0	1	1	7.7

Optimal solution:
$p = 7.7$; $x = 1.1, y = 0, z = 2.2, w = 0, v = 4.4$.

Applications

21. Unknowns: $x =$ # calculus texts, $y =$ # history texts, $z =$ # marketing texts

Maximize $p = 10x + 4y + 8z$ subject to $x + y + z \le 650$, $2x + y + 3z \le 1,000$, $x \ge 0$, $y \ge 0$, $z \ge 0$.

	x	y	z	s	t	p		
s	1	1	1	1	0	0	650	$2R_1 - R_2$
t	2	1	3	0	1	0	1000	
p	−10	−4	−8	0	0	1	0	$R_3 + 5R_2$

	x	y	z	s	t	p	
s	0	1	−1	2	−1	0	300
x	2	1	3	0	1	0	1000
p	0	1	7	0	5	1	5000

Optimal solution:

$p = 5,000/1 = 5,000$; $x = 1,000/2 = 500, y = 0, z = 0$.

You should purchase 500 calculus texts, no history texts, and no marketing texts. The maximum profit is $5,000 per semester.

23. Unknowns: $x = $ # gallons of PineOrange, $y = $ # gallons of PineKiwi, $z = $ # gallons of OrangeKiwi

Maximize $p = x + 2y + z$ subject to $2x + 3y \le 800$, $2x + 3z \le 650$, $y + z \le 350, x \ge 0, y \ge 0, z \ge 0$.

	x	y	z	s	t	u	p		
s	2	3	0	1	0	0	0	800	
t	2	0	3	0	1	0	0	650	
u	0	1	1	0	0	1	0	350	$3R_3 - R_1$
p	−1	−2	−1	0	0	0	1	0	$3R_4 + 2R_1$

	x	y	z	s	t	u	p		
y	2	3	0	1	0	0	0	800	
t	2	0	3	0	1	0	0	650	$R_2 - R_3$
u	−2	0	3	−1	0	3	0	250	
p	1	0	−3	2	0	0	3	1600	$R_4 + R_3$

	x	y	z	s	t	u	p		
y	2	3	0	1	0	0	0	800	$2R_1 - R_2$
t	4	0	0	1	1	−3	0	400	
z	−2	0	3	−1	0	3	0	250	$2R_3 + R_2$
p	−1	0	0	1	0	3	3	1850	$4R_4 + R_2$

	x	y	z	s	t	u	p	
y	0	6	0	1	−1	3	0	1200
x	4	0	0	1	1	−3	0	400
z	0	0	6	−1	1	3	0	900
p	0	0	0	5	1	9	12	7800

Optimal solution:

$p = 7,800/12 = 650$; $x = 400/4 = 100, y = 1,200/6 = 200, z = 900/6 = 150$.

The company makes a maximum profit of $650 by making 100 gallons of PineOrange, 200 gallons of PineKiwi, and 150 gallons of OrangeKiwi.

25. Unknowns:

$x = $ # sections of Ancient History, $y = $ # sections of Medieval History, $z = $ # sections of Modern History

Maximize $p = x + 2y + 3z$ (in tens of thousands of dollars) subject to

$x + y + z \leq 45$, $10x + 5y + 20z \leq 500$, $x + y + 2z \leq 60$, $x \geq 0$, $y \geq 0$, $z \geq 0$.

	x	y	z	s	t	u	p		
s	1	1	1	1	0	0	0	45	$20R_1 - R_2$
t	10	5	20	0	1	0	0	500	
u	1	1	2	0	0	1	0	60	$10R_3 - R_2$
p	-1	-2	-3	0	0	0	1	0	$20R_4 + 3R_2$

	x	y	z	s	t	u	p		
s	10	15	0	20	-1	0	0	400	$R_1 - 3R_3$
z	10	5	20	0	1	0	0	500	$R_2 - R_3$
u	0	5	0	0	-1	10	0	100	
p	10	-25	0	0	3	0	20	1500	$R_4 + 5R_3$

	x	y	z	s	t	u	p		
s	10	0	0	20	2	-30	0	100	
z	10	0	20	0	2	-10	0	400	$R_2 - R_1$
y	0	5	0	0	-1	10	0	100	$2R_3 + R_1$
p	10	0	0	0	-2	50	20	2000	$R_4 + R_1$

	x	y	z	s	t	u	p	
t	10	0	0	20	2	-30	0	100
z	0	0	20	-20	0	20	0	300
y	10	10	0	20	0	-10	0	300
p	20	0	0	20	0	20	20	2100

Optimal solution:
$p = 2,100/20 = 105$; $x = 0$, $y = 300/10 = 30$, $z = 300/20 = 15$.
The department should offer no Ancient History, 30 sections of Medieval History, and 15 sections of Modern History, for a profit of $\$105 \times 10,000 = \$1,050,000$.
Answers to additional question: The values of the slack variables are:

$t = 1,000/2 = 500$, meaning that there will be 500 students without classes.

$s = u = 0$, meaning that all time slots and professors are used.

27. Unknowns:

$x = $ # acres of tomatoes, $y = $ # acres of lettuce, $z = $ # acres of carrots

Maximize $p = 20x + 15y + 5z$ subject to $x + y + z \leq 100$, $5x + 4y + 2z \leq 400$, $4x + 2y + 2z \leq 500$, $x \geq 0$, $y \geq 0$,

$z \geq 0$. (p is measured in hundreds of dollars.)

	x	y	z	s	t	u	p		
s	1	1	1	1	0	0	0	100	$5R_1 - R_2$
t	5	4	2	0	1	0	0	400	
u	4	2	2	0	0	1	0	500	$5R_3 - 4R_2$
p	−20	−15	−5	0	0	0	1	0	$R_4 + 4R_2$

	x	y	z	s	t	u	p	
s	0	1	3	5	−1	0	0	100
x	5	4	2	0	1	0	0	400
u	0	−6	2	0	−4	5	0	900
p	0	1	3	0	4	0	1	1600

Optimal solution:

$p = 1,600/1 = 1,600$; $x = 400/5 = 80$, $y = 0$, $z = 0$.

Plant 80 acres of tomatoes, and no lettuce or carrots, for a maximum profit of $160,000.

The slack variable corresponding to the number of acres available is s. The value of s is $s = 100/5 = 20$, meaning that you will leave 20 acres unplanted.

29. Unknowns:

$x = $ # servings of granola, $y = $ # servings of nutty granola, $z = $ # servings of nuttiest granola

Maximize $p = 6x + 8y + 3z$ subject to

$x + y + 5z \le 1,500$, $4x + 8y + 8z \le 10,000$, $2x + 4y + 8z \le 4,000$, $x \ge 0$, $y \ge 0$, $z \ge 0$.

	x	y	z	s	t	u	p		
s	1	1	5	1	0	0	0	1500	$4R_1 - R_3$
t	4	8	8	0	1	0	0	10000	$R_2 - 2R_3$
u	2	4	8	0	0	1	0	4000	
p	−6	−8	−3	0	0	0	1	0	$R_4 + 2R_3$

	x	y	z	s	t	u	p		
s	2	0	12	4	0	−1	0	2000	
t	0	0	−8	0	1	−2	0	2000	
y	2	4	8	0	0	1	0	4000	$R_3 - R_1$
p	−2	0	13	0	0	2	1	8000	$R_4 + R_1$

	x	y	z	s	t	u	p	
x	2	0	12	4	0	−1	0	2000
t	0	0	−8	0	1	−2	0	2000
y	0	4	−4	−4	0	2	0	2000
p	0	0	25	4	0	1	1	10000

Optimal solution:

$p = 10,000/1 = 10,000$; $x = 2,000/2 = 1,000$, $y = 2,000/4 = 500$, $z = 0$.

The Choral Society can make a profit of $10,000 by selling 1,000 servings of granola, 500 servings of nutty granola, and no nuttiest granola.

To obtain the ingredients left over, look at the values of the slack variables in the final tableau: $s = 0$, so there are no toasted oats left over. $t = 2,000/1 = 2,000$, so there are 2,000 oz of almonds left over. $u = 0$, so there are no raisins left over.

31. Unknowns:

$x = $ # axes, $y = $ # maces, $z = $ # spears.

Maximize $p = 6x + 6y + 8z$ subject to

$8x + 5y + 2z \le 50,000, \ 2x + 4y + 6z \le 40,000, \ z - x \le 0, x \ge 0, \ y \ge 0, \ z \ge 0.$

	x	y	z	s	t	u	p		
s	8	5	2	1	0	0	0	50000	$R_1 - 2R_3$
t	2	4	6	0	1	0	0	40000	$R_2 - 6R_3$
u	−1	0	1	0	0	1	0	0	
p	−6	−6	−8	0	0	0	1	0	$R_4 + 8R_3$

	x	y	z	s	t	u	p		
s	10	5	0	1	0	−2	0	50000	
t	8	4	0	0	1	−6	0	40000	$5R_2 - 4R_1$
z	−1	0	1	0	0	1	0	0	$10R_3 + R_1$
p	−14	−6	0	0	0	8	1	0	$5R_4 + 7R_1$

	x	y	z	s	t	u	p	
x	10	5	0	1	0	−2	0	50000
t	0	0	0	−4	5	−22	0	0
z	0	5	10	1	0	8	0	50000
p	0	5	0	7	0	26	5	350000

Optimal solution:
$p = 350,000/5 = 70,000; \ x = 50,000/10 = 5,000, y = 0, z = 50,000/10 = 5,000.$
Achlúk can inflict a maximum of 70,000 units of damage using an arsenal of 5,000 axes, no maces, and 5,000 spears.

33. Unknowns:

$x = $ millions of gallons of oil allocated to process A, $y = $ millions of gallons of oil allocated to process B, $z = $ millions of gallons of oil allocated to process C

Maximize $p = 4(0.60)x + 4(0.55)y + 4(0.50)z = 2.4x + 2.2y + 2.0z$, which we change to $p = 24x + 22y + 20z$ to work with integers, subject to $x + y + z \le 50$, $150x + 100y + 50z \le 3,000$ or $3x + 2y + z \le 60, x \ge 0, \ y \ge 0, \ z \ge 0.$

	x	y	z	s	t	p		
s	1	1	1	1	0	0	50	$3R_1 - R_2$
t	3	2	1	0	1	0	60	
p	−24	−22	−20	0	0	1	0	$R_3 + 8R_2$

	x	y	z	s	t	p		
s	0	1	2	3	−1	0	90	
x	3	2	1	0	1	0	60	$2R_2 - R_1$
p	0	−6	−12	0	8	1	480	$R_3 + 6R_1$

	x	y	z	s	t	p	
z	0	1	2	3	−1	0	90
x	6	3	0	−3	3	0	30
p	0	0	0	18	2	1	1020

Optimal solution:
$p = 1,020/1 = 1,020$; $x = 30/6 = 5$, $y = 0$, $z = 90/2 = 45$.
Allocate 5 million gallons to process A and 45 million gallons to process C.
Another solution: Allocate 10 million gallons to process B and 40 million gallons to process C.

35. Unknowns: $x = $ # servings of Xtend, $y = $ # servings of Gainz, $z = $ # servings of Strongevity

Maximize $p = 7x + 6y$ subject to $2y + 2.5z \leq 40$, $2.5x + 3y + z \leq 60$, $x - z \leq 0$, $x \geq 0$, $y \geq 0$, $z \geq 0$.

	x	y	z	s	t	u	p		
s	0	2	2.5	1	0	0	0	40	$2R\_1$
t	2.5	3	1	0	1	0	0	60	$2R\_2$
u	1	0	−1	0	0	1	0	0	
p	−7	−6	0	0	0	0	1	0	

	x	y	z	s	t	u	p		
s	0	4	5	2	0	0	0	80	
t	5	6	2	0	2	0	0	120	$R_2 - 5R_3$
u	1	0	−1	0	0	1	0	0	
p	−7	−6	0	0	0	0	1	0	$R_4 + 7R_3$

	x	y	z	s	t	u	p		
s	0	4	5	2	0	0	0	80	
t	0	6	7	0	2	−5	0	120	$5R_2 - 7R_1$
x	1	0	−1	0	0	1	0	0	$5R_3 + R_1$
p	0	−6	−7	0	0	7	1	0	$5R_4 + 7R_1$

	x	y	z	s	t	u	p		
z	0	4	5	2	0	0	0	80	
t	0	2	0	−14	10	−25	0	40	$2R_2 - R_1$
x	5	4	0	2	0	5	0	80	$R_3 - R_1$
p	0	−2	0	14	0	35	5	560	$2R_4 + R_1$

	x	y	z	s	t	u	p	
y	0	4	5	2	0	0	0	80
t	0	0	-5	-30	20	-50	0	0
x	5	0	-5	0	0	5	0	0
p	0	0	5	30	0	70	10	1200

Optimal solution:
$p = 1,200/10 = 120$; $x = 0/5 = 0, y = 80/4 = 20, z = 0.$
Use 20 servings of Gainz and none of the others for 120 g of BCAAs.

37. Unknowns: $x = $ # DUK shares, $y = $ # DTV shares, $z = $ # OCR shares

Maximize $p = 4x + 10y + 30z$ subject to $80x + 100y + 90z \le 90,000$, $(0.04)(80x) + (0.01)(90z) \le 900$ or

$32x + 9z \le 9,000, \ x \ge 0, \ y \ge 0, \ z \ge 0.$

	x	y	z	s	t	p		
s	80	100	90	1	0	0	90000	
t	32	0	9	0	1	0	9000	$10R_2 - R_1$
p	-4	-10	-30	0	0	1	0	$3R_3 + R_1$

	x	y	z	s	t	p	
z	80	100	90	1	0	0	90000
t	240	-100	0	-1	10	0	0
p	68	70	0	1	0	3	90000

Optimal solution:
$p = 90,000/3 = 30,000$; $x = 0, y = 0, z = 90,000/90 = 1,000.$
Buy 1,000 shares of OCR and no others. The broker is wrong.

39. Unknowns: $x = $ amount allocated to automobile loans (in $ millions), $y = $ amount allocated to furniture loans (in $ millions), $z = $ amount allocated to signature loans (in $ millions), $w = $ amount allocated to other secured loans

Maximize $p = 8x + 10y + 12z + 10w$ (100 times the return) subject to

$x + y + z + w \le 5, \ -x - y + 9z - w \le 0, \ -x + y + w \le 0, \ -2x + w \le 0, \ x \ge 0, \ y \ge 0, \ z \ge 0, \ w \ge 0.$

	x	y	z	w	s	t	u	v	p		
s	1	1	1	1	1	0	0	0	0	5	$9R_1 - R_2$
t	-1	-1	9	-1	0	1	0	0	0	0	
u	-1	1	0	1	0	0	1	0	0	0	
v	-2	0	0	1	0	0	0	1	0	0	
p	-8	-10	-12	-10	0	0	0	0	1	0	$3R_5 + 4R_2$

	x	y	z	w	s	t	u	v	p		
s	10	10	0	10	9	−1	0	0	0	45	$R_1 - 10R_3$
z	−1	−1	9	−1	0	1	0	0	0	0	$R_2 + R_3$
u	−1	1	0	1	0	0	1	0	0	0	
v	−2	0	0	1	0	0	0	1	0	0	
p	−28	−34	0	−34	0	4	0	0	3	0	$R_5 + 34R_3$

	x	y	z	w	s	t	u	v	p		
s	20	0	0	0	9	−1	−10	0	0	45	
z	−2	0	9	0	0	1	1	0	0	0	$10R_2 + R_1$
y	−1	1	0	1	0	0	1	0	0	0	$20R_3 + R_1$
v	−2	0	0	1	0	0	0	1	0	0	$10R_4 + R_1$
p	−62	0	0	0	0	4	34	0	3	0	$10R_5 + 31R_1$

	x	y	z	w	s	t	u	v	p	
x	20	0	0	0	9	−1	−10	0	0	45
z	0	0	90	0	9	9	0	0	0	45
y	0	20	0	20	9	−1	10	0	0	45
v	0	0	0	10	9	−1	−10	10	0	45
p	0	0	0	0	279	9	30	0	30	1395

Optimal solution:

$p = 1,395/30 = 93/2$; $x = 45/20 = 9/4$, $y = 45/20 = 9/4$, $z = 45/90 = 1/2$, $w = 0$.

Allocate \$2,250,000 to automobile loans, \$2,250,000 to furniture loans, and \$500,000 to signature loans. Another optimal solution (pivot in the w-column): $p = 8,370/180 = 46.5$; $x = 45/20 = 2.25$, $y = 0$, $z = 45/90 = 0.5$,

$w = 45/20 = 2.25$. Allocate \$2,250,000 to automobile loans, \$500,000 to signature loans, and \$2,250,000 to other secured loans.

In general, Allocate \$2,250,000 to automobile loans, \$500,000 to signature loans, and \$2,250,000 to any combination of furniture loans and other secured loans.

41. Unknowns: $x =$ amount invested in Warner, $y =$ amount invested in Universal, $z =$ amount invested in Sony,

$w =$ amount invested in EMI

Maximize $p = 8x + 20y + 10z + 15w$ subject to

$0.12x + 0.20y + 0.20z + 0.15w \le 15,000$ or $12x + 20y + 20z + 15w \le 1,500,000$, $x − 4y + z + w \le 0$,

$x + y + z + w \le 100,000$, $x \ge 0$, $y \ge 0$, $z \ge 0$, $w \ge 0$.

	x	y	z	w	s	t	r	p		
s	12	20	20	15	1	0	0	0	1500000	
t	1	−4	1	1	0	1	0	0	0	$5R_2 + R_1$
r	1	1	1	1	0	0	1	0	100000	$20R_3 − R_1$
p	−8	−20	−10	−15	0	0	0	1	0	$R_4 + R_1$

	x	y	z	w	s	t	r	p	
y	12	20	20	15	1	0	0	0	1500000
t	17	0	25	20	1	5	0	0	1500000
r	8	0	0	5	−1	0	20	0	500000
p	4	0	10	0	1	0	0	1	1500000

Optimal solution:
$p = 1,500,000/1 = 1,500,000$; $x = 0, y = 1,500,000/20 = 75,000, z = 0, w = 0$.
Invest \$75,000 in Universal, none in the rest. Another optimal solution is: Invest \$18,750 in Universal, and \$75,000 in EMI.

43. Unknowns:

$x = $ # boards sent from Tucson to Honolulu

$y = $ # boards sent from Tucson to Venice Beach

$z = $ # boards sent from Toronto to Honolulu

$w = $ # boards sent from Toronto to Venice Beach

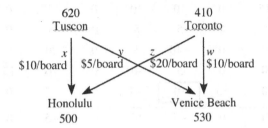

Maximize $p = x + y + z + w$ subject to

$x + y \leq 620,\ z + w \leq 410,\ x + z \leq 500,\ y + w \leq 530,\ 10x + 5y + 20z + 10w \leq 6,550,\ x \geq 0,\ y \geq 0,\ z \geq 0,\ w \geq 0.$

	x	y	z	w	s	t	u	v	r	p		
s	1	1	0	0	1	0	0	0	0	0	620	$R_1 - R_3$
t	0	0	1	1	0	1	0	0	0	0	410	
u	1	0	1	0	0	0	1	0	0	0	500	
v	0	1	0	1	0	0	0	1	0	0	530	
r	10	5	20	10	0	0	0	0	1	0	6550	$R_5 - 10R_3$
p	−1	−1	−1	−1	0	0	0	0	0	1	0	$R_6 + R_3$

	x	y	z	w	s	t	u	v	r	p		
s	0	1	−1	0	1	0	−1	0	0	0	120	
t	0	0	1	1	0	1	0	0	0	0	410	
x	1	0	1	0	0	0	1	0	0	0	500	
v	0	1	0	1	0	0	0	1	0	0	530	$R_4 - R_1$
r	0	5	10	10	0	0	−10	0	1	0	1550	$R_5 - 5R_1$
p	0	−1	0	−1	0	0	1	0	0	1	500	$R_6 + R_1$

	x	y	z	w	s	t	u	v	r	p		
y	0	1	-1	0	1	0	-1	0	0	0	120	$15R_1 + R_5$
t	0	0	1	1	0	1	0	0	0	0	410	$15R_2 - R_5$
x	1	0	1	0	0	0	1	0	0	0	500	$15R_3 - R_5$
v	0	0	1	1	-1	0	1	1	0	0	410	$15R_4 - R_5$
r	0	0	**15**	10	-5	0	-5	0	1	0	950	
p	0	0	-1	-1	1	0	0	0	0	1	620	$15R_6 + R_5$

	x	y	z	w	s	t	u	v	r	p		
y	0	15	0	10	10	0	-20	0	1	0	2750	$R_1 - R_5$
t	0	0	0	5	5	15	5	0	-1	0	5200	$2R_2 - R_5$
x	15	0	0	-10	5	0	20	0	-1	0	6550	$R_3 + R_5$
v	0	0	0	5	-10	0	20	15	-1	0	5200	$2R_4 - R_5$
z	0	0	15	**10**	-5	0	-5	0	1	0	950	
p	0	0	0	-5	10	0	-5	0	1	15	10250	$2R_6 + R_5$

	x	y	z	w	s	t	u	v	r	p		
y	0	15	-15	0	15	0	-15	0	0	0	1800	$3R_1 + R_4$
t	0	0	-15	0	15	30	15	0	-3	0	9450	$3R_2 - R_4$
x	15	0	15	0	0	0	15	0	0	0	7500	$3R_3 - R_4$
v	0	0	-15	0	-15	0	**45**	30	-3	0	9450	
w	0	0	15	10	-5	0	-5	0	1	0	950	$9R_5 + R_4$
p	0	0	15	0	15	0	-15	0	3	30	21450	$3R_6 + R_4$

	x	y	z	w	s	t	u	v	r	p	
y	0	45	-60	0	30	0	0	30	-3	0	14850
t	0	0	-30	0	60	90	0	-30	-6	0	18900
x	45	0	60	0	15	0	0	-30	3	0	13050
u	0	0	-15	0	-15	0	45	30	-3	0	9450
w	0	0	120	90	-60	0	0	30	6	0	18000
p	0	0	30	0	30	0	0	30	6	90	73800

Optimal solution:
$p = 73,800/90 = 820$; $x = 13,050/45 = 290$, $y = 14,850/45 = 330$, $z = 0$, $w = 18,000/90 = 200$.
Make the following shipments: Tucson to Honolulu: 290 boards; Tucson to Venice Beach: 330 boards; Toronto to Honolulu: 0 boards; Toronto to Venice Beach: 200 boards, giving 820 boards shipped.

45. Unknowns: $x =$ # people you fly from Chicago to Los Angeles, $y =$ # people you fly from Chicago to New York,

$z =$ # people you fly from Denver to Los Angeles, $w =$ # people you fly from Denver to New York

Maximize $p = x + y + z + w$ subject to

$195x + 182y + 395z + 166w \leq 4,520$, $x + y \leq 20$, $z + w \leq 10$, $x + z \leq 10$, $y + w \leq 15$, $x \geq 0$, $y \geq 0$, $z \geq 0$,

$w \geq 0$.

	x	y	z	w	s	t	u	v	r	p	
s	195	182	395	166	1	0	0	0	0	0	4520
t	1	1	0	0	0	1	0	0	0	0	20
u	0	0	1	1	0	0	1	0	0	0	10
v	1	0	1	0	0	0	0	1	0	0	10
r	0	1	0	1	0	0	0	0	1	0	15
p	−1	−1	−1	−1	0	0	0	0	0	1	0

	x	y	z	w	s	t	u	v	r	p	
s	0	182	200	166	1	0	0	−195	0	0	2570
t	0	1	−1	0	0	1	0	−1	0	0	10
u	0	0	1	1	0	0	1	0	0	0	10
x	1	0	1	0	0	0	0	1	0	0	10
r	0	1	0	1	0	0	0	0	1	0	15
p	0	−1	0	−1	0	0	0	1	0	1	10

	x	y	z	w	s	t	u	v	r	p	
s	0	0	382	166	1	−182	0	−13	0	0	750
y	0	1	−1	0	0	1	0	−1	0	0	10
u	0	0	1	1	0	0	1	0	0	0	10
x	1	0	1	0	0	0	0	1	0	0	10
r	0	0	1	1	0	−1	0	1	1	0	5
p	0	0	−1	−1	0	1	0	0	0	1	20

	x	y	z	w	s	t	u	v	r	p	
z	0	0	1	0.43	0	−0.48	0	−0.03	0	0	1.96
y	0	1	0	0.43	0	0.52	0	−1.03	0	0	11.96
u	0	0	0	0.57	0	0.48	1	0.03	0	0	8.04
x	1	0	0	−0.43	0	0.48	0	1.03	0	0	8.04
r	0	0	0	0.57	0	−0.52	0	1.03	1	0	3.04
p	0	0	0	−0.57	0	0.52	0	−0.03	0	1	21.96

	x	y	z	w	s	t	u	v	r	p	
w	0	0	2.3	1	0.01	−1.1	0	−0.08	0	0	4.52
y	0	1	−1	0	0	1	0	−1	0	0	10
u	0	0	−1.3	0	−0.01	1.1	1	0.08	0	0	5.48
x	1	0	1	0	0	0	0	1	0	0	10
r	0	0	−1.3	0	−0.01	0.1	0	1.08	1	0	0.48
p	0	0	1.3	0	0.01	−0.1	0	−0.08	0	1	24.52

	x	y	z	w	s	t	u	v	r	p	
w	0	0	1	1	0	0	1	0	0	0	10
y	0	1	0.19	0	0.01	0	−0.91	−1.07	0	0	5
t	0	0	−1.19	0	−0.01	1	0.91	0.07	0	0	5
x	1	0	1	0	0	0	0	1	0	0	10
r	0	0	−1.19	0	−0.01	0	−0.09	$\boxed{1.07}$	1	0	0
p	0	0	1.19	0	0.01	0	0.09	−0.07	0	1	25

	x	y	z	w	s	t	u	v	r	p	
w	0	0	1	1	0	0	1	0	0	0	10
y	0	1	−1	0	0	0	−1	0	1	0	5
t	0	0	−1.11	0	−0.01	1	0.92	0	−0.07	0	5
x	1	0	2.11	0	0.01	0	0.08	0	−0.93	0	10
v	0	0	−1.11	0	−0.01	0	−0.08	1	0.93	0	0
p	0	0	1.11	0	0.01	0	0.08	0	0.07	1	25

Optimal solution:
$p = 25$; $x = 10, y = 5, z = 0, w = 10$.
Fly 10 people from Chicago to Los Angeles, 5 people from Chicago to New York, and 10 people from Denver to New York.

Communication and reasoning exercises

47. Yes; the given problem can be stated as: Maximize $p = 3x - 2y$ subject to $-x + y - z \leq 0$, $x - y - z \leq 6$, $x \geq 0$,

$y \geq 0$, $z \geq 0$.

49. The graphical method applies only to LP problems in two unknowns, whereas the simplex method can be used to solve LP problems with any number of unknowns.

51. She is correct. Since there are only two constraints, there can only be two active variables, giving two or fewer nonzero values for the unknowns at each stage.

53. A basic solution to a system of linear equations is a solution in which all the nonpivotal variables are taken to be zero; that is, all variables whose values are arbitrary are assigned the value zero. To obtain a basic solution for a given system of linear equations, one can row-reduce the associated augmented matrix, write down the general solution, and then set all the parameters (variables with "arbitrary" values) equal to zero.

55. No. Let us assume for the sake of simplicity that all the pivots are 1s. (They may certainly be changed to 1s without affecting the value of any of the variables.) Since the entry at the bottom of the pivot column is negative, the bottom row gets replaced by itself plus a positive multiple of the pivot row. The value of the objective function (bottom right entry) is thus replaced by itself plus a positive multiple of the nonnegative rightmost entry of the pivot row. Therefore, it cannot decrease.

Section 5.4

1. Introduce slack variables and rewrite the constraints and objective function in standard form, and set up the first tableau:

$x + 2y - s = 6$
$-x + y + t = 4$
$2x + y + u = 8$
$-x - y + p = 0$

	x	y	s	t	u	p	
*s	1	2	−1	0	0	0	6
t	−1	1	0	1	0	0	4
u	2	1	0	0	1	0	8
p	−1	−1	0	0	0	1	0

The first starred row is the s-row, and its largest positive entry is the 2 in the y-column. Thus, we use this column as the pivot column. The test ratios are: s: 6/2, t: 4/1, u: 8/1. The smallest test ratio is s: 6/2. Thus we pivot on the 2 in the s-row.

	x	y	s	t	u	p		
*s	1	2	−1	0	0	0	6	
t	−1	1	0	1	0	0	4	$2R_2 - R_1$
u	2	1	0	0	1	0	8	$2R_3 - R_1$
p	−1	−1	0	0	0	1	0	$2R_4 + R_1$

	x	y	s	t	u	p	
y	1	2	−1	0	0	0	6
t	−3	0	1	2	0	0	2
u	3	0	1	0	2	0	10
p	−1	0	−1	0	0	2	6

As there are no more starred rows, we go to Phase 2, and do the standard simplex method.

	x	y	s	t	u	p		
y	1	2	−1	0	0	0	6	$3R_1 - R_3$
t	−3	0	1	2	0	0	2	$R_2 + R_3$
u	3	0	1	0	2	0	10	
p	−1	0	−1	0	0	2	6	$3R_4 + R_3$

	x	y	s	t	u	p		
y	0	6	−4	0	−2	0	8	$R_1 + 2R_2$
t	0	0	2	2	2	0	12	
x	3	0	1	0	2	0	10	$2R_3 - R_2$
p	0	0	−2	0	2	6	28	$R_4 + R_2$

	x	y	s	t	u	p	
y	0	6	0	4	2	0	32
s	0	0	2	2	2	0	12
x	6	0	0	-2	2	0	8
p	0	0	0	2	4	6	40

Optimal solution:
$p = 40/6 = 20/3$; $x = 8/6 = 4/3$, $y = 32/6 = 16/3$.

3. Introduce slack variables and rewrite the constraints and objective function in standard form, and set up the first tableau:
$x + y + s = 25$
$x - t = 10$
$-x + 2y - u = 0$
$-12x - 10y + p = 0$

	x	y	s	t	u	p	
s	1	1	1	0	0	0	25
*t	1	0	0	-1	0	0	10
*u	-1	2	0	0	-1	0	0
p	-12	-10	0	0	0	1	0

The first starred row is the t-row, and its largest positive entry is the 1 in the x-column. Thus, we use this column as the pivot column. The test ratios are: s: 25/1, t: 10/1. The smallest test ratio is t: 10/1. Thus we pivot on the 1 in the t-row.

	x	y	s	t	u	p		
s	1	1	1	0	0	0	25	$R_1 - R_2$
*t	1	0	0	-1	0	0	10	
*u	-1	2	0	0	-1	0	0	$R_3 + R_2$
p	-12	-10	0	0	0	1	0	$R_4 + 12R_2$

	x	y	s	t	u	p	
s	0	1	1	1	0	0	15
x	1	0	0	-1	0	0	10
*u	0	2	0	-1	-1	0	10
p	0	-10	0	-12	0	1	120

The first starred row is the u-row, and its largest positive entry is the 2 in the y-column. Thus, we use this column as the pivot column. The test ratios are: s: 15/1, u: 10/2. The smallest test ratio is u: 10/2. Thus we pivot on the 2 in the u-row.

	x	y	s	t	u	p		
s	0	1	1	1	0	0	15	$2R_1 - R_3$
x	1	0	0	-1	0	0	10	
*u	0	2	0	-1	-1	0	10	
p	0	-10	0	-12	0	1	120	$R_4 + 5R_3$

	x	y	s	t	u	p	
s	0	0	2	3	1	0	20
x	1	0	0	−1	0	0	10
y	0	2	0	−1	−1	0	10
p	0	0	0	−17	−5	1	170

As there are no more starred rows, we go to Phase 2, and do the standard simplex method.

	x	y	s	t	u	p		
s	0	0	2	3	1	0	20	
x	1	0	0	−1	0	0	10	$3R_2 + R_1$
y	0	2	0	−1	−1	0	10	$3R_3 + R_1$
p	0	0	0	−17	−5	1	170	$3R_4 + 17R_1$

	x	y	s	t	u	p	
t	0	0	2	3	1	0	20
x	3	0	2	0	1	0	50
y	0	6	2	0	−2	0	50
p	0	0	34	0	2	3	850

Optimal solution:
$p = 850/3; \ x = 50/3, y = 50/6 = 25/3.$

5. Introduce slack variables and rewrite the constraints and objective function in standard form, and set up the first tableau:
$x + y + z + s = 150$
$x + y + z - t = 100$
$-2x - 5y - 3z + p = 0$

	x	y	z	s	t	p	
s	1	1	1	1	0	0	150
*t	1	1	1	0	−1	0	100
p	−2	−5	−3	0	0	1	0

The first starred row is the t-row, and its largest positive entry is the 1 in the x-column. Thus, we use this column as the pivot column. The test ratios are: s: 150/1, t: 100/1. The smallest test ratio is t: 100/1. Thus we pivot on the 1 in the t-row.

	x	y	z	s	t	p		
s	1	1	1	1	0	0	150	$R_1 - R_2$
*t	1	1	1	0	−1	0	100	
p	−2	−5	−3	0	0	1	0	$R_3 + 2R_2$

	x	y	z	s	t	p	
s	0	0	0	1	1	0	50
x	1	1	1	0	−1	0	100
p	0	−3	−1	0	−2	1	200

As there are no more starred rows, we go to Phase 2, and do the standard simplex method.

	x	y	z	s	t	p		
s	0	0	0	1	1	0	50	
x	1	1	1	0	−1	0	100	
p	0	−3	−1	0	−2	1	200	$R_3 + 3R_2$

	x	y	z	s	t	p		
s	0	0	0	1	1	0	50	
y	1	1	1	0	−1	0	100	$R_2 + R_1$
p	3	0	2	0	−5	1	500	$R_3 + 5R_1$

	x	y	z	s	t	p	
t	0	0	0	1	1	0	50
y	1	1	1	1	0	0	150
p	3	0	2	5	0	1	750

Optimal solution:
$p = 750/1 = 750$; $x = 0, y = 150/1 = 150, z = 0$.

7. Introduce slack variables and rewrite the constraints and objective function in standard form, and set up the first tableau:
$$x + 2y + z + s = 40$$
$$2y − z − t = 10$$
$$2x − y + z − u = 20$$
$$−10x − 20y − 15z + p = 0$$

	x	y	z	s	t	u	p	
s	1	2	1	1	0	0	0	40
*t	0	2	−1	0	−1	0	0	10
*u	2	−1	1	0	0	−1	0	20
p	−10	−20	−15	0	0	0	1	0

The first starred row is the t-row, and its largest positive entry is the 2 in the y-column. Thus, we use this column as the pivot column. The test ratios are: s: 40/2, t: 10/2. The smallest test ratio is t: 10/2. Thus we pivot on the 2 in the t-row.

	x	y	z	s	t	u	p		
s	1	2	1	1	0	0	0	40	$R_1 - R_2$
*t	0	2	-1	0	-1	0	0	10	
*u	2	-1	1	0	0	-1	0	20	$2R_3 + R_2$
p	-10	-20	-15	0	0	0	1	0	$R_4 + 10R_2$

	x	y	z	s	t	u	p	
s	1	0	2	1	1	0	0	30
y	0	2	-1	0	-1	0	0	10
*u	4	0	1	0	-1	-2	0	50
p	-10	0	-25	0	-10	0	1	100

The first starred row is the u-row, and its largest positive entry is the 4 in the x-column. Thus, we use this column as the pivot column. The test ratios are: s: 30/1, u: 50/4. The smallest test ratio is u: 50/4. Thus we pivot on the 4 in the u-row.

	x	y	z	s	t	u	p		
s	1	0	2	1	1	0	0	30	$4R_1 - R_3$
y	0	2	-1	0	-1	0	0	10	
*u	4	0	1	0	-1	-2	0	50	
p	-10	0	-25	0	-10	0	1	100	$2R_4 + 5R_3$

	x	y	z	s	t	u	p	
s	0	0	7	4	5	2	0	70
y	0	2	-1	0	-1	0	0	10
x	4	0	1	0	-1	-2	0	50
p	0	0	-45	0	-25	-10	2	450

As there are no more starred rows, we go to Phase 2, and do the standard simplex method.

	x	y	z	s	t	u	p		
s	0	0	7	4	5	2	0	70	
y	0	2	-1	0	-1	0	0	10	$7R_2 + R_1$
x	4	0	1	0	-1	-2	0	50	$7R_3 - R_1$
p	0	0	-45	0	-25	-10	2	450	$7R_4 + 45R_1$

	x	y	z	s	t	u	p	
z	0	0	7	4	5	2	0	70
y	0	14	0	4	-2	2	0	140
x	28	0	0	-4	-12	-16	0	280
p	0	0	0	180	50	20	14	6300

Optimal solution:

$p = 6,300/14 = 450;\ x = 280/28 = 10,\ y = 140/14 = 10,\ z = 70/7 = 10.$

9. Introduce slack variables and rewrite the constraints and objective function in standard form, and set up the first tableau:

$x + y + z + w + s = 40$
$2x + y - z - w - t = 10$
$x + y + z + w - r = 10$
$-x - y - 3z - w + p = 0$

	x	y	z	w	s	t	r	p	
s	1	1	1	1	1	0	0	0	40
*t	2	1	-1	-1	0	-1	0	0	10
*r	1	1	1	1	0	0	-1	0	10
p	-1	-1	-3	-1	0	0	0	1	0

The first starred row is the t-row, and its largest positive entry is the 2 in the x-column. Thus, we use this column as the pivot column. The test ratios are: s: 40/1, t: 10/2, r: 10/1. The smallest test ratio is t: 10/2. Thus we pivot on the 2 in the t-row.

	x	y	z	w	s	t	r	p		
s	1	1	1	1	1	0	0	0	40	$2R_1 - R_2$
*t	2	1	-1	-1	0	-1	0	0	10	
*r	1	1	1	1	0	0	-1	0	10	$2R_3 - R_2$
p	-1	-1	-3	-1	0	0	0	1	0	$2R_4 + R_2$

	x	y	z	w	s	t	r	p	
s	0	1	3	3	2	1	0	0	70
x	2	1	-1	-1	0	-1	0	0	10
*r	0	1	3	3	0	1	-2	0	10
p	0	-1	-7	-3	0	-1	0	2	10

The first starred row is the r-row, and its largest positive entry is the 3 in the z-column. Thus, we use this column as the pivot column. The test ratios are: s: 70/3, r: 10/3. The smallest test ratio is r: 10/3. Thus we pivot on the 3 in the r-row.

	x	y	z	w	s	t	r	p		
s	0	1	3	3	2	1	0	0	70	$R_1 - R_3$
x	2	1	-1	-1	0	-1	0	0	10	$3R_2 + R_3$
*r	0	1	3	3	0	1	-2	0	10	
p	0	-1	-7	-3	0	-1	0	2	10	$3R_4 + 7R_3$

	x	y	z	w	s	t	r	p	
s	0	0	0	0	2	0	2	0	60
x	6	4	0	0	0	-2	-2	0	40
z	0	1	3	3	0	1	-2	0	10
p	0	4	0	12	0	4	-14	6	100

As there are no more starred rows, we go to Phase 2, and do the standard simplex method.

	x	y	z	w	s	t	r	p		
s	0	0	0	0	2	0	2	0	60	
x	6	4	0	0	0	−2	−2	0	40	$R_2 + R_1$
z	0	1	3	3	0	1	−2	0	10	$R_3 + R_1$
p	0	4	0	12	0	4	−14	6	100	$R_4 + 7R_1$

	x	y	z	w	s	t	r	p	
r	0	0	0	0	2	0	2	0	60
x	6	4	0	0	2	−2	0	0	100
z	0	1	3	3	2	1	0	0	70
p	0	4	0	12	14	4	0	6	520

Optimal solution:
$p = 520/6 = 260/3$; $x = 100/6 = 50/3$, $y = 0$, $z = 70/3$, $w = 0$.

11. Convert the given problem into a maximization problem:

Maximize $p = -6x - 6y$ subject to $x + 2y \geq 20$, $2x + y \geq 20$, $x \geq 0$, $y \geq 0$.

Introduce slack variables and rewrite the constraints and objective function in standard form, and set up the first tableau:
$x + 2y - s = 20$
$2x + y - t = 20$
$6x + 6y + p = 0$

	x	y	s	t	p	
$*s$	1	2	−1	0	0	20
$*t$	2	1	0	−1	0	20
p	6	6	0	0	1	0

The first starred row is the s-row, and its largest positive entry is the 2 in the y-column. Thus, we use this column as the pivot column. The test ratios are: s: 20/2, t: 20/1. The smallest test ratio is s: 20/2. Thus we pivot on the 2 in the s-row.

	x	y	s	t	p		
$*s$	1	2	−1	0	0	20	
$*t$	2	1	0	−1	0	20	$2R_2 - R_1$
p	6	6	0	0	1	0	$R_3 - 3R_1$

	x	y	s	t	p	
y	1	2	−1	0	0	20
$*t$	3	0	1	−2	0	20
p	3	0	3	0	1	−60

The first starred row is the t-row, and its largest positive entry is the 3 in the x-column. Thus, we use this column as the pivot column. The test ratios are: y: 20/1, t: 20/3. The smallest test ratio is t: 20/3. Thus we pivot on the 3 in the t-row.

	x	y	s	t	p		
y	1	2	−1	0	0	20	$3R_1 − R_2$
*t	3	0	1	−2	0	20	
p	3	0	3	0	1	−60	$R_3 − R_2$

	x	y	s	t	p	
y	0	6	−4	2	0	40
x	3	0	1	−2	0	20
p	0	0	2	2	1	−80

As there are no more starred rows, we go to Phase 2, and do the standard simplex method.
However, there are no negative numbers in the bottom row, so there are no pivot steps to do in Phase 2. Therefore, we are done. Optimal solution: $p = -80/1 = -80$; $x = 20/3$, $y = 40/6 = 20/3$.
Since $c = -p$, the minimum value of c is 80.

13. Convert the given problem into a maximization problem:

Maximize $p = -2x - y - 3z$ subject to $x + y + z \geq 100$, $2x + y \geq 50$, $y + z \geq 50$, $x \geq 0$, $y \geq 0$, $z \geq 0$.

Introduce slack variables and rewrite the constraints and objective function in standard form, and set up the first tableau:

$x + y + z - s = 100$
$2x + y - t = 50$
$y + z - u = 50$
$2x + y + 3z + p = 0$

	x	y	z	s	t	u	p	
*s	1	1	1	−1	0	0	0	100
*t	2	1	0	0	−1	0	0	50
*u	0	1	1	0	0	−1	0	50
p	2	1	3	0	0	0	1	0

The first starred row is the s-row, and its largest positive entry is the 1 in the x-column. Thus, we use this column as the pivot column. The test ratios are: s: 100/1, t: 50/2. The smallest test ratio is t: 50/2. Thus we pivot on the 2 in the t-row.

	x	y	z	s	t	u	p		
*s	1	1	1	−1	0	0	0	100	$2R_1 − R_2$
*t	2	1	0	0	−1	0	0	50	
*u	0	1	1	0	0	−1	0	50	
p	2	1	3	0	0	0	1	0	$R_4 − R_2$

	x	y	z	s	t	u	p	
*s	0	1	2	-2	1	0	0	150
x	2	1	0	0	-1	0	0	50
*u	0	1	1	0	0	-1	0	50
p	0	0	3	0	1	0	1	-50

The first starred row is the s-row, and its largest positive entry is the 2 in the z-column. Thus, we use this column as the pivot column. The test ratios are: s: 150/2, u: 50/1. The smallest test ratio is u: 50/1. Thus we pivot on the 1 in the u-row.

	x	y	z	s	t	u	p		
*s	0	1	2	-2	1	0	0	150	$R_1 - 2R_3$
x	2	1	0	0	-1	0	0	50	
*u	0	1	[1]	0	0	-1	0	50	
p	0	0	3	0	1	0	1	-50	$R_4 - 3R_3$

	x	y	z	s	t	u	p	
*s	0	-1	0	-2	1	2	0	50
x	2	1	0	0	-1	0	0	50
z	0	1	1	0	0	-1	0	50
p	0	-3	0	0	1	3	1	-200

The first starred row is the s-row, and its largest positive entry is the 2 in the u-column. Thus, we use this column as the pivot column. The only positive entry in this column is the 2 in the s-row, so we pivot on that entry.

	x	y	z	s	t	u	p		
*s	0	-1	0	-2	1	[2]	0	50	
x	2	1	0	0	-1	0	0	50	
z	0	1	1	0	0	-1	0	50	$2R_3 + R_1$
p	0	-3	0	0	1	3	1	-200	$2R_4 - 3R_1$

	x	y	z	s	t	u	p	
u	0	-1	0	-2	1	2	0	50
x	2	1	0	0	-1	0	0	50
z	0	1	2	-2	1	0	0	150
p	0	-3	0	6	-1	0	2	-550

As there are no more starred rows, we go to Phase 2, and do the standard simplex method.

	x	y	z	s	t	u	p		
u	0	-1	0	-2	1	2	0	50	$R_1 + R_2$
x	2	[1]	0	0	-1	0	0	50	
z	0	1	2	-2	1	0	0	150	$R_3 - R_2$
p	0	-3	0	6	-1	0	2	-550	$R_4 + 3R_2$

	x	y	z	s	t	u	p		
u	2	0	0	-2	0	2	0	100	
y	2	1	0	0	-1	0	0	50	$2R_2 + R_3$
z	-2	0	2	-2	[2]	0	0	100	
p	6	0	0	6	-4	0	2	-400	$R_4 + 2R_3$

	x	y	z	s	t	u	p	
u	2	0	0	-2	0	2	0	100
y	2	2	2	-2	0	0	0	200
t	-2	0	2	-2	2	0	0	100
p	2	0	4	2	0	0	2	-200

Optimal solution: $p = -200/2 = -100$; $x = 0, y = 200/2 = 100, z = 0$.
Since $c = -p$, the minimum value of c is 100.

15. Convert the given problem into a maximization problem:

Maximize $p = -50x - 50y - 11z$ subject to $2x + z \geq 3$, $2x + y - z \geq 2$, $3x + y - z \leq 3$, $x \geq 0$, $y \geq 0$, $z \geq 0$.

Introduce slack variables and rewrite the constraints and objective function in standard form, and set up the first tableau:
$2x + z - s = 3$
$2x + y - z - t = 2$
$3x + y - z + u = 3$
$50x + 50y + 11z + p = 0$

	x	y	z	s	t	u	p	
*s	2	0	1	-1	0	0	0	3
*t	2	1	-1	0	-1	0	0	2
u	3	1	-1	0	0	1	0	3
p	50	50	11	0	0	0	1	0

The first starred row is the s-row, and its largest positive entry is the 2 in the x-column. Thus, we use this column as the pivot column. The test ratios are: s: 3/2, t: 2/2, u: 3/3. The smallest test ratio is t: 2/2. Thus we pivot on the 2 in the t-row.

	x	y	z	s	t	u	p		
*s	2	0	1	-1	0	0	0	3	$R_1 - R_2$
*t	[2]	1	-1	0	-1	0	0	2	
u	3	1	-1	0	0	1	0	3	$2R_3 - 3R_2$
p	50	50	11	0	0	0	1	0	$R_4 - 25R_2$

	x	y	z	s	t	u	p	
*s	0	−1	2	−1	1	0	0	1
x	2	1	−1	0	−1	0	0	2
u	0	−1	1	0	3	2	0	0
p	0	25	36	0	25	0	1	−50

The first starred row is the s-row, and its largest positive entry is the 2 in the z-column. Thus, we use this column as the pivot column. The test ratios are: s: 1/2, u: 0/1. The smallest test ratio is u: 0/1. Thus we pivot on the 1 in the u-row.

	x	y	z	s	t	u	p		
*s	0	−1	2	−1	1	0	0	1	$R_1 - 2R_3$
x	2	1	−1	0	−1	0	0	2	$R_2 + R_3$
u	0	−1	[1]	0	3	2	0	0	
p	0	25	36	0	25	0	1	−50	$R_4 - 36R_3$

	x	y	z	s	t	u	p	
*s	0	1	0	−1	−5	−4	0	1
x	2	0	0	0	2	2	0	2
z	0	−1	1	0	3	2	0	0
p	0	61	0	0	−83	−72	1	−50

The first starred row is the s-row, and its largest positive entry is the 1 in the y-column. Thus, we use this column as the pivot column. The only positive entry in this column is the 1 in the s-row, so we pivot on that entry.

	x	y	z	s	t	u	p		
*s	0	[1]	0	−1	−5	−4	0	1	
x	2	0	0	0	2	2	0	2	
z	0	−1	1	0	3	2	0	0	$R_3 + R_1$
p	0	61	0	0	−83	−72	1	−50	$R_4 - 61R_1$

	x	y	z	s	t	u	p	
y	0	1	0	−1	−5	−4	0	1
x	2	0	0	0	2	2	0	2
z	0	0	1	−1	−2	−2	0	1
p	0	0	0	61	222	172	1	−111

As there are no more starred rows, we go to Phase 2, and do the standard simplex method.
However, there are no negative numbers in the bottom row, so there are no pivot steps to do in Phase 2. Therefore, we are done. Optimal solution: $p = −111/1 = −111$; $x = 2/2 = 1$, $y = 1/1 = 1$, $z = 1/1 = 1$.
Since $c = −p$, the minimum value of c is 111.

17. Convert the given problem into a maximization problem: Maximize $p = −x − y − z − w$ subject to

$5x − y + w \geq 1{,}000$, $z + w \leq 2{,}000$, $x + y \leq 500$, $x \geq 0$, $y \geq 0$, $z \geq 0$, $w \geq 0$.

Introduce slack variables and rewrite the constraints and objective function in standard form, and set up the first tableau:

$5x - y + w - s = 1,000$
$z + w + t = 2,000$
$x + y + u = 500$
$x + y + z + w + p = 0$

	x	y	z	w	s	t	u	p	
*s	5	−1	0	1	−1	0	0	0	1000
t	0	0	1	1	0	1	0	0	2000
u	1	1	0	0	0	0	1	0	500
p	1	1	1	1	0	0	0	1	0

The first starred row is the s-row, and its largest positive entry is the 5 in the x-column. Thus, we use this column as the pivot column. The test ratios are: s: 1000/5, u: 500/1. The smallest test ratio is s: 1000/5. Thus we pivot on the 5 in the s-row.

	x	y	z	w	s	t	u	p		
*s	5	−1	0	1	−1	0	0	0	1000	
t	0	0	1	1	0	1	0	0	2000	
u	1	1	0	0	0	0	1	0	500	$5R_3 - R_1$
p	1	1	1	1	0	0	0	1	0	$5R_4 - R_1$

	x	y	z	w	s	t	u	p	
x	5	−1	0	1	−1	0	0	0	1000
t	0	0	1	1	0	1	0	0	2000
u	0	6	0	−1	1	0	5	0	1500
p	0	6	5	4	1	0	0	5	−1000

As there are no more starred rows, we go to Phase 2, and do the standard simplex method.
However, there are no negative numbers in the bottom row, so there are no pivot steps to do in Phase 2. Therefore, we are done. Optimal solution: $p = -1,000/5 = -200$; $x = 1,000/5 = 200, y = 0, z = 0, w = 0$. Since $c = -p$, the minimum value of c is 200.

19. You can use the online Pivot and Gauss-Jordan Tool in decimal mode to do the pivoting, or use the online Simplex Method Tool. When entering problems in the Simplex Method Tool there is no need to enter the inequalities $x \geq 0$, $y \geq 0$, etc.

	x	y	z	w	s	t	u	p	
s	1.2	1	1	1	1	0	0	0	40.5
*t	2.2	1	−1	−1	0	−1	0	0	10
*u	1.2	1	1	1.2	0	0	−1	0	10.5
p	−2	−3	−1.1	−4	0	0	0	1	0

	x	y	z	w	s	t	u	p	
s	0	0.45	1.55	1.55	1	0.55	0	0	35.05
x	1	0.45	−0.45	−0.45	0	−0.45	0	0	4.55
*u	0	0.45	1.55	1.75	0	0.55	−1	0	5.05
p	0	−2.09	−2.01	−4.91	0	−0.91	0	1	9.09

	x	y	z	w	s	t	u	p	
s	0	0.05	0.18	0	1	0.06	0.89	0	30.58
x	1	0.57	−0.05	0	0	−0.31	−0.26	0	5.86
w	0	0.26	0.89	1	0	0.31	−0.57	0	2.89
p	0	−0.81	2.34	0	0	0.63	−2.81	1	23.28

	x	y	z	w	s	t	u	p	
u	0	0.06	0.2	0	1.13	0.07	1	0	34.54
x	1	0.59	0	0	0.29	−0.29	0	0	14.85
w	0	0.29	1	1	0.65	0.35	0	0	22.68
p	0	−0.65	2.9	0	3.18	0.82	0	1	120.41

	x	y	z	w	s	t	u	p	
u	−0.1	0	0.2	0	1.1	0.1	1	0	33.05
y	1.7	1	0	0	0.5	−0.5	0	0	25.25
w	−0.5	0	1	1	0.5	0.5	0	0	15.25
p	1.1	0	2.9	0	3.5	0.5	0	1	136.75

Optimal solution:
$p = 136.75$; $x = 0, y = 25.25, z = 0, w = 15.25$.

21. First convert the problem to a maximization problem as in Example 3 by taking p to be the negative of c. Then use the online Pivot and Gauss-Jordan Tool in decimal mode to do the pivoting, or use the online Simplex Method Tool. When entering problems in the Simplex Method Tool there is no need to enter the inequalities $x \geq 0$, $y \geq 0$, etc.

	x	y	z	s	t	u	p	
*s	1	1.5	1.2	−1	0	0	0	100
*t	2	1.5	0	0	−1	0	0	50
*u	0	1.5	1.1	0	0	−1	0	50
p	2.2	1	3.3	0	0	0	1	0

	x	y	z	s	t	u	p	
*s	−1	0	1.2	−1	1	0	0	50
y	1.33	1	0	0	−0.67	0	0	33.33
*u	−2	0	1.1	0	1	−1	0	0
p	0.87	0	3.3	0	0.67	0	1	−33.33

	x	y	z	s	t	u	p	
*s	1.18	0	0	−1	−0.09	1.09	0	50
y	1.33	1	0	0	−0.67	0	0	33.33
z	−1.82	0	1	0	0.91	−0.91	0	0
p	6.87	0	0	0	−2.33	3	1	−33.33

	x	y	z	s	t	u	p	
*s	0	−0.89	0	−1	0.5	1.09	0	20.45
x	1	0.75	0	0	−0.5	0	0	25
z	0	1.36	1	0	0	−0.91	0	45.45
p	0	−5.15	0	0	1.1	3	1	−205

	x	y	z	s	t	u	p	
u	0	−0.81	0	−0.92	0.46	1	0	18.75
x	1	0.75	0	0	−0.5	0	0	25
z	0	0.62	1	−0.83	0.42	0	0	62.5
p	0	−2.71	0	2.75	−0.27	0	1	−261.25

	x	y	z	s	t	u	p	
u	1.08	0	0	−0.92	−0.08	1	0	45.83
y	1.33	1	0	0	−0.67	0	0	33.33
z	−0.83	0	1	−0.83	0.83	0	0	41.67
p	3.62	0	0	2.75	−2.08	0	1	−170.83

	x	y	z	s	t	u	p	
u	1	0	0.1	−1	0	1	0	50
y	0.67	1	0.8	−0.67	0	0	0	66.67
t	−1	0	1.2	−1	1	0	0	50
p	1.53	0	2.5	0.67	0	0	1	−66.67

Optimal solution: $p = −66.67$; $x = 0, y = 66.67, z = 0$.
Since $c = −p$, the minimum value of c is 66.67.

23. First convert the problem to a maximization problem as in Example 3 by taking p to be the negative of c. Then use the online Pivot and Gauss-Jordan Tool in decimal mode to do the pivoting, or use the online Simplex Method Tool. When entering problems in the Simplex Method Tool there is no need to enter the inequalities $x \geq 0$, $y \geq 0$, etc.

	x	y	z	w	s	t	u	p	
s	5.12	−1	0	1	1	0	0	0	1000
*t	0	0	1	1	0	−1	0	0	2000
u	1.22	1	0	0	0	0	1	0	500
p	1.1	1	1.5	−1	0	0	0	1	0

	x	y	z	w	s	t	u	p	
s	5.12	-1	0	1	1	0	0	0	1000
z	0	0	1	1	0	-1	0	0	2000
u	1.22	1	0	0	0	0	1	0	500
p	1.1	1	0	-2.5	0	1.5	0	1	-3000

	x	y	z	w	s	t	u	p	
w	5.12	-1	0	1	1	0	0	0	1000
z	-5.12	1	1	0	-1	-1	0	0	1000
u	1.22	1	0	0	0	0	1	0	500
p	13.9	-1.5	0	0	2.5	1.5	0	1	-500

	x	y	z	w	s	t	u	p	
w	6.34	0	0	1	1	0	1	0	1500
z	-6.34	0	1	0	-1	-1	-1	0	500
y	1.22	1	0	0	0	0	1	0	500
p	15.73	0	0	0	2.5	1.5	1.5	1	250

Optimal solution: $p = 250$; $x = 0, y = 500, z = 500, w = 1,500$.
Since $c = -p$, the minimum value of c is -250.

Applications

25. Unknowns: $x = $ # acres of tomatoes, $y = $ # acres of lettuce, $z = $ # acres of carrots

Maximize $p = 20x + 15y + 5z$ (in hundreds of dollars) subject to

$x + y + z \leq 100, \ 5x + 4y + 2z \geq 400, \ 4x + 2y + 2z \leq 500, \ x \geq 0, \ y \geq 0, \ z \geq 0.$

	x	y	z	s	t	u	p		
s	1	1	1	1	0	0	0	100	$5R_1 - R_2$
*t	5	4	2	0	-1	0	0	400	
u	4	2	2	0	0	1	0	500	$5R_3 - 4R_2$
p	-20	-15	-5	0	0	0	1	0	$R_4 + 4R_2$

	x	y	z	s	t	u	p		
s	0	1	3	5	1	0	0	100	
x	5	4	2	0	-1	0	0	400	$R_2 + R_1$
u	0	-6	2	0	4	5	0	900	$R_3 - 4R_1$
p	0	1	3	0	-4	0	1	1600	$R_4 + 4R_1$

	x	y	z	s	t	u	p	
t	0	1	3	5	1	0	0	100
x	5	5	5	5	0	0	0	500
u	0	−10	−10	−20	0	5	0	500
p	0	5	15	20	0	0	1	2000

Optimal solution:

$p = 2,000/1 = 2,000$; $x = 500/5 = 100, y = 0, z = 0$.

Plant 100 acres of tomatoes and no other crops. This will give you a profit of $200,000. Since the slack variable corresponding to farm area is $s = 0$, you will be using all 100 acres of farm.

27. Unknowns: $x = $ # mailings to the East Coast, $y = $ # mailings to the Midwest, $z = $ # mailings to the West Coast

Minimize $c = 40x + 60y + 50z$ subject to $100x + 100y + 50z \geq 1,500$, $50x + 100y + 100z \geq 1,500, x \geq 0, y \geq 0$,

$z \geq 0$.

	x	y	z	s	t	p		
*s	100	100	50	−1	0	0	1500	
*t	50	100	100	0	−1	0	1500	$2R_2 - R_1$
p	40	60	50	0	0	1	0	$5R_3 - 2R_1$

	x	y	z	s	t	p		
x	100	100	50	−1	0	0	1500	$3R_1 - R_2$
*t	0	100	150	1	−2	0	1500	
p	0	100	150	2	0	5	−3000	$R_3 - R_2$

	x	y	z	s	t	p	
x	300	200	0	−4	2	0	3000
z	0	100	150	1	−2	0	1500
p	0	0	0	1	2	5	−4500

Optimal solution: $p = -4,500/5 = -900$; $x = 3,000/300 = 10, y = 0, z = 1,500/150 = 10$.

Since $c = -p$, the minimum value of c is 900.

Send 10 mailings to the East Coast, none to the Midwest, and 10 to the West Coast. Cost: $900. Another solution resulting in the same cost (pivot on the y-column in the last tableau above) is no mailings to the East Coast, 15 to the Midwest, None to the West Coast.

29. Unknowns: $x = $ # quarts orange juice, $y = $ # quarts orange concentrate

Minimize $c = 0.5x + 2y$ subject to $x \geq 10,000$, $y \geq 1,000$, $10x + 50y \geq 200,000, x \geq 0, y \geq 0$.

	x	y	s	t	u	p		
*s	1	0	−1	0	0	0	10000	
*t	0	1	0	−1	0	0	1000	
*u	10	50	0	0	−1	0	200000	
p	0.5	2	0	0	0	1	0	2R_4

	x	y	s	t	u	p		
*s	1	0	-1	0	0	0	10000	
*t	0	1	0	-1	0	0	1000	
*u	10	50	0	0	-1	0	200000	$R_3 - 10R_1$
p	1	4	0	0	0	2	0	$R_4 - R_1$

	x	y	s	t	u	p		
x	1	0	-1	0	0	0	10000	
*t	0	1	0	-1	0	0	1000	
*u	0	50	10	0	-1	0	100000	$R_3 - 50R_2$
p	0	4	1	0	0	2	-10000	$R_4 - 4R_2$

	x	y	s	t	u	p		
x	1	0	-1	0	0	0	10000	
y	0	1	0	-1	0	0	1000	$50R_2 + R_3$
*u	0	0	10	50	-1	0	50000	
p	0	0	1	4	0	2	-14000	$25R_4 - 2R_3$

	x	y	s	t	u	p	
x	1	0	-1	0	0	0	10000
y	0	50	10	0	-1	0	100000
t	0	0	10	50	-1	0	50000
p	0	0	5	0	2	50	-450000

Optimal solution: $p = -450,000/50 = -9,000$; $x = 10,000/1 = 10,000$, $y = 100,000/50 = 2,000$.
Since $c = -p$, the minimum value of c is 9,000.
Succulent Citrus should produce 10,000 quarts of orange juice and 2,000 quarts of orange concentrate.

31. Unknowns: $x = $ # regional music albums, $y = $ # pop/rock music albums, $z = $ # tropical music albums

Maximize $p = 5x + 4y + 6z$ subject to

$x - 4z \geq 0$, $x + y + z \leq 40,000$, $y \geq 10,000$, $x \geq 0$, $y \geq 0$, $z \geq 0$.

	x	y	z	s	t	u	p		
*s	1	0	-4	-1	0	0	0	0	
t	1	1	1	0	1	0	0	40000	$R_2 - R_1$
*u	0	1	0	0	0	-1	0	10000	
p	-5	-4	-6	0	0	0	1	0	$R_4 + 5R_1$

	x	y	z	s	t	u	p		
x	1	0	-4	-1	0	0	0	0	
t	0	1	5	1	1	0	0	40000	$R_2 - R_3$
*u	0	1	0	0	0	-1	0	10000	
p	0	-4	-26	-5	0	0	1	0	$R_4 + 4R_3$

	x	y	z	s	t	u	p		
x	1	0	-4	-1	0	0	0	0	$5R_1 + 4R_2$
t	0	0	5	1	1	1	0	30000	
y	0	1	0	0	0	-1	0	10000	
p	0	0	-26	-5	0	-4	1	40000	$5R_4 + 26R_2$

	x	y	z	s	t	u	p	
x	5	0	0	-1	4	4	0	120000
z	0	0	5	1	1	1	0	30000
y	0	1	0	0	0	-1	0	10000
p	0	0	0	1	26	6	5	980000

Optimal solution:

$p = 980,000/5 = 196,000$; $x = 120,000/5 = 24,000$, $y = 10,000/1 = 10,000$, $z = 30,000/5 = 6,000$.
You should sell 24,000 regional music albums, 10,000 pop/rock music albums, and 6,000 tropical music albums per day for a maximum revenue of $196,000.

33. Unknowns:

$x = $ # axes, $y = $ # maces, $z = $ # spears.

Maximize $p = 6x + 6y + 8z$ subject to

$8x + 5y + 2z \leq 600$, $x + y + 2z \geq 200$, $-y + 2z \leq 0, x \geq 0$, $y \geq 0$, $z \geq 0$.

	x	y	z	s	t	u	p		
s	8	5	2	1	0	0	0	600	$R_1 - R_3$
*t	1	1	2	0	-1	0	0	200	$R_2 - R_3$
u	0	-1	2	0	0	1	0	0	
p	-6	-6	-8	0	0	0	1	0	$R_4 + 4R_3$

	x	y	z	s	t	u	p		
s	8	6	0	1	0	-1	0	600	$R_1 - 3R_2$
*t	1	2	0	0	-1	-1	0	200	
z	0	-1	2	0	0	1	0	0	$2R_3 + R_2$
p	-6	-10	0	0	0	4	1	0	$R_4 + 5R_2$

	x	y	z	s	t	u	p		
s	5	0	0	1	3	2	0	0	
y	1	2	0	0	-1	-1	0	200	$3R_2 + R_1$
z	1	0	4	0	-1	1	0	200	$3R_3 + R_1$
p	-1	0	0	0	-5	-1	1	1000	$3R_4 + 5R_1$

	x	y	z	s	t	u	p	
t	5	0	0	1	3	2	0	0
y	8	6	0	1	0	−1	0	600
z	8	0	12	1	0	5	0	600
p	22	0	0	5	0	7	3	3000

Optimal solution:

$p = 3,000/3 = 1,000$; $x = 0, y = 600/6 = 100, z = 600/12 = 50$.

Achlúk can inflict a maximum of 1,000 units of damage using an arsenal of no axes, 100 maces, and 50 spears.

35. Unknowns:

$x = $ # axes, $y = $ # maces, $z = $ # spears.

Minimize $c = 8x + 5y + 2z$ subject to

$3x + 3y + 4z \geq 1,000, \; x + y + 2z \geq 200, \; x + 2y + 3z \leq 500, \; x \geq 0, \; y \geq 0, \; z \geq 0.$

	x	y	z	s	t	u	p		
*s	3	3	4	−1	0	0	0	1000	$R_1 - 2R_2$
*t	1	1	2	0	−1	0	0	200	
u	1	2	3	0	0	1	0	500	$2R_3 - 3R_2$
p	8	5	2	0	0	0	1	0	$R_4 - R_2$

	x	y	z	s	t	u	p		
*s	1	1	0	−1	2	0	0	600	$3R_1 - 2R_3$
z	1	1	2	0	−1	0	0	200	$3R_2 + R_3$
u	−1	1	0	0	3	2	0	400	
p	7	4	0	0	1	0	1	−200	$3R_4 - R_3$

	x	y	z	s	t	u	p		
*s	5	1	0	−3	0	−4	0	1000	
z	2	4	6	0	0	2	0	1000	$5R_2 - 2R_1$
t	−1	1	0	0	3	2	0	400	$5R_3 + R_1$
p	22	11	0	0	0	−2	3	−1000	$5R_4 - 22R_1$

	x	y	z	s	t	u	p	
x	5	1	0	−3	0	−4	0	1000
z	0	18	30	6	0	18	0	3000
t	0	6	0	−3	15	6	0	3000
p	0	33	0	66	0	78	15	−27000

Optimal solution: $p = −27,000/15 = −1,800$; $x = 1,000/5 = 200, y = 0, z = 3,000/30 = 100$.

Since $c = −p$, the minimum value of c is 1,800.

Use 200 axes, no maces, and 100 spears for a minimum cost of 1,800 gold pieces.

37. Unknowns: $x =$ # servings of cereal, $y =$ # servings of dessert, $z =$ # servings of juice

Minimize $c = 10x + 53y + 27z$ subject to $60x + 80y + 60z \geq 120$ or $3x + 4y + 3z \geq 6$, $45y + 120z \geq 120$ or

$3y + 8z \geq 8$, $x \geq 0$, $y \geq 0$, $z \geq 0$.

You can use the online Pivot and Gauss-Jordan Tool in decimal mode do the pivoting.

	x	y	z	s	t	p	
*s	3	4	3	−1	0	0	6
*t	0	3	8	0	−1	0	8
p	10	53	27	0	0	1	0

	x	y	z	s	t	p	
y	0.75	1	0.75	−0.25	0	0	1.5
*t	−2.25	0	5.75	0.75	−1	0	3.5
p	−29.75	0	−12.75	13.25	0	1	−79.5

	x	y	z	s	t	p	
y	1.04	1	0	−0.35	0.13	0	1.04
z	−0.39	0	1	0.13	−0.17	0	0.61
p	−34.74	0	0	14.91	−2.22	1	−71.74

	x	y	z	s	t	p	
x	1	0.96	0	−0.33	0.13	0	1
z	0	0.38	1	0	−0.12	0	1
p	0	33.29	0	3.33	2.13	1	−37

Optimal solution: $p = -37$; $x = 1, y = 0, z = 1$.
Since $c = -p$, the minimum value of c is 37.
Provide one serving of cereal, one serving of juice, and no dessert!

39. Unknowns: $x =$ # bundles from Nadir, $y =$ # bundles from Blunt, $z =$ # bundles from Sonny

Minimize $c = 3x + 4y + 5z$ (in thousands of dollars) subject to

$5x + 10y + 15z \geq 150$, $10x + 10y + 10z \geq 200$, $15x + 10y + 10z \geq 150$, $x \geq 0$, $y \geq 0$, $z \geq 0$.

	x	y	z	s	t	u	p		
*s	5	10	15	−1	0	0	0	150	
*t	10	10	10	0	−1	0	0	200	$3R_2 - 2R_1$
*u	15	10	10	0	0	−1	0	150	$3R_3 - 2R_1$
p	3	4	5	0	0	0	1	0	$3R_4 - R_1$

	x	y	z	s	t	u	p		
z	5	10	15	−1	0	0	0	150	$7R_1 - R_3$
*t	20	10	0	2	−3	0	0	300	$7R_2 - 4R_3$
*u	35	10	0	2	0	−3	0	150	
p	4	2	0	1	0	0	3	−150	$35R_4 - 4R_3$

	x	y	z	s	t	u	p		
z	0	60	105	−9	0	3	0	900	
*t	0	30	0	6	−21	12	0	1500	$2R_2 - R_1$
x	35	10	0	2	0	−3	0	150	$6R_3 - R_1$
p	0	30	0	27	0	12	105	−5850	$2R_4 - R_1$

	x	y	z	s	t	u	p		
y	0	60	105	−9	0	3	0	900	$7R_1 + 3R_3$
*t	0	0	−105	21	−42	21	0	2100	$R_2 - R_3$
x	210	0	−105	21	0	−21	0	0	
p	0	0	−105	63	0	21	210	−12600	$R_4 - 3R_3$

	x	y	z	s	t	u	p		
y	630	420	420	0	0	−42	0	6300	$R_1 + R_2$
*t	−210	0	0	0	−42	42	0	2100	
s	210	0	−105	21	0	−21	0	0	$2R_3 + R_2$
p	−630	0	210	0	0	84	210	−12600	$R_4 - 2R_2$

	x	y	z	s	t	u	p		
y	420	420	420	0	−42	0	0	8400	$R_1 - 2R_3$
u	−210	0	0	0	−42	42	0	2100	$R_2 + R_3$
s	210	0	−210	42	−42	0	0	2100	
p	−210	0	210	0	84	0	210	−16800	$R_4 + R_3$

	x	y	z	s	t	u	p	
y	0	420	840	−84	42	0	0	4200
u	0	0	−210	42	−84	42	0	4200
x	210	0	−210	42	−42	0	0	2100
p	0	0	0	42	42	0	210	−14700

Optimal solution: $p = -14,700/210 = -70$; $x = 2,100/210 = 10$, $y = 4,200/420 = 10$, $z = 0$.
Since $c = -p$, the minimum value of c is 70.
Buy 10 bundles from Nadir, 10 from Blunt, and none from Sonny. Cost: $70,000. Another solution resulting in the same cost (pivot on the z-column in the last tableau) is 15 bundles from Nadir, none from Blunt, and 5 from Sonny.

41. Unknowns: $x =$ # servings of Xtend, $y =$ # servings of Gainz, $z =$ # servings of Strongevity

Minimize $c = x + 1.1y + 1.3z$ subject to $2y + 2.5z \geq 70$, $2.5x + 3y + z \geq 50$, $7x + 6y \geq 60$, $x \geq 0$, $y \geq 0$, $z \geq 0$.
You can use the online Pivot and Gauss-Jordan Tool in decimal mode do the pivoting.

	x	y	z	s	t	u	p		
*s	0	2	2.5	-1	0	0	0	70	$2R_1$
*t	7	6	0	0	-1	0	0	60	
*u	2.5	3	1	0	0	-1	0	50	$2R_3$
p	1	1.1	1.3	0	0	0	1	0	$10R_4$

	x	y	z	s	t	u	p		
*s	0	4	5	-2	0	0	0	140	
*t	7	6	0	0	-1	0	0	60	
*u	5	6	2	0	0	-2	0	100	$5R_3 - 2R_1$
p	10	11	13	0	0	0	10	0	$5R_4 - 13R_1$

	x	y	z	s	t	u	p		
z	0	4	5	-2	0	0	0	140	
*t	7	6	0	0	-1	0	0	60	
*u	25	22	0	4	0	-10	0	220	$7R_3 - 25R_2$
p	50	3	0	26	0	0	50	-1820	$7R_4 - 50R_2$

	x	y	z	s	t	u	p		
z	0	4	5	-2	0	0	0	140	$14R_1 + R_3$
x	7	6	0	0	-1	0	0	60	
*u	0	4	0	28	25	-70	0	40	
p	0	-279	0	182	50	0	350	-15740	$2R_4 - 13R_3$

	x	y	z	s	t	u	p		
z	0	60	70	0	25	-70	0	2000	$R_1 - 10R_2$
x	7	6	0	0	-1	0	0	60	
s	0	4	0	28	25	-70	0	40	$3R_3 - 2R_2$
p	0	-610	0	0	-225	910	700	-32000	$3R_4 + 305R_2$

	x	y	z	s	t	u	p		
z	-70	0	70	0	35	-70	0	1400	$11R_1 - 5R_3$
y	7	6	0	0	-1	0	0	60	$77R_2 + R_3$
s	-14	0	0	84	77	-210	0	0	
p	2135	0	0	0	-980	2730	2100	-77700	$11R_4 + 140R_3$

	x	y	z	s	t	u	p	
z	-700	0	770	-420	0	280	0	15400
y	525	462	0	84	0	-210	0	4620
t	-14	0	0	84	77	-210	0	0
p	21525	0	0	11760	0	630	23100	-854700

Optimal solution: $p = -854,700/23,100 = -37$; $x = 0$, $y = 4,620/462 = 10$, $z = 15,400/770 = 20$.
Since $c = -p$, the minimum value of c is 37.
Use no Xtend, 10 servings of Gainz, and 20 servings of Strongevity for a total cost of $37.

43. Unknowns: $x = $ # convention-style hotels, $y = $ # vacation-style hotels, and $z = $ # small motels

a. Minimize $c = 100x + 20y + 4z$ subject to $500x + 200y + 50z \geq 1,400$, $x \geq 1$, $z \leq 2$, $x \geq 0$, $y \geq 0$, $z \geq 0$.

	x	y	z	s	t	u	p		
$*s$	500	200	50	-1	0	0	0	1400	$R_1 - 500R_2$
$*t$	$\boxed{1}$	0	0	0	-1	0	0	1	
u	0	0	1	0	0	1	0	2	
p	100	20	4	0	0	0	1	0	$R_4 - 100R_2$

	x	y	z	s	t	u	p		
$*s$	0	200	50	-1	$\boxed{500}$	0	0	900	
x	1	0	0	0	-1	0	0	1	$500R_2 + R_1$
u	0	0	1	0	0	1	0	2	
p	0	20	4	0	100	0	1	-100	$5R_4 - R_1$

	x	y	z	s	t	u	p		
t	0	$\boxed{200}$	50	-1	500	0	0	900	
x	500	200	50	-1	0	0	0	1400	$R_2 - R_1$
u	0	0	1	0	0	1	0	2	
p	0	-100	-30	1	0	0	5	-1400	$2R_4 + R_1$

	x	y	z	s	t	u	p		
y	0	200	50	-1	500	0	0	900	$R_1 - 50R_3$
x	500	0	0	0	-500	0	0	500	
u	0	0	$\boxed{1}$	0	0	1	0	2	
p	0	0	-10	1	500	0	10	-1900	$R_4 + 10R_3$

	x	y	z	s	t	u	p	
y	0	200	0	-1	500	-50	0	800
x	500	0	0	0	-500	0	0	500
z	0	0	1	0	0	1	0	2
p	0	0	0	1	500	10	10	-1880

Optimal solution: $p = -1,880/10 = -188$; $x = 500/500 = 1, y = 800/200 = 4, z = 2/1 = 2$.
Since $c = -p$, the minimum value of c is 188.
Build 1 convention-style hotel, 4 vacation-style hotels, and 2 small motels. The total cost will amount to $188 million.
b. Since 20% of the $188 million cost is $37.6 million, you will still be covered by the $50 million subsidy.

45. Unknowns:

$x = $ # boards sent from Tucson to Honolulu

$y = $ # boards sent from Tucson to Venice Beach

$z = $ # boards sent from Toronto to Honolulu

$w = $ # boards sent from Toronto to Venice Beach

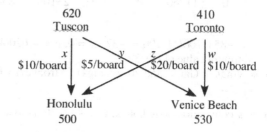

Minimize $c = 10x + 5y + 20z + 10w$ subject to

$x + y \leq 620$, $z + w \leq 410$, $x + z \geq 500$, $y + w \geq 530$, $x \geq 0$, $y \geq 0$, $z \geq 0$, $w \geq 0$.

	x	y	z	w	s	t	u	v	p		
s	1	1	0	0	1	0	0	0	0	620	$R_1 - R_3$
t	0	0	1	1	0	1	0	0	0	410	
*u	1	0	1	0	0	0	-1	0	0	500	
*v	0	1	0	1	0	0	0	-1	0	530	
p	10	5	20	10	0	0	0	0	1	0	$R_5 - 10R_3$

	x	y	z	w	s	t	u	v	p		
s	0	1	-1	0	1	0	1	0	0	120	
t	0	0	1	1	0	1	0	0	0	410	
x	1	0	1	0	0	0	-1	0	0	500	
*v	0	1	0	1	0	0	0	-1	0	530	$R_4 - R_1$
p	0	5	10	10	0	0	10	0	1	-5000	$R_5 - 5R_1$

	x	y	z	w	s	t	u	v	p		
y	0	1	-1	0	1	0	1	0	0	120	$R_1 + R_4$
t	0	0	1	1	0	1	0	0	0	410	$R_2 - R_4$
x	1	0	1	0	0	0	-1	0	0	500	$R_3 - R_4$
*v	0	0	1	1	-1	0	-1	-1	0	410	
p	0	0	15	10	-5	0	5	0	1	-5600	$R_5 - 15R_4$

	x	y	z	w	s	t	u	v	p		
y	0	1	0	1	0	0	0	-1	0	530	$R_1 - R_4$
t	0	0	0	0	1	1	1	1	0	0	
x	1	0	0	-1	1	0	0	1	0	90	$R_3 + R_4$
z	0	0	1	1	-1	0	-1	-1	0	410	
p	0	0	0	-5	10	0	20	15	1	-11750	$R_5 + 5R_4$

	x	y	z	w	s	t	u	v	p	
y	0	1	−1	0	1	0	1	0	0	120
t	0	0	0	0	1	1	1	1	0	0
x	1	0	1	0	0	0	−1	0	0	500
w	0	0	1	1	−1	0	−1	−1	0	410
p	0	0	5	0	5	0	15	10	1	−9700

Optimal solution: $p = -9,700/1 = -9,700$; $x = 500/1 = 500$, $y = 120/1 = 120$, $z = 0$, $w = 410/1 = 410$.
Since $c = -p$, the minimum value of c is 9,700.
Make the following shipments: Tucson to Honolulu: 500 boards per week; Tucson to Venice Beach: 120 boards per week; Toronto to Honolulu: 0 boards per week; Toronto to Venice Beach: 410 boards per week. Minimum weekly cost is $9,700.

47. Unknowns: $x =$ amount overdrawn from Congressional Integrity Bank, $y =$ amount from Citizens' Trust, $z =$ amount from Checks R Us

Maximize $p = x$ subject to

$x + y + z \le 10,000$, $x + y \le \frac{1}{4}(x + y + z)$ or $3x + 3y - z \le 0$, $x \ge 2,500$, $x \ge 0$, $y \ge 0$, $z \ge 0$.

	x	y	z	s	t	u	p		
s	1	1	1	1	0	0	0	10000	$3R_1 - R_2$
t	3	3	−1	0	1	0	0	0	
*u	1	0	0	0	0	−1	0	2500	$3R_3 - R_2$
p	−1	0	0	0	0	0	1	0	$3R_4 + R_2$

	x	y	z	s	t	u	p		
s	0	0	4	3	−1	0	0	30000	$R_1 - 4R_3$
x	3	3	−1	0	1	0	0	0	$R_2 + R_3$
*u	0	−3	1	0	−1	−3	0	7500	
p	0	3	−1	0	1	0	3	0	$R_4 + R_3$

	x	y	z	s	t	u	p		
s	0	12	0	3	3	12	0	0	
x	3	0	0	0	0	−3	0	7500	$4R_2 + R_1$
z	0	−3	1	0	−1	−3	0	7500	$4R_3 + R_1$
p	0	0	0	0	0	−3	3	7500	$4R_4 + R_1$

	x	y	z	s	t	u	p	
u	0	12	0	3	3	12	0	0
x	12	12	0	3	3	0	0	30000
z	0	0	4	3	−1	0	0	30000
p	0	12	0	3	3	0	12	30000

Optimal solution:
$p = 30,000/12 = 2,500$; $x = 30,000/12 = 2,500$, $y = 0$, $z = 30,000/4 = 7,500$.

Withdraw \$2,500 from Congressional Integrity Bank, \$0 from Citizens' Trust, \$7,500 from Checks R Us.

49. Unknowns: $x =$ # people you fly from Chicago to Los Angeles, $y =$ # people you fly from Chicago to New York,

$z =$ # people you fly from Denver to Los Angeles, $w =$ # people you fly from Denver to New York

Minimize $c = 200x + 125y + 225z + 280w$ subject to

$x + y \leq 20$, $z + w \leq 10$, $x + z \geq 10$, $y + w \geq 15$, $x \geq 0$, $y \geq 0$, $z \geq 0$, $w \geq 0$.

	x	y	z	w	s	t	u	v	p		
s	1	1	0	0	1	0	0	0	0	20	$R_1 - R_3$
t	0	0	1	1	0	1	0	0	0	10	
*u	1	0	1	0	0	0	-1	0	0	10	
*v	0	1	0	1	0	0	0	-1	0	15	
p	200	125	225	280	0	0	0	0	1	0	$R_5 - 200R_3$

	x	y	z	w	s	t	u	v	p		
s	0	1	-1	0	1	0	1	0	0	10	
t	0	0	1	1	0	1	0	0	0	10	
x	1	0	1	0	0	0	-1	0	0	10	
*v	0	1	0	1	0	0	0	-1	0	15	$R_4 - R_1$
p	0	125	25	280	0	0	200	0	1	-2000	$R_5 - 125R_1$

	x	y	z	w	s	t	u	v	p		
y	0	1	-1	0	1	0	1	0	0	10	$R_1 + R_4$
t	0	0	1	1	0	1	0	0	0	10	$R_2 - R_4$
x	1	0	1	0	0	0	-1	0	0	10	$R_3 - R_4$
*v	0	0	1	1	-1	0	-1	-1	0	5	
p	0	0	150	280	-125	0	75	0	1	-3250	$R_5 - 150R_4$

	x	y	z	w	s	t	u	v	p	
y	0	1	0	1	0	0	0	-1	0	15
t	0	0	0	0	1	1	1	1	0	5
x	1	0	0	-1	1	0	0	1	0	5
z	0	0	1	1	-1	0	-1	-1	0	5
p	0	0	0	130	25	0	225	150	1	-4000

Optimal solution: $p = -4,000/1 = -4,000$; $x = 5/1 = 5$, $y = 15/1 = 15$, $z = 5/1 = 5$, $w = 0$.
Since $c = -p$, the minimum value of c is 4,000.
Fly 5 people from Chicago to LA, 15 from Chicago to New York, 5 from Denver to LA, and none from Denver to New York at a total cost of \$4,000.

51. Unknowns: $x =$ # cardiologists hired, $y =$ # rehabilitation specialists hired, $z =$ # infectious disease specialists hired

Maximize $p = 12x + 19y + 14z$ (thousands of dollars) subject to

$x + y + 3z \geq 27$, $10x + 10y + 10z \leq 200 - 30 = 170$, $x \geq 0$, $y \geq 0$, $z \geq 0$.

	x	y	z	s	t	p		
*s	1	1	[3]	−1	0	0	27	
t	1	1	1	0	1	0	17	$3R_2 - R_1$
p	−12	−19	−14	0	0	1	0	$3R_3 + 14R_1$

	x	y	z	s	t	p		
z	1	1	3	−1	0	0	27	$2R_1 - R_2$
t	2	[2]	0	1	3	0	24	
p	−22	−43	0	−14	0	3	378	$2R_3 + 43R_2$

	x	y	z	s	t	p	
z	0	0	6	−3	−3	0	30
y	2	2	0	1	3	0	24
p	42	0	0	15	129	6	1788

Optimal solution:
$p = 1{,}788/6 = 298$; $x = 0, y = 24/2 = 12, z = 30/6 = 5$.
Hire no more cardiologists, 12 rehabilitation specialists, and 5 infectious disease specialists.

Communication and reasoning exercises

53. In a general linear programming problem, the solution $x = 0$, $y = 0$, ... represented by the initial tableau may not be feasible (feasible solutions must satisfy all the constraints), and this shows up as a basic solution with some negative values for surplus variables. In order for a basic solution to be feasible, none of the basic variables (including surplus and slack variables) can be negative. In phase I we use pivoting to arrive at a basic solution where no basic variables are negative.

55. The basic solution corresponding to the initial tableau has all the unknowns equal to zero, and this is not a feasible solution because it does not satisfy the given inequality.

57. We can rewrite the given problem as a standard LP problem by multiplying both sides of the first constraint by −1:

Maximize $p = x + y$ subject to $-x + 2y \le 0$, $2x + y \le 10$, $x \ge 0$, $y \ge 0$. Therefore, this problem can be solved using the techniques of either section (choice (C)).

59. Answers may vary. Examples are Exercises 1 and 2.

61. A simple example is: Maximize $p = x + y$ subject to $x + y \le 10$, $x + y \ge 20$, $x \ge 0$, $y \ge 0$.

	x	y	s	t	p		
s	[1]	1	1	0	0	10	
*t	1	1	0	−1	0	20	$R_2 - R_1$
p	−1	−1	0	0	1	0	$R_3 + R_1$

	x	y	s	t	p	
x	1	1	1	0	0	10
*t	0	0	-1	-1	0	10
p	0	0	1	0	1	10

We now find it impossible to find a feasible solution, since there are no positive entries in the starred row. This problem has an empty feasible region, and hence no optimal solution.

Section 5.5

1. Maximize $p = 2x + y$

subject to

$x + 2y \le 6$

$-x + y \le 2$

$x \ge 0, \; y \ge 0.$

$$\begin{bmatrix} 1 & 2 & 6 \\ -1 & 1 & 2 \\ 2 & 1 & 0 \end{bmatrix} \to \begin{bmatrix} 1 & -1 & 2 \\ 2 & 1 & 1 \\ 6 & 2 & 0 \end{bmatrix}$$

Minimize $c = 6s + 2t$

subject to

$s - t \ge 2$

$2s + t \ge 1$

$s \ge 0, \; t \ge 0.$

3. Minimize $c = 2s + t + 3u$

subject to

$s + t + u \ge 100$

$2s + t \ge 50$

$s \ge 0, \; t \ge 0, \; u \ge 0.$

$$\begin{bmatrix} 1 & 1 & 1 & 100 \\ 2 & 1 & 0 & 50 \\ 2 & 1 & 3 & 0 \end{bmatrix} \to \begin{bmatrix} 1 & 2 & 2 \\ 1 & 1 & 1 \\ 1 & 0 & 3 \\ 100 & 50 & 0 \end{bmatrix}$$

Maximize $p = 100x + 50y$

subject to

$x + 2y \le 2$

$x + y \le 1$

$x \le 3x \ge 0, \; y \ge 0.$

5. Maximize

$p = x + y + z + w$

subject to

$x + y + z \le 3$

$y + z + w \le 4$

$x + z + w \le 5$

$x + y + w \le 6$

$x \ge 0, \; y \ge 0$

$z \ge 0, \; w \ge 0.$

$$\begin{bmatrix} 1 & 1 & 1 & 0 & 3 \\ 0 & 1 & 1 & 1 & 4 \\ 1 & 0 & 1 & 1 & 5 \\ 1 & 1 & 0 & 1 & 6 \\ 1 & 1 & 1 & 1 & 0 \end{bmatrix} \to \begin{bmatrix} 1 & 0 & 1 & 1 & 1 \\ 1 & 1 & 0 & 1 & 1 \\ 1 & 1 & 1 & 0 & 1 \\ 0 & 1 & 1 & 1 & 1 \\ 3 & 4 & 5 & 6 & 0 \end{bmatrix}$$

Minimize

$c = 3s + 4t + 5u + 6v$

subject to

$s + u + v \ge 1$

$s + t + v \ge 1$

$s + t + u \ge 1$

$t + u + v \ge 1$

$s \ge 0, \; t \ge 0$

$u \ge 0, \; v \ge 0.$

7. Minimize $c = s + 3t + u$

subject to

$5s - t + v \ge 1,000$

$u - v \ge 2,000$

$s + t \ge 500$

$s \ge 0, \; t \ge 0$

$u \ge 0, \; v \ge 0.$

$$\begin{bmatrix} 5 & -1 & 0 & 1 & 1,000 \\ 0 & 0 & 1 & -1 & 2,000 \\ 1 & 1 & 0 & 0 & 500 \\ 1 & 3 & 1 & 0 & 0 \end{bmatrix} \to \begin{bmatrix} 5 & 0 & 1 & 1 \\ -1 & 0 & 1 & 3 \\ 0 & 1 & 0 & 1 \\ 1 & -1 & 0 & 0 \\ 1,000 & 2,000 & 500 & 0 \end{bmatrix}$$

Maximize

$p = 1,000x + 2,000y + 500z$

subject to

$5x + z \le 1$

$-x + z \le 3$

$y \le 1$

$x - y \le 0$

$x \ge 0, \; y \ge 0, \; z \ge 0.$

9. Minimize $c = s + t$

subject to

$s + 2t \ge 6$

$2s + t \ge 6$

$s \ge 0, \; t \ge 0.$

$$\begin{bmatrix} 1 & 2 & 6 \\ 2 & 1 & 6 \\ 1 & 1 & 0 \end{bmatrix} \to \begin{bmatrix} 1 & 2 & 1 \\ 2 & 1 & 1 \\ 6 & 6 & 0 \end{bmatrix}$$

Maximize $p = 6x + 6y$

subject to

$x + 2y \le 1$

$2x + y \le 1$

$x \ge 0, \; y \ge 0.$

Solve the dual problem using the standard simplex method:

	x	y	s	t	p		
s	1	2	1	0	0	1	$2R_1 - R_2$
t	2	1	0	1	0	1	
p	−6	−6	0	0	1	0	$R_3 + 3R_2$

	x	y	s	t	p		
s	0	3	2	−1	0	1	
x	2	1	0	1	0	1	$3R_2 - R_1$
p	0	−3	0	3	1	3	$R_3 + R_1$

	x	y	s	t	p	
y	0	3	2	−1	0	1
x	6	0	−2	4	0	2
p	0	0	2	2	1	4

The solution to the primal problem is $c = 4/1 = 4$; $s = 2/1 = 2$, $t = 2/1 = 2$.

11. Minimize $c = 6s + 6t$ Maximize $p = 20x + 20y$

subject to   subject to

$$\begin{aligned} s + 2t &\geq 20 \\ 2s + t &\geq 20 \\ s &\geq 0,\ t \geq 0. \end{aligned} \quad \begin{bmatrix} 1 & 2 & 20 \\ 2 & 1 & 20 \\ 6 & 6 & 0 \end{bmatrix} \rightarrow \begin{bmatrix} 1 & 2 & 6 \\ 2 & 1 & 6 \\ 20 & 20 & 0 \end{bmatrix} \quad \begin{aligned} x + 2y &\leq 6 \\ 2x + y &\leq 6 \\ x &\geq 0,\ y \geq 0. \end{aligned}$$

Solve the dual problem using the standard simplex method:

	x	y	s	t	p		
s	1	2	1	0	0	6	$2R_1 - R_2$
t	2	1	0	1	0	6	
p	−20	−20	0	0	1	0	$R_3 + 10R_2$

	x	y	s	t	p		
s	0	3	2	−1	0	6	
x	2	1	0	1	0	6	$3R_2 - R_1$
p	0	−10	0	10	1	60	$3R_3 + 10R_1$

	x	y	s	t	p	
y	0	3	2	−1	0	6
x	6	0	−2	4	0	12
p	0	0	20	20	3	240

The solution to the primal problem is $c = 240/3 = 80$; $s = 20/3$, $t = 20/3$.

13. Minimize $c = 0.2s + 0.3t$

subject to

$2s + t \geq 10$

$s + 2t \geq 10$

$s + t \geq 8$

$s \geq 0, \, t \geq 0.$

$$\begin{bmatrix} 2 & 1 & 10 \\ 1 & 2 & 10 \\ 1 & 1 & 8 \\ 0.2 & 0.3 & 0 \end{bmatrix} \rightarrow \begin{bmatrix} 2 & 1 & 1 & 0.2 \\ 1 & 2 & 1 & 0.3 \\ 10 & 10 & 8 & 0 \end{bmatrix}$$

Maximize $p = 10x + 10y + 8z$

subject to

$2x + y + z \leq 0.2$

$x + 2y + z \leq 0.3$

$x \geq 0, \, y \geq 0, \, z \geq 0.$

Solve the dual problem using the standard simplex method. (Note: Do not rewrite the constraints without decimals at this stage—clear them in the first step of the simplex method.)

	x	y	z	s	t	p		
s	2	1	1	1	0	0	0.2	5R_1
t	1	2	1	0	1	0	0.3	10R_2
p	−10	−10	−8	0	0	1	0	

	x	y	z	s	t	p		
s	10	5	5	5	0	0	1	
t	10	20	10	0	10	0	3	$R_2 - R_1$
p	−10	−10	−8	0	0	1	0	$R_3 + R_1$

	x	y	z	s	t	p		
x	10	5	5	5	0	0	1	$3R_1 - R_2$
t	0	15	5	−5	10	0	2	
p	0	−5	−3	5	0	1	1	$3R_3 + R_2$

	x	y	z	s	t	p		
x	30	0	10	20	−10	0	1	
y	0	15	5	−5	10	0	2	$2R_2 - R_1$
p	0	0	−4	10	10	3	5	$5R_3 + 2R_1$

	x	y	z	s	t	p	
z	30	0	10	20	−10	0	1
y	−30	30	0	−30	30	0	3
p	60	0	0	90	30	15	27

The solution to the primal problem is $c = 27/15 = 1.8$; $s = 90/15 = 6$, $t = 30/15 = 2$.

15. Minimize $c = 2s + t$ subject to

$$3s + t \geq 30$$
$$s + t \geq 20$$
$$s + 3t \geq 30$$
$$s \geq 0, \ t \geq 0.$$

$$\begin{bmatrix} 3 & 1 & 30 \\ 1 & 1 & 20 \\ 1 & 3 & 30 \\ 2 & 1 & 0 \end{bmatrix} \rightarrow \begin{bmatrix} 3 & 1 & 1 & 2 \\ 1 & 1 & 3 & 1 \\ 30 & 20 & 30 & 0 \end{bmatrix}$$

Maximize $p = 30x + 20y + 30z$
subject to

$$3x + y + z \leq 2$$
$$x + y + 3z \leq 1$$
$$x \geq 0, \ y \geq 0, \ z \geq 0.$$

Solve the dual problem using the standard simplex method:

	x	y	z	s	t	p		
s	3	1	1	1	0	0	2	
t	1	1	3	0	1	0	1	$3R_2 - R_1$
p	−30	−20	−30	0	0	1	0	$R_3 + 10R_1$

	x	y	z	s	t	p		
x	3	1	1	1	0	0	2	$8R_1 - R_2$
t	0	2	8	−1	3	0	1	
p	0	−10	−20	10	0	1	20	$2R_3 + 5R_2$

	x	y	z	s	t	p		
x	24	6	0	9	−3	0	15	$R_1 - 3R_2$
z	0	2	8	−1	3	0	1	
p	0	−10	0	15	15	2	45	$R_3 + 5R_2$

	x	y	z	s	t	p	
x	24	0	−24	12	−12	0	12
y	0	2	8	−1	3	0	1
p	0	0	40	10	30	2	50

The solution to the primal problem is $c = 50/2 = 25$; $s = 10/2 = 5$, $t = 30/2 = 15$.

17. Minimize $c = s + 2t + 3u$
subject to

$$3s + 2t + u \geq 60$$
$$2s + t + 3u \geq 60$$
$$s \geq 0, \ t \geq 0, \ u \geq 0.$$

$$\begin{bmatrix} 3 & 2 & 1 & 60 \\ 2 & 1 & 3 & 60 \\ 1 & 2 & 3 & 0 \end{bmatrix} \rightarrow \begin{bmatrix} 3 & 2 & 1 \\ 2 & 1 & 2 \\ 1 & 3 & 3 \\ 60 & 60 & 0 \end{bmatrix}$$

Maximize $p = 60x + 60y$
subject to

$$3x + 2y \leq 1$$
$$2x + y \leq 2$$
$$x + 3y \leq 3$$
$$x \geq 0, \ y \geq 0.$$

Solve the dual problem using the standard simplex method:

	x	y	s	t	u	p		
s	3	2	1	0	0	0	1	
t	2	1	0	1	0	0	2	$3R_2 - 2R_1$
u	1	3	0	0	1	0	3	$3R_3 - R_1$
p	−60	−60	0	0	0	1	0	$R_4 + 20R_1$

	x	y	s	t	u	p		
x	3	2	1	0	0	0	1	
t	0	−1	−2	3	0	0	4	$2R_2 + R_1$
u	0	7	−1	0	3	0	8	$2R_3 - 7R_1$
p	0	−20	20	0	0	1	20	$R_4 + 10R_1$

	x	y	s	t	u	p		
y	3	2	1	0	0	0	1	
t	3	0	−3	6	0	0	9	
u	−21	0	−9	0	6	0	9	
p	30	0	30	0	0	1	30	

The solution to the primal problem is $c = 30/1 = 30$; $s = 30/1 = 30$, $t = 0$, $u = 0$.

19. Minimize $c = 2s + t + 3u$ Maximize $p = 100x + 50y + 50z$
  subject to   subject to

$$s + t + u \geq 100$$
$$2s + t \geq 50$$
$$t + u \geq 50$$
$$s \geq 0, \, t \geq 0, \, u \geq 0.$$

$$\begin{bmatrix} 1 & 1 & 1 & 100 \\ 2 & 1 & 0 & 50 \\ 0 & 1 & 1 & 50 \\ 2 & 1 & 3 & 0 \end{bmatrix} \rightarrow \begin{bmatrix} 1 & 2 & 0 & 2 \\ 1 & 1 & 1 & 1 \\ 1 & 0 & 1 & 3 \\ 100 & 50 & 50 & 0 \end{bmatrix}$$

$$x + 2y \leq 2$$
$$x + y + z \leq 1$$
$$x + z \leq 3$$
$$x \geq 0, \, y \geq 0, \, z \geq 0.$$

Solve the dual problem using the standard simplex method:

	x	y	z	s	t	u	p		
s	1	2	0	1	0	0	0	2	$R_1 - R_2$
t	1	1	1	0	1	0	0	1	
u	1	0	1	0	0	1	0	3	$R_3 - R_2$
p	−100	−50	−50	0	0	0	1	0	$R_4 + 100R_2$

	x	y	z	s	t	u	p		
s	0	1	−1	1	−1	0	0	1	
x	1	1	1	0	1	0	0	1	
u	0	−1	0	0	−1	1	0	2	
p	0	50	50	0	100	0	1	100	

The solution to the primal problem is $c = 100/1 = 100$; $s = 0$, $t = 100/1 = 100$, $u = 0$.

21. Minimize $c = s + t + u$ Maximize $p = 60x + 60y + 60z$
   subject to   subject to

$$3s + 2t + u \geq 60$$

$$2s + t + 3u \geq 60$$

$$\begin{bmatrix} 3 & 2 & 1 & 60 \\ 2 & 1 & 3 & 60 \\ 1 & 3 & 2 & 60 \\ 1 & 1 & 1 & 0 \end{bmatrix} \rightarrow \begin{bmatrix} 3 & 2 & 1 & 1 \\ 2 & 1 & 3 & 1 \\ 1 & 3 & 2 & 1 \\ 60 & 60 & 60 & 0 \end{bmatrix}$$

$$s + 3t + 2u \geq 60$$

$$s \geq 0, \ t \geq 0, \ u \geq 0.$$

$$3x + 2y + z \leq 1$$

$$2x + y + 3z \leq 1$$

$$x + 3y + 2z \leq 1$$

$$x \geq 0, \ y \geq 0, \ z \geq 0.$$

Solve the dual problem using the standard simplex method:

	x	y	z	s	t	u	p		
s	3	2	1	1	0	0	0	1	
t	2	1	3	0	1	0	0	1	$3R_2 - 2R_1$
u	1	3	2	0	0	1	0	1	$3R_3 - R_1$
p	-60	-60	-60	0	0	0	1	0	$R_4 + 20R_1$

	x	y	z	s	t	u	p		
x	3	2	1	1	0	0	0	1	$7R_1 - R_2$
t	0	-1	7	-2	3	0	0	1	
u	0	7	5	-1	0	3	0	2	$7R_3 - 5R_2$
p	0	-20	-40	20	0	0	1	20	$7R_4 + 40R_2$

	x	y	z	s	t	u	p		
x	21	15	0	9	-3	0	0	6	$18R_1 - 5R_3$
z	0	-1	7	-2	3	0	0	1	$54R_2 + R_3$
u	0	54	0	3	-15	21	0	9	
p	0	-180	0	60	120	0	7	180	$3R_4 + 10R_3$

	x	y	z	s	t	u	p	
x	378	0	0	147	21	-105	0	63
z	0	0	378	-105	147	21	0	63
y	0	54	0	3	-15	21	0	9
p	0	0	0	210	210	210	21	630

The solution to the primal problem is $c = 630/21 = 30$; $s = 210/21 = 10$, $t = 210/21 = 10$, $u = 210/21 = 10$.

23. Add $k = 2$ to each entry and put 1s to the right and below:

$$\begin{bmatrix} 1 & 3 & 4 & 1 \\ 4 & 1 & 0 & 1 \\ 1 & 1 & 1 & 0 \end{bmatrix}$$

This gives the following LP problem:

Maximize $p = x + y + z$
subject to

$$x + 3y + 4z \leq 1$$

$$4x + y \leq 1$$

$$x \geq 0, \ y \geq 0, \ z \geq 0.$$

Here are the tableaux:

	x	y	z	s	t	p		
s	1	3	4	1	0	0	1	$4R_1 - R_2$
t	4	1	0	0	1	0	1	
p	−1	−1	−1	0	0	1	0	$4R_3 + R_2$

	x	y	z	s	t	p		
s	0	11	16	4	−1	0	3	
x	4	1	0	0	1	0	1	
p	0	−3	−4	0	1	4	1	$4R_3 + R_1$

	x	y	z	s	t	p		
z	0	11	16	4	−1	0	3	
x	4	1	0	0	1	0	1	$11R_2 - R_1$
p	0	−1	0	4	3	16	7	$11R_3 + R_1$

	x	y	z	s	t	p		
y	0	11	16	4	−1	0	3	
x	44	0	−16	−4	12	0	8	
p	0	0	16	48	32	176	80	

The solution to the primal problem is $p = 80/176 = 5/11$; $x = 8/44 = 2/11$, $y = 3/11$, $z = 0$. So, the column player's

optimal strategy is $C = [2/5 \ 3/5 \ 0]^T$ and the value of the game is $e = 11/5 - 2 = 1/5$.

The solution to the dual problem is $s = 48/176 = 3/11$, $t = 32/176 = 2/11$, so the row player's optimal strategy is

$R = [3/5 \ 2/5]$.

25. Add $k = 2$ to each entry and put 1s to the right and below:

$$\begin{bmatrix} 1 & 3 & 4 & 1 \\ 4 & 1 & 0 & 1 \\ 3 & 4 & 2 & 1 \\ 1 & 1 & 1 & 0 \end{bmatrix}$$

This gives the following LP problem:

Maximize $p = x + y + z$

subject to

$$x + 3y + 4z \leq 1$$

$$4x + y \leq 1$$

$$3x + 4y + 2z \leq 1$$

$$x \geq 0, \ y \geq 0, \ z \geq 0.$$

Here are the tableaux:

	x	y	z	s	t	u	p		
s	1	3	4	1	0	0	0	1	$4R_1 - R_2$
t	4	1	0	0	1	0	0	1	
u	3	4	2	0	0	1	0	1	$4R_3 - 3R_2$
p	-1	-1	-1	0	0	0	1	0	$4R_4 + R_2$

	x	y	z	s	t	u	p		
s	0	11	16	4	-1	0	0	3	$R_1 - 2R_3$
x	4	1	0	0	1	0	0	1	
u	0	13	8	0	-3	4	0	1	
p	0	-3	-4	0	1	0	4	1	$2R_4 + R_3$

	x	y	z	s	t	u	p		
s	0	-15	0	4	5	-8	0	1	
x	4	1	0	0	1	0	0	1	$5R_2 - R_1$
z	0	13	8	0	-3	4	0	1	$5R_3 + 3R_1$
p	0	7	0	0	-1	4	8	3	$5R_4 + R_1$

	x	y	z	s	t	u	p	
t	0	-15	0	4	5	-8	0	1
x	20	20	0	-4	0	8	0	4
z	0	20	40	12	0	-4	0	8
p	0	20	0	4	0	12	40	16

The solution to the primal problem is $p = 16/40 = 2/5$; $x = 4/20 = 1/5$, $y = 0$, $z = 8/40 = 1/5$. So, the column

player's optimal strategy is $C = [1/2 \ \ 0 \ \ 1/2]^T$ and the value of the game is $e = 5/2 - 2 = 1/2$.

The solution to the dual problem is $s = 4/40 = 1/10$, $t = 0$, $u = 12/40 = 3/10$, so the row player's optimal strategy is

$R = [1/4 \ \ 0 \ \ 3/4]$.

27. The first row is dominated by the last, so we can remove it. That is as far as we can reduce by dominance. Add

$k = 3$ to each entry and put 1s to the right and below:

$$\begin{bmatrix} 5 & 2 & 1 & 0 & 1 \\ 4 & 5 & 3 & 4 & 1 \\ 3 & 5 & 6 & 6 & 1 \\ 1 & 1 & 1 & 1 & 0 \end{bmatrix}.$$

This gives the following LP problem:

Maximize $p = x + y + z + w$

subject to

$$5x + 2y + z \leq 1$$

$$4x + 5y + 3z + 4w \leq 1$$

$$3x + 5y + 6z + 6w \leq 1$$

$$x \geq 0,\ y \geq 0,\ z \geq 0,\ w \geq 0.$$

Here are the tableaux:

	x	y	z	w	s	t	u	p		
s	5	2	1	0	1	0	0	0	1	
t	4	5	3	4	0	1	0	0	1	$5R_2 - 4R_1$
u	3	5	6	6	0	0	1	0	1	$5R_3 - 3R_1$
p	−1	−1	−1	−1	0	0	0	1	0	$5R_4 + R_1$

	x	y	z	w	s	t	u	p		
x	5	2	1	0	1	0	0	0	1	
t	0	17	11	20	−4	5	0	0	1	
u	0	19	27	30	−3	0	5	0	2	$2R_3 - 3R_2$
p	0	−3	−4	−5	1	0	0	5	1	$4R_4 + R_2$

	x	y	z	w	s	t	u	p		
x	5	2	1	0	1	0	0	0	1	$21R_1 - R_3$
w	0	17	11	20	−4	5	0	0	1	$21R_2 - 11R_3$
u	0	−13	21	0	6	−15	10	0	1	
p	0	5	−5	0	0	5	0	20	5	$21R_4 + 5R_3$

	x	y	z	w	s	t	u	p	
x	105	55	0	0	15	15	−10	0	20
w	0	500	0	420	−150	270	−110	0	10
z	0	−13	21	0	6	−15	10	0	1
p	0	40	0	0	30	30	50	420	110

The solution to the primal problem is $p = 110/420 = 11/42$; $x = 20/105 = 4/21$, $y = 0$, $z = 1/21$,

$w = 10/420 = 1/42$. So, the column player's optimal strategy is $C = [8/11 \ 0 \ 2/11 \ 1/11]^T$ and the value of the game

is $e = 42/11 - 3 = 9/11$.

The solution to the dual problem is $s = 30/420 = 1/14$, $t = 30/420 = 1/14$, $u = 50/420 = 5/42$, so the row player's

optimal strategy is $R = [0 \ 3/11 \ 3/11 \ 5/11]$. (Note that we put in 0 for the first row that we removed at the

beginning.)

Applications

29. Unknowns: $s = $ # ounces of fish, $t = $ # ounces of cornmeal

Minimize $c = 5s + 5t$ subject to $8t + 4t \geq 48$, $4s + 8t \geq 48$, $s \geq 0$, $t \geq 0$.

$$\text{Dualize: } \begin{bmatrix} 8 & 4 & 48 \\ 4 & 8 & 48 \\ 5 & 5 & 0 \end{bmatrix} \rightarrow \begin{bmatrix} 8 & 4 & 5 \\ 4 & 8 & 5 \\ 48 & 48 & 0 \end{bmatrix}.$$

Dual problem: Maximize $p = 48x + 48y$ subject to $8x + 4y \le 5$, $4x + 8y \le 5$, $x \ge 0$, $y \ge 0$.

Solve the dual problem:

	x	y	s	t	p	
s	8	4	1	0	0	5
t	4	8	0	1	0	5	$2R_2 - R_1$
p	-48	-48	0	0	1	0	$R_3 + 6R_1$

	x	y	s	t	p	
x	8	4	1	0	0	5	$3R_1 - R_2$
t	0	12	-1	2	0	5	
p	0	-24	6	0	1	30	$R_3 + 2R_2$

	x	y	s	t	p	
x	24	0	4	-2	0	10
y	0	12	-1	2	0	5
p	0	0	4	4	1	40

The solution to the primal problem is $c = 40/1 = 40$, $s = 4/1 = 4$, $t = 4/1 = 4$. Meow should use 4 ounces each of fish and cornmeal, for a total cost of $40¢$ per can. The shadow costs are the values of x and y in the final tableau: $x = 10/24 = 5/12¢$ per gram of protein, $y = 5/12¢$ per gram of fat.

31. Unknowns: $s = $ # ounces of chicken, $t = $ # ounces of grain

Minimize $c = 10s + t$ subject to $10s + 2t \ge 200$, $5s + 2t \ge 150$, $s \ge 0$, $t \ge 0$.

$$\text{Dualize: } \begin{bmatrix} 10 & 2 & 200 \\ 5 & 2 & 150 \\ 10 & 1 & 0 \end{bmatrix} \rightarrow \begin{bmatrix} 10 & 5 & 10 \\ 2 & 2 & 1 \\ 200 & 150 & 0 \end{bmatrix}$$

Dual problem: Maximize $p = 200x + 150y$ subject to $10x + 5y \le 10$, $2x + 2y \le 1$, $x \ge 0$, $y \ge 0$.

Solve the dual problem:

	x	y	s	t	p	
s	10	5	1	0	0	10	$R_1 - 5R_2$
t	2	2	0	1	0	1	
p	-200	-150	0	0	1	0	$R_3 + 100R_2$

	x	y	s	t	p	
s	0	-5	1	-5	0	5
x	2	2	0	1	0	1
p	0	50	0	100	1	100

The solution to the primal problem is $c = 100/1 = 100$; $s = 0$, $t = 100/1 = 100$. Use no chicken and 100 oz of grain for a total cost of $1. The shadow costs are the values of x and y in the final tableau: $x = 1/2¢$ per gram of protein, $y = 0¢$ per gram of fat.

33. Unknowns: $s = $ # servings of cereal, $t = $ # servings of dessert, $u = $ # servings of juice

Minimize $c = 10s + 53t + 27u$ subject to $60s + 80t + 60u \geq 120$, $45t + 120u \geq 120$, $s \geq 0$, $t \geq 0$, $u \geq 0$.

$$\text{Dualize: } \begin{bmatrix} 60 & 80 & 60 & 120 \\ 0 & 45 & 120 & 120 \\ 10 & 53 & 27 & 0 \end{bmatrix} \rightarrow \begin{bmatrix} 60 & 0 & 10 \\ 80 & 45 & 53 \\ 60 & 120 & 27 \\ 120 & 120 & 0 \end{bmatrix}.$$

Dual problem: Maximize $p = 120x + 120y$ subject to $60x \leq 10$, $80x + 45y \leq 53$, $60x + 120y \leq 27$, $x \geq 0$, $y \geq 0$.
Solve the dual problem:

	x	y	s	t	u	p		
s	60	0	1	0	0	0	10	
t	80	45	0	1	0	0	53	$3R_2 - 4R_1$
u	60	120	0	0	1	0	27	$R_3 - R_1$
p	-120	-120	0	0	0	1	0	$R_4 + 2R_1$

	x	y	s	t	u	p		
x	60	0	1	0	0	0	10	
t	0	135	-4	3	0	0	119	$8R_2 - 9R_3$
u	0	120	-1	0	1	0	17	
p	0	-120	2	0	0	1	20	$R_4 + R_3$

	x	y	s	t	u	p	
x	60	0	1	0	0	0	10
t	0	0	-23	24	-9	0	799
y	0	120	-1	0	1	0	17
p	0	0	1	0	1	1	37

The solution to the primal problem is $c = 37/1 = 37$; $s = 1/1 = 1$, $t = 0$, $u = 1$. Prepare one serving of cereal, one serving of juice, and no dessert! for a total cost of $37¢$. The shadow costs are the values of x and y in the final tableau: $x = 10/60 = 1/6¢$ per calorie and $y = 17/120¢$ per % U.S. RDA of vitamin C.

35. Unknowns: $s = $ # mailings to the East Coast, $t = $ # mailings to the Midwest, $u = $ # mailings to the West Coast

Minimize $c = 40s + 60t + 50u$ subject to

$100s + 100t + 50u \geq 1,500$, $50s + 100t + 100u \geq 1,500$, $s \geq 0$, $t \geq 0$, $u \geq 0$.

Dualize: $\begin{bmatrix} 100 & 100 & 50 & 1{,}500 \\ 50 & 100 & 100 & 1{,}500 \\ 40 & 60 & 50 & 0 \end{bmatrix} \rightarrow \begin{bmatrix} 100 & 50 & 40 \\ 100 & 100 & 60 \\ 50 & 100 & 50 \\ 1{,}500 & 1{,}500 & 0 \end{bmatrix}.$

Dual problem: Maximize $p = 1{,}500x + 1{,}500y$ subject to

$100x + 50y \leq 40$; $100x + 100y \leq 60$, $50x + 100y \leq 50, x \geq 0, \; y \geq 0$.

Solve the dual problem:

	x	y	s	t	u	p		
s	100	50	1	0	0	0	40	
t	100	100	0	1	0	0	60	$R_2 - R_1$
u	50	100	0	0	1	0	50	$2R_3 - R_1$
p	−1500	−1500	0	0	0	1	0	$R_4 + 15R_1$

	x	y	s	t	u	p		
x	100	50	1	0	0	0	40	$R_1 - R_2$
t	0	50	−1	1	0	0	20	
u	0	150	−1	0	2	0	60	$R_3 - 3R_2$
p	0	−750	15	0	0	1	600	$R_4 + 15R_2$

	x	y	s	t	u	p	
x	100	0	2	−1	0	0	20
y	0	50	−1	1	0	0	20
u	0	0	2	−3	2	0	0
p	0	0	0	15	0	1	900

The solution to the primal problem is $c = 900/1 = 900$; $s = 0$, $t = 15/1 = 15$, $u = 0$. Send 15 mailings to the Midwest and none to the East or West Coasts. Cost: \$900. The shadow costs are the values of x and y in the final tableau:

$x = 20/100 = 1/5 = 20¢$ per Democrat and $y = 20/50 = 2/5 = 40¢$ per Republican.

Another solution comes from pivoting on the 150 in the y-column in the second tableau instead of the 50: $s = 10$, $t = 0$, $u = 10$. Send 10 mailings to the East Coast, none to the Midwest, 10 to the West Coast.

37. Unknowns: $x = $ # full-page ads in *Sports Illustrated*, $y = $ # full-page ads in *GQ*.
Expressing costs and readers in thousands, the primal problem is:
Minimize $c = 2s + t$ subject to

$600s + 150t \geq 9{,}000$, $s \geq 6$, $t \geq 8$, $s \geq 0, t \geq 0$.

Dualize: $\begin{bmatrix} 600 & 150 & 9{,}000 \\ 1 & 0 & 6 \\ 0 & 1 & 8 \\ 2 & 1 & 0 \end{bmatrix} \rightarrow \begin{bmatrix} 600 & 1 & 0 & 2 \\ 150 & 0 & 1 & 1 \\ 9{,}000 & 6 & 8 & 0 \end{bmatrix}.$

Dual problem: Maximize $p = 9{,}000x + 6y + 8z$ subject to

$600x + y \le 2$; $150x + z \le 1$, $x \ge 0$, $y \ge 0$, $z \ge 0$.

Solve the dual problem:

	x	y	z	s	t	p		
s	600	1	0	1	0	0	2	
t	150	0	1	0	1	0	1	$4R_2 - R_1$
p	-9000	-6	-8	0	0	1	0	$R_3 + 15R_1$

	x	y	z	s	t	p		
x	600	1	0	1	0	0	2	
t	0	-1	4	-1	4	0	2	
p	0	9	-8	15	0	1	30	$R_3 + 2R_2$

	x	y	z	s	t	p	
x	600	1	0	1	0	0	2
z	0	-1	4	-1	4	0	2
p	0	7	0	13	8	1	34

The solution to the primal problem is $c = 34/1 = 34$; $s = 13/1 = 13$, $t = 8/1 = 8$. Place 13 ads in Sports Illustrated and 8 in GQ. Cost: \$34,000

39. Unknowns: $s = $ # sleep spells, $t = $ # shock spells

Minimize $c = 50s + 20t$ subject to

$3s + t \ge 48$, $3s + 2t \ge 52$, $-s + t \ge 0$, $s \ge 0$, $t \ge 0$.

Dualize: $\begin{bmatrix} 3 & 1 & 48 \\ 3 & 2 & 52 \\ -1 & 1 & 0 \\ 50 & 20 & 0 \end{bmatrix} \rightarrow \begin{bmatrix} 3 & 3 & -1 & 50 \\ 1 & 2 & 1 & 20 \\ 48 & 52 & 0 & 0 \end{bmatrix}$.

Dual problem: Maximize $p = 48x + 52y$ subject to

$3x + 3y - z \le 50$, $x + 2y + z \le 20$, $x \ge 0$, $y \ge 0$, $z \ge 0$.

Solve the dual problem:

	x	y	z	s	t	p		
s	3	3	-1	1	0	0	50	$2R_1 - 3R_2$
t	1	2	1	0	1	0	20	
p	-48	-52	0	0	0	1	0	$R_3 + 26R_2$

	x	y	z	s	t	p		
s	3	0	-5	2	-3	0	40	
y	1	2	1	0	1	0	20	$3R_2 - R_1$
p	-22	0	26	0	26	1	520	$3R_3 + 22R_1$

	x	y	z	s	t	p		
x	3	0	-5	2	-3	0	40	$8R_1 + 5R_2$
y	0	6	8	-2	6	0	20	
p	0	0	-32	44	12	3	2440	$R_3 + 4R_2$

	x	y	z	s	t	p	
x	24	30	0	6	6	0	420
z	0	6	8	-2	6	0	20
p	0	24	0	36	36	3	2520

The solution to the primal problem is $c = 2{,}520/3 = 840$; $s = 36/3 = 12$, $t = 36/3 = 12$. Gillian should use 12 sleep spells and 12 shock spells, costing 840 therms of energy.

41. We observe that the first column dominates the last, so we can eliminate the last. In the matrix that remains, the second row dominates the third, so we eliminate the third. We cannot reduce any further. Add $k = 500$ to each entry and put 1s to the right and below:

$$\begin{bmatrix} 300 & 200 & 1 \\ 0 & 1{,}000 & 1 \\ 1 & 1 & 0 \end{bmatrix}$$

This gives the following LP problem:

Maximize $p = x + y$ subject to $300x + 200y \le 1$, $1{,}000y \le 1$, $x \ge 0$, $y \ge 0$.

Here are the tableaux:

	x	y	s	t	p		
s	300	200	1	0	0	1	
t	0	1000	0	1	0	1	
p	-1	-1	0	0	1	0	$300R_3 + R_1$

	x	y	s	t	p		
x	300	200	1	0	0	1	$5R_1 - R_2$
t	0	1000	0	1	0	1	
p	0	-100	1	0	300	1	$10R_3 + R_2$

	x	y	s	t	p	
x	1500	0	5	-1	0	4
y	0	1000	0	1	0	1
p	0	0	10	1	3000	11

The solution to the primal problem is $p = 11/3{,}000$; $x = 4/1{,}500 = 1/375$, $y = 1/1{,}000$. So, T. N. Spend's optimal strategy is $C = [8/11 \ \ 3/11 \ \ 0]^T$ and the value of the game is $e = 3{,}000/11 - 500 = -2{,}500/11 \approx -227$.

The solution to the dual problem is $s = 10/3{,}000$, $t = 1/3{,}000$, so T. L. Down's optimal strategy is

$R = [10/11 \ \ 1/11 \ \ 0]$.

280

T. N. Spend should spend about 73% of the days in Littleville, 27% in Metropolis, and skip Urbantown. T. L. Down should spend about 91% of the days in Littleville, 9% in Metropolis, and skip Urbantown. The expected outcome is that T. L. Down will lose about 227 votes per day of campaigning.

43. With A the row player and B the column player, the payoff matrix is

$$P = \begin{bmatrix} 2 & -1 & 0 \\ -1 & 4 & -1 \\ 0 & -1 & 6 \end{bmatrix}.$$

This game does not reduce. Add $k = 1$ to each entry and put 1s to the right and below:

$$\begin{bmatrix} 3 & 0 & 1 & 1 \\ 0 & 5 & 0 & 1 \\ 1 & 0 & 7 & 1 \\ 1 & 1 & 1 & 0 \end{bmatrix}.$$

This gives the following LP problem:

Maximize $p = x + y + z$ subject to $3x + z \le 1$, $5y \le 1$, $x + 7z \le 1$, $x \ge 0$, $y \ge 0$, $z \ge 0$.

Here are the tableaus:

	x	y	z	s	t	u	p		
s	3	0	1	1	0	0	0	1	
t	0	5	0	0	1	0	0	1	
u	1	0	7	0	0	1	0	1	$3R_3 - R_1$
p	-1	-1	-1	0	0	0	1	0	$3R_4 + R_1$

	x	y	z	s	t	u	p		
x	3	0	1	1	0	0	0	1	
t	0	5	0	0	1	0	0	1	
u	0	0	20	-1	0	3	0	2	
p	0	-3	-2	1	0	0	3	1	$5R_4 + 3R_2$

	x	y	z	s	t	u	p		
x	3	0	1	1	0	0	0	1	$20R_1 - R_3$
y	0	5	0	0	1	0	0	1	
u	0	0	20	-1	0	3	0	2	
p	0	0	-10	5	3	0	15	8	$2R_4 + R_3$

	x	y	z	s	t	u	p	
x	60	0	0	21	0	-3	0	18
y	0	5	0	0	1	0	0	1
z	0	0	20	-1	0	3	0	2
p	0	0	0	9	6	3	30	18

The solution to the primal problem is $p = 18/30 = 3/5$; $x = 18/60 = 3/10$, $y = 1/5$, $z = 2/20 = 1/10$. So, player B's

optimal strategy is $C = [1/2 \ \ 1/3 \ \ 1/6]^T$ and the value of the game is $e = 5/3 - 1 = 2/3$.

The solution to the dual problem is $s = 9/30 = 3/10$, $t = 6/30 = 1/5$, $u = 3/30 = 1/10$, so player A's optimal strategy is $R = [1/2 \; 1/3 \; 1/6]$.

Each player should show one finger with probability 1/2, two fingers with probability 1/3, and three fingers with probability 1/6. The expected outcome is that player A will win 2/3 point per round, on average.

45. Let Colonel Blotto be the row player and Captain Kije the column player. Label possible moves by the number of regiments sent to the first of the two locations; the remaining regiments are sent to the other location. The payoff matrix is then the following:

		0	1	2	3
				Captain Kije	
	0	4	2	1	0
	1	1	3	0	−1
Colonel Blotto	2	−2	2	2	−2
	3	−1	0	3	1
	4	0	1	2	4

We cannot reduce this game. Add $k = 2$ to each entry and put 1s to the right and below:

$$\begin{bmatrix} 6 & 4 & 3 & 2 & 1 \\ 3 & 5 & 2 & 1 & 1 \\ 0 & 4 & 4 & 0 & 1 \\ 1 & 2 & 5 & 3 & 1 \\ 2 & 3 & 4 & 6 & 1 \\ 1 & 1 & 1 & 1 & 0 \end{bmatrix}.$$

This gives the following LP problem:

Maximize $p = x + y + z + w$ subject to

$$6x + 4y + 3z + 2w \le 1$$

$$3x + 5y + 2z + w \le 1$$

$$4y + 4z \le 1$$

$$x + 2y + 5z + 3w \le 1$$

$$2x + 3y + 4z + 6w \le 1$$

$$x \ge 0, \; y \ge 0, \; z \ge 0, \; w \ge 0.$$

Here are the tableaux:

	x	y	z	w	s	t	u	v	r	p		
s	6	4	3	2	1	0	0	0	0	0	1	
t	3	5	2	1	0	1	0	0	0	0	1	$2R_2 - R_1$
u	0	4	4	0	0	0	1	0	0	0	1	
v	1	2	5	3	0	0	0	1	0	0	1	$6R_4 - R_1$
r	2	3	4	6	0	0	0	0	1	0	1	$3R_5 - R_1$
p	−1	−1	−1	−1	0	0	0	0	0	1	0	$6R_6 + R_1$

	x	y	z	w	s	t	u	v	r	p		
x	6	4	3	2	1	0	0	0	0	0	1	$8R_1 - R_5$
t	0	6	1	0	-1	2	0	0	0	0	1	
u	0	4	4	0	0	0	1	0	0	0	1	
v	0	8	27	16	-1	0	0	6	0	0	5	$R_4 - R_5$
r	0	5	9	[16]	-1	0	0	0	3	0	2	
p	0	-2	-3	-4	1	0	0	0	0	6	1	$4R_6 + R_5$

	x	y	z	w	s	t	u	v	r	p		
x	48	27	15	0	9	0	0	0	-3	0	6	$2R_1 - 9R_2$
t	0	[6]	1	0	-1	2	0	0	0	0	1	
u	0	4	4	0	0	0	1	0	0	0	1	$3R_3 - 2R_2$
v	0	3	18	0	0	0	0	6	-3	0	3	$2R_4 - R_2$
w	0	5	9	16	-1	0	0	0	3	0	2	$6R_5 - 5R_2$
p	0	-3	-3	0	3	0	0	0	3	24	6	$2R_6 + R_2$

	x	y	z	w	s	t	u	v	r	p		
x	96	0	21	0	27	-18	0	0	-6	0	3	$10R_1 - 21R_3$
y	0	6	1	0	-1	2	0	0	0	0	1	$10R_2 - R_3$
u	0	0	[10]	0	2	-4	3	0	0	0	1	
v	0	0	35	0	1	-2	0	12	-6	0	5	$2R_4 - 7R_3$
w	0	0	49	96	-1	-10	0	0	18	0	7	$10R_5 - 49R_3$
p	0	0	-5	0	5	2	0	0	6	48	13	$2R_6 + R_3$

	x	y	z	w	s	t	u	v	r	p		
x	960	0	0	0	228	-96	-63	0	-60	0	9	
y	0	60	0	0	-12	24	-3	0	0	0	9	
z	0	0	10	0	2	-4	3	0	0	0	1	
v	0	0	0	0	-12	24	-21	24	-12	0	3	
w	0	0	0	960	-108	96	-147	0	180	0	21	
p	0	0	0	0	12	0	3	0	12	96	27	

The solution to the primal problem is

$p = 27/96 = 9/32$; $x = 9/960 = 3/320$, $y = 9/60 = 3/20$, $z = 1/10$, $w = 21/960 = 7/320$.

So, Captain Kije's optimal strategy is $C = [1/30 \ 8/15 \ 16/45 \ 7/90]^T$ and the value of the game is

$e = 32/9 - 2 = 14/9$. In fact, this is only one of several optimal strategies for Captain Kije; which you find depends on the choices of pivots you make in the simplex method. Other optimal strategies are the one given above, in reverse order, and $[1/18 \ 4/9 \ 4/9 \ 1/18]^T$.

The solution to the dual problem (which is unique) is

$s = 12/96 = 1/8$, $t = 0$, $u = 3/96 = 1/32$, $v = 0$, $r = 12/96 = 1/8$,

so Colonel Blotto's optimal strategy is $R = [4/9 \ 0 \ 1/9 \ 0 \ 4/9]$.

Write moves as (x, y), where x represents the number of regiments sent to the first location and y represents the number sent to the second location. Colonel Blotto should play $(0, 4)$ with probability 4/9, $(2, 2)$ with probability 1/9, and $(4, 0)$ with probability 4/9. Captain Kije has several optimal strategies, one of which is to play $(0, 3)$ with probability 1/30, $(1, 2)$ with probability 8/15, $(2, 1)$ with probability 16/45, and $(3, 0)$ with probability 7/90. The expected outcome is that Colonel Blotto will win 14/9 points on average.

Communication and reasoning exercises

47. Two variables and three constraints: In the matrix formulation of an LP problem, the number of rows is one more than the number of constraints, and the number of columns is one more than the number of variables. Thus, the matrix form of the primal problem has three rows and four columns, and so its transpose has 4 rows and 3 columns, translating to three constraints and two variables.

49. The dual of a standard minimization problem satisfying the nonnegative objective condition is a standard maximization problem, which can be solved by using the standard simplex algorithm, thus avoiding the need to do Phase I.

51. Answers will vary. We use a minimization problem that does not satisfy the nonnegative objective condition. An example is: Minimize $c = x - y$ subject to $x - y \geq 100$, $x + y \geq 200$, $x \geq 0$, $y \geq 0$. This problem can be solved using the techniques in the preceding section.

53. The dual problem is a nonstandard maximization problem, because the right-hand sides of its constraints are the entries in the bottom row of the matrix representation of the primal problem, and at least one of those entries is negative.

55. If the given problem is a standard minimization problem satisfying the nonnegative objective condition, its dual is a standard maximization problem and so can be solved using a single-phase simplex method. Otherwise, dualizing may not save any labor, since the dual will not be a standard maximization problem.

Chapter 5 Review

1. $2x - 3y \leq 12$

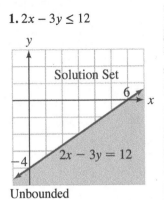

Unbounded

3. $x + 2y \leq 20$

$3x + 2y \leq 30$

$x \geq 0, \ y \geq 0$

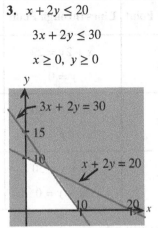

Bounded; Corner points: $(0,0), \ (0,10), \ (5, 15/2), \ (10,0)$

5. Maximize $p = 2x + y$ subject to $3x + y \leq 30, \ x + y \leq 12, \ x + 3y \leq 30, \ x \geq 0, \ y \geq 0$.

Corner Point	Lines through Point	Coordinates	$p = 2x + y$
A	$x = 0, \ y = 0$	$(0,0)$	0
B	$3x + y = 30$ $y = 0$	$(10,0)$	20
C	$3x + y = 30$ $x + y = 12$	$(9,3)$	$\boxed{21}$
D	$x + y = 12$ $x + 3y = 30$	$(3,9)$	15
E	$x + 3y = 30$ $x = 0$	$(0,10)$	10

Maximum value occurs at C: $p = 21$; $x = 9, \ y = 3$.

7. Minimize $c = 2x + y$ subject to $3x + y \geq 30$, $x + 2y \geq 20$, $2x - y \geq 0$, $x \geq 0$, $y \geq 0$.

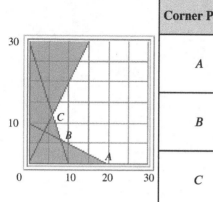

Corner Point	Lines through Point	Coordinates	$c = 2x + y$
A	$x + 2y = 20$ $y = 0$	$(20, 0)$	40
B	$3x + y = 30$ $x + 2y = 20$	$(8, 6)$	$\boxed{22}$
C	$3x + y = 30$ $2x - y = 0$	$(6, 12)$	24

Minimum value occurs at B: $c = 22$; $x = 8$, $y = 6$.

9. Introduce slack variables and rewrite the constraints and objective function in standard form, and set up the first tableau:

$x + 2y + 2z + s = 60$
$2x + y + 3z + t = 60$
$-x - y - 2z + p = 0$

	x	y	z	s	t	p	
s	1	2	2	1	0	0	60
t	2	1	3	0	1	0	60
p	-1	-1	-2	0	0	1	0

The most negative entry in the bottom row is the -2 in the z-column, so we use this column as the pivot column. The test ratios are: s: 60/2, t: 60/3. The smallest test ratio is t: 60/3. Thus we pivot on the 3 in the t-row.

	x	y	z	s	t	p		
s	1	2	2	1	0	0	60	$3R_1 - 2R_2$
t	2	1	$\boxed{3}$	0	1	0	60	
p	-1	-1	-2	0	0	1	0	$3R_3 + 2R_2$

	x	y	z	s	t	p	
s	-1	4	0	3	-2	0	60
z	2	1	3	0	1	0	60
p	1	-1	0	0	2	3	120

The most negative entry in the bottom row is the -1 in the y-column, so we use this column as the pivot column. The test ratios are: s: 60/4, z: 60/1. The smallest test ratio is s: 60/4. Thus we pivot on the 4 in the s-row.

	x	y	z	s	t	p		
s	-1	$\boxed{4}$	0	3	-2	0	60	
z	2	1	3	0	1	0	60	$4R_2 - R_1$
p	1	-1	0	0	2	3	120	$4R_3 + R_1$

	x	y	z	s	t	p	
y	−1	4	0	3	−2	0	60
z	9	0	12	−3	6	0	180
p	3	0	0	3	6	12	540

As there are no more negative numbers in the bottom row, we are done, and read off the solution:
Optimal solution:
$p = 540/12 = 45$; $x = 0$, $y = 60/4 = 15$, $z = 180/12 = 15$.

11. Introduce slack variables and rewrite the constraints and objective function in standard form, and set up the first tableau:
$x + y + z − s = 100$
$y + z + t = 80$
$x + z + u = 80$
$−x − y − 3z + p = 0$

	x	y	z	s	t	u	p	
*s	1	1	1	−1	0	0	0	100
t	0	1	1	0	1	0	0	80
u	1	0	1	0	0	1	0	80
p	−1	−1	−3	0	0	0	1	0

The first starred row is the s-row, and its largest positive entry is the 1 in the x-column. Thus, we use this column as the pivot column. The test ratios are: s: $100/1$, u: $80/1$. The smallest test ratio is u: $80/1$. Thus we pivot on the 1 in the u-row.

	x	y	z	s	t	u	p		
*s	1	1	1	−1	0	0	0	100	$R_1 − R_3$
t	0	1	1	0	1	0	0	80	
u	1	0	1	0	0	1	0	80	
p	−1	−1	−3	0	0	0	1	0	$R_4 + R_3$

	x	y	z	s	t	u	p	
*s	0	1	0	−1	0	−1	0	20
t	0	1	1	0	1	0	0	80
x	1	0	1	0	0	1	0	80
p	0	−1	−2	0	0	1	1	80

The first starred row is the s-row, and its largest positive entry is the 1 in the y-column. Thus, we use this column as the pivot column. The test ratios are: s: $20/1$, t: $80/1$. The smallest test ratio is s: $20/1$. Thus we pivot on the 1 in the s-row.

	x	y	z	s	t	u	p	
*s	0	1	0	-1	0	-1	0	20
t	0	1	1	0	1	0	0	80
x	1	0	1	0	0	1	0	80
p	0	-1	-2	0	0	1	1	80

$R_2 - R_1$

$R_4 + R_1$

	x	y	z	s	t	u	p	
y	0	1	0	-1	0	-1	0	20
t	0	0	1	1	1	1	0	60
x	1	0	1	0	0	1	0	80
p	0	0	-2	-1	0	0	1	100

As there are no more starred rows, we go to Phase 2, and do the standard simplex method.

	x	y	z	s	t	u	p	
y	0	1	0	-1	0	-1	0	20
t	0	0	1	1	1	1	0	60
x	1	0	1	0	0	1	0	80
p	0	0	-2	-1	0	0	1	100

$R_3 - R_2$

$R_4 + 2R_2$

	x	y	z	s	t	u	p	
y	0	1	0	-1	0	-1	0	20
z	0	0	1	1	1	1	0	60
x	1	0	0	-1	-1	0	0	20
p	0	0	0	1	2	2	1	220

Optimal solution:
$p = 220/1 = 220$; $x = 20/1 = 20$, $y = 20/1 = 20$, $z = 60/1 = 60$.

13. Introduce slack variables and rewrite the constraints and objective function in standard form, and set up the first tableau:
$3x + 2y + z - s = 60$
$2x + y + 3z - t = 60$
$x + 2y + 3z + p = 0$

	x	y	z	s	t	p	
*s	3	2	1	-1	0	0	60
*t	2	1	3	0	-1	0	60
p	1	2	3	0	0	1	0

The first starred row is the s-row, and its largest positive entry is the 3 in the x-column. Thus, we use this column as the pivot column. The test ratios are: s: 60/3, t: 60/2. The smallest test ratio is s: 60/3. Thus we pivot on the 3 in the s-row.

	x	y	z	s	t	p		
*s	3	2	1	-1	0	0	60	
*t	2	1	3	0	-1	0	60	$3R_2 - 2R_1$
p	1	2	3	0	0	1	0	$3R_3 - R_1$

	x	y	z	s	t	p	
x	3	2	1	-1	0	0	60
*t	0	-1	7	2	-3	0	60
p	0	4	8	1	0	3	-60

The first starred row is the t-row, and its largest positive entry is the 7 in the z-column. Thus, we use this column as the pivot column. The test ratios are: x: 60/1, t: 60/7. The smallest test ratio is t: 60/7. Thus we pivot on the 7 in the t-row.

	x	y	z	s	t	p		
x	3	2	1	-1	0	0	60	$7R_1 - R_2$
*t	0	-1	7	2	-3	0	60	
p	0	4	8	1	0	3	-60	$7R_3 - 8R_2$

	x	y	z	s	t	p	
x	21	15	0	-9	3	0	360
z	0	-1	7	2	-3	0	60
p	0	36	0	-9	24	21	-900

As there are no more starred rows, we go to Phase 2, and do the standard simplex method.

	x	y	z	s	t	p		
x	21	15	0	-9	3	0	360	$2R_1 + 9R_2$
z	0	-1	7	2	-3	0	60	
p	0	36	0	-9	24	21	-900	$2R_3 + 9R_2$

	x	y	z	s	t	p	
x	42	21	63	0	-21	0	1260
s	0	-1	7	2	-3	0	60
p	0	63	63	0	21	42	-1260

Optimal solution: $p = -1,260/42 = -30$; $x = 1,260/42 = 30$, $y = 0$, $z = 0$.
Since $c = -p$, the minimum value of c is 30.

15. Since technology is indicated, we use the online Simplex Method Tool. When entering the problems in this tool there is no need to enter the inequalities $x \geq 0$, $y \geq 0$, etc.

```
Tableau #1
x      y      z      s1     s2     s3     -c
3      2      -1     -1     0      0      0      10
2      1      3      0      -1     0      0      20
1      3      -2     0      0      -1     0      30
1      -2     4      0      0      0      1      0
```

...

```
Tableau #6
x      y      z      s1     s2     s3     -c
2      1      3      0      -1     0      0      20
5      0      11     0      -3     1      0      30
1      0      7      1      -2     0      0      30
5      0      10     0      -2     0      1      40
```

Looking at the sixth tableau, we find that there is no possible pivot above the most negative entry in the bottom row, indicating that the feasible region is unbounded (and no optimal solution exists).

17. Introduce slack variables and rewrite the constraints and objective function in standard form, and set up the first tableau:
$$x + y - s = 30$$
$$x + z - t = 20$$
$$x + y - w + u = 10$$
$$y + z - w + v = 10$$
$$x + y + z + w + p = 0$$

	x	y	z	w	s	t	u	v	p	
*s	1	1	0	0	-1	0	0	0	0	30
*t	1	0	1	0	0	-1	0	0	0	20
u	1	1	0	-1	0	0	1	0	0	10
v	0	1	1	-1	0	0	0	1	0	10
p	1	1	1	1	0	0	0	0	1	0

The first starred row is the s-row, and its largest positive entry is the 1 in the x-column. Thus, we use this column as the pivot column. The test ratios are: s: 30/1, t: 20/1, u: 10/1. The smallest test ratio is u: 10/1. Thus we pivot on the 1 in the u-row.

	x	y	z	w	s	t	u	v	p		
*s	1	1	0	0	-1	0	0	0	0	30	$R_1 - R_3$
*t	1	0	1	0	0	-1	0	0	0	20	$R_2 - R_3$
u	1	1	0	-1	0	0	1	0	0	10	
v	0	1	1	-1	0	0	0	1	0	10	
p	1	1	1	1	0	0	0	0	1	0	$R_5 - R_3$

	x	y	z	w	s	t	u	v	p	
*s	0	0	0	1	-1	0	-1	0	0	20
*t	0	-1	1	1	0	-1	-1	0	0	10
x	1	1	0	-1	0	0	1	0	0	10
v	0	1	1	-1	0	0	0	1	0	10
p	0	0	1	2	0	0	-1	0	1	-10

The first starred row is the s-row, and its largest positive entry is the 1 in the w-column. Thus, we use this column as the pivot column. The test ratios are: s: 20/1, t: 10/1. The smallest test ratio is t: 10/1. Thus we pivot on the 1 in the t-row.

	x	y	z	w	s	t	u	v	p		
*s	0	0	0	1	-1	0	-1	0	0	20	$R_1 - R_2$
*t	0	-1	1	[1]	0	-1	-1	0	0	10	
x	1	1	0	-1	0	0	1	0	0	10	$R_3 + R_2$
v	0	1	1	-1	0	0	0	1	0	10	$R_4 + R_2$
p	0	0	1	2	0	0	-1	0	1	-10	$R_5 - 2R_2$

	x	y	z	w	s	t	u	v	p	
*s	0	1	-1	0	-1	1	0	0	0	10
w	0	-1	1	1	0	-1	-1	0	0	10
x	1	0	1	0	0	-1	0	0	0	20
v	0	0	2	0	0	-1	-1	1	0	20
p	0	2	-1	0	0	2	1	0	1	-30

The first starred row is the s-row, and its largest positive entry is the 1 in the y-column. Thus, we use this column as the pivot column. The only positive entry in this column is the 1 in the s-row, so we pivot on that entry.

	x	y	z	w	s	t	u	v	p		
*s	0	[1]	-1	0	-1	1	0	0	0	10	
w	0	-1	1	1	0	-1	-1	0	0	10	$R_2 + R_1$
x	1	0	1	0	0	-1	0	0	0	20	
v	0	0	2	0	0	-1	-1	1	0	20	
p	0	2	-1	0	0	2	1	0	1	-30	$R_5 - 2R_1$

	x	y	z	w	s	t	u	v	p	
y	0	1	-1	0	-1	1	0	0	0	10
w	0	0	0	1	-1	0	-1	0	0	20
x	1	0	1	0	0	-1	0	0	0	20
v	0	0	2	0	0	-1	-1	1	0	20
p	0	0	1	0	2	0	1	0	1	-50

As there are no more starred rows, we go to Phase 2, and do the standard simplex method.
However, there are no negative numbers in the bottom row, so there are no pivot steps to do in Phase 2. Therefore, we are done. Optimal solution: $p = -50/1 = -50$; $x = 20/1 = 20$, $y = 10/1 = 10$, $z = 0$, $w = 20/1 = 20$.

Since $c = -p$, the minimum value of c is 50. Another solution is: $x = 30$, $y = 0$, $z = 0$, $w = 20$.

19. Minimize $c = 2s + t$ subject to $3s + 2t \geq 60$, $2s + t \geq 60$, $s + 3t \geq 60$, $s \geq 0$, $t \geq 0$.

Dualize: $\begin{bmatrix} 3 & 2 & 60 \\ 2 & 1 & 60 \\ 1 & 3 & 60 \\ 2 & 1 & 0 \end{bmatrix} \rightarrow \begin{bmatrix} 3 & 2 & 1 & 2 \\ 2 & 1 & 3 & 1 \\ 60 & 60 & 60 & 0 \end{bmatrix}$.

Dual problem: Maximize $p = 60x + 60y + 60z$ subject to $3x + 2y + z \leq 2$, $2x + y + 3z \leq 1$, $x \geq 0$, $y \geq 0$, $z \geq 0$.
Solve the dual problem:

	x	y	z	s	t	p		
s	3	2	1	1	0	0	2	$2R_1 - 3R_2$
t	2	1	3	0	1	0	1	
p	−60	−60	−60	0	0	1	0	$R_3 + 30R_2$

	x	y	z	s	t	p		
s	0	1	−7	2	−3	0	1	
x	2	1	3	0	1	0	1	$R_2 - R_1$
p	0	−30	30	0	30	1	30	$R_3 + 30R_1$

	x	y	z	s	t	p		
y	0	1	−7	2	−3	0	1	$10R_1 + 7R_2$
x	2	0	10	−2	4	0	0	
p	0	0	−180	60	−60	1	60	$R_3 + 18R_2$

	x	y	z	s	t	p	
y	14	10	0	6	−2	0	10
z	2	0	10	−2	4	0	0
p	36	0	0	24	12	1	60

Solution to the primal problem: $c = 60$; $s = 24$, $t = 12$. Another possible solution: $c = 60$; $s = 0$, $t = 60$ (pivot on the 1 in the second row in the second tableau). Since the original unknowns were x and y, we write these optimal solutions as:

$c = 60$; $x = 24$, $y = 12$ or $c = 60$; $x = 0$, $y = 60$.

21. First rewrite the given problem as a standard minimization problem with all the constraints using "$\geq$":

Minimize $c = 2s + t$ subject to $3s + 2t \geq 10$, $-2s + t \geq -30$, $s + 3t \geq 60$, $s \geq 0$, $t \geq 0$.

Dualize: $\begin{bmatrix} 3 & 2 & 10 \\ -2 & 1 & -30 \\ 1 & 3 & 60 \\ 2 & 1 & 0 \end{bmatrix} \rightarrow \begin{bmatrix} 3 & -2 & 1 & 2 \\ 2 & 1 & 3 & 1 \\ 10 & -30 & 60 & 0 \end{bmatrix}$.

Dual problem: Maximize $p = 10x - 30y + 60z$ subject to $3x - 2y + z \leq 2$, $2x + y + 3z \leq 1$, $x \geq 0$, $y \geq 0$, $z \geq 0$.

Solve the dual problem:

	x	y	z	s	t	p		
s	3	-2	1	1	0	0	2	$3R_1 - R_2$
t	2	1	3	0	1	0	1	
p	-10	30	-60	0	0	1	0	$R_3 + 20R_2$

	x	y	z	s	t	p	
s	7	-7	0	3	-1	0	5
z	2	1	3	0	1	0	1
p	30	50	0	0	20	1	20

Solution to the primal problem: $c = 20/1 = 20$; $s = 0$, $t = 20/1 = 20$. Since the original unknowns were x and y, we write the optimal solution as: $c = 20$; $x = 0$, $y = 20$.

23. The first column is dominated by the last, so we can eliminate it. We cannot reduce any further. Add $k = 2$ to each entry and put 1s to the right and below:

$$\begin{bmatrix} 4 & 1 & 1 \\ 0 & 3 & 1 \\ 1 & 2 & 1 \\ 1 & 1 & 0 \end{bmatrix}.$$

This gives the following LP problem:

Maximize $p = x + y$ subject to $4x + y \leq 1$, $3y \leq 1$, $x + 2y \leq 1$, $x \geq 0$, $y \geq 0$.

Here are the tableaux:

	x	y	s	t	u	p		
s	4	1	1	0	0	0	1	
t	0	3	0	1	0	0	1	
u	1	2	0	0	1	0	1	$4R_3 - R_1$
p	-1	-1	0	0	0	1	0	$4R_4 + R_1$

	x	y	s	t	u	p		
x	4	1	1	0	0	0	1	$3R_1 - R_2$
t	0	3	0	1	0	0	1	
u	0	7	-1	0	4	0	3	$3R_3 - 7R_2$
p	0	-3	1	0	0	4	1	$R_4 + R_2$

	x	y	s	t	u	p	
x	12	0	3	-1	0	0	2
y	0	3	0	1	0	0	1
u	0	0	-3	-7	12	0	2
p	0	0	1	1	0	4	2

The solution to the primal problem is $p = 2/4 = 1/2$; $x = 2/12 = 1/6$, $y = 1/3$. So, the column player's optimal

strategy is $C = [0 \ \ 1/3 \ \ 2/3]^T$ and the value of the game is $e = 2 - 2 = 0$.

The solution to the dual problem is $s = 1/4$, $t = 1/4$, $u = 0$, so the row player's optimal strategy is $R = [1/2 \ \ 1/2 \ \ 0]$.

25. This game cannot be reduced. Add $k = 3$ to each entry and put 1s to the right and below:

$$\begin{bmatrix} 0 & 1 & 6 & 1 \\ 4 & 3 & 3 & 1 \\ 1 & 5 & 4 & 1 \\ 1 & 1 & 1 & 0 \end{bmatrix}.$$

This gives the following LP problem:

Maximize $p = x + y + z$ subject to $y + 6z \le 1$, $4x + 3y + 3z \le 1$, $x + 5y + 4z \le 1$, $x \ge 0$, $y \ge 0$, $z \ge 0$.

Here are the tableaux:

	x	y	z	s	t	u	p		
s	0	1	6	1	0	0	0	1	
t	4	3	3	0	1	0	0	1	
u	1	5	4	0	0	1	0	1	$4R_3 - R_2$
p	-1	-1	-1	0	0	0	1	0	$4R_4 + R_2$

	x	y	z	s	t	u	p		
s	0	1	6	1	0	0	0	1	$17R_1 - R_3$
x	4	3	3	0	1	0	0	1	$17R_2 - 3R_3$
u	0	17	13	0	-1	4	0	3	
p	0	-1	-1	0	1	0	4	1	$17R_4 + R_3$

	x	y	z	s	t	u	p		
s	0	0	89	17	1	-4	0	14	
x	68	0	12	0	20	-12	0	8	$89R_2 - 12R_1$
y	0	17	13	0	-1	4	0	3	$89R_3 - 13R_1$
p	0	0	-4	0	16	4	68	20	$89R_4 + 4R_1$

	x	y	z	s	t	u	p	
z	0	0	89	17	1	-4	0	14
x	6052	0	0	-204	1768	-1020	0	544
y	0	1513	0	-221	-102	408	0	85
p	0	0	0	68	1428	340	6052	1836

The solution to the primal problem is

$p = 1,836/6,052 = 27/89$; $x = 544/6,052 = 8/89$, $y = 85/1,513 = 5/89$, $z = 14/89$.

So, the column player's optimal strategy is $C = [8/27 \ \ 5/27 \ \ 14/27]^T$ and the value of the game is

$e = 89/27 - 3 = 8/27.$

The solution to the dual problem is $s = 68/6{,}052 = 1/89$, $t = 1{,}428/6{,}052 = 21/89$, $u = 340/6{,}052 = 5/89$, so the row player's optimal strategy is $R = [1/27 \; 21/27 \; 5/27] = [1/27 \; 7/9 \; 5/27]$.

27. Apply the simplex method to the given problem:

	x	y	s	t	p		
*s	-2	$\boxed{1}$	-1	0	0	1	
*t	1	-2	0	-1	0	1	$R_2 + 2R_1$
p	1	2	0	0	1	0	$R_3 - 2R_1$

	x	y	s	t	p	
y	-2	1	-1	0	0	1
*t	-3	0	-2	-1	0	3
p	5	0	2	0	1	-2

As there are no positive entries in the t-row (other than the right-hand side), we cannot go to the next step. Thus, it is impossible to get into the feasible region by completing Phase 1. In other words, there are no feasible solutions (choice A). Another way of seeing this would be to graph the constraints. This would show an empty feasible region.

29. First note that the objective function, $Z = x_1 + 4x_2 + 2x_3 - 10$, is not linear because of the "-10". However, maximizing $p = x_1 + 4x_2 + 2x_3$ will also maximize Z, provided we remember to subtract 10 from the optimal value of the objective function when we are done. Thus, we solve the problem using $p = x_1 + 4x_2 + 2x_3$ as our objective function:

	x_1	x_2	x_3	s_1	s_2	z		
s_1	4	1	1	1	0	0	45	$R_1 - R_2$
s_2	-1	$\boxed{1}$	2	0	1	0	0	
z	-1	-4	-2	0	0	1	0	$R_3 + 4R_2$

	x_1	x_2	x_3	s_1	s_2	z		
s_1	$\boxed{5}$	0	-1	1	-1	0	45	
x_2	-1	1	2	0	1	0	0	$5R_2 + R_1$
z	-5	0	6	0	4	1	0	$R_3 + R_1$

	x_1	x_2	x_3	s_1	s_2	z	
x_1	5	0	-1	1	-1	0	45
x_2	0	5	9	1	4	0	45
z	0	0	5	1	3	1	45

Optimal Solution: $p = 45$; $x_1 = 45/5 = 9$, $x_2 = 45/5 = 9$, $x_3 = 0$. Since the maximum value of p is 45, the maximum value of Z is $45 - 10 = 35$.

31. Unknowns: $x =$ # packages from Duffin House, $y =$ # packages from Higgins Press

Minimize $c = 50x + 80y$ subject to $5x + 5y \geq 4,000$, $5x + 10y \geq 6,000$, $x \geq 0$, $y \geq 0$.

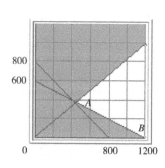

Corner Point	Lines through Point	Coordinates	$c = 50x + 80y$
A	$5x + 5y = 4,000$ $x = 0$	$(0, 800)$	$64,000$
B	$5x + 5y = 4,000$ $5x + 10y = 6,000$	$(400, 400)$	$\boxed{52,000}$
C	$5x + 10y = 6,000$ $y = 0$	$(1,200, 0)$	$60,000$

Purchase 400 packages from each for a minimum cost of $52,000.

33. a. Since the lowest cost is given by ordering the same number of packages from each supplier (point B in the feasible region in Exercise 31), changing that by ordering at least 20% more packages from Duffin as from Higgins will result in a higher cost, since the solution will no longer be point B but another point in the feasible region (choice B), assuming it is possible to meet all the conditions. On the other hand, it is conceivable that the extra constraint will result in an empty feasible region (choice D). Thus, the correct answers are (B) and (D).

b. Unknowns: as in Exercise 31. Minimize $c = 50x + 80y$ subject to $5x + 5y \geq 4,000$, $5x + 10y \geq 6,000$, $x \geq 1.2y$ or $x - 1.2y \geq 0$, $x \geq 0$, $y \geq 0$.

Corner Point	Lines through Point	Coordinates	$c = 50x + 80y$
A	$5x + 10y = 6,000$ $x - 1.2y = 0$	$(450, 375)$	$\boxed{52,500}$
B	$5x + 10y = 6,000$ $y = 0$	$(1,200, 0)$	$60,000$

Solution: Purchase 450 packages from Duffin House, 375 from Higgins Press for a minimum cost of $52,500.

35. Unknowns: $x =$ # shares in EEE, $y =$ # shares in RRR

Minimize $c = 2.0x + 3.0y$ subject to $50x + 55y \leq 12,100$, $2.25x + 2.75y \geq 550$, $x \geq 0$, $y \geq 0$.

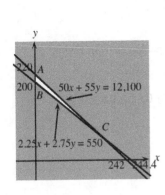

Corner Point	Lines through Point	Coordinates	$c = 2.0x + 3.0y$
A	$x = 0$ $50x + 55y = 12{,}100$	$(0, 220)$	660
B	$x = 0$ $2.25x + 2.75y = 550$	$(0, 200)$	600
C	$50x + 55y = 12{,}100$ $2.25x + 2.75y = 550$	$(220, 20)$	500

Solution: Buy 220 shares of EEE and 20 shares of RRR. The minimum total risk index is $c = 500$.

37. Take $x = $ # Sprinkles, $y = $ # Storms, $z = $ # Hurricanes.

Maximize $p = x + y + 2z$ subject to

$x + 2y + 2z \le 600$, $2x + y + 3z \le 600$, $x + 3y + 6z \le 600$, $x \ge 0$, $y \ge 0$, $z \ge 0$.

	x	y	z	s	t	u	p		
s	1	2	2	1	0	0	0	600	$3R_1 - R_3$
t	2	1	3	0	1	0	0	600	$2R_2 - R_3$
u	1	3	$\boxed{6}$	0	0	1	0	600	
p	-1	-1	-2	0	0	0	1	0	$3R_4 + R_3$

	x	y	z	s	t	u	p		
s	2	3	0	3	0	-1	0	1200	$3R_1 - 2R_2$
t	$\boxed{3}$	-1	0	0	2	-1	0	600	
z	1	3	6	0	0	1	0	600	$3R_3 - R_2$
p	-2	0	0	0	0	1	3	600	$3R_4 + 2R_2$

	x	y	z	s	t	u	p		
s	0	11	0	9	-4	-1	0	2400	$10R_1 - 11R_3$
x	3	-1	0	0	2	-1	0	600	$10R_2 + R_3$
z	0	$\boxed{10}$	18	0	-2	4	0	1200	
p	0	-2	0	0	4	1	9	3000	$5R_4 + R_3$

	x	y	z	s	t	u	p	
s	0	0	-198	90	-18	-54	0	10800
x	30	0	18	0	18	-6	0	7200
y	0	10	18	0	-2	4	0	1200
p	0	0	18	0	18	9	45	16200

Optimal solution:

$p = 16,200/45 = 360$; $x = 7,200/30 = 240$, $y = 1,200/10 = 120$, $z = 0$.
Make 240 Sprinkles, 120 Storms, and no Hurricanes.

39. Unknowns: $x =$ # packages from Duffin House, $y =$ # packages from Higgins Press, $z =$ # packages from Ewing

Books. Minimize $c = 50x + 150y + 100z$ subject to

$5x + 10y + 5z \geq 4,000$, $2x + 10y + 5z \geq 6,000$, $y \geq 0.5(x + y + z)$ or $x - y + z \leq 0$, $x \geq 0$, $y \geq 0$, $z \geq 0z \geq 0$.
Convert the given problem to a maximization problem:

Maximize $p = -50x - 150y - 100z$ subject to

$5x + 10y + 5z \geq 4,000$, $2x + 10y + 5z \geq 6,000$, $x - y + z \leq 0$, $x \geq 0$, $y \geq 0$, $z \geq 0$.

	x	y	z	s	t	u	p		
*s	5	10	5	-1	0	0	0	4000	
*t	2	10	5	0	-1	0	0	6000	$R_2 - R_1$
u	1	-1	1	0	0	1	0	0	$10R_3 + R_1$
p	50	150	100	0	0	0	1	0	$R_4 - 15R_1$

	x	y	z	s	t	u	p		
y	5	10	5	-1	0	0	0	4000	$R_1 + R_2$
*t	-3	0	0	1	-1	0	0	2000	
u	15	0	15	-1	0	10	0	4000	$R_3 + R_2$
p	-25	0	25	15	0	0	1	-60000	$R_4 - 15R_2$

	x	y	z	s	t	u	p	
y	2	10	5	0	-1	0	0	6000
s	-3	0	0	1	-1	0	0	2000
u	12	0	15	0	-1	10	0	6000
p	20	0	25	0	15	0	1	-90000

Optimal solution: $p = -90,000/1 = -90,000$; $x = 0$, $y = 6,000/10 = 600$, $z = 0$.
Since $c = -p$, the minimum value of c is 90,000.
Order 600 packages from Higgins and none from the others, for a total cost of $90,000.

41. Take $x =$ the number of credits of Sciences, $y =$ the number of credits of Fine Arts, $z =$ the number of credits of

Liberal Art, $w =$ the number of credits of Mathematics.
Given information:

(1) The total number of credits is at least 120: $\quad x + y + z + w \geq 120$.

(2) At least as many Science credits as Fine Arts credits: $\quad x \geq y$, or $x - y \geq 0$.

(3) At most twice as many Mathematics credits as Science credits. Rephrasing: The number of Mathematics credits is at most twice the number of Science credits:

$\quad w \leq 2x \Rightarrow -2x + w \leq 0$.

(4) Liberal Arts credits exceed Mathematics credits by no more than one third of the number of Fine Arts credits. Rephrasing: The number of Liberal Arts credits minus the number of Mathematics credits is at most one third of the number of Fine Arts credits:

$\quad z - w \leq \frac{1}{3}y \Rightarrow 3z - 3w \leq y \Rightarrow -y + 3z - 3w \leq 0$

Thus, the linear programming problem is: Minimize $c = 300x + 300y + 200z + 200w$ subject to

$x + y + z + w \geq 120, \; x - y \geq 0, \; -2x + w \leq 0, \; -y + 3z - 3w \leq 0, \; x \geq 0, \; y \geq 0, \; z \geq 0, \; w \geq 0.$

b. Using technology (Website → On Line Utilities → Simplex Method Tool) We obtain the following solution:

$c = 26{,}400; \; x = 24, \; y = 0, \; z = 48, \; w = 48.$

Billy-Sean should take the following combination: Sciences: 24 credits, Fine Arts: no credits, Liberal Arts: 48 credits, Mathematics: 48 credits, for a total cost of $26,400.

43. Unknowns: $x = $ # packages from New York to O'Hagan.com, $y = $ # packages form New York to

FantasyBooks.com, $z = $ # packages from Illinois to O'Hagan.com, $w = $ # packages form Illinois to

FantasyBooks.com

Minimize $c = 20x + 50y + 30z + 40w$ subject to

$x + y \leq 600, \; z + w \leq 300, \; x + z \geq 600, \; y + w \geq 200, x \geq 0, \; y \geq 0, \; z \geq 0, \; w \geq 0.$

	x	y	z	w	s	t	u	v	p		
s	1	1	0	0	1	0	0	0	0	600	$R_1 - R_3$
t	0	0	1	1	0	1	0	0	0	300	
*u	1	0	1	0	0	0	-1	0	0	600	
*v	0	1	0	1	0	0	0	-1	0	200	
p	20	50	30	40	0	0	0	0	1	0	$R_5 - 20R_3$

	x	y	z	w	s	t	u	v	p		
s	0	1	-1	0	1	0	1	0	0	0	
t	0	0	1	1	0	1	0	0	0	300	
x	1	0	1	0	0	0	-1	0	0	600	
*v	0	1	0	1	0	0	0	-1	0	200	$R_4 - R_1$
p	0	50	10	40	0	0	20	0	1	-12000	$R_5 - 50R_1$

	x	y	z	w	s	t	u	v	p		
y	0	1	-1	0	1	0	1	0	0	0	$R_1 + R_4$
t	0	0	1	1	0	1	0	0	0	300	$R_2 - R_4$
x	1	0	1	0	0	0	-1	0	0	600	$R_3 - R_4$
*v	0	0	1	1	-1	0	-1	-1	0	200	
p	0	0	60	40	-50	0	-30	0	1	-12000	$R_5 - 60R_4$

	x	y	z	w	s	t	u	v	p		
y	0	1	0	1	0	0	0	-1	0	200	
t	0	0	0	0	1	1	1	1	0	100	
x	1	0	0	-1	1	0	0	1	0	400	$R_3 + R_1$
z	0	0	1	1	-1	0	-1	-1	0	200	$R_4 - R_1$
p	0	0	0	-20	10	0	30	60	1	-24000	$R_5 + 20R_1$

	x	y	z	w	s	t	u	v	p	
w	0	1	0	1	0	0	0	-1	0	200
t	0	0	0	0	1	1	1	1	0	100
x	1	1	0	0	1	0	0	0	0	600
z	0	-1	1	0	-1	0	-1	0	0	0
p	0	20	0	0	10	0	30	40	1	-20000

Optimal solution: $p = -20,000/1 = -20,000$; $x = 600/1 = 600$, $y = 0$, $z = 0/1 = 0$, $w = 200/1 = 200$. Since $c = -p$, the minimum value of c is 20,000.

45. First reduce the game: We can eliminate O'Hagan's option of offering no promotion, and then we can eliminate FantasyBook's no promotion option. The entries in the remaining payoff matrix are nonnegative, so we put 1s to the right and below:

$$\begin{bmatrix} 20 & 10 & 15 & 1 \\ 0 & 15 & 10 & 1 \\ 1 & 1 & 1 & 0 \end{bmatrix}.$$

This gives the following LP problem:

Maximize $p = x + y + z$ subject to $20x + 10y + 15z \le 1$, $15y + 10z \le 1$, $x \ge 0$, $y \ge 0$, $z \ge 0$.

Here are the tableaus:

	x	y	z	s	t	p		
s	**20**	10	15	1	0	0	1	
t	0	15	10	0	1	0	1	
p	-1	-1	-1	0	0	1	0	$20R_3 + R_1$

	x	y	z	s	t	p		
x	20	10	15	1	0	0	1	$3R_1 - 2R_2$
t	0	**15**	10	0	1	0	1	
p	0	-10	-5	1	0	20	1	$3R_3 + 2R_2$

	x	y	z	s	t	p	
x	60	0	25	3	-2	0	1
y	0	15	10	0	1	0	1
p	0	0	5	3	2	60	5

The solution to the primal problem is $p = 5/60 = 1/12$; $x = 1/60$, $y = 1/15$, $z = 0$. So, FantasyBook's optimal strategy is $C = [0 \ 1/5 \ 4/5 \ 0]^T$ and the value of the game is $e = 12$.

The solution to the dual problem is $s = 3/60 = 1/20$, $t = 2/60 = 1/30$, so O'HaganBook's optimal strategy is $R = [0 \ 3/5 \ 2/5]$.

Section 6.1

1. The elements of F are the four seasons: spring, summer, fall, winter. Thus, $F = \{\text{spring, summer, fall, winter}\}$.

3. The elements of I are all the positive integers no greater than 6, namely $1, 2, 3, 4, 5, 6$. Thus, $I = \{1, 2, 3, 4, 5, 6\}$.

5. $A = \{n | n \text{ is a positive integer and } 0 \le n \le 3\} = \{1, 2, 3\}$ (Note that 0 is not positive, so we exclude it.)

7. $B = \{n | n \text{ is an even positive integer and } 0 \le n \le 8\} = \{2, 4, 6, 8\}$

9. a. If the coins are distinguishable, $S = \{(H, H), (H, T), (T, H), (T, T)\}$.
(Note that (H, T) and (T, H) are different outcomes, since the first coin is distinguished from the second.)
b. If the coins are indistinguishable, then (H, T) and (T, H) are the same outcome, and so
$\quad S = \{(H, H), (H, T), (T, T)\}$.

11. If the dice are distinguishable, then the outcomes can be thought of as ordered pairs. Thus, since the numbers add to 6, $S = \{(1, 5), (2, 4), (3, 3), (4, 2), (5, 1)\}$.

13. If the dice are indistinguishable, then the outcomes are characterized only by the numbers that come up, and not by the order. Thus, since the numbers add to 6, $S = \{(1, 5), (2, 4), (3, 3)\}$.

15. As the numbers facing up can never add to 13 (the largest sum is 12), there are no such outcomes. In other words, $S = \varnothing$.

17.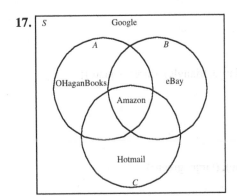

19.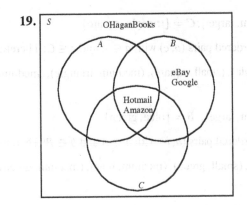

21. $A = \{\text{June, Janet, Jill, Justin, Jeffrey, Jello}\}$, $B = \{\text{Janet, Jello, Justin}\}$
$A \cup B$ is the set of all elements that are in A or B (or both): June, Janet, Jill, Justin, Jeffrey, Jello. Thus,
$\quad A \cup B = \{\text{June, Janet, Jill, Justin, Jeffrey, Jello }\} = A$.

23. $A = \{\text{June, Janet, Jill, Justin, Jeffrey, Jello}\}$
$A \cup \varnothing$ is the set of all elements that are in A or $\varnothing$ (or both). Since $\varnothing$ has no elements, $A \cup \varnothing = A$ for every set A.

25. $A = \{\text{June, Janet, Jill, Justin, Jeffrey, Jello}\}$, $B = \{\text{Janet, Jello, Justin}\}$, $C = \{\text{Sally, Solly, Molly, Jolly, Jello}\}$
$\quad B \cup C = \{\text{Janet, Jello, Justin}\} \setminus cup \{\text{Sally, Solly, Molly, Jolly, Jello}\}$

301

= {Janet, Justin, Sally, Solly, Molly, Jolly, Jello}

Therefore,

$A \cup (B \cup C)$ = {June, Janet, Jill, Justin, Jeffrey, Jello }∪{Janet, Justin, Sally, Solly, Molly, Jolly, Jello} = {June, Janet, Jill, Justin, Jeffrey, Jello, Sally, Solly, Molly, Jolly}.

27. B = {Janet, Jello, Justin}, C = {Sally, Solly, Molly, Jolly, Jello}

$C \cap B$ is the set of all elements that are common to both C and B. Thus, $C \cap$ = B{Jello}

29. A = {June, Janet, Jill, Justin, Jeffrey, Jello}

$A \cap \emptyset$ is the set of all elements that are common to both A and $\emptyset$. Since $\emptyset$ has no elements, $A \cap \emptyset = \emptyset$ for every set A.

31. A = {June, Janet, Jill, Justin, Jeffrey, Jello}, B = {Janet, Jello, Justin}, C = {Sally, Solly, Molly, Jolly, Jello}

$A \cap B$ = {Janet, Jello, Justin}

Therefore, $(A \cap B) \cap C$ = {Janet, Jello, Justin } ∩ {Sally, Solly, Molly, Jolly, Jello } = {Jello}.

33. A = {June, Janet, Jill, Justin, Jeffrey, Jello}, B = {Janet, Jello, Justin}, C = {Sally, Solly, Molly, Jolly, Jello}

$A \cap B$ = {Janet, Jello, Justin}

Therefore, $(A \cap B) \cup C$ = {Janet, Jello, Justin }∪{Sally, Solly, Molly, Jolly, Jello}

= {Janet, Justin, Jello, Sally, Solly, Molly, Jolly}.

35. A = {small, medium, large }, C = {triangle, square}

$A \times C$ is the set of all ordered pairs (a, c) with $a \in A$ and $c \in C$: Therefore,

$A \times C$ = {(small, triangle), (small, square), (medium, triangle), (medium, square), (large, triangle), (large, square)}.

37. A = {small, medium, large}, B = {blue, green}

$A \times B$ is the set of all ordered pairs (a, b) with $a \in A$ and $b \in B$: Therefore,

$A \times B$ = {(small, blue), (small, green), (medium, blue), (medium, green), (large, blue), (large, green)}.

39. B = {blue, green}, C = {triangle, square}

To represent $B \times C$ we use the elements of B = {blue, green} for the row headings, and the elements of C = {triangle, square} for the column headings:

	A	B	C
1		Triangle	Square
2	Blue	Blue Triangle	Blue Square
3	Green	Green Triangle	Green Square

41. A = {small, medium, large}, B = {blue, green}

To represent $A \times B$ we use the elements of A = {small, medium, large} for the row headings, and the elements of

302

$B = \{$blue, green$\}$ for the column headings:

◿	A	B	C
1		Blue	Green
2	Small	Small Blue	Small Green
3	Medium	Medium Blue	Medium Green
4	Large	Large Blue	Large Green

43. If a die is rolled and a coin is tossed, each outcome is a pair (b, a), where b is an outcome when a die is rolled and a is an outcome when a coin is tossed. Thus, the set of outcomes is $B \times A = \{$1H, 1T, 2H, 2T, 3H, 3T, 4H, 4T, 5H, 5T, 6H, 6T$\}$.

45. If a coin is tossed 3 times, each outcome is a triple (a_1, a_2, a_3), where a_1, a_2, and a_3 are outcomes when a coin is tossed. Thus, the set of outcomes is $A \times A \times A = \{$HHH, HHT, HTH, HTT, THH, THT, TTH, TTT$\}$.

47. E is the set of outcomes in which at least one die shows an even number. Thus, E' is the set of outcomes in which *both* dice show an odd number: $E' = \{(1, 1), (1, 3), (1, 5), (3, 1), (3, 3), (3, 5), (5, 1), (5, 3), (5, 5)\}$.

49. $E \cup F$ is the set of outcomes in which either at least one die is even, or at least one is odd. Since this includes *all* the outcomes (in every outcome there is either an odd or even outcome), its complement is empty: $(E \cup F)' = \varnothing$.

51. By Exercise 47, E' is the set of outcomes in which *both* dice show an odd number: By Exercise 48, F' is the set of outcomes in which both dice show an even number. Thus, $E' \cup F'$ is the set of outcomes in which either both are even, or both are odd:

$$E' \cup F' = \{(1, 1), (1, 3), (1, 5), (3, 1), (3, 3), (3, 5), (5, 1), (5, 3), (5, 5), (2, 2), (2, 4), (2, 6),$$
$$(4, 2), (4, 4), (4, 6), (6, 2), (6, 4), (6, 6)\}.$$

53. $(A \cup B)'$ is the region outside $A \cup B$, while $A' \cap B'$ is the overlap of the region outside A with that outside B (the gray region in the diagram below):

$(A \cup B)' = A' \cap B'$

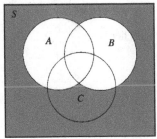

55. $(A \cap B) \cap C$ is the overlap of $A \cap B$ and C, which is the same as the overlap of all three sets: A, B, C. Similarly for $A \cap (B \cap C)$ (the gray region in the diagram below):

$(A \cap B) \cap C = A \cap (B \cap C)$

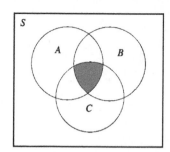

57. $A \cup (B \cap C)$ is the union of A with $B \cap C$, and so consists of all elements in A together with those in both B and C. This is the same as $(A \cup B) \cap (A \cup C)$ (the gray region in the diagram below):

$A \cup (B \cap C) = (A \cup B) \cap (A \cup C)$

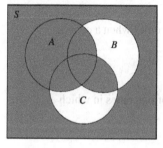

59. $S' = \varnothing$

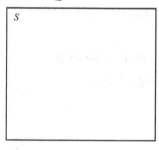

Applications

61. The set of clients who owe her money is

　　$A = \{\text{Acme, Crafts, Effigy, Global}\}$.

The set of clients who have done at least \$10,000 worth of business with her is

　　$B = \{\text{Acme, Brothers, Crafts, Dion}\}$.

Therefore, the set of clients who owe her money and have done at least \$10,000 worth of business with her is

　　$A \cap B = \{\text{Acme, Crafts}\}$.

63. The set of clients who have done at least \$10,000 worth of business with her is

　　$B = \{\text{Acme, Brothers, Crafts, Dion}\}$.

The set of clients who have employed her in the last year is

　　$C = \{\text{Acme, Crafts, Dion, Effigy, Global, Hilbert}\}$.

Therefore, the set of clients who have done at least \$10,000 worth of business with her or have employed her in the last year is

　　$B \cup C = \{\text{Acme, Brothers, Crafts, Dion, Effigy, Global, Hilbert}\}$.

65. The set of clients who do not owe her money is

$A' = \{$Brothers, Dion, Floyd, Hilbert$\}$.

The set of clients who have employed her in the last year is

$C = \{$Acme, Crafts, Dion, Effigy, Global, Hilbert$\}$.

Therefore, the set of clients who do not owe her money and have employed her in the last year is

$A' \cap C = \{$Dion, Hilbert$\}$.

67. The clients who owe her money is

$A = \{$Acme, Crafts, Effigy, Global$\}$.

The set of clients who have not done at least $10,000 worth of business with her is

$B' = \{$Effigy, Floyd, Global, Hilbert$\}$.

The set of clients who have not employed her in the last year is

$C' = \{$Brothers, Floyd$\}$.

Therefore, the set of clients who owe her money, have not done at least $10,000 worth of business with her, and have not employed her in the last year is

$A \cap B' \cap C' = \emptyset$.

69. You can organize the spreadsheet as follows: For the row headings, use the years 2003, 2004, 2005, and 2006. For the column headings, use Sailboats, Motor Boats, and Yachts:

	A	B	C	D
1		**Sailboats**	**Motor Boats**	**Yachts**
2	**2003**	(2003 Sailboats)	(2003 Motor Boats)	(2003 Yachts)
3	**2004**	(2004 Sailboats)	(2004 Motor Boats)	(2004 Yachts)
4	**2005**	(2005 Sailboats)	(2005 Motor Boats)	(2005 Yachts)
5	**2006**	(2006 Sailboats)	(2006 Motor Boats)	(2006 Yachts)

This setup gives a tabular representation of the Cartesian product

$\{2003, 2004, 2005, 2006\} \times \{$Sailboats, Motor boats, Yachts$\}$.

Alternatively, you could use the years for the column headings and $\{$Sailboats, Motor Boats, Yachts$\}$ for the row headings and obtain a representation of

$\{$Sailboats, Motor boats, Yachts$\} \times \{2003, 2004, 2005, 2006\}$.

Communication and reasoning exercises

71. The collection of all iPads and jPads combined is the set of all items that are either iPads (that is, in the set I) or jPads (that is, in the set J), and is therefore the union of the two sets: $I \cup J$.

73. Techno music that is neither European nor Dutch:

Techno AND NOT (European OR Dutch)

Techno $\cap$ (European $\cup$ Dutch)$'$ (Choice B).

75. Let $A = \{1\}$, $B = \{2\}$, and $C = \{1, 2\}$. Then $(A \cap B) \cup C = \{1, 2\}$ but $A \cap (B \cup C) = \{1\}$. In general, $A \cap (B \cup C)$ must be a subset of A, but $(A \cap B) \cup C$ need not be; also, $(A \cap B) \cup C$ must contain C as a subset, but $A \cap (B \cup C)$ need not.

77. A universal set is a set containing all "things" currently under consideration. When discussing sets of positive integers, the universe might be the set of all positive integers, or the set of all integers (positive, negative, and 0), or any other set containing the set of all positive integers.

79. $A \cap (B \cup C')$ means A, and either B or not C. Thus, for instance, take A as the set of suppliers who deliver components on time, B as the set of suppliers whose components are known to be of high quality, and C as the set of suppliers who do not promptly replace defective components. Then selecting suppliers in $A \cap (B \cup C')$ means selecting those who deliver components on time and are either companies whose components are of high quality or are suppliers who promptly replace defective components.

81. Let $A = \{$movies that are violent$\}$, $B = \{$movies that are shorter than 2 hours$\}$, $C = \{$movies that have a tragic ending$\}$, and $D = \{$movies that have an unexpected ending$\}$. The given sentence can be rewritten as "She prefers movies in $A' \cap B \cap (C \cup D)'$." It can also be rewritten as "She prefers movies in $A' \cap B \cap C' \cap D'$."

83. Removing the comma would cause the statement to be ambiguous, as it could then correspond to either WWII $\cup$ (Comix $\cap$ Aliens$'$) or to (WWII $\cup$ Comix) $\cap$ Aliens$'$. (See Exercise 76.)

Section 6.2

1. $n(A)$ = Number of elements in $A = 4, n(B) = $ Number of elements in $B = 5$. Therefore, $n(A) + n(B) = 4 + 5 = 9$.

3. $A \cup B = \{$Dirk, Johan, Frans, Sarie, Tina, Klaas, Henrika$\}$, and so $n(A \cup B) = 7$.

5. $B \cap C = \{$Frans$\}$, $A \cup (B \cap C) = \{$Dirk, Johan, Frans, Sarie$\}$, and so $n(A \cup (B \cap C)) = 4$.

7. From Exercise 3, we know that $n(A \cup B) = 7$. On the other hand, $n(A) + n(B) - n(A \cap B) = 4 + 5 - 2 = 7$. Therefore, $n(A \cup B) = n(A) + n(B) - n(A \cap B)$.

9. $n(A \times A) = n(A)n(A) = 2 \times 2 = 4$

11. $n(B \times C) = n(B)n(C) = 6 \times 3 = 18$

13. $n(A \times B \times B) = n(A)n(B \times B) = n(A)n(B)n(B) = 2 \times 6 \times 6 = 72$

15. $n(A \cup B) = n(A) + n(B) - n(A \cap B) = 43 + 20 - 3 = 60$

17. $n(A \cup B) = n(A) + n(B) - n(A \cap B) \Rightarrow 100 = 60 + 60 - n(A \cap B)$, and so $n(A \cap B) = 60 + 60 - 100 = 20$.

19. $n(A') = n(S) - n(A) = 10 - 4 = 6$

21. $n((A \cap B)') = n(S) - n(A \cap B) = 10 - 1 = 9$

23. Use $S = \{$BA, MU, SO, SU, LI, MS, WA, RT, DU, LY$\}$, $A = \{$SO, LI, MS, RT$\}$, $B = \{$BA, MU, SO$\}$.
$A' = \{$BA, MU, SU, WA, DU, LY$\}$, $B' = \{$SU, LI, MS, WA, RT, DU, LY$\}$ $\Rightarrow A' \cap B' = \{SU, WA, DU, LY\}$ and so $n(A' \cap B') = 4$.

25. $n((A \cap B)') = 9$ from Exercise 21, and so $n(A') + n(B') - n((A \cup B)') = 6 + 7 - 4 = 9$.
Therefore, $n((A \cap B)') = n(A') + n(B') - n((A \cup B)')$.

27. Assign labels to the regions of the diagram where the quantities are unknown:

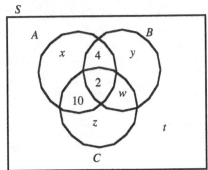

$n(A) = 20 \Rightarrow x + 16 = 20 \Rightarrow x = 4$

$n(B \cap C) = 8 \Rightarrow 2 + w = 8 \Rightarrow w = 6$

$n(B) = 20 \Rightarrow 6 + y + w = 20 \Rightarrow 6 + y + 6 = 20 \Rightarrow y = 8$

$n(C) = 28 \Rightarrow 12 + z + w = 28 \Rightarrow 12 + z + 6 = 28 \Rightarrow z = 10$

$n(S) = 50 \Rightarrow x + y + z + w + t + 16 = 50 \Rightarrow 4 + 8 + 10 + 6 + t + 16 = 50 \Rightarrow t = 6$

Thus, the completed diagram is

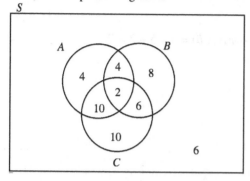

29. Assign labels to the regions of the diagram where the quantities are unknown:

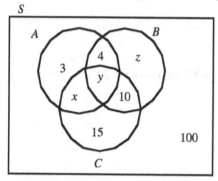

$n(A) = 10 \Rightarrow x + y + 7 = 10 \Rightarrow x + y = 3$

$n(B) = 19 \Rightarrow y + z + 14 = 19 \Rightarrow y + z = 5$

$n(S) = 140 \Rightarrow x + y + z + 132 = 140 \Rightarrow x + y + z = 8$

This is a system of 3 linear equations in 3 unknowns:

$x + y = 3 \qquad (1) \quad y + z = 5 \quad (2) \quad x + y + z = 8. \quad (3)$

To solve, we can substitute (1) into (3), giving

$3 + z = 8 \Rightarrow z = 5$

(2) now gives $y + 5 = 5 \Rightarrow y = 0$.

(1) now gives $x + 0 = 3 \Rightarrow x = 3$.

Thus, the solution is $(x, y, z) = (3, 0, 5)$. The completed diagram is:

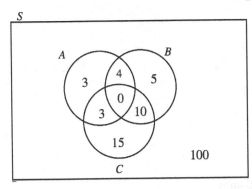

Applications

31. Let A be the set of Web sites containing "asteroid" and let C be the set of Web sites containing the word "comet."

We are told that $n(A) = 25.0, n(C) = 93.5, n(A \cap C) = 3.1$. Therefore,

$$n(A \cup C) = n(A) + n(C) - n(A \cap C) = 25.0 + 93.5 - 3.1 = 115.4 \text{ million sites.}$$

33. Let B be the set of people who had black hair, and let R be the set of people who had a whole row to themselves.

We are told that $n(B \cup R) = 37, n(B) = 33,$ and $n(R) = 6.$ We are asked to find $n(B \cap R)$.

$$n(B \cup R) = n(B) + n(R) - n(B \cap R) \Rightarrow 37 = 33 + 6 - n(B \cap R)$$

So, $n(B \cap R) = 33 + 6 - 37 = 2.$

35. Let A be the set of gamers who used smartphones, and let B be the set of gamers who used tablets. We are told that

$n(A \cup B) = 132, \ n(A) = 123,$ and $n(B) = 44$ (working in millions of gamers). We are asked to find the number who

used tablets excluding those who used both: $n(B) - n(A \cap B)$, so we first calculate $n(A \cap B)$:

$$n(A \cup B) = n(A) + n(B) - n(A \cap B) \Rightarrow 132 = 123 + 44 - n(A \cap B)$$

So, $n(A \cap B) = 123 + 44 - 132 = 35,$ so 35 million gamers used both. Therefore,

$$n(B) - n(A \cap B) = 44 - 35 = 9 \text{ million gamers.}$$

37. $C \cap N$ is the set of authors who are both successful and new.

$C \cup N$ is the set of authors who are either successful or new (or both).

$$n(C) = 30, n(N) = 20, n(C \cap N) = 5, n(C \cup N) = 45$$

$$n(C \cup N) = n(C) + n(N) - n(C \cap N) \Rightarrow 45 = 30 + 20 - 5 \checkmark$$

39. $C \cap N'$ is the set of authors who are successful but not new—in other words, the set of authors who are successful

and established. $n(C \cap N') = 25$

41. Of the 80 established authors, 25 are successful. Thus, the percentage of established authors who are successful is

$$\frac{25}{80} \approx 0.3125, \text{ or } 31.25\%.$$

Of the 30 successful authors, 25 are established. Thus, the percentage of successful authors who are established is

$$\frac{25}{30} \approx 0.8333, \text{ or } 83.33\%.$$

43. The set of housing starts in the Midwest in 2014 is $M \cap C$.

$n(M \cap C) = 110$ thousand units

45. The set of housing starts in 2013 excluding housing starts in the South is $B \cap T'$.

$n(B \cap T') = 620 - 330 = 290$ thousand units

47. The set of housing starts in 2013 in the West and Midwest is $B \cap (W \cup M)$.

$n(B \cap (W \cup M)) = n((B \cap W) \cup (B \cap M)) = 100 + 130 - 0 = 230$ thousand units

49. The set of non-manufacturing industries that increased is $X \cap M'$.

	X	Y	Z	Totals
F	3	4	1	8
M	8	3	3	14
T	6	1	0	7
H	4	1	1	6
U	3	1	1	5
Totals	24	10	6	40

$n(X \cap M') = 24 - 8 = 16$

51. $n(H' \cup Z)$ is the number of industries that either were not in the health-care sector or were unchanged in value (or both).

	X	Y	Z	Totals
F	3	4	1	8
M	8	3	3	14
T	6	1	0	7
H	4	1	1	6
U	3	1	1	5
Totals	24	10	6	40

$n(H' \cup Z) = 40 - (4 + 1) = 35$

53. $n(T \cap Y) = 1; n(Y) = 10$

	X	Y	Z	Totals
F	3	4	1	8
M	8	3	3	14
T	6	1	0	7
H	4	1	1	6
U	3	1	1	5
Totals	24	10	6	40

$\dfrac{n(T \cap Y)}{n(Y)} = \dfrac{1}{10}$. This is the fraction of industries that decreased that were from the information technology sector.

55. Let S be the set of all children in the study, let R be the set of children who had rickets, and let U be the set of urban children. We are told that $n(S) = 1556, n(R) = 1024, n(U) = 243, n(U \cap R) = 93$.
Represent the given information in a Venn diagram:

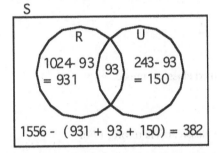

a. From the diagram, $n(R \cap U') = 931$.

b. From the diagram, $n(U' \cap R') = 382$.

57. a. Following is a Venn diagram with most of the unknowns marked with labels:

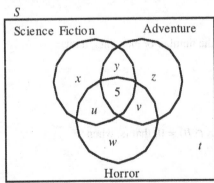

15 had seen a science fiction movie and a horror movie $\Rightarrow 5 + u = 15 \Rightarrow u = 10$

5 had seen an adventure movie and a horror movie $\Rightarrow 5 + v = 5 \Rightarrow v = 0$

25 had seen a science fiction movie and an adventure movie $\Rightarrow 5 + y = 25 \Rightarrow y = 20$

35 had seen a horror movie 35 had seen a horror movie

$$\Rightarrow 5 + u + v + w = 35 \Rightarrow 5 + 10 + 0 + w = 35 \Rightarrow w = 20$$

55 had seen an adventure movie $\Rightarrow 5 + y + z + v = 55 \Rightarrow 5 + 20 + z + 0 = 55 \Rightarrow z = 30$

40 had seen a science fiction movie $\Rightarrow 5 + x + y + u = 40 \Rightarrow 5 + x + 20 + 10 = 40 \Rightarrow x = 5$

Finally, the total number of people in the survey was 100:

$$\Rightarrow 5 + x + y + z + u + v + w + t = 100$$

$$\Rightarrow 5 + 5 + 20 + 30 + 10 + 0 + 20 + t = 100$$

$$\Rightarrow t = 10$$

Here is the completed diagram:

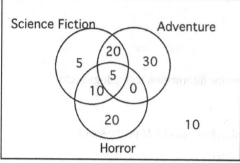

b. Of the 40 people who had seen science fiction, 15 saw a horror movie. Therefore, the percentage of science fiction movie fans who are also horror movie fans can be estimated as

$$\frac{15}{40} = 0.375, \text{ or } 37.5\%.$$

59. Let R be the set of students who liked rock music, and let C be the set of those who liked classical music.

$$n(R) = 22, n(R \cap C) = 5$$

We are also given other information that we do not need.

Since 5 of the 22 people who liked rock also liked classical, the other 17 did not like classical. Therefore, 17 of those that enjoyed rock did not enjoy classical music.

Communication and reasoning exercises

61. Since every element of A is in B, and since B contains at least one more element than A (otherwise they would be the same set), $n(A) < n(B)$.

63. The number of elements in the Cartesian product of two finite sets is the product of the number of elements in the two sets.

65. Answers will vary.

67. Since $n(A \cup B) = n(A) + n(B) - n(A \cap B)$, we get $n(A \cup B) \neq n(A) + n(B)$ when $n(A \cap B) \neq 0$; that is, when $A \cap B \neq \varnothing$.

69. Since $A \subseteq A \cup B$, the only way they can have the same number of elements is if $A = A \cup B$; that is, when $B \subseteq A$.

71. $n(A \cup B \cup C) = n(A) + n(B) + n(C) - n(A \cap B) - n(B \cap C) - n(A \cap C) + n(A \cap B \cap C)$

Section 6.3

1. Alternative 1: 2 outcomes

Alternative 2: 3 outcomes

Alternative 3: 5 outcomes

Total number of outcomes: $2 + 3 + 5 = 10$

3. Step 1: 2 outcomes

Step 2: 3 outcomes

Step 3: 5 outcomes

Total number of outcomes: $2 \times 3 \times 5 = 30$

5. Alternative 1: $1 \times 2 = 2$ outcomes

Alternative 2: $2 \times 2 = 4$ outcomes

Total number of outcomes: $2 + 4 = 6$

7. Step 1: $1 + 2 = 3$ outcomes

Step 2: $2 + 2 + 1 = 5$ outcomes

Total number of outcomes: $3 \times 5 = 15$

9. Alternative 1: $(3 + 1) \times 2 = 8$ outcomes

Alternative 2: 5 outcomes

Total number of outcomes: $8 + 5 = 13$

11. Step 1: $(3 \times 1) + 2 = 6$ outcomes

Step 2: 5 outcomes

Total number of outcomes: $6 \times 5 = 30$

13. Decision algorithm: Start with 4 empty slots, and select slots in which to place the letters.
Step 1: Select a slot to place the b: 4 choices.
Step 2: Place the "a"s in the remaining slots: 1 choice.

Total number of outcomes: $4 \times 1 = 4$

Applications

15. Decision algorithm:
Step 1: Select a flavor: 31 choices.
Step 2: Select cone, cup, or sundae: 3 choices.

Total number of outcomes: $31 \times 3 = 93$

17. Decision algorithm:
Step 1: Select the first bit: 2 choices.
Step 2: Select the second bit: 2 choices.
Step 3: Select the third bit: 2 choices.
Step 4: Select the fourth bit: 2 choices.

Total number of outcomes: $2 \times 2 \times 2 \times 2 = 16$

19. Decision algorithm: Start with 6 empty slots, and select slots in which to place the ternary digits.
Step 1: Select a slot to place the 1: 6 choices.
Step 2: Select a slot to place the 2: 5 choices.
Step 3: Place 0s in the remaining slots: 1 choice.

Total number of outcomes: $6 \times 5 \times 1 = 30$

21. Alternative 1 (Gummy candy):
Step 1: Select a size: 3 choices.
Step 2: Select a shape: 3 choices.
Alternative 2 (Licorice nibs):
Step 1: Select a size: 2 choices.
Step 2: Select a color: 2 choices.

Total number of outcomes: $3 \times 3 + 2 \times 2 = 13$

23. Alternative 1: Single disc:
Step 1: Select a type: 2 choices.
Step 2: Select a color: 5 choices.
Alternative 2: Spindle of discs:
Step 1: Select spindle size: 2 choices.
Step 2: Select a type: 2 choices.
Step 3: Select a color: 2 choices.

Total number of choices: $(2 \times 5) + (2 \times 2 \times 2) = 18$

25. Step 1: Answer 1st t/f question: 2 choices.
Step 2: Answer 2nd t/f question: 2 choices.
. . .
Step 10: Answer 10th t/f question: 2 choices.
Step 11: Answer 1st multiple choice question: 5 choices.
Step 12: Answer 2nd multiple choice question: 5 choices.

Total number of choices: $(2 \times 2 \times \ldots \times 2) \times (5 \times 5) = 2^{10} \times 5^2 = 25,600$

27. Alternative 1: Do Part A:

Steps 1–8: Answer the 8 t/f questions; 2 choices each, giving 2^8 possible choices.

Alternative 2: Answer Part B:

Steps 1-5: Answer 5 multiple choice questions with 5 choices each, giving 5^5 possible choices.

Total number of choices: $2^8 + 5^5 = 3,381$

29. a. Step 1: Select a mutual fund: 4 choices.
Step 2: Select a muni. bond fund: 3 choices.
Step 3: Select a stock: 8 choices.
Step 4: Select a precious metal: 3 choices.

Total number of choices: $4 \times 3 \times 8 \times 3 = 288$

b. Step 1: Select 3 mutual funds: 4 choices (select which one to leave out).
Step 2: Select 2 municipal bond funds: 3 choices (select which one to leave out).
Step 3: Select one stock: 8 choices.
Step 4: Select 2 precious metals 3 choices (select which one to leave out).

Total number of choices: $4 \times 3 \times 8 \times 3 = 288$

31. The number of possible characters that can be represented equals the number of possible bytes:

Steps 1–8: Select 0 or 1 for each bit: $2 \times 2 \times \ldots \times 2 = 28$ choices.

Thus, the number of possible characters is $2^8 = 256$.

33. Decision algorithm to obtain a symmetry of the five-pointed star:
Alternative 1: Pure rotation: 5 choices
Alternative 2: Rotation followed by a flip: 5 choices

Total number of symmetries: $5 + 5 = 10$

35. As the variable name must end in a digit, it cannot have length 1 (as that would be a single letter only).
Alternative 1: Length two: Letter, digit:
Step 1: Choose a letter (uppercase or lowercase): 52 choices.
Step 2: Choose a digit: 10 choices.
Alternative 2: Length three: Letter, (letter or digit), digit
Step 1: Choose a letter: 52 choices.
Step 3: Choose a letter or digit: 62 choices.
Step 2: Choose a digit: 10 choices.

Total number of variables $= 52 \times 10 + 52 \times 62 \times 10 = 32,760$

37. Step 1: Select a winner of the North Carolina-Central Connecticut game: 2 choices\gap[40] Step 2: Select a winner of the Virginia-Syracuse game: 1 choice (we know that Syracuse is the winner).\gap[40] Step 3: Select the overall winner: 2 choices.Total number of choices: $2 \times 1 \times 2 = 4$

39. a. Step 1: Choose the first digit: 8 choices

Steps 2–7: Choose the remaining 6 digits: 10^6 choices

Total number: $8 \times 10^6 = 8{,}000{,}000$

b. To count the numbers beginning with 463, 460, or 400, use the following decision algorithm:

Alternative 1: Start with 463.

Steps 1–4: Choose the remaining 4 digits: 10^4 choices.

Alternative 2: Start with 460.

Steps 1–4: Choose the remaining 4 digits: 10^4 choices.

Alternative 3: Start with 400.

Steps 1–4: Choose the remaining 4 digits: 10^4 choices.

Total number of choices: $10^4 + 10^4 + 10^4 = 30{,}000$

c. Use the following decision algorithm:
Step 1: Select the 1st digit (other than 0 or 1): 8 choices.
Step 2: Select a 2nd digit different from the 1st digit: 9 choices.
Step 2: Select a 3rd digit different from the 2nd digit: 9 choices.

. . .

Step 7: Select a 7th digit different from the 6th digit: 9 choices.

Total number of choices: $8 \times 9^6 = 4{,}251{,}528$

41. a. Use the following decision algorithm:
Step 1: Select the first digit: 2 choices.
Step 2 Select digits 2 through 6: 1 choice (a single issuer)

Step 3: Select digits 7 through 15 (customer id): 10^9 choices.

Step 4: Select the check digit: 1 choice (as it is determined by the digits that precede it).

Total number of choices: $2 \times 10^9 = 2$ billion possible card numbers

b. Use the following decision algorithm:
Step 1: Select the first digit: 1 choice (6 for Discover)
Step 2 Select digits 2 through 6: 1 choice (a single issuer)

Step 3: Select digits 7 through 15 (customer id): 10^9 choices.

Step 4: Select the check digit: 9 choices (exclude the correct check digit).

Total number of choices: $10^9 \times 9 = 9$ billion possible card numbers

43. a. Steps 1–3: Choose three bases (4 choices each): $4^3 = 64$ choices

b. Steps 1–n: Choose n bases (4 choices each): 4^n choices

c. From part (b) with $n = 2.1 \times 10^{10}$, we obtain a total of $4^n = 4^{2.1 \times 10^{10}}$ possible DNA chains.

45. a. Steps 1–6: Select 6 hexadecimal digits (16 choices per digit): $16^6 = 16{,}777{,}216$ possible colors
b. Step 1: Choose the 1st repeating pair of digits: 16 choices.

Step 2: Choose the 2nd repeating pair of digits: 16 choices.

Step 3: Choose the 3rd repeating pair of digits: 16 choices.

Total number of colors: $16^3 = 4{,}096$

c. Step 1: Choose the digit x: 16 choices.

Step 2: Choose the digit y: 16 choices.

Total number of grayscale shades: $16^2 = 256$

d. Step 1: Choose the position for the sequence xy: 3 choices.

Step 2: Choose the values of x and y: 16^{2} choices.

Total number of choices: 3×16^2 choices. However, the above decision algorithm leads to the sequence 000000 in three ways. The number of pure colors is therefore $3 \times 16^2 - 2 = 766$.

47. Step 1: Choose a male actor to play Escalus: 10 choices.
Step 2: Choose a male actor to play Paris: 9 choices.
...
Step 6: Choose a male actor to play Tybalt: 5 choices.
Step 7: Choose a male actor to play Friar Lawrence: 4 choices.
Step 8: Choose a female actor to play Lady Montague: 8 choices.
Step 9: Choose a female actor to play Lady Capulet: 7 choices.
Step 10: Choose a female actor to play Juliet: 6 choices.
Step 11: Choose a female actor to play Juliet's Nurse: 5 choices.
Total number of casts: $(10 \times 9 \times 8 \times 7 \times 6 \times 5 \times 4) \times (8 \times 7 \times 6 \times 5) = 1,016,064,000$

49. a. Steps 1–3: Choose 3 letters: 26^3 choices.

Steps 4–6: Choose 3 digits: 10^3 choices.

Total number of license plates: $26^3 \times 10^3 = 17,576,000$

b. Steps 1–2 Choose 2 letters: 26^2 choices.

Step 3: Choose a letter other than I, O, or Q: 23 choices.

Steps 4–6: Choose 3 digits: 10^3 choices.

Total number of license plates: $26^2 \times 23 \times 10^3 = 15,548,000$

c. A decision algorithm for the number of reserved plates:
Step 1: Choose VET, MDZ, or DPZ: 3 choices.

Step 2–4: Choose 3 digits: 10^3 choices.

Total number of reserved plates: 3×10^3

From part (b), the total number of possible plates is 15,548,000. Therefore, the number of unreserved plates is
$15,548,000 - 3 \times 10^3 = 15,545,000$.

51. a. Start with 4 empty slots in which to place the letters R and D.
Step 1: Select a slot for the single D: 4 choices.
Step 2: Place Rs in the remaining slots: 1 choice.

Total number of sequences: $4 \times 1 = 4$

b. Each route in the maze is a sequence of 4 moves: either right (R) or down (D). Since only one down move is possible, the number of such routes equals the number of four-letter sequences that contain only the letters R and D, with D occurring only once: 4 possibilities, by part (a).

c. If we allowed left and/or up moves, we could get an unlimited number of routes, such as
RRRD, RLRRRD, RLRLRRRD, ...

53. a. Step 1: Choose a 1st cylinder: 6 choices.
Step 2: Choose a 2nd cylinder on the opposite side: 3 choices.
Step 3: Choose a 3rd cylinder on the same side as the first: 2 choices.
Step 4: Chose a 4th cylinder on the same side as the second: 2 choices.
Step 5: Chose a 5th cylinder on the same side as the first: 1 choice.
Step 6: Chose a 6th cylinder on the same side as the second: 1 choice.

Total number of firing sequences: $6 \times 3 \times 2 \times 2 \times 1 \times 1 = 72$

b. Use the same decision algorithm as for part (a), except that, in Step 1, we only have 3 choices (the cylinders on the left).

Total number of firing sequences: $3 \times 3 \times 2 \times 2 \times 1 \times 1 = 36$

55. Decision algorithm for producing a painting:

    Steps 1–5: Select blue or gray for each of the odd-numbered lines: 2^{5} choices.

    Step 6: Select a single color for the remaining lines: 3 choices.

Total number of paintings: $2^5 \times 3 = 96$

57. a. Decision algorithm for incorrect codes that will open the lock:Step 1: Decide which digit will be the wrong one: 4 choices.

    Step 2: Choose an incorrect digit in that place: 9 choices (one is correct, leaving 9 incorrect ones).

Total number of incorrect codes that will open the lock: $4 \times 9 = 36$

b. From part (a), there are 36 incorrect codes that will open the lock. Also, there is only 1 correct code. Therefore, the total number of codes that will open the lock is $36 + 1 = 37$.

59. Decision algorithm for creating a calendar:

    Step 1: Choose a day of the week for Jan 1: 7 choices.

    Step 2: Decide whether or not it is a leap year: 2 choices.

Total number of possible calendars: $7 \times 2 = 14$

61. Decision algorithm to select a particular iteration:

    Step 1: Choose a value for i: 10 choices.

    Step 2: Choose a value for j: 19 choices (19 integers in the range 2–20).

    Step 3: Choose a value for k: 10 choices.

Total number of iterations: $10 \times 19 \times 10 = 1,900$

63. Decision algorithm to place a single 1×1 block:

    Step 1: Choose a position in the left-right direction: m choices.

    Step 2: Choose a position in the front-back direction: n choices.

    Step 3: Choose a position in the up-down direction: r choices.

Total number of possible solids: mnr

65. The number of possible sequences of length 1 is 2 (either a dot: $\cdot$ or a dash: $-$)

The number of possible sequences of length 2 is $2 \times 2 = 4$ (2 steps, 2 choices per step):

    $\cdot\,\cdot,\ \cdot\,-,\ -\,\cdot,\ -\,-$

The number of possible sequences of length 3 is $2^3 = 8$ (3 steps, 2 choices per step).

The number of possible sequences of length 4 is $2^4 = 16$ (3 steps, 2 choices per step).

So far, using different lengths from 1 to 4, we can encode $1 + 2 + 4 + 8 + 16 = 31$ possible letters, which is enough to include the whole alphabet. (Using lengths 1–3 will not work, since that will give only $1 + 2 + 4 + 8 = 15$ possible letters.) Thus, we need to use sequences of up to 4 dots and dashes.

Communication and reasoning exercises

67. The multiplication principle is based on the cardinality of the Cartesian product of two sets.

69. The decision algorithm produces every pair of shirts twice, first in one order and then in the other. Therefore, it is not valid. The actual number of pairs of shirts is half of the number computed by the algorithm: $90/2 = 45$.

71. Think of placing the five squares in a row of five empty slots. Step 1: Choose a slot for the blue square 5 choices. Step 2: Choose a slot for the green square: 4 choices. Step 3: Choose the remaining 3 slots for the yellow squares: 1 choice. Hence, there are 20 possible five-square sequences.

Section 6.4

1. $6! = 6 \times 5 \times 4 \times 3 \times 2 \times 1 = 720$

3. $\dfrac{8!}{6!} = \dfrac{8 \times 7 \times \cancel{6 \times 5 \times 4 \times 3 \times 2 \times 1}}{\cancel{6 \times 5 \times 4 \times 3 \times 2 \times 1}} = 56$

5. $P(6,4) = 6 \times 5 \times 4 \times 3 = 360$

7. $\dfrac{P(6,4)}{4!} = \dfrac{6 \times 5 \times 4 \times 3}{4 \times 3 \times 2 \times 1} = 15$

9. $C(3,2) = \dfrac{3 \times 2}{2 \times 1} = 3$

Alternatively, $C(3,2) = C(3, 3-2) = C(3,1) = 3$

11. $C(10,8) = C(10,2)$

(because $10 - 8 = 2$)

$= \dfrac{10 \times 9}{2 \times 1} = 45$

13. $C(20,1) = \dfrac{20}{1} = 20$

15. $C(100,98) = C(100,2)$

(because $100 - 98 = 2$)

$= \dfrac{100 \times 99}{2 \times 1} = 4{,}950$

17. The number of ordered lists of 4 items chosen from 6 is $P(6,4) = 6 \times 5 \times 4 \times 3 = 360$.

19. The number of unordered lists of 3 items chosen from 7 is $C(7,3) = \dfrac{7 \times 6 \times 3}{3 \times 2 \times 1} = 35$.

21. Each 5-letter sequence containing the letters b, o, g, e, y is a list of 5 letters chosen from the above 5 (or a permutation of the 5 letters) and the number of these is

$P(5,5) = 5! = 5 \times 4 \times 3 \times 2 \times 1 = 120.$

23. Each 3-letter sequence is a list of 3 letters chosen from the given 6, and so the number of 3-letter sequences is

$P(6,3) = 6 \times 5 \times 4 = 120.$

25. Each 3-letter unordered set is a list of 3 letters chosen from the given 6, and so the number of 3-letter sets is

$C(6,3) = \dfrac{6 \times 5 \times 4}{3 \times 2 \times 1} = 20.$

27. Since there are repeated letters, we use a decision algorithm to construct such a sequence. Start with 6 empty slots.
Step 1: Choose a slot for the k: 6 choices.

Step 2: Choose 3 slots from the remaining 5 for the a's: $C(5,3)$ choices.

Step 3: Choose 2 slots from the remaining 2 for the u's: $C(2,2)$ slots.

Total number of sequences: $6 \times C(5,3) \times C(2,2) = 60$

29. There are a total of 10 marbles in the bag. The number of sets of 4 chosen from 10 marbles is $C(10,4) = 210$.

31. Decision algorithm for assembling a collection of 4 marbles that includes all the red ones:
Step 1: Choose 3 red marbles: $C(3,3) = 1$ choice.

Step 2: Choose 1 non-red marble: $C(7, 1) = 7$ choices. (There are 7 non-red marbles to choose from.)

Total number of sets $= C(3, 3)C(7, 1) = 7$

33. Decision algorithm for assembling a collection of 4 marbles that includes no red ones:

Step 1: Choose 4 non-red marbles: $C(7, 4) = 35$ choices. (There are 7 non-red marbles to choose from.)

Total number of sets $= C(7, 4) = 35$.

35. Decision algorithm for assembling a collection as specified:
Step 1: Choose 1 red marble: 3 choices.
Step 2: Choose 1 green marble: 2 choices.
Step 3: Choose 1 yellow marble: 2 choices.
Step 4: Choose 1 orange marble: 2 choices.

Total number of sets: $3 \times 2 \times 2 \times 2 = 24$

37. A set containing at least 2 red marbles must contain either 2 or 3 red marbles.
Alternative 1: Decide on 2 red marbles.

Step 1: Choose 2 red marbles: $C(3, 2)$ choices.

Step 2: Choose 3 non-red marbles: $C(7, 3)$ choices.

Alternative 3: Decide on 3 red marbles.

Step 1: Choose 3 red marbles: $C(3, 3)$ choices.

Step 2: Choose 2 non-red marbles: $C(7, 2)$ choices.

Total number of choices: $C(3, 2)C(7, 3) + C(3, 3)C(7, 2) = 126$

39. A set containing at most 1 yellow marbles must contain either 0 or 1 yellow marbles.
Alternative 1: Decide on 0 yellow marbles.

Step 1: Choose 5 non-yellow marbles: $C(8, 5)$ choices.

Alternative 2: Decide on 1 yellow marble.

Step 1: Choose 1 yellow marble: $C(2, 1)$ choices.

Step 2: Choose 4 non-yellow marbles: $C(8, 4)$ choices.

Total number of choices: $C(8, 5) + C(2, 1)C(8, 4) = 196$

41. Decision algorithm for assembling a collection as specified:
Alternative 1: Use the lavender marble but no yellow ones.

Step 1: Select the lavender marble: $C(1, 1) = 1$ choice.

Step 2: Select 4 non-lavender non-yellow marbles: $C(7, 4)$ choices.

Alternative 2: Use a yellow marble but no lavender ones.

Step 1: Select 1 yellow marble: $C(2, 1)$ choices.

Step 2: Select 4 non-lavender non-yellow marbles: $C(7, 4)$ choices.

Total number of sets: $C(1, 1)C(7, 4) + C(2, 1)C(7, 4) = 105$

43. Think of the outcome of a sequence of 30 dice throws as a sequence of "words" of length 30 using the "letters" 1, 2, 3, ... , 6. Using this interpretation, a sequence with five 1s is a 30-letter word containing five 1s. For the decision algorithm, start with 30 empty slots and choose numbers to fill the slots:

Step 1: Choose five slots for the 1s: $C(30, 5)$ choices.

Steps 2–26: Choose one of 2, 3, 4, 5, 6 to fill each of the remaining 25 slots: $5 \times 5 \times ... \times 5 = 5^{25}$ choices.

Total number of sequences with five 1s: $C(30, 5) \times 5^{25}$

Since there are 6^{30} different sequences possible, the fraction of sequences with five 1s is

$$\frac{C(30, 5) \times 5^{25}}{6^{30}} \approx 0.192.$$

45. For the decision algorithm, start with 30 empty slots and choose numbers to fill the slots:

Step 1: Choose 15 slots for the even numbers: $C(30, 15)$ choices.

Next 15 steps: Choose $2, 4,$ or 6 for each of these 15 slots: 3^{15} choices.

Next 15 steps: Choose $1, 3,$ or 5 for each of the remaining 15 slots: 3^{15} choices.

Since there are 6^{30} different sequences possible, the fraction of sequences with exactly 15 even numbers is

$$\frac{C(30, 15) \times 3^{15} \times 3^{15}}{6^{30}} \approx 0.144.$$

47. Start with 11 empty slots. Decision algorithm:

Step 1: Select a slot for the m: $C(11, 1) = 11$ choices.

Step 2: Select 4 slots for the i's: $C(10, 4)$ choices.

Step 3: Select 6 slots for the s's: $C(6, 4)$ choices.

Step 4: Select 2 slots for the p's: $C(2, 2) = 1$ choice.

Total number of sequences: $C(11, 1)C(10, 4)C(6, 4)C(2, 2)$

49. Start with 11 empty slots. Decision algorithm:

Step 1: Select 2 slots for the m's: $C(11, 2)$ choices.

Step 2: Select 1 slot for the e: $C(9, 1) = 9$ choices.

Step 3: Select 1 slot for the g: $C(8, 1) = 8$ choices.

Step 4: Select 3 slots for the a's: $C(7, 3)$ choices.

Step 5: Select 1 slot for the l: $C(4, 1) = 4$ choices.

Step 6: Select 1 slot for the o: $C(3, 1) = 3$ choices.

Step 7: Select 1 slot for the n: $C(2, 1) = 2$ choices.

Step 8: Select 1 slot for the i: $C(1, 1) = 1$ choice.

Total number of sequences: $C(11, 2)C(9, 1)C(8, 1)C(7, 3)C(4, 1)C(3, 1)C(2, 1)C(1, 1)$

51. Start with 10 empty slots. Decision algorithm:

Step 1: Select 2 slots for the c's: $C(10, 2)$ choices.

Step 2: Select 4 slots for the a's: $C(8, 4)$ choices.

Step 3: Select 1 slot for the s: $C(4, 1) = 4$ choices.

Step 4: Select 1 slot for the b: $C(3, 1) = 3$ choices.

Step 5: Select 1 slot for the l: $C(2, 1) = 2$ choices.

Step 6: Select 1 slot for the n: $C(1, 1) = 1$ choice.

Total number of sequences: $C(10, 2)C(8, 4)C(4, 1)C(3, 1)C(2, 1)C(1, 1)$

Applications

53. Each itinerary is a list of 4 venues chosen from 4. Thus, the number of itineraries is

$P(4, 4) = 4! = 4 \times 3 \times 2 \times 1 = 24$.

55. Decision algorithm for constructing a two pair hand:

Step 1: Select 2 denominations for the pairs: $C(13, 2)$ choices.

Step 2: Select two cards of the lowest-ranked denomination above: $C(4, 2)$ choices.

Step 3: Select two cards of the other denomination above: $C(4, 2)$ choices.

Step 4: Select a single card that belongs to neither of the two denominations: $C(44, 1) = 44$ choices. $(52 - 8 = 44)$

Total number of two pair hands: $C(13, 2)C(4, 2)C(4, 2)C(44, 1) = 123,552$

57. Decision algorithm for constructing a two of a kind hand:

Step 1: Select a denomination for the two cards: $C(13, 1) = 13$ choices.

Step 2: Select 2 cards of the above denomination: $C(4, 2)$ choices.

Step 3: Select 3 denominations for the singles: $C(12, 3)$ choices.

Step 4: Select 1 card of the lowest-ranked denomination above: $C(4, 1) = 4$ choices.

Step 5: Select 1 card of the next-ranked denomination: $C(4, 1) = 4$ choices.

Step 6: Select 1 card of the highest-ranked: $C(4, 1) = 4$ choices.

Total number of two of a kind hands: $13 \times C(4, 2)C(12, 3) \times 4 \times 4 \times 4 = 1,098,240$

59. A straight is a run of 5 cards of consecutive denominations: A, 2, 3, 4, 5 up through 10, J, Q, K, A, but not all of the same suit. If we ignore the restriction about the suits, we get the following decision algorithm:

Step 1: Select a starting denomination for the straight: (A, 2, ..., 10): 10 choices.

Steps 2–6: Select a card of each of the above denominations: $4 \times 4 \times 4 \times 4 \times 4 = 4^5$ choices.

Total number of runs of 5 cards: $10(4^5$

However, these runs include hands in which all 5 cards are of the same suit (straight flushes). There are 10×4 of these (Step 1: Select a starting card; Step 2: Select a single suit). Excluding the straight flushes gives

Total number of straights: $10 \times 4^5 - 10 \times 4 = 10,200$.

61. a. Since a portfolio consists of a set of 5 stocks chosen from the 10, the number of possible portfolios is

$C(10, 5) = 252$ portfolios.

b. Decision algorithm for assembling a portfolio as per part (b):

Step 1: Choose VZ and MCD: $C(2, 2) = 1$ choice

Step 2: Choose 3 more stocks from 6 (VZ, MCD, KO, XOM excluded): $C(6, 3) = 20$

Total number of choices: $1 \times 20 = 20$ portfolios

c. Five of the listed stocks have yields above 3.5%. Decision algorithm for assembling a portfolio as per part (c):

Alternative 1: Exactly 4 stocks have yields above 3.5%:

Step 1: Choose 4 stocks with yields above 3.5%: $C(5, 4) = 5$ choices.

Step 2: Choose 1 stock with a yield not above 3.5%: $C(5, 1) = 5$ choices.

Total number of choices for Alternative 1: $5 \times 5 = 25$ choices.

Alternative 2: All 5 stocks have yields above 3.5%:

Step 1: Choose 5 stocks with yields above 3.5%: $C(5, 5) = 1$ choice.

Total number of choices for Alternative 2: 1 choice.

Total number of portfolios $= 25 + 1 = 26$

63. a. Decision algorithm for assembling a collection of stocks:

Step 1: Choose 3 tech stocks: $C(6, 3) = 20$ choices.

Step 2: Choose 2 non-tech stocks: $C(6, 2) = 15$ choices.

Total number of choices: $20 \times 15 = 300$ collections

b. Decision algorithm for assembling a collection of stocks that declined in value (there were 3 stocks in each category that declined):

Step 1: Choose 3 declining tech stocks: $C(3, 3) = 1$ choice.

Step 2: Choose 2 declining non-tech stocks: $C(3, 2) = 3$ possible choices.

Total number of choices: $1 \times 3 = 3$ collections

c. Since only 3 collections of out a possible 300 consist entirely of stocks that declined in value, the chances of selecting one of those collections is 3 in 300; that is, 1 in 100, or .01.

65. a. As a seeding consists of a list of 16 teams, the number of seedings is $P(16, 16) = 16!$

b. To choose a seeding of the designated type, we have the following decision algorithm:

Choose the order in which the games 1 vs 16, 2 vs 15, etc. appear: $P(8, 8) = 8!$

In each of these 8 games, choose the order of the teams: 2^8.

So, the number is $P(8, 8) \times 2^8 = 8! \times 2^8$.

67. There are 32 first-round games, 16 second-round games, 8 third round games, 4 fourth-round games, 2 fifth-round games, and one final, giving a total of 63 games.

Decision algorithm for picking the winners with 15 upsets in the first four rounds:

Step 1: Choose 15 games in which the upsets occur: $C(60, 15)$ choices

Step 2: Choose the winners for each of the three games in the last two rounds: $2 \times 2 \times 2 = 8$.

Total number of possiblties: $C(60, 15) \times 8 = 425{,}552{,}713{,}541{,}760$.

69. a. The number of groups of 4 chosen from 10 is $C(10, 4) = 210$.

b. Decision algorithm for assembling a group of 4 movies that satisfies the Lara twins:

Alternative 1: Exactly one of "Captain America" or "Thor"

Step 1: Choose 1 out of "Captain America" and "Thor" : $C(2, 1) = 2$ choices.

Step 2: Choose 3 from the remaining 7 ("Bridesmaids" is excluded): $C(7, 3) = 35$ choices.

Total number of groups in Alternative 1 is therefore $2 \times 35 = 70$.

Alternative 2: Both "Captain America" and "Thor"

Step 1: Choose 2 out of "Captain America" and "Thor" : $C(2, 2) = 1$ choice.

Step 2: Choose 2 from the remaining 7 ("Captain America", "Thor", and "Bridesmaids" excluded).

$C(7, 2) = 21$ choices.

Total number of groups in Alternative 2 is therefore $1 \times 21 = 21$.

Total number of groups that make the Lara twins happy is therefore $70 + 21 = 91$.

c. As 91 of the 210 groups (less than half) make the Lara twins happy, they are less likely than not to be satisfied with a random selection.

71. a. Each itinerary is a permutation of the 23 listed cities: 23! of them altogether.

b. Once the first five stops are determined, a decision algorithm for completing the itinerary is:

Steps 1–18: Select a different city from the remaining 18 for each remaining stop:

$18 \times 17 \times \ldots \times 2 \times 1$ choices.

Total number of itineraries: $18 \times 17 \times \ldots \times 2 \times 1 = 18!$

c. Decision algorithm for constructing an itinerary of the required type (think of making an itinerary as filling a sequence of 23 empty slots with different cities):

Step 1: Choose a starting point in the itinerary to insert the sequence of 5 named cities: 1 through 19: 19 choices.

Next 18 steps: Select a different city from the remaining 18 for each remaining slot: $18 \times 17 \times \ldots \times 2 \times 1$ choices.

Total number of itineraries: $19 \times 18!$

73. We wish to compute the number of groups of 4 chosen from 5 contestants (since we are excluding Ben and Ann): $C(5, 4) = 5$ possible groups. (Choice A).

75. Each handshake corresponds to a pair of people, so the question is really asking how many pairs of people there are in a group of 10: $C(10, 2) = 45$ (Choice D).

77. a. Note: In a set of 3 numbers, all 3 are different. The number of combinations = the number of sets of 3 numbers chosen from 40: $C(40, 3) = 9,880$.

b. Decision algorithm for constructing a combination in which a number appears twice:
Step 1: Select the number that appears twice: 40 choices.
Step 3: Select another number: 39 choices.

Total number of new combinations: $40 \times 39 = 1,560$

c. Decision algorithm for constructing a combination in which a number appears 3 times:
Step 1: Select the number that appears 3 times: 40 choices.
This gives 40 new combinations.

Total number of combinations: $9,880 + 1,560 + 40 = 11,480$

79. a. The number of combinations of 2 equations chosen from the 20 constraints is $C(20, 2) = 190$ systems of equations to solve.

b. Replace 20 by n, getting $C(n, 2)$.

Communication and reasoning exercises

81. You should choose the multiplication principle because the multiplication principle can be used to solve all problems that call for the formulas for permutations, as well as others.

83. A permutation is an ordered list, or sequence, and only choices (A) and (D) can be represented as ordered lists. (In a presidential cabinet, each portfolio is distinct.)

85. A permutation. Changing the order in a list of driving instructions can result in a different outcome; for instance, "1. Turn left. 2. Drive one mile." and "1. Drive one mile. 2. Turn left." will take you to different locations.

87. Urge your friend not to focus on formulas but instead to learn to formulate decision algorithms and use the principles of counting.

89. It is ambiguous on the following point: Are the three students to play different characters, or are they to play a group of three, such as "three guards"? This should be made clear in the exercise.

Chapter 6 Review

1. The negative integers greater than or equal to -3 are $-3, -2,$ and -1. Thus, $N = \{-3, -2, -1\}$

3. If the dice are distinguishable, then the outcomes can be thought of as ordered pairs. Thus, since the numbers are different, S is the set of all ordered pairs of distinct numbers in the range 1–6:

$S = \{(1,2), (1,3), (1,4), (1,5), (1,6), (2,1), (2,3), (2,4), (2,5), (2,6), (3,1), (3,2), (3,4), (3,5), (3,6),$

$(4,1), (4,2), (4,3), (4,5), (4,6), (5,1), (5,2), (5,3), (5,4), (5,6), (6,1), (6,2), (6,3), (6,4), (6,5)\}$

5. $A = \{a, b\}, B = \{b, c\}, S = \{a, b, c, d\}, B' = \{a, d\}$

$\quad A \cup B' = \{a, b\} \cup \{a, d\} = \{a, b, d\}$

$\quad A \times B' = \{(a, a), (a, d), (b, a), (b, d)\}$

7. The set of outcomes when a day in August and a time of that day are selected is the set of pairs (day in August, time). That is, it is the set $A \times B$.

9. The set of all integers that are not odd perfect squares is the set of integers other than those in E' ($E' = $ the odd integers) and Q: $(E' \cap Q)'$ or $E \cup Q'$

11. Let S be the set of all novels in your home; $n(S) = 400$.

Let A be the event that you have read a novel: $n(A) = 150$.

Let B be the event that Roslyn has read a novel: $n(A) = 200$.

We are also told that $n(A \cap B) = 50$.

To compute $n(A \cup B)$ we use

$\quad n(A \cup B) = n(A) + n(B) - n(A \cap B) = 150 + 200 - 50 = 300.$

The event that neither of you has read a novel is $(A \cup B)'$, and its cardinality is given by

$\quad n[(A \cup B)'] = n(S) - n(A \cup B) = 400 - 300 = 100.$

Note that the formula used for the last calculation is $n(C') = n(S) - n(C)$.

13. The number of outcomes from rolling the dice is $n(A(B))$, where A is the set of outcomes of the red dice and B is the set of outcomes of the green one.

$\quad n(A(B)) = n(A)n(B) = 6 \cdot 6 = 36$

The set of losing combinations has the form $A \cup B$, where here A is the set of doubles and B is the set of outcomes in which the green die shows an odd number and the red die shows an even number. To compute its cardinality, we use

$\quad n(A \cup B) = n(A) + n(B) - n(A \cap B) = 6 + 3 \times 3 - 0 = 15.$

The set of winning combinations is the complement of the set C of losing combinations and is given by

$\quad n(C') = n(S) - n(C) = 36 - 15 = 21.$

15. Decision algorithm for constructing a two of a kind hand with no Aces:

$\quad$ Step 1: Select a denomination other than Ace for the pair: $C(12, 1)$ choices.

$\quad$ Step 2: Select 2 cards of that denomination: $C(4, 2)$ choices.

Step 3: Select 3 denominations (other than Ace or the one already selected above) for the remaining singles: $C(11, 3)$ choices.

Step 4: Select a suit for the highest-ranked denomination just chosen: $C(4, 1)$ choices.

Step 5: Select a suit for the next highest-ranked denomination just chosen: $C(4, 1)$ choices.

Step 6: Select a suit for the lowest-ranked denomination just chosen: $C(4, 1)$ choices.

Total number of hands: $C(12, 1)C(4, 2)C(11, 3)C(4, 1)C(4, 1)C(4, 1)$

17. Decision algorithm for constructing a straight flush:

Step 1: Select a suit for the flush: $C(4, 1)$ choices.

Step 2: Select a starting denomination for the consecutive run: $C(10, 1)$ choices.

(A, 2, 3, 4, 5 up through 10, J, Q, K, A)

Total number of hands: $C(4, 1)C(10, 1)$

19. There are a total of $4 + 2 = 6$ red and green marbles. Thus, there are $C(6, 5) = 6$ possible sets of such marbles.

21. Decision algorithm for constructing a set of five marbles that include all the red ones:

Step 1: Select 4 red marbles: $C(4, 4)$ choices.

Step 2: Select 1 non-red marble: $C(8, 1)$ choices.

Total number of choices: $C(4, 4)C(8, 1) = 8$

23. Decision algorithm for constructing a set of 5 marbles that include at least 2 yellow ones:
Alternative 1: Use 2 yellow ones:

Step 1: Select 2 yellow marbles: $C(3, 2)$ choices.

Step 2: Select 3 non-yellow marbles: $C(9, 3)$ choices.

Alternative 2: Use 3 yellow ones:

Step 1: Select 3 yellow marbles: $C(3, 3)$ choices.

Step 2: Select 2 non-yellow marbles: $C(9, 2)$ choices.

Total number of choices: $C(3, 2)C(9, 3) + C(3, 3)C(9, 2) = 288$

25. $S \cup T$ is the set of books that are either sci fi or stored in Texas (or both);

$n(S \cup T) = n(S) + n(T) - n(S \cap T) = 33,000 + 94,000 - 15,000 = 112,000$

27. $C \cup S'$ is the set of books that are either stored in California or not sci fi. To compute its cardinality, add the corresponding entries (shown below in thousands):

	S	H	R	O	Total
W	10	12	12	30	64
C	8	12	6	16	42
T	15	15	20	44	94
Total	33	39	38	90	20

$n(C \cup S') = 175,000$

29. $R \cap (T \cup H)$ is the set of romance books that are also horror books or stored in Texas.

	S	H	R	O	Total
W	10	12	12	30	64
C	8	12	6	16	42
T	15	15	20	44	94
Total	33	39	38	90	20

$n(R \cap (T \cup H)) = 20,000$

31. Let H denote the set of OHaganBooks.com customers, J the set of JungleBooks.com customers, and F the set of FarmerBooks.com customers.

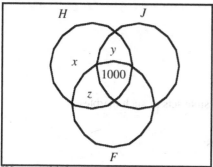

We are told that

OHaganBooks.com had 3500 customers: $x + y + z + 1,000 = 3,500.$

2,000 customers were shared with JungleBooks.com: $y + 1,000 = 2,000, \Rightarrow y = 1,000.$

1,500 customers were shared with FarmerBooks.com: $z + 1,000 = 1,500 \Rightarrow z = 500.$

We can now compute x from the first equation: $x + 1,000 + 500 + 1,000 = 3,500 \Rightarrow x = 1,000.$

We are asked for the number that are exclusive OHaganBooks.com customers:$x = 1,000.$

33. Let H denote the set of OHaganBooks.com customers, J the set of JungleBooks.com customers, and F the set of FarmerBooks.com customers.
Let us start with the information from Exercise 31 filled in:

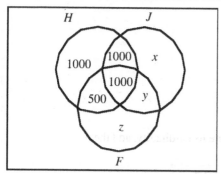

JungleBooks.com has a total of 3,600 customers: $x + y + 2,000 = 3,600 \Rightarrow x + y = 1,600.$

FarmerBooks.com has 3,400 customers: $y + z + 1,500 = 3,400 \Rightarrow y + z = 1,900.$

JungleBooks.com and FarmerBooks.com share 1,100 customers between them:

$y + 1,000 = 1,100 \Rightarrow y = 100$

giving

$x + 100 = 1,600 \Rightarrow x = 1,500$

$100 + z = 1,900 \Rightarrow z = 1,800.$

Thus, OHaganBooks has 1,000 exclusive customers, JungleBooks has $x = 1,500$ exclusive customers, and FarmerBooks has $z = 1,800$ exclusive customers—the most.

35. Let H denote the set of OHaganBooks.com customers, J the set of JungleBooks.com customers, and F the set of FarmerBooks.com customers.
Here is the complete data from Exercise 31 above:

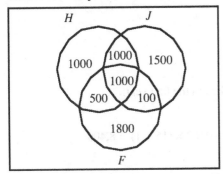

$n(H \cup J) = 3,500 + 1,500 + 100 = 5,100$

$n(H \cup F) = 3,500 + 100 + 1,800 = 5,400$

Thus, OHaganBooks should merge with FarmerBooks.com for a combined customer base of 5\,400.

37. Decision algorithm for constructing a 3-letter code:
    Step 1: Choose the 1st letter: 26 choices.
    Step 2: Choose the 2nd letter: 26 choices.
    Step 3: Choose the 3rd letter: 26 choices.
Total number of codes: $26 \times 26 \times 26 = 26^3 = 17,576$

39. Decision algorithm for constructing a code:
    Step 1: Choose the 1st letter: 26 choices.
    Step 2: Choose the 2nd letter: 25 choices.
    Step 3: Choose the 1st digit: 9 choices.
    Step 4: Choose the 2nd digit: 10 choices.
Total number of codes: $26 \times 25 \times 9 \times 10 = 58,500$

41. Decision algorithm for constructing a course schedule meeting the minimum requirements:

    Step 1: Select 3 liberal arts courses: $C(6, 3)$ choices.

    Step 2: Select 3 math courses: $C(6, 3)$ choices.

        (Note: At this stage, you have must select a minimum of 4 courses to make up the required 10, and so you must select 2 sciences and 2 fine arts.)

    Step 3: Select 2 science courses: $C(5, 2)$ choices.

    Step 4: Select 2 fine arts courses: $C(6, 2)$ choices.

Total number of schedules: $C(6, 3)C(6, 3)C(5, 2)C(6, 2) = 20 \times 20 \times 10 \times 15 = 60,000$

43. Decision algorithm for constructing a course schedule meeting all of the above requirements, and the physics complication:
    Alternative 1: Take neither Physics I nor Physics II.

        Step 1: Select 3 liberal arts courses: $C(6, 3)$ choices.

        Step 2: Select Calc I: 1 choice.

        Step 3: Select 2 other math courses: $C(5, 2)$ choices.

Step 4: Select 2 science courses other than Physics I and II: $C(3, 2)$ choices.

Step 5: Select 2 fine arts courses other than the single pair that cannot be taken in the first year:

$C(6, 2) - 1$ choices

Total number of schedules for Alternative 1: $C(6, 3)C(5, 2)C(3, 2)[C(6, 2) - 1] = 20 \times 10 \times 3 \times (15 - 1) = 8{,}400$

Alternative 2: Take Physics I but not Physics II.

Step 1: Select 3 liberal arts courses: $C(6, 3)$ choices.

Step 2: Select Calc I: 1 choice.

Step 3: Select 2 other math courses: $C(5, 2)$ choices.

Step 4: Select Physics I: 1 choice.

Step 5: Select 1 science course other than Physics I and II: $C(3, 1)$ choices.

Step 6: Select 2 fine arts courses other than the single pair that cannot be taken in the first year:

$C(6, 2) - 1$ choices.

Total number of schedules for Alternative 1: $C(6, 3)C(5, 2)C(3, 1)[C(6, 2) - 1] = 20 \times 10 \times 3 \times (15 - 1) = 8{,}400$

Alternative 3: Take Physics I and Physics II.

Step 1: Select 3 liberal arts courses: $C(6, 3)$ choices.

Step 2: Select Calc I: 1 choice.

Step 3: Select 2 other math courses: $C(5, 2)$ choices.

Step 4: Select Physics I & II: 1 choice.

Step 5: Select 2 fine arts courses other than the single pair that cannot be taken in the first year:

$C(6, 2) - 1$ choices.

Total number of schedules for Alternative 1: $C(6, 3)C(5, 2)[C(6, 2) - 1] = 20 \times 10 \times (15 - 1) = 2{,}800$

Total number of course schedules overall: $8{,}400 + 8{,}400 + 2{,}800 = 19{,}600$

Section 7.1

1. $S = \{HH, HT, TH, TT\}$; $E = \{HH, HT, TH\}$

3. $S = \{HHH, HHT, HTH, HTT, THH, THT, TTH, TTT\}$; $E = \{HTT, THT, TTH, TTT\}$

5. $S = \left\{\begin{array}{cccccc} (1,1), & (1,2), & (1,3), & (1,4), & (1,5), & (1,6), \\ (2,1), & (2,2), & (2,3), & (2,4), & (2,5), & (2,6), \\ (3,1), & (3,2), & (3,3), & (3,4), & (3,5), & (3,6), \\ (4,1), & (4,2), & (4,3), & (4,4), & (4,5), & (4,6), \\ (5,1), & (5,2), & (5,3), & (5,4), & (5,5), & (5,6), \\ (6,1), & (6,2), & (6,3), & (6,4), & (6,5), & (6,6) \end{array}\right\}$ $E = \{(1,4), (2,3), (3,2), (4,1)\}$

7. $S = \left\{\begin{array}{cccccc} (1,1), & (1,2), & (1,3), & (1,4), & (1,5), & (1,6), \\ & (2,2), & (2,3), & (2,4), & (2,5), & (2,6), \\ & & (3,3), & (3,4), & (3,5), & (3,6), \\ & & & (4,4), & (4,5), & (4,6), \\ & & & & (5,5), & (5,6), \\ & & & & & (6,6) \end{array}\right\}$ $E = \{(1,3), (2,2)\}$

9. S same as Exercise 7; $E = \{(2,2), (2,3), (2,5), (3,3), (3,5), (5,5)\}$

11. $S = \{m, o, z, a, r, t\}$; $E = \{o, a\}$

13. There are $4 \times 3 = 12$ possible sequences of 2 different letters chosen from *sore*:

$S = \{(s,o), (s,r), (s,e), (o,s), (o,r), (o,e), (r,s), (r,o), (r,e), (e,s), (e,o), (e,r)\}$;

$E = \{(o,s), (o,r), (o,e), (e,s), (e,o), (e,r)\}$.

15. There are $5 \times 4 = 20$ possible sequences of 2 different digits chosen from 0–4:

$S = \{01, 02, 03, 04, 10, 12, 13, 14, 20, 21, 23, 24, 30, 31, 32, 34, 40, 41, 42, 43\}$

$E = \{10, 20, 21, 30, 31, 32, 40, 41, 42, 43\}$.

17. $S = \{\text{domestic car, imported car, van, antique car, antique truck}\}$; $E = \{\text{van, antique truck}\}$

19. a. The sample space is the set of all sets of 4 gummy candies chosen from the packet of 12.

b. The event that April will get the combination she desires is the set of all sets of 4 gummy candies in which two are strawberry and two are black currant.

21. a. The sample space is the set of all lists of 15 people chosen from 20.
b. The event that Hillary Clinton is the Secretary of State is the set of all lists of 15 people chosen from 20, in which Hillary Clinton occupies the eleventh (Secretary of State) position.

23. A: The red die shows 1; B: The numbers add to 4.

"The red die shows 1 and the numbers add to 4" is the event $A \cap B$.   And = ∩

There is only one possible outcome in $A \cap B$: $A \cap B = \{(1, 3)\}$.

Therefore, $n(A \cap B) = 1$.

25. B: The numbers add to 4.

"The numbers do not add to 4" is the event B'.

$B = \{(1, 3), (2, 2), (3, 1)\}; \; n(B') = n(S) - n(B) = 36 - 3 = 33$

27. "The numbers do not add to 4" is the event B'. "The numbers add to 11" is the event D'.

Therefore, "The numbers do not add to 4 *but* they do add to 11" is the event $B' \cap D'$.

$B' \cap D' = \{(6, 5), (5, 6)\}$, and so $n(B' \cap D') = 2$

29. "At least one of the numbers is 1" is the event C. "The numbers add to 4" is the event B.

Therefore, "At least one of the numbers is 1 or the numbers add to 4" is the event $C \cup B$.

$C \cup B = \{(1, 1), (1, 2), (1, 3), (1, 4), (1, 5), (1, 6), (6, 1), (5, 1), (4, 1), (3, 1), (2, 1), (2, 2)\}$

$n(C \cup B) = 12$

31. W: You will use the Website tonight. I: Your math grade will improve.

Therefore, "You will use the Website tonight *and* your math grade will improve." is the event $W \cap I$.

And =∩

33. E: You will use the Website every night. I: Your math grade will improve.

Therefore, "Either you will use the Website every night, *or* your math grade will not improve." is the event $E \cup I'$.

Or = ∪

35. I: Your math grade will improve. W: You will use the Website tonight. E: You will use the Website every night.

The given statement is: "Either your math grade will improve, or you will use the Website tonight *but* you will not use it every night."

The comma after "improve" breaks the statement into two pieces:

     Either

    your math grade will improve: I

      or

    you will use the Website tonight *but* you will not use it every night: $(W \cap E')$.

Therefore, the given statement is the event $I \cup (W \cap E')$.

37. *I*: Your math grade will improve. *W*: You will use the Website tonight. *E*: You will use the Website every night.

The given statement is: "Either your math grade will improve *or* you will use the Website tonight, *but* you will not use it not every night."

The comma after "tonight" breaks the statement into two pieces:

Either your math grade will improve *or* you will use the Website tonight, $(I \cup W)$

but

you will not use it every night: E'.

Therefore, the given statement is the event $(I \cup W) \cap E'$.

39. The sample space consists of all sets of 3 marbles chosen from 8. $n(S) = C(8, 3) = 56$

The event consists of all sets of 3 non-red marbles chosen from 4. $n(E) = C(4, 3) = 4$

41. Using the multiplication principle: $n(E) = C(4, 1) \times C(2, 1) \times C(2, 1) = 16$

Applications

43. Looking at the map, we find the following regions that saw an increase in housing prices of 6% or more: Pacific, Mountain, West South Central, South Atlantic. Therefore, $E = \{$Pacific, Mountain, West South Central, South Atlantic$\}$.

45. *E*: The region you choose saw an increase in housing prices of 6% or more.

F: The region you choose is on the east coast.

$E \cup F$ translates to *E or F*: You choose a region that saw an increase in housing prices of 6% or more *or* is on the east coast.

$E \cup F = \{$Pacific, Mountain, West South Central, New England, Middle Atlantic, South Atlantic$\}$

$E \cap F$ translates to *E and F*: You choose a region that saw an increase in housing prices of 6% or more *and* is on the east coast.

$E \cap F = \{$South Atlantic$\}$

47. Two events are mutually exclusive if they have no outcomes in common.

a. *E*: You choose a region from among the two with the highest percentage increase in housing prices.

   $E = \{$Mountain, Pacific$\}$

F: You choose a region that is not on the east or west coast.

$F = \{$Mountain, West North Central, West South Central, East North Central, East South Central$\}$

Since $E \cap F = \{$Mountain$\}$, $E \cap F \neq \emptyset$, they are not mutually exclusive.

b. *E*: You choose a region from among the two with the highest increase in housing prices.

   $E = \{$Mountain, Pacific$\}$

F: You choose a region that is on the east coast.

   $F = \{$New England, Middle Atlantic, South Atlantic$\}$

Since $E \cap F = \emptyset$, they are mutually exclusive.

49. *S*: an author is successful.   *N*: an author is new.

$S \cap N$ is the event that an author is successful and new.

$S \cup N$ is the event that an author is either successful or new

$n(S \cap N) = 5$ (see table)

$n(S \cup N) = 45$ (see table)

	N	E	Total
S	5	25	30
U	15	55	70
Total	20	80	100

	N	E	Total
S	5	25	30
U	15	55	70
Total	20	80	100

51. Since N and E have no outcomes in common (an author cannot be both new and established), they are mutually disjoint.

N and S are not mutually disjoint, since $n(N \cap S) = 5$ (see table).

S and E are not mutually disjoint, since $n(S \cap E) = 25$ (see table).

	N	E	Total
S	5	25	30
U	15	55	70
Total	20	80	100

53. S: an author is successful. N: an author is new.

$S \cap N'$ is the event that an author is successful but not a new author.

$n(S \cap N') = 25$ (see table)

	N	$E = N'$	Total
S	5	25	30
U	15	55	70
Total	20	80	100

55. The total number of established authors is 80. Of these, 25 are successful. Thus, the percentage that are successful is

$$\frac{25}{80} = 31.25\%.$$

The total number of successful authors is 30. Of these, 25 are established. Thus, the percentage that are established is

$$\frac{25}{30} \approx 83.33\%.$$

57. "An industry increased in value but was not in the manufacturing sector" is the event $X \cap M'$.

	X	Y	Z	**Totals**
F	3	4	1	8
M	8	3	3	14
T	6	1	0	7
H	4	1	1	6
U	3	1	1	5
Totals	24	10	6	40

$$n(X \cap M') = 24 - 8 = 16$$

59. $H' \cup Z = \{\text{Health care}\}' \cup \{\text{Unchanged in value}\}$; the event that an industry was either not in the health care sector or was unchanged in value (or both).

$$n(H' \cup Z) = 40 - (4 + 1) = 35 \text{ (see table)}$$

	X	Y	Z	**Totals**
F	3	4	1	8
M	8	3	3	14
T	6	1	0	7
H	4	1	1	6
U	3	1	1	5
Totals	24	10	6	40

61. For a pair of events to be mutually exclusive, they must have no outcomes in common. Three kinds can be identified:

(1) Pairs of events corresponding to separate rows: F and M, F and T, M and T

(2) Pairs of events corresponding to separate columns: X and Y, Y and Z, X and Z

(3) Pairs corresponding to a row and column whose intersection is empty: T and Z

63. E: The dog's flight drive is strongest. H: The dog's fight drive is weakest.

 G: The dog's fight drive is strongest.

a. "The dog's flight drive is not strongest and its fight drive is weakest" is the event $E' \cap H$.

b. "The dog's flight drive is strongest or its fight drive is weakest" is the event $E \cup H$.

c. "Neither the dog's flight drive nor fight drive are strongest" is the event $E' \cap G' = (E \cup G)'$.

65. a. The dog's fight and flight drives are both strongest: Intersection of rightmost column and bottom row: {9}
b. The dog's fight drive is strongest, but its flight drive is neither weakest nor strongest: Intersection of rightmost column and middle row: {6}

67. a. {1, 4, 7} This is the leftmost column: The dog's fight drive is weakest.
b. {1, 9}: The dog's fight and flight drives are either both strongest or both weakest.
c. {3, 6, 7, 8, 9} This is the union of the rightmost column and the bottom row: Either the dog's fight drive is strongest, or its flight drive is strongest.

69. S is the set of sets of 4 gummy bears chosen from 6; $n(S) = C(6, 4) = 15$.

Decision algorithm for constructing a set including the raspberry gummy bear:

    Step 1: Select the raspberry one: $C(1, 1) = 1$ choice

    Step 2: Select 3 non-raspberry ones: $C(5, 3) = 10$ choices

Total number of choices: $n(E) = 1 \times 10 = 10$

71. a. Decision algorithm for constructing a finish (winner, second place and third place) for the race:
    Step 1: Select the winner: 7 choices
    Step 2: Select second place: 6 choices
    Step 3: Select third place: 5 choices

Total number of finishes: $n(S) = 7 \times 6 \times 5 = 210$

b. E: Electoral College is in second or third place. F: Celera is the winner.

Therefore, $E \cap F$ is the event that Celera wins and Electoral College is in second or third place. In other words, it is the set of all lists of three horses in which Celera is first and Electoral College is second or third.

Decision algorithm for constructing a finish in $E \cap F$:

    Alternative 1: Electoral College is second:
        Step 1: Select Celera as the winner: 1 choice
        Step 2: Select Electoral College for second place: 1 choice
        Step 3: Select third place: 5 choices

    Total number of finishes for Alternative 1: $1 \times 1 \times 5 = 5$

    Alternative 2: Electoral College is third:
        Step 1: Select Celera as the winner: 1 choice
        Step 2: Select Electoral College for third place: 1 choice
        Step 3: Select second place: 5 choices

Total number of finishes for Alternative 2: $1 \times 1 \times 5 = 5$

Therefore, $n(E \cap F) = 5 + 5 = 10$.

Communication and reasoning exercises

73. An event is a <u>subset of the sample space.</u>

75. $(E \cap F)'$ is the complement of the event $E \cap F$. Since $E \cap F$ is the event that both E and F occur, $(E \cap F)'$ is the event that <u>E and F do not both occur.</u>

77. Choice (B): The event should be a subset of the sample space, and only (B) has the property (the collection of tall, dark strangers you meet is a subset of the set of people you meet).

79. True; Consider the following experiment: Select an element of the set S at random. Then the sample space is the set of elements of S. In other words, the sample space is S.

81. Answers may vary. Cast a die, and record the remainder when the number facing up is divided by 2.

83. Yes. For instance, $E = \{(2,5),(5,1)\}$ and $F = \{(4,3)\}$ are two such events.

Section 7.2

1. $P(E) = \dfrac{fr(E)}{N} = \dfrac{40}{100} = .4$

3. $N = 800$, $fr(E) = 640$; $P(E) = \dfrac{fr(E)}{N} = \dfrac{640}{800} = .8$

5. $E = \{1, 2, 3, 4\}$; $P(E) = \dfrac{fr(E)}{N} = \dfrac{8 + 8 + 8 + 12}{60} = \dfrac{36}{60} = .6$

7. We obtain the relative frequency distribution by dividing the frequencies by $N = 4{,}000$:

Outcome	Frequency	Relative Frequency
HH	1,100	$\dfrac{1{,}100}{4{,}000} = .275$
HT	950	$\dfrac{950}{4{,}000} = .2375$
TH	1,200	$\dfrac{1{,}200}{4{,}000} = .3$
TT	750	$\dfrac{750}{4{,}000} = .1875$

9. Refer to the relative frequency distribution in the solution to Exercise 7. The event the that the second coin lands with heads up is

$E = \{HH, TH\} \Rightarrow P(E) = .275 + .3 = .575.$

11. From the solution to Exercise 9, the second coin comes up heads approximately 58% of the time. Therefore, the second coin *seems* slightly biased in favor of heads, especially in view of the fact that the coin was tossed 4,000 times. On the other hand, it is conceivable that the coin is fair and that heads came up 58% of the time purely by chance. Deciding which conclusion is more reasonable requires some knowledge of inferential statistics.

13. Yes: Since the relative frequencies are between 0 and 1 (inclusive) and add up to 1, the given distribution can be a relative frequency distribution.

15. No: Relative frequencies cannot be negative.

17. Yes: Since the relative frequencies are between 0 and 1 (inclusive) and add up to 1, the given distribution can be a relative frequency distribution.

19. Relative frequency distribution:

Outcome	1	2	3	4	5
Rel. Frequency	.2	.3	.1	.1	x

Since the probabilities of all the outcomes add to 1,

$.2 + .3 + .1 + .1 + x = 1 \Rightarrow .7 + x = 1 \Rightarrow x = .3.$

a. From the properties of relative frequency distributions,

$P(\{1,3,5\}) = P(1) + P(3) + P(5) = .2 + .1 + .3 = .6.$

b. Since $E = \{1,2,3\}$, $E' = \{4,5\}$,

$P(E') = P(\{4,5\}) = P(4) + P(5) = .1 + .3 = .4$

21. Use the following formulas to generate binary digits (0 represents heads, say, and 1 represents tails):
TI-83/84 Plus: `randInt(0,1)` To obtain `randInt`, follow [MATH] → `PRB`.
Spreadsheet: `=RANDBETWEEN(0,1)`
Answers will vary. The estimated probability that heads comes up should be around .5.

23. Simulate tossing two coins by generating two random binary digits (0 represents heads, say, and 1 represents tails). If the outcome is one head and one tail, the digits should add to 1. Otherwise, they will add to 0 or 2. The sum of these digits can be obtained as follows:

TI-83/84 Plus: `randInt(0,1)+randInt(0,1)`
To obtain `randInt`, follow [MATH] → `PRB`.
Spreadsheet: `=RANDBETWEEN(0,1)+RANDBETWEEN(0,1)`

Then count how many 1s you get.
Answers will vary. The estimated probability that the outcome is one head and one tail should be around .5.

Applications

25. a. $P(E) = \dfrac{fr(E)}{N} = \dfrac{270}{500} = .54$

b. $P(E) = \dfrac{fr(E)}{N} = \dfrac{70 + 10}{500} = \dfrac{80}{500} = .16$

c. $P(E) = \dfrac{fr(E)}{N} = \dfrac{500 - 10}{500} = \dfrac{490}{500} = .98$

27. a. We obtain the relative frequency distribution by dividing the frequencies by $N = 134 + 52 + 9 + 5 = 200$:

Outcome	Frequency	Rel. Frequency
Current	134	$\dfrac{134}{200} = .67$
Past Due	52	$\dfrac{52}{200} = .26$
In Foreclosure	9	$\dfrac{9}{200} = .045$
Repossessed	5	$\dfrac{5}{200} = .025$

b. Take E to be the event that a randomly selected subprime mortgage in Texas was not current.

$E = \{$Past Due, In Foreclosure, Repossessed$\}$; $P(E) = .26 + .045 + .025 = .33$

29. a. Using the result of Quick Example #3, the relative frequency distribution is obtained by converting the percentages into decimals:

Age	0–14	15–29	30–64	> 64
Percentage	.30	.27	.37	.06

b. The event E that a resident of Mexico is *not* between 15 and 64 is shown by the shaded parts of the distribution:

Age	0–14	15–29	30–64	> 64
Percentage	.30	.27	.37	.06

Thus the relative frequency is $P(E) = .30 + .06 = .36$.

31. a. We obtain the relative frequency distribution by dividing the frequencies by $N = 1 + 4 + 4 + 1 = 10$:

Outcome	Frequency	Rel. Frequency
3	1	$\dfrac{1}{10} = .1$
2	4	$\dfrac{4}{10} = .4$
1	4	$\dfrac{4}{10} = .4$
0	1	$\dfrac{1}{10} = .1$

b. Take E to be the event that a randomly selected small SUV will have a crash test rating of "Acceptable" (2) or better.

$E = \{2, 3\}; P(E) = .4 + .1 = .5$

33. The number of households is $N = 2,000$. To calculate $fr(E)$ in each case, we use

$$P(E) = \frac{fr(E)}{N}, \quad \text{giving } fr(E) = P(E) \times N.$$

Here is a chart showing the frequencies and their calculation:

Outcome	Dial-up	Cable Modem	DSL	Other
Rel. Frequency	.628	.206	.151	.015
Frequency	$.628 \times 2,000 = 1,256$	$.206 \times 2,000 = 412$	$.151 \times 2,000 = 302$	$.015 \times 2,000 = 30$

35. Using the given classification of the data, we get the following:

Outcome	Surge	Plunge	Steady
Frequency	4	6	10
Rel. Frequency	$\dfrac{4}{20} = .2$	$\dfrac{6}{20} = .3$	$\dfrac{10}{20} = .5$

37. $P(E \cap S) = \dfrac{fr(E \cap S)}{N} = \dfrac{25}{100} = .25$

	N	E	Total
S	5	25	30
U	15	55	70
Total	20	80	100

39. $P(N) = \dfrac{fr(N)}{N} = \dfrac{20}{100} = .2$

	N	E	Total
S	5	25	30
U	15	55	70
Total	20	80	100

41. $P(U) = \dfrac{fr(U)}{N} = \dfrac{70}{100} = .7$

	N	E	Total
S	5	25	30
U	15	55	70
Total	20	80	100

43. Restrict attention to the successful authors only:

$P(\text{Successful author is established}) = \dfrac{25}{30} = \dfrac{5}{6}$

	N	E	Total
S	5	25	30

45. Restrict attention to the established authors only:

$P(\text{Established author is successful}) = \dfrac{25}{80} = \dfrac{5}{16}$

	E
S	25
U	55
Total	80

47. $P(\text{Contaminated}) = .8$

Therefore, $P(U) = 1 - .8 = .2$.

Of the contaminated chicken, 20% had the strain resistant to antibiotics. Therefore,

$\qquad P(R) = .2 \times .8 = .16 \qquad$ (20% of 80%).

The other 80% of contaminated chicken is not resistant. Therefore,

$\qquad P(C) = .8 \times .8 = .64 \qquad$ (80% of 80%).

This gives the following relative frequency distribution:

Outcome	U	C	R
Rel. Freq.	.2	.64	.16

49. Conventionally grown produce:

The events are: No pesticide (NP), Single pesticide (SP), Multiple pesticides (MP).

We are told that 73% of conventionally grown foods had residues from at least one pesticide. Therefore,

$P(NP) = 1 - .73 = .27.$

We are also told that conventionally grown foods were 6 times as likely to contain multiple pesticides as organic foods, and that 10% of organic foods had multiple pesticides. Therefore,

$P(MP) = 6 \times .10 = .60.$

This leaves

$P(SP) = 1 - (.27 + .60) = .13.$

Probability distribution for conventionally grown produce:

Outcome	NP	SP	MP
Probability	.27	.13	.60

Organic produce:

We are told that $P(MP) = .10$ and that 23% had either single or multiple pesticide residues. Therefore,

$P(SP) = .23 - .10 = .13.$

This leaves

$P(NP) = 1 - (.10 + .13) = .77.$

Probability distribution for organic produce:

Outcome	NP	SP	MP
Probability	.77	.13	.10

51. $P(\text{False negative}) = \dfrac{10}{400} = .025$ $P(\text{False positive}) = \dfrac{10}{200} = .05$

53. Answers will vary.

Communication and reasoning exercises

55. The estimated probability of an event E is defined to be <u>the fraction of times E occurs</u>.

57. 101; $fr(E)$ can be any number beween 0 and 100 inclusive, so the possible answers are $0/100 = 0$, $1/100 = .01$, $2/100 = .02$, ..., $99/100 = .99$, $100/100 = 1$.

59. Wrong. For a pair of fair dice, the probability of a pair of matching numbers is 1/6, as Ruth says. However, it is quite possible, although not very likely, that if you cast a pair of fair dice 20 times, you will never obtain a matching pair. (In fact, there is approximately a 2.6% chance that this will happen.) In general, a nontrivial claim about theoretical probability can never be absolutely validated or refuted experimentally. All we can say is that the evidence suggests that the dice are not fair.

61. For a (large) number of days, record the temperature prediction for the next day, and then check the actual high temperature the next day. Record whether the prediction was accurate (within, say, 2°F of the actual temperature). The fraction of times the prediction was accurate is the relative frequency.

Section 7.3

1.

Outcome	a	b	c	d	e
Probability	.1	.05	.6	.05	x

Since the probabilities of all the outcomes add to 1,

$$.1 + .05 + .6 + .05 + x = 1 \Rightarrow .8 + x = 1 \Rightarrow x = .2.$$

a. $P(\{a, c, e\}) = P(a) + P(c) + P(e) = .1 + .6 + .2 = .9$

b. $P(E \cup F) = P(\{a, b, c, e\}) = P(a) + P(b) + P(c) + P(e) = .1 + .05 + .6 + .2 = .95$

c. $P(E') = P(\{b, d\}) = P(b) + P(d) = .05 + .05 = .1$

d. $P(E \cap F) = P\{(c, e)\} = P(c) + P(e) = .6 + .2 = .8$

3. $P(E) = \dfrac{n(E)}{n(S)} = \dfrac{5}{20} = \dfrac{1}{4}$

5. $P(E) = \dfrac{n(E)}{n(S)} = \dfrac{10}{10} = 1$

7. $n(S) = 4, \; n(E) = 3 \Rightarrow P(E) = \dfrac{n(E)}{n(S)} = \dfrac{3}{4}$

9. $S = \{HH, HT, TH, TT\}; \; n(S) = 4; \; E = \{HH, HT, TH\}; \; n(E) = 3 \Rightarrow P(E) = \dfrac{n(E)}{n(S)} = \dfrac{3}{4}$

11. $S = \{HHH, HHT, HTH, HTT, THH, THT, TTH, TTT\}; \; n(S) = 8; \; E = \{HTT, THT, TTH, TTT\}; \; n(E) = 4$

$$P(E) = \frac{n(E)}{n(S)} = \frac{4}{8} = \frac{1}{2}$$

13. $n(S) = 36, \; n(E) = 4 \Rightarrow P(E) = \dfrac{n(E)}{n(S)} = \dfrac{4}{36} = \dfrac{1}{9}$

15. $n(S) = 36, \; n(E) = 0 \Rightarrow P(E) = \dfrac{n(E)}{n(S)} = \dfrac{0}{36} = 0$

17. $n(S) = 36, \; E = \{(2, 2), (2, 3), (2, 5), (3, 2), (3, 3), (3, 5), (5, 2), (5, 3), (5, 5)\}; \; n(E) = 9$

$$P(E) = \frac{n(E)}{n(S)} = \frac{9}{36} = \frac{1}{4}$$

19. $E = \{(4, 4), (2, 3)\}$

The outcomes for a pair of indistinguishable dice are not equally likely, so we cannot use $P(E) = n(E)/n(S)$. Using the Example in the textbook,

$$P(4, 4) = \frac{1}{36}, \text{ and } P(2, 3) = \frac{1}{18}.$$

Therefore, $P(E) = \dfrac{1}{36} + \dfrac{1}{18} = \dfrac{1}{12}.$

The corresponding event for distinguishable dice is $\{(4, 4), (2, 3), (3, 2)\}$.

(We can also use this to compute $P(E)$: $P(E) = \dfrac{n(E)}{n(S)} = \dfrac{3}{36} = \dfrac{1}{12}$.)

21. Start with

Outcome	1	2	3	4	5	6
Probability	x	$2x$	x	$2x$	x	$2x$

Since the sum of the probabilities of the outcomes must be 1, we get

$$9x = 1, \text{ and so } x = \frac{1}{9}.$$

This gives:

Outcome	1	2	3	4	5	6
Probability	1/9	2/9	1/9	2/9	1/9	2/9

$$P(\{1, 2, 3\}) = \frac{1}{9} + \frac{2}{9} + \frac{1}{9} = \frac{4}{9}$$

23. Start with

Outcome	1	2	3	4
Probability	$8x$	$4x$	$2x$	x

Since the sum of the probabilities of the outcomes must be 1, we get

$$15x = 1, \text{ and so } x = \frac{1}{15}.$$

This gives:

Outcome	1	2	3	4
Probability	8/15	4/15	2/15	1/15

25. $P(A) = .1$, $P(B) = .6$, $P(A \cap B) = .05$

$P(A \cup B) = P(A) + P(B) - P(A \cap B) = .1 + .6 - .05 = .65$

27. $A \cap B = \varnothing$, $P(A) = .3$, $P(A \cup B) = .4$

$\quad P(A \cup B) = P(A) + P(B)$ (Mutually exclusive)

$\quad .4 = .3 + P(B) \Rightarrow P(B) = .4 - .3 = .1$

29. $A \cap B = \varnothing$, $P(A) = .3$, $P(B) = .4$

$\quad P(A \cup B) = P(A) + P(B) = .3 + .4 = .7$ (Mutually exclusive)

31. $P(A \cup B) = .9$, $P(B) = .6$, $P(A \cap B) = .1$

$\quad P(A \cup B) = P(A) + P(B) - P(A \cap B)$

$\quad .9 = P(A) + .6 - .1 \Rightarrow P(A) = .9 - .6 + .1 = .4$

33. $P(A) = .75 \Rightarrow P(A') = 1 - P(A) = 1 - .75 = .25$

35. $P(A) = .3, P(B) = .4, P(C) = .3$

Since $A, B,$ and C are mutually exclusive, $P(A \cup B \cup C) = P(A) + P(B) + P(C) = .3 + .4 + .3 = 1.$

37. $P(A) = .3, P(B) = .4$

Since A and B are mutually exclusive,

$\quad P(A \cup B) = P(A) + P(B) = .3 + .4 = .7 \Rightarrow P((A \cup B)') = 1 - P(A \cup B) = 1 - .7 = .3.$

39. Since $A \cup B = S, P(A \cup B) = 1.$

Since $A \cap B = \varnothing, A$ and B are mutually exclusive, so $P(A \cup B) = P(A) + P(B).$

Substituting $P(A \cup B) = 1$ gives $1 = P(A) + P(B).$

41. $P(A) = .2, P(B) = .1; P(A \cup B) = .4$

$P(A \cup B)$ is more than $P(A) + P(B)$. But, since $P(A \cup B) = P(A) + P(B) - P(A \cap B), P(A \cup B)$ cannot be more than

$P(A) + P(B)$. Therefore, the given information does not describe a probability distribution.

43. $P(A) = .2, P(B) = .4; P(A \cap B) = .2$

Since $P(A \cap B)$ is $\leq$ both $P(A)$ and $P(B),$ the given information is consistent with a probability distribution.

45. $P(A) = .1, P(B) = 0; P(A \cup B) = 0$

No; $P(A \cup B)$ should be $\geq P(A),$ since $A \cup B \supseteq A.$

Applications

47. a. $E = \{\text{Current, Past Due}\}; n(E) = 136{,}330 + 53{,}310 = 189{,}640$

$\quad P(E) = \dfrac{n(E)}{n(S)} = \dfrac{189{,}640}{203{,}480} \approx .93$

b. $E = \{\text{Past Due, In Foreclosure, Repossessed}\}; n(E) = 203{,}480 - 136{,}330 = 67{,}150$

$\quad P(E) = \dfrac{n(E)}{n(S)} = \dfrac{67{,}150}{203{,}480} \approx .33$

49. The probability distribution is obtained by converting the percentages to decimals:

Outcome	Hispanic or Latino	White (not Hispanic)	African American	Asian	Other
Probability	.48	.29	.08	.08	.07

$P(\text{neither White nor Asian}) = 1 - .29 - .08 = .63$

51. a. $S = \{\text{Stock market success, Sold to other concern, Fail}\}$

b. $P(\text{Stock market success}) = \dfrac{2}{10} = .2 \qquad P(\text{Sold to other concerns}) = \dfrac{3}{10} = .3 \qquad P(\text{Fail}) = 1 - (.2 + .3) = .5$

c. To realize profits for early investors, a start-up venture must be either a stock market success or sold to another concern, so

$$P(\text{Profit}) = .3 + .2 = .5.$$

53. The outcomes of the experiment are: SUV, pickup, passenger car, and minivan. The given information tells us that $P(\text{SUV}) = .25$ and $P(\text{pickup}) = .15$. We are also told that $P(\text{passenger car})$ is five times $P(\text{minivan})$. Take $P(\text{minivan})$ $= x$. Then we have:

Outcome	SUV	Pickup	Passenger Car	Minivan
Probability	.25	.15	$5x$	x

Since the sum of the probabilities must be 1, we get

$$.25 + .15 + 5x + x = 1 \Rightarrow .40 + 6x = 1 \Rightarrow 6x = 1 - .40 = .60 \Rightarrow x = .10.$$

So, $P(\text{minivan}) = .10$, and $P(\text{passenger car}) = 5x = .50$. This gives the required distribution:

Outcome	SUV	Pickup	Passenger Car	Minivan
Probability	.25	.15	.50	.10

55. Start with

Outcome	1	2	3	4	5	6
Probability	0	x	x	x	x	0

$4x = 1$, so $x = 1/4 = .25$, giving

Outcome	1	2	3	4	5	6
Probability	0	.25	.25	.25	.25	0

$P(\text{odd}) = 0 + .25 + .25 = .5$

57. Start with

Outcome	1	2	3	4	5	6
Probability	x	$2x$	$2x$	$2x$	$2x$	x

$10x = 1$, so $x = .1$, giving

Outcome	1	2	3	4	5	6
Probability	.1	.2	.2	.2	.2	.1

$P(\text{odd}) = .1 + .2 + .2 = .5$

59. Let $x = P(\text{matching numbers})$. Then $P(\text{Mismatching numbers}) = 2x$.

There are 6 matching pairs and 30 mismatching ones. Since the probabilities add to 1,

$$6x + 30(2x) = 1 \Rightarrow 66x = 1, \text{ so } x = \frac{1}{66}.$$

This gives:

$$P(1,1) = P(2,2) = \ldots = P(6,6) = \frac{1}{66}$$

$$P(1,2) = \ldots = P(6,5) = \frac{2}{66} = \frac{1}{33}.$$

To obtain an odd sum, the pair must consist of an even and odd number, and there are nine such (mismatching) pairs:

$$P(\text{odd sum}) = 9 \times \frac{1}{33} = \frac{6}{11}.$$

61. Take $P(2) = 15x$. This gives

Outcome	1	2	3	4	5	6
Probability	$5x$	$15x$	$5x$	$3x$	$5x$	$5x$

$38x = 1$, so $x = \dfrac{1}{38}$, giving

Outcome	1	2	3	4	5	6
Probability	$\dfrac{5}{38}$	$\dfrac{15}{38}$	$\dfrac{5}{38}$	$\dfrac{3}{38}$	$\dfrac{5}{38}$	$\dfrac{5}{38}$

$$P(\text{odd}) = \frac{5}{38} + \frac{5}{38} + \frac{5}{38} = \frac{15}{38}$$

63. Take D: I will meet a tall dark stranger; $P(D) = 1/3$.

T: I will travel; $P(T) = 2/3$.

Also, $P(D \cap T) = 1/6$.

$$P(D \cup T) = P(D) + P(T) - P(D \cap T) = \frac{1}{3} + \frac{2}{3} - \frac{1}{6} = \frac{5}{6}$$

65. Take A: A randomly selected person polled ranked jobs or health care as the top domestic priority.

We are told that $P(A) = .61$, and we are asked to find $P(A')$.

$$P(A') = 1 - P(A) = 1 - .61 = .39$$

67. Take T: A randomly chosen electric car was manufactured by Tesla; $P(T) = .30$.

N: It was manufactured by Nissan; $P(N) = .35$.

The events T and N are mutually exclusive, so

$$P(T \cup N) = P(T) + P(N) = .30 + .35 = .65.$$

We are asked to find $P[(T \cup N)'] = 1 - P(T \cup N) = 1 - .65 = .35$.

69. S = set of all applicants; $n(S) = 70{,}139$

E = set of admitted applicants; $n(E) = 13{,}692$

$$P(E) = \frac{n(E)}{n(S)} = \frac{13,692}{70,139} \approx .20$$

71. S = set of all applicants; $n(S) = 70,139$

E = set of admitted applicants who had a Math SAT below 400; $n(E) = 9$

$$P(E) = \frac{n(E)}{n(S)} = \frac{9}{70,139} \approx .00013 \approx .00$$

73. S = set of all applicants; $n(S) = 70,139$

E = set of applicants not admitted; $n(E) = 56,447$

$$P(E) = \frac{n(E)}{n(S)} = \frac{56,447}{70,139} \approx .80$$

75. R: An applicant had a Math SAT in the range 500–599. A: An applicant was admitted.

$$P(R) = \frac{14,529}{70,139} \approx .207 \qquad P(A) = \frac{13,692}{70,139} \approx .195 \qquad P(R \cap A) = \frac{1,410}{70,139} \approx .020$$

$$P(R \cup A) = P(R) + P(A) - P(R \cap A) \approx .207 + .195 - .020 \approx .38$$

77. R: An applicant had a Math SAT in the range 500–599. A: An applicant was admitted.

We are asked to find $P[(R \cup A)']$. In Exercise 75 we computed $P(R \cup A) \approx .38$.

So $\qquad P[(R \cup A)'] = 1 - P(R \cup A) \approx 1 - .38 = .62.$

79. S = set of all applicants who had a Math SAT below 400; $n(S) = 1,661$

E = set of admitted applicants who had a Math SAT below 400; $n(E) = 9$

$$P(E) = \frac{n(E)}{n(S)} = \frac{9}{1,661} \approx .01$$

81. S = set of all admitted applicants; $n(S) = 13,692$

E = set of admitted applicants who had a math SAT of 700 or above; $n(E) = 8,398$

$$P(E) = \frac{n(E)}{n(S)} = \frac{8,398}{13,692} \approx .61$$

83. S = set of all rejected applicants; $n(S) = 56,447$

E = set of rejected applicants who had a math SAT below 600; $n(E) = 13,119 + 6,714 + 1,652 = 21,485$

$$P(E) = \frac{n(E)}{n(S)} = \frac{21,485}{56,447} \approx .38$$

85. Use the following events:

E: Agree that it should be the government's responsibility to provide a decent standard of living for the elderly.

F: Agree that it would be a good idea to invest part of their Social Security taxes on their own.

$$P(E \cup F) = P(E) + P(F) - P(E \cap F)$$

$P(E \cup F) = .79 + .43 - P(E \cap F) \Rightarrow P(E \cup F) = 1.22 - P(E \cap F)$ ... (1)

Since $P(E \cup F)$ must be less than or equal to 1, $P(E \cap F)$ must be at least .22, so the smallest percentage of people that could have agreed with both statements is 22%.

For the second part of the question, we ask how large $P(E \cap F)$ can be for the formula (1) to make sense. The largest conceivable value for $P(E \cap F)$ is .43 (because $E \cap F$ is a subset of E and F and hence its probability cannot exceed either .79 or .43). Thus, the largest percentage of people that could have agreed with both statements is 43%.

87. Use the following events: B: Failed it backward; $P(B) = \frac{2}{3}$; D: Failed it sideways; $P(D) = \frac{3}{4}$.

We are also told that $P(B \cap D) = \frac{5}{12}$.

$P(B \cup D) = P(B) + P(D) - P(B \cap D) = \frac{2}{3} + \frac{3}{4} - \frac{5}{12} = 1$

Therefore, all of them failed it either backwards or sideways, so all were disqualified.

89. Use the following events: V: Vaccinated; $P(V) = .80$; F: Gets flu; $P(F) = .10$.
We are also told that 2% of the vaccinated population gets the flu. Therefore, the percentage that are vaccinated and also get the flu is

$P(V \cap F) = .02P(V) = .02 \times .80 = .016$.

Now

$P(V \cup F) = P(V) + P(F) - P(V \cap F) = .80 + .10 - .016 = .884$.

Thus, 88.4% of the population either gets vaccinated or gets the disease.

Communication and reasoning exercises

91. Here is one possible experiment: Roll a die, and observe which of the following outcomes occurs: Outcome A: 1 or 2 facing up; $P(A) = 1/3$, Outcome B: 3 or 4 facing up; $P(B) = 1/3$, Outcome C: 5 or 6 facing up; $P(C) = 1/3$.

93. He is wrong. It is possible to have a run of losses of any length. Each time he plays the game, the chances of losing are the same, regardless of his history of wins or losses. Tony may have grounds to *suspect* that the game is rigged, but he has no proof.

95. The probability of the union of two events is the sum of the probabilities of the two events if <u>they are mutually exclusive.</u>

97. Wrong. For example, the modeled probability of winning a state lottery is small but nonzero. However, the vast majority of people who play the lottery every day of their lives never win, no matter how frequently they play, so the relative frequency is zero for these people.

99. When $A \cap B = \varnothing$ we have $P(A \cap B) = P(\varnothing) = 0$, so

$P(A \cup B) = P(A) + P(B) - P(A \cap B) = P(A) + P(B) - 0 = P(A) + P(B)$.

101. Zero. According to the assumption, no matter how many thunderstorms occur, lightning cannot strike a given spot more than once, so, after n trials the estimated probability will never exceed $1/n$ and so will approach zero as the number of trials gets large. Since the modeled probability models the limit of relative frequencies as the number of trials gets large, it must therefore be zero.

103. $P(A \cup B \cup C) = P(A) + P(B) + P(C) - P(A \cap B) - P(A \cap C) - P(B \cap C) + P(A \cap B \cap C)$

Section 7.4

1. $n(S) = C(10, 5) = 252$ $n(E) = C(4, 4)C(6, 1) = 6$ (4 red, 1 non-red)

$$P(E) = \frac{n(E)}{n(S)} = \frac{6}{252} = \frac{1}{42}$$

3. $n(S) = C(10, 5) = 252$

E: At least 1 white marble

The complementary event is F: No white ones.

$$n(F) = C(8, 5) = 56 \qquad (5 \text{ non-white})$$

$$P(F) = \frac{n(F)}{n(S)} = \frac{56}{252} = \frac{2}{9}$$

Therefore,

$$P(E) = 1 - P(F) = 1 - \frac{2}{9} = \frac{7}{9}.$$

5. $n(S) = C(10, 5) = 252$

$n(E) = C(4, 2)C(3, 1)C(2, 1)C(1, 1) = 36$ (2 red, 1 green, 1 white, 1 purple)

$$P(E) = \frac{n(E)}{n(S)} = \frac{36}{252} = \frac{1}{7}$$

7. $n(S) = C(10, 5) = 252$

$n(E) = C(7, 5) + C(3, 1)C(7, 4) = 21 + 105 = 126$ (5 non-green or 1 green, 4 non-green)

$$P(E) = \frac{n(E)}{n(S)} = \frac{126}{252} = \frac{1}{2}$$

9. E: She does not have all the red ones.

The complementary event is F: She has all the red ones. In Exercise 1, we computed this probability:

$$P(F) = \frac{1}{42}.$$

Therefore,

$$P(E) = 1 - \frac{1}{42} = \frac{41}{42}.$$

11. $n(S) = C(10, 2) = 45$

There are only 3 stocks listed with yields of 3.75% or more. Therefore,

$$n(E) = C(3, 2) = 3 \Rightarrow P(E) = \frac{n(E)}{n(S)} = \frac{3}{45} = \frac{1}{15}.$$

13. $n(S) = C(10, 4) = 210$; $n(E) = C(1, 1)C(8, 3) = 56$

(Choose T and then 3 out of the 8 that remain when you exclude T and KO.)

$$P(E) = \frac{n(E)}{n(S)} = \frac{56}{210} = \frac{4}{15}$$

15. $n(S) = C(10, 2) = 45$

Ending up with 200 shares of PFE means that PFE was one of the stocks you chose:

$n(E) = C(1, 1)C(9, 1) = 9$ (Choose PFE and then 1 out of the 9 that remain.)

$$P(E) = \frac{n(E)}{n(S)} = \frac{9}{45} = \frac{1}{5}.$$

17. Number of possible completed tests: $n(S) = 2^8 \times 5^5 \times 5!$ (8 true/false, 5 multiple choice, 5 matching)

Number of correct answers: $n(E) = 1$

$$P(E) = \frac{n(E)}{n(S)} = \frac{1}{2^8 \times 5^5 \times 5!}$$

19. $n(S) = C(52, 5) = 2{,}598{,}960$; $n(E) = 1{,}098{,}240$ (6.4 Exercise 57)

$$P(E) = \frac{n(E)}{n(S)} = \frac{1{,}098{,}240}{2{,}598{,}960} \approx .4226$$

21. $n(S) = C(52, 5) = 2{,}598{,}960$; $n(E) = 123{,}552$ (6.4 Exercise 55)

$$P(E) = \frac{123{,}552}{2{,}598{,}960} \approx .0475$$

23. $n(S) = C(52, 5) = 2{,}598{,}960$

Decision algorithm for computing the number of flushes:
First compute the number of hands in which all 5 cards are of the same suit:

Step 1: Select a suit: 4 choices.

Step 2: Select 5 cards of that suit: $C(13, 5)$ choices.

Total number of choices: $4 \times C(13, 5)$

Among these are hands in which the 5 cards are consecutive: A, 2, 3, 4, 5 up through 10, J, Q, K, A (straight or royal flushes). The number of straight or royal flushes is 4×10 (Step 1: Select a suit; Step 2: Select a starting card for the run). Excluding these gives
Total number of flushes:

$n(E) = 4 \times C(13, 5) - 4 \times 10 = 5{,}108$

$$P(E) = \frac{n(E)}{n(S)} = \frac{5{,}108}{2{,}598{,}960} \approx .0020.$$

25. $n(S) = 27^{39}$; $n(E) = 1$ (The correct sequence)

$$P(E) = \frac{n(E)}{n(S)} = \frac{1}{27^{39}}$$

27. $n(S) = C(7, 4) = 35$; $n(E) = C(5, 4) = 5 \Rightarrow P(E) = \frac{n(E)}{n(S)} = \frac{5}{35} = \frac{1}{7}$

29. $n(S) = C(50, 5) = 2{,}118{,}760$

Big Winner: $n(E) = 1 \Rightarrow P(E) = \frac{n(E)}{n(S)} = \frac{1}{2{,}118{,}760} \approx .000000472$

Small-Fry Winner: $n(F) = C(5, 4)C(45, 1) = 225$ (4 winning numbers & 1 losing number)

$$P(F) = \frac{n(F)}{n(S)} = \frac{225}{2,118,760} \approx .000106194$$

$$P(E \cup F) = P(E) + P(F) = \frac{226}{2,118,760} \approx .000106666 \qquad \text{(Mutually exclusive)}$$

31. $n(S) = C(700, 400)$

a. $n(E) = C(100, 100)C(600, 300) = C(600, 300)$ $\qquad$ 100 managers, 300 non-managers

$$P(E) = \frac{n(E)}{n(S)} = \frac{C(600, 300)}{C(700, 400)}$$

b. $n(E) = C(1, 1)C(699, 399) = C(699, 399)$ $\qquad$ You, 399 others

$$P(E) = \frac{n(E)}{n(S)} = \frac{C(699, 399)}{C(700, 400)}$$

Note that

$$\frac{C(699, 399)}{C(700, 400)} = \frac{699!/(399! \times 300!)}{700!/(400! \times 300!)} = \frac{699! \times 400!}{700! \times 399!} = \frac{400}{700}.$$

33. $n(S) = 10 \times 10 \times 10 = 10^3$; $n(E) = 10 \times 9 \times 8 = P(10, 3) = 720$

$$P(E) = \frac{n(E)}{n(S)} = \frac{720}{1,000} = .72$$

35. The number of possible outcomes is $n(S) = 2 \times 2 \times 2 = 8$. $\qquad$ (Select a winner 3 times)

The only way North Carolina (NC) will beat Central Connecticut (CC) but lose to Virginia is if NC beats CC, Virginia beats Syracuse, and CC loses to Virginia. Therefore,

$$n(E) = 1 \Rightarrow P(E) = \frac{n(E)}{n(S)} = \frac{1}{8}.$$

37. As a seeding consists of a list of 16 teams, $n(S) = $ the number of possible seedings $= P(16, 16) = 16!$
To choose a seeding of the designated type, we have the following decision algorithm:

Choose the order in which the games 1 vs 16, 2 vs 15, etc. appear: $P(8, 8) = 8!$

In each of these 8 games, choose the order of the teams: 2^8.

So, $n(E) = P(8, 8) \times 2^8 = 8! \times 2^8$

$$P(E) = \frac{n(E)}{n(S)} = \frac{8! \times 2^8}{16!}$$

39. There are 32 first-round games, 16 second-round games, 8 third round games, and 4 fourth-round games, giving a total of 60 games in the first four rounds. There are two more rounds to complete the tournament (called "the final four" and the finals): 2 fifth-round games, and one final. Thus, there are $n(S) = 2 \times 2 \times \cdots \times 2 = 2^{63}$ possible brackets.

a. Because only one of these brackets is the correct one, $n(E) = 1$, so $P(E) = 1/2^{63}$.

b. Decision algorithm for picking the winners with 15 upsets in the first four rounds:

Step 1: Choose 15 games in which the upsets occur: $C(60, 15)$ choices

Step 2: Choose the winners for each of the three games in the last two rounds: $2 \times 2 \times 2 = 8$.

Total number of possiblties: $n(E) = C(60, 15) \times 8 = 425,552,713,541,760$ possibilities

$$P(E) = \frac{n(E)}{n(S)} = \frac{C(60, 15) \times 8}{2^{63}} = \frac{425,552,713,541,760}{9,223,372,036,854,775,808}$$

41. A perfect progression is a sequence of 8 "Won" scores in the range 0–7 that are all different. S is the set of all sequences of 8 digits in the range 0–7.

$$n(S) = 8^8; \, n(E) = 8! \Rightarrow P(E) = \frac{n(E)}{n(S)} = \frac{8!}{8^8}$$

43. Each of the two random moves from the starting position can be up, down, left, or right. Since this gives 4 choices per move, there are

$$n(S) = 4 \times 4 = 16$$

possible sequences of two moves. Only two of these sequences will get you to the Finish node: right + down and down + right. Therefore,

$$n(E) = 2 \Rightarrow P(E) = \frac{n(E)}{n(S)} = \frac{2}{16} = \frac{1}{8}.$$

45. The number of possible sequences of 4 digits in the range 0–9 is $n(S) = 10 \times 10 \times 10 \times 10 = 10,000$.

Number of correct codes: 1
Number of codes that are correct except for a single digit:
    Select a slot for the incorrect digit: 4 choices.
    Select the incorrect digit: 9 choices.
    Fill in the remaining slots with the correct digits: 1 choice.

Total number of incorrect codes: $4 \times 9 \times 1 = 36$

Therefore, $n(E) = 1 + 36 = 37 \Rightarrow P(E) = \dfrac{n(E)}{n(S)} = \dfrac{37}{10,000} = .0037.$

47. a. Decision algorithm for forming a committee:

    Select a chief investigator: $C(6, 1)$ choices.

    Select an assistant investigator: $C(6, 1)$ choices.

    Select 2 at-large investigators: $C(10, 2)$ choices.

    Select 5 ordinary members: $C(8, 5)$ choices.

This gives

    $n(S) = C(6, 1)C(6, 1)C(10, 2)C(8, 5) = 90,720.$

b. Decision algorithm to make Larry happy:
    Alternative 1: Larry is chief and Otis is assistant:
        Select Larry for chief and Otis for assistant: 1 choice.

        Select 2 at-large investigators: $C(10, 2)$ choices.

        Select 5 ordinary members: $C(8, 5)$ choices.

This gives $C(10, 2)C(8, 5)$ choices for this alternative.

    Alternative 2: Larry is not on the committee:

        Select a chief investigator: $C(5, 1)$ choices.

        Select an assistant investigator: $C(6, 1)$ choices.

        Select 2 at-large investigators: $C(9, 2)$ choices.

        Select 5 ordinary members: $C(7, 5)$ choices.

This gives $C(5, 1)C(6, 1)C(9, 2)C(7, 5)$ choices for this alternative.

Adding gives a total of

$$n(E) = C(10, 2)C(8, 5) + C(5, 1)C(6, 1)C(9, 2)C(7, 5) = 25{,}200.$$

c. $P(E) = \dfrac{n(E)}{n(S)} = \dfrac{25{,}200}{90{,}720} \approx .28$

Communication and reasoning exercises

49. The four outcomes listed are not equally likely; for example, (red, blue) can occur in four ways. The methods of this section yield a probability for (red, red) of $C(2, 2)/C(4, 2) = 1/6$.

51. No. If we do not pay attention to order, the probability is $C(5, 2)/C(9, 2) = 10/36 = 5/18$. If we do pay attention to order, the probability is $P(5, 2)/P(9, 2) = 20/72 = 5/18$ again. The difference between permutations and combinations cancels when we compute the probability.

53. Answers will vary.

Section 7.5

1. $P(A|B) = \dfrac{P(A \cap B)}{P(B)} = \dfrac{.2}{.5} = .4$

3. $P(A \cap B) = P(A|B)P(B) = (.2)(.4) = .08$

5. $P(A \cap B) = P(A|B)P(B)$

$.3 = .4P(B) \Rightarrow P(B) = \dfrac{.3}{.4} = .75$

7. If A and B are independent, then $P(A \cap B) = P(A)P(B) = (.5)(.4) = .2$.

9. If A and B are independent, then $P(A|B) = P(A) = .5$.

11. 10% of all Anchovians detest anchovies (D).

Rewording: The probability that an Anchovian detests anchovies is $.10: P(D) = .10$.

30% of all married Anchovians (M) detest anchovies.
Rewording this as in Example 2, we get:
The probability that an Anchovian detests anchovies, given that he or she is married, is $.30$; $P(D|M) = .30$.

13. 30% of all lawyers who lost clients (L) were antitrust lawyers (A).
Rewording this as in Example 2, we get:
The probability that a lawyer was an antitrust lawyer, given that she lost clients, is $.30$; $P(A|L) = .30$.
10% of all antitrust lawyers lost clients.
Rewording this as in Example 2, we get:
The probability that a lawyer lost clients, given that she was an antitrust lawyer, was $.10$; $P(L|A) = .10$.

15. 55% of those who go out in the midday sun (M) are Englishmen (E).
Rewording this as in Example 2, we get:
The probability that someone is an Englishman, given that he goes out in the midday sun, is $.55$; $P(E|M) = .55$.

5% of those who do not go out in the midday sun are Englishmen.
Rewording this as in Example 2, we get:
The probability that someone is an Englishman, given that he does not go out in the midday sun, is $.05$;

$P(E|M') = .05$.

17. A: The sum is 5. B: The green one is not a 1.

$P(A \cap B) = P\{(3,2),(2,3),(1,4)\} = \dfrac{3}{36} = \dfrac{1}{12}$ $P(B) = \dfrac{30}{36} = \dfrac{5}{6}$ $P(A|B) = \dfrac{P(A \cap B)}{P(B)} = \dfrac{1/12}{5/6} = \dfrac{1}{10}$

19. A: The red one is 5. B: The sum is 6.

$P(A \cap B) = P\{(5,1)\} = \dfrac{1}{36}$

$P(B) = P\{(1,5),(2,4),(3,3),(4,2),(5,1)\} = \dfrac{5}{36}$ $P(A|B) = \dfrac{P(A \cap B)}{P(B)} = \dfrac{1/36}{5/36} = \dfrac{1}{5}$

21. *A*: The sum is 5. *B*: The dice have opposite parity.

$$P(A \cap B) = P\{(4,1),(3,2),(2,3),(1,4)\} = \frac{4}{36} = \frac{1}{9} \qquad P(B) = \frac{18}{36} = \frac{1}{2} \qquad P(A|B) = \frac{P(A \cap B)}{P(B)} = \frac{1/9}{1/2} = \frac{2}{9}$$

23. *A*: She gets all 3 red ones. *B*: She gets the fluorescent pink one.

$n(S) = C(10,4) = 210$

$$P(A \cap B) = \frac{C(1,1)C(3,3)}{210} = \frac{1}{210} \qquad P(B) = \frac{C(1,1)C(9,3)}{210} = \frac{84}{210}$$

$$P(A|B) = \frac{P(A \cap B)}{P(B)} = \frac{1/210}{84/210} = \frac{1}{84}$$

25. *A*: She gets no red ones. *B*: She gets the fluorescent pink one.

$n(S) = C(10,4) = 210$

$$P(A \cap B) = \frac{C(1,1)C(6,3)}{210} = \frac{20}{210} \qquad P(B) = \frac{C(1,1)C(9,3)}{210} = \frac{84}{210}$$

$$P(A|B) = \frac{P(A \cap B)}{P(B)} = \frac{20/210}{84/210} = \frac{5}{21}$$

27. *A*: She gets one of each color other than fluorescent pink. *B*: She gets at least one red one.

Notice that $A \cap B = A$. $n(S) = C(10,4) = 210$

$$P(A \cap B) = P(A) = \frac{C(3,1)C(2,1)C(2,1)C(2,1)}{210} = \frac{24}{210} \qquad P(B) = 1 - \frac{C(7,4)}{210} = 1 - \frac{35}{210} = \frac{175}{210}$$

$$P(A|B) = \frac{P(A \cap B)}{P(B)} = \frac{24/210}{175/210} = \frac{24}{175}$$

29

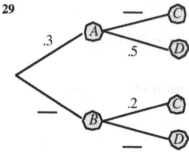

Since the probabilities leaving each branching point must add to 1, we can fill in the missing probabilities:

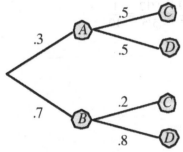

The probability of each outcome is obtained by multiplying the probabilities along the branches leading to the corresponding node:

356

$P(A \cap C) = .3 \times .5 = .15$

$P(A \cap D) = .3 \times .5 = .15$

$P(B \cap C) = .7 \times .2 = .14$

$P(B \cap D) = .7 \times .8 = .56.$

31

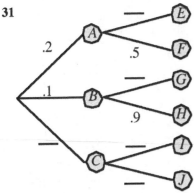

Since the probabilities leaving each branching point must add to 1, we can fill in some of the missing probabilities (below left):

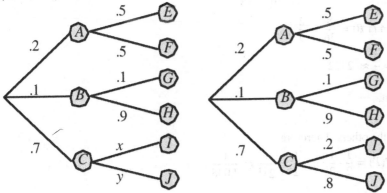

To get the remaining two (above right), use the given information about the probabilities of the outcomes:

$$P(C \cap I) = .7 \times x = .14 \Rightarrow x = \frac{.14}{.7} = .2 \Rightarrow y = 1 - .2 = .8.$$

Finally, the probability of each outcome is obtained by multiplying the probabilities along the branches leading to the corresponding node:

$P(A \cap E) = .2 \times .5 = .10 \qquad P(A \cap F) = .2 \times .5 = .10$

$P(B \cap G) = .1 \times .1 = .01 \qquad P(B \cap H) = .1 \times .9 = .09$

$P(C \cap I) = .7 \times .2 = .14 \qquad P(C \cap J) = .7 \times .8 = .56.$

33. *A*: Your new skateboard design is a success.   *B*: Your new skateboard design is a failure.

If your new skateboard design is a success, it cannot be a failure, and *vice versa*. Therefore, *A* and *B* are mutually exclusive.

35. *A*: Your new skateboard design is a success.   *B*: Your competitor's new skateboard design is a failure.

Since the likelihood of *A* does affect the likelihood of *B*, the events are not independent. Since it is possible for *A* and *B* to occur together, they are not mutually exclusive. Therefore, the correct choice is "neither."

37. A: The red die is 1, 2, or 3; $P(A) = \frac{1}{2}$ B: The green die is even; $P(B) = \frac{1}{2}$

$A \cap B$: The red die is 1, 2, or 3 and the green one is even; $P(A \cap B) = \frac{9}{36} = \frac{1}{4}$

$P(A)P(B) = \frac{1}{2} \times \frac{1}{2} = \frac{1}{4}$ $P(A \cap B) = \frac{1}{4}$

Since $P(A \cap B) = P(A)P(B)$, A and B are independent.

39. A: Exactly one die is 1; $P(A) = \frac{10}{36} = \frac{5}{18}$ B: The sum is even; $P(B) = \frac{1}{2}$

$A \cap B$: Exactly one die is 1 and the sum is even; $P(A \cap B) = \frac{4}{36} = \frac{1}{9}$

$P(A)P(B) = \frac{5}{18} \times \frac{1}{2} = \frac{5}{36}$ $P(A \cap B) = \frac{1}{9}$

Since $P(A \cap B) \neq P(A)P(B)$, A and B are dependent.

41. A: Neither die is 1; $P(A) = \frac{25}{36}$ B: Exactly one die is 2; $P(B) = \frac{10}{36} = \frac{5}{18}$

$A \cap B$: Neither die is 1 and exactly one is 2; $P(A \cap B) = \frac{8}{36} = \frac{2}{9}$

$P(A)P(B) = \frac{25}{36} \times \frac{5}{18} \approx .1929$ $P(A \cap B) = \frac{2}{9} \approx .2222$

Since $P(A \cap B) \neq P(A)P(B)$, A and B are dependent.

43. The outcome of each coin toss is independent of the others. Therefore,

$$P(HTTHHHTHHTT) = P(H)P(T)P(T)...P(T) = \frac{1}{2} \cdot \frac{1}{2} \cdot ... \cdot \frac{1}{2} = \frac{1}{2^{11}} = \frac{1}{2,048}.$$

Applications

45. The two events we are interested in are:

B: A person in the U.S. declared personal bankruptcy. E: A person in the U.S. recently experienced a "big three" event.

We are told that:

The probability that a person in the U.S. would declare personal bankruptcy was .006. That is, $P(B) = .006$.

The probability that a person in the U.S. would declare personal bankruptcy *and* had recently experienced a "big three" event was .005. That is, $P(E \cap B) = .005$.

We are asked to find the probability that a person had recently experienced one of the "big three" events given that she had declared personal bankruptcy. That is, we are asked to find $P(E|B)$.

$$P(E|B) = \frac{P(E \cap B)}{P(B)} = \frac{.005}{.006} \approx .8$$

47. W: A home sale took place in the West. F: A home sale took place in April 2015.

a. We are asked for $P(W|F)$.

$W \cap F$ is the event that a home sale took place in the West and in April 2015.

$$P(W \cap F) = \frac{110}{5,000}; \; P(F) = \frac{450}{5,000} \Rightarrow P(W|F) = \frac{110/5,000}{450/5,000} = \frac{110}{450} \approx .24$$

b. We are asked for $P(F|W)$.

$$P(W \cap F) = \frac{110}{5,000}; \; P(W) = \frac{1,200}{5,000} \Rightarrow P(W|F) = \frac{110/5,000}{1,200/5,000} = \frac{110}{1,200} \approx .092$$

49. The two events are

A: Agreed that it should be the government's responsibility to provide a decent standard of living for the elderly.

$P(A) = .79$

B: Agreed that it would be a good idea to invest part of their Social Security taxes on their own. $P(B) = .43$

By independence, $P(A \cap B) = P(A)P(B) = (.79)(.43) \approx .34$.

51. Take X: Used Brand X; $P(X) = .40$; G: Gave up doing laundry; $P(G) = .05$

$$P(X \cap G) = .04, \, P(X)P(G) = .40 \times .05 = .02 \ne P(X \cap G).$$

Therefore, X and G are not independent.

$$P(G|X) = \frac{P(G \cap X)}{P(X)} = \frac{.04}{.40} = .1$$

which is larger than $P(G)$. Therefore, a user of Brand X is more likely to give up doing laundry than a randomly chosen person.

53. Use the following events: D: Involved in deadly accident. T: Involved in tire-related accident.

We are told that $P(D \cap T) = .000003$.

We are also told that the probability that a tire-related accident would prove deadly was .02. Rephrasing this,

The probability that an accident is deadly, *given that* it is tire related, is .02: $P(D|T) = .02$.

We are asked to find $P(T)$. Start with the formula:

$$P(D|T) = \frac{P(D \cap T)}{P(T)} \Rightarrow .02 = \frac{.000003}{P(T)} \Rightarrow .02 P(T) = .000003 \Rightarrow P(T) = \frac{.000003}{.02} = .00015.$$

55. $P(E|S) = \dfrac{P(E \cap S)}{P(S)} = \dfrac{.25}{.30} = \dfrac{5}{6}$ or $P(E|S) = \dfrac{fr(E \cap S)}{fr(S)} = \dfrac{25}{30} = \dfrac{5}{6}$

	N	E	**Total**
S	5	25	30
U	15	55	70
Total	20	80	100

57. $P(U|N) = \dfrac{P(U \cap N)}{P(N)} = \dfrac{.15}{.20} = .75$ or $P(U|N) = \dfrac{fr(U \cap N)}{fr(N)} = \dfrac{15}{20} = .75$

	N	*E*	**Total**
S	5	25	30
U	15	55	70
Total	20	80	100

59. $P(U|E) = \dfrac{P(U \cap E)}{P(E)} = \dfrac{.55}{.80} = \dfrac{11}{16}$ or $P(U|E) = \dfrac{fr(U \cap E)}{fr(E)} = \dfrac{55}{80} = \dfrac{11}{16}$

	N	*E*	**Total**
S	5	25	30
U	15	55	70
Total	20	80	100

61. $P(\text{An unsuccessful author is established}) = P(E|U)$

$P(E|U) = \dfrac{P(E \cap U)}{P(U)} = \dfrac{.55}{.70} = \dfrac{11}{14}$ or $P(E|U) = \dfrac{fr(E \cap U)}{fr(U)} = \dfrac{55}{70} = \dfrac{11}{14}$

	N	*E*	**Total**
S	5	25	30
U	15	55	70
Total	20	80	100

63

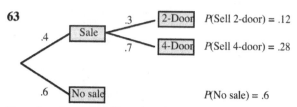

$P(\text{Sell 2-door}) = .12$

$P(\text{Sell 4-door}) = .28$

$P(\text{No sale}) = .6$

Note that the probability of each outcome was obtained by multiplying the probabilities of the corresponding edges.

65. Notice that the total number of vehicles adds up to 100, so that the numbers give the probabilities for the first branches of the tree:

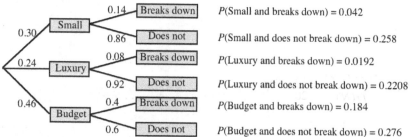

$P(\text{Small and breaks down}) = 0.042$

$P(\text{Small and does not break down}) = 0.258$

$P(\text{Luxury and breaks down}) = 0.0192$

$P(\text{Luxury and does not break down}) = 0.2208$

$P(\text{Budget and breaks down}) = 0.184$

$P(\text{Budget and does not break down}) = 0.276$

Note that the probability of each outcome was obtained by multiplying the probabilities of the corresponding edges.

67

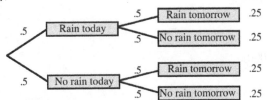

Note that the probability of each outcome was obtained by multiplying the probabilities of the corresponding edges. From the tree, P(No rain today or tomorrow) = .25.

69. Use the following events: A: Employed; B: Bachelor's degree or higher

$$P(A|B) = \frac{fr(A \cap B)}{fr(B)} = \frac{48.8}{67.3} \approx .73.$$

71. Use the following events: A: Bachelor's degree or higher; B: Employed

$$P(A|B) = \frac{fr(A \cap B)}{fr(B)} = \frac{48.8}{127.9} \approx .38.$$

73. Rephrasing the question, we want to find the probability that someone was not in the labor force, given that he or she had not completed a Bachelor's degree or higher.

A: Not in labor force; B: Not completed a Bachelor's degree or higher

$$P(A|B) = \frac{fr(A \cap B)}{fr(B)} = \frac{74.4 - 16.9}{209.1 - 67.3} \approx .41$$

75. Rephrasing the question, we want to find the probability that someone was employed, given that he or she had completed a Bachelor's degree or higher and was in the labor force.

A: Employed; B: Bachelor's degree or higher and in the labor force

$$P(A|B) = \frac{fr(A \cap B)}{fr(B)} = \frac{48.8}{67.3 - 16.9} \approx .97$$

77. We compare P(High school diploma only|Unemployed) with P(High school diploma only|Employed).

$$P(\text{High school diploma only}|\text{Unemployed}) = \frac{2.2}{6.8} \approx .32$$

$$P(\text{High school diploma only}|\text{Employed}) = \frac{33.9}{127.9} \approx .27$$

Your friend is right: The probability that an unemployed person has a high school diploma only is .32, while the corresponding figure for an employed person is .27.

79. $P(K|D) = 1.31 P(K|D')$

81. a. If the course has a positive effect on productivity, then the probability of a productivity increase is greater for a person who took the course. That is, $P(I|T) > P(I)$.

b. If T and I are independent, then $P(I|T) = P(I)$ so that taking the course had no effect on productivity: The course was ineffective.

83. a. Use the following events: A: The person was an Internet user B: The person's family income was at least

$35,000.

$$P(A|B) = \frac{P(A \cap B)}{P(B)} = \frac{.35}{.35 + .24} \approx .59$$

b. We compare $P(\text{Internet user} \mid \geq \$35,000)$ with $P(\text{Internet user} \mid < \$35,000)$.

By part (a), $P(\text{Internet user} \mid \geq \$35,000) = .59$.

By a similar calculation, $P(\text{Internet user} \mid < \$35,000) = \frac{.11}{.11 + .30} \approx .27$.

Therefore, a person was more likely to be an Internet user if his or her family income was $35,000 or more.

85. From the table, the probability that a Jeep Wrangler was reported stolen was .0170. In other words,
The probability that a vehicle was reported stolen given that it was a Jeep Wrangler was .0170;

$$P(R|J) = .0170.$$

87. From the table, the probability that a Toyota Land Cruiser was reported stolen was .0143. In other words,
The probability that a vehicle was reported stolen given that it was a Toyota Land Cruiser was .0143.
This is exactly what Choice (D) says.

89. Consider the following events:

B: My BMW 3-series will be stolen; $P(B) = .0077$.

L: My Lexus GS300 will be stolen; $P(L) = .0074$.

a. As the events are independent, $P(B \cap L) = P(B)P(L) = .0077 \times .0074 \approx .000057$.

b. $P(B \cup L) = P(B) + P(L) - P(B \cap L) = .0077 + .0074 - .000057 = .015043$

91. Use the following events: V: Tests positive; U: Use steroids.

$$P(U|V) = .90,\ P(U \cap V) = .10;\ P(U|V) = \frac{P(U \cap V)}{P(V)}$$

$$.90 = \frac{.10}{P(V)} \Rightarrow .90P(V) = .10 \Rightarrow P(V) = \frac{.10}{.90} \approx .11 \text{ or } 11\%$$

93. Use the following events: A: Contaminated by *Salmonella*; B: Contaminated by a strain of *Salmonella* resistant to at least three antibiotics.

We are told that $P(A) = .20$, and also that the probability that a *Salmonella*-contaminated sample was contaminated

by a strain resistant to at least three antibiotics was .53. Rewriting this gives:

The probability that a sample was contaminated by a strain resistant to at least three antibiotics *given that* it was
contaminated by *Salmonella* is .53: $P(B|A) = .53$.

We are asked to find $P(B \cap A)$ (which is actually the same as $P(B)$), so we use the formula:

$$P(B|A) = \frac{P(B \cap A)}{P(A)} \Rightarrow .53 = \frac{P(B \cap A)}{.20}$$

and obtain $P(B \cap A) = .53 \times .20 = .106$.

95. Take A: Contaminated by *Salmonella*

$B1$: Contaminated by a strain of *Salmonella* resistant to at least one antibiotic

$B3$: Contaminated by a strain of *Salmonella* resistant to at least three antibiotics

We are told:

$P(A) = .20, P(B1|A) = .84, P(B3|A) = .53.$

We are asked to find $P(B3|B1)$.

$$P(B3|B1) = \frac{P(B3 \cap B1)}{P(B1)} = \frac{P(B3)}{P(B1)} = \frac{P(B3 \cap A)}{P(B1 \cap A)}$$

because $B3 \cap B1 = B3 = B3 \cap A$, and $B1 = B1 \cap A$. But we have

$$P(B1|A) = \frac{P(B1 \cap A)}{P(A)} = .84$$

$$P(B3|A) = \frac{P(B3 \cap A)}{P(A)} = .53.$$

Taking their ratio gives

$$\frac{P(B3 \cap A)}{P(B1 \cap A)} = \frac{.53}{.84} \approx .631.$$

Therefore,

$$P(B3|B1) = \frac{P(B3 \cap A)}{P(B1 \cap A)} \approx .631.$$

97. Here is the survey information with the event names as in the text, and with the unknown x used for the event $A' \cap B'$ as shown:

	Saw Ad (B)	Did Not See Ad (B')	Total
Purchased Game (A)	20	40	60
Did Not Purchase Game (A')	180	x	$180 + x$
Total	200	$40 + x$	$240 + x$

As the survey showed the ad had no effect on on sales, the estimated probabilities $P(A)$ and $P(A|B)$ are equal. So,

$$\frac{60}{240 + x} = \frac{20}{200} = \frac{1}{10}$$

Cross-multiplying gives

$$600 = 240 + x \implies x = 360.$$

Thus, the entire table is

	Saw Ad	Did Not See Ad	Total
Purchased Game	20	40	60
Did Not Purchase Game	180	360	540
Total	200	400	600

Communication and reasoning exercises

99. Let W be the event that someone misspells "Waner", and let C be the event that someone misspells "Costenoble."
As these events are independent, the probability of misspelling both names is

$$P(W \cap C) = P(W)P(C) = p \times p = p^2$$

Thus the percentage of people who misspell both names is expected to be $100p^2$ percent.

101. Answers will vary. Here is a simple one: E: The first toss is a head, F: The second toss is a head, G: The third toss is a head.

103. The probability you seek is $P(E|F)$, or should be. If, for example, you were going to place a wager on whether E occurs or not, it is crucial to know that the sample space has been reduced to F. (You know that F did occur.) If you base your wager on $P(E)$ rather than $P(E|F)$, you will misjudge your likelihood of winning.

105. You might explain that the conditional probability of E is not the *a priori* probability of E, but it is the probability of E in a hypothetical world in which the outcomes are restricted to be what is given. In the example she is citing, yes, the probability of throwing a double-six is 1/36 in the absence of any other knowledge. However, by the "conditional probability" of throwing a double-six given that the sum is larger than 7, we might mean the probability of a double-six in a situation in which the dice have already been thrown, but all we know is that the sum is greater than 7. Since there are only 15 ways in which that can happen, the conditional probability is 1/15. For a more extreme case, consider the conditional probability of throwing a double-six given that the sum is 12.

107. If $A \subseteq B$, then $A \cap B = A$, so $P(A \cap B) = P(A)$ and $P(A|B) = \dfrac{P(A \cap B)}{P(B)} = \dfrac{P(A)}{P(B)}$.

109. Your friend is correct. If A and B are mutually exclusive, then $P(A \cap B) = 0$. On the other hand, if A and B are independent, then $P(A \cap B) = P(A)P(B)$. Thus, $P(A)P(B) = 0$. If a product is 0, then one of the factors must be 0, so either $P(A) = 0$ or $P(B) = 0$. Thus, it cannot be true that A and B are mutually exclusive, have nonzero probabilities, and are independent all at the same time.

111. Suppose that A and B are independent, so that

$$P(A \cap B) = P(A)P(B).$$

Then

$$P(A' \cap B') = 1 - P(A \cup B) = 1 - [P(A) + P(B) - P(A \cap B)] = 1 - [P(A) + P(B) - P(A)P(B)]$$

[since $P(A \cap B) = P(A)P(B)$]

$$= (1 - P(A))(1 - P(B)) = P(A')P(B').$$

Therefore, $P(A' \cap B') = P(A')P(B')$, and so A' and B' are independent.

Section 7.6

1. $P(A|B) = .8$, $P(B) = .2$, $P(A|B') = .3$, $P(B') = 1 - P(B) = 1 - .2 = .8$

$$P(B|A) = \frac{P(A|B)P(B)}{P(A|B)P(B) + P(A|B')P(B')} = \frac{(.8)(.2)}{(.8)(.2) + (.3)(.8)} = \frac{.16}{.16 + .24} = .4$$

3. $P(X|Y) = .8$, $P(Y') = .3$, $P(X|Y') = .5$, $P(Y) = 1 - P(Y') = 1 - .3 = .7$,

$$P(Y|X) = \frac{P(X|Y)P(Y)}{P(X|Y)P(Y) + P(X|Y')P(Y')} = \frac{(.8)(.7)}{(.8)(.7) + (.5)(.3)} = \frac{.56}{.56 + .15} \approx .7887$$

5. $P(X|Y_1) = .4$, $P(X|Y_2) = .5$, $P(X|Y_3) = .6$, $P(Y_1) = .8$, $P(Y_2) = .1$, $P(Y_3) = 1 - (.8 + .1) = .1$

$$P(Y_1|X) = \frac{P(X|Y_1)P(Y_1)}{P(X|Y_1)P(Y_1) + P(X|Y_2)P(Y_2) + P(X|Y_3)P(Y_3)} = \frac{(.4)(.8)}{(.4)(.8) + (.5)(.1) + (.6)(.1)}$$

$$= \frac{.32}{.32 + .05 + .06} \approx .7442$$

7. $P(X|Y_1) = .4$, $P(X|Y_2) = .5$, $P(X|Y_3) = .6$, $P(Y_1) = .8$, $P(Y_2) = .1$, $P(Y_3) = 1 - (.8 + .1) = .1$

$$P(Y_2|X) = \frac{P(X|Y_2)P(Y_2)}{P(X|Y_1)P(Y_1) + P(X|Y_2)P(Y_2) + P(X|Y_3)P(Y_3)} = \frac{(.5)(.1)}{(.4)(.8) + (.5)(.1) + (.6)(.1)}$$

$$= \frac{.05}{.32 + .05 + .06} \approx .1163$$

Applications

9. Use the following events: D: Decreased spending on music; I: Internet user

$P(D|I) = .11$, $P(I) = .40$, $P(D|I') = .2$. We are asked to compute $P(I|D)$.

$$P(I|D) = \frac{P(D|I)P(I)}{P(D|I)P(I) + P(D|I')P(I')} = \frac{(.11)(.4)}{(.11)(.4) + (.2)(.6)} = \frac{.044}{.044 + .12} \approx .268, \text{ or } 26.8\%$$

11. Use the following events: A: Snows in Greenland; B: Glaciers grow;

$P(A) = \frac{1}{25} = .04$, $P(B|A) = .20$, $P(B|A') = .04$. We are asked to compute $P(A|B)$.

$$P(A|B) = \frac{P(B|A)P(A)}{P(B|A)P(A) + P(B|A')P(A')} = \frac{(.2)(.04)}{(.2)(.04) + (.04)(.96)} = \frac{.008}{.008 + .0384} \approx .1724$$

13. Use the following events: A: Admitted; T: SAT of 700 or more.

$P(T) = .36$, $P(A|T) = .34$, $P(A|T') = .12$. We are asked to compute $P(T|A)$.

$$P(T|A) = \frac{P(A|T)P(T)}{P(A|T)P(T) + P(A|T')P(T')} = \frac{(.34)(.36)}{(.34)(.36) + (.12)(.64)} = \frac{.1224}{.1224 + .0768} \approx .61$$

15. Use the following events: C: Driving a car; D: Driver dies (in a severe side-impact).

We are given $P(C) = .454$. Also, the information in the table tells us that $P(D|C) = 1, P(D|C') = .3$. We are asked to

find $P(C|D)$. Bayes' theorem states

$$P(C|D) = \frac{P(D|C)P(C)}{P(D|C)P(C) + P(D|C')P(C')} = \frac{.454}{.454 + (.3)(1 - .454)} \approx .73.$$

17. Use the following events: A: Fit enough to play; B: Failed the fitness test and therefore dropped from the team.

$P(B|A) = .5$, $P(B|A') = 1$, $P(A) = .45$, so $P(A') = .55$

We are asked to compute the probability that Mona was justifiably dropped, which is $P(A'|B)$.

$$P(A'|B) = \frac{P(B|A')P(A')}{P(B|A)P(A) + P(B|A')P(A')} = \frac{(1)(.55)}{(.5)(.45) + (1)(.55)} = \frac{.55}{.225 + .55} \approx .71$$

19. Use the following events: L: Driving a light truck; V: Driving an SUV; C: Driving a car; D: Driver dies (in a severe side-impact).

We are given the following information: $P(L) = .273$, $P(V) = .273$, $P(C) = .454$. Also, the information in the table

tells us that $P(D|T) = .210$, $P(D|V) = .371$, $P(D|C) = 1$.

We are asked to find the probability that a driver was driving an SUV, given that the driver died (in a severe side-impact); that is, $P(V|D)$. Bayes' theorem states

$$P(V|D) = \frac{P(D|V)P(V)}{P(D|L)P(L) + P(D|V)P(V) + P(D|C)P(C)} = \frac{(.371)(.273)}{(.210)(.273) + (.371)(.273) + (1)(.454)} \approx .165.$$

21. Use the following events: A: Admitted into UCLA; C: California applicant; U: Applicant from another U.S. state; I: International applicant.

$P(A|C) = .22$, $P(A|U) = .28$, $P(A|I) = .22$, $P(C) = .84$, $P(U) = .10$, $P(I) = .06$

We are asked to compute $P(C|A)$.

$$P(C|A) = \frac{P(A|C)P(C)}{P(A|C)P(C) + P(A|U)P(U) + P(A|I)P(I)} = \frac{(.22)(.84)}{(.22)(.84) + (.28)(.10) + (.22)(.06)} \approx .82,$$

or approximately 82%

23. Use the following events: I: Used the Internet for e-mail; C: Caucasian; A: African American; H: Hispanic; R: Other.

$P(I|C) = .86$, $P(I|A) = .77$, $P(I|H) = .77$, $P(I|R) = .85$, $P(C) = .69$, $P(A) = .12$, $P(H) = .13$,

$P(R) = 1 - (.69 + .12 + .13) = .06$

We are asked to compute $P(H|I)$.

$$P(H|I) = \frac{P(I|H)P(H)}{P(I|C)P(C) + P(I|A)P(A) + P(I|H)P(H) + P(I|R)P(R)}$$

$$= \frac{(.77)(.13)}{(.86)(.69) + (.77)(.12) + (.77)(.13) + (.85)(.06)}$$

$\approx .1196$, or approximately 12%

25. Use the following events: M: Married; L: Have pool.

$P(M|L) = .86$, $P(L) = .15$

a. Given that $P(M|L') = .90$, we are asked to find $P(L|M)$.

$$P(L|M) = \frac{P(M|L)P(L)}{P(M|L)P(L) + P(M|L')P(L')} = \frac{(.86)(.15)}{(.86)(.15) + (.9)(.85)} = \frac{.129}{.129 + .765} \approx .1443, \text{ or } 14.43\%$$

b. From part (a), 14.43% of married couples have pools. The percentage of single people with pools is obtained as follows: Take S: Single; L: Have pool.

$$P(L) = .15, \quad P(S|L) = 1 - .86 = .14, \quad P(S|L') = 1 - .90 = .10 \text{ (From part (b))}$$

$$P(L|S) = \frac{P(S|L)P(L)}{P(S|L)P(L) + P(S|L')P(L')} = \frac{(.14)(.15)}{(.14)(.15) + (.1)(.85)} = \frac{.021}{.021 + .085} \approx .1981, \text{ or } 19.81\%$$

Therefore, 19.81% of single homeowners have pools. Thus, pool manufacturers should go after the single homeowners.

27. Use the following events: A: Former student of Prof. A.; C: Earned C– or lower.

All of Professor A's former students wound up with a C– or lower. In other words, $P(C|A) = 1$.

Two thirds of students not from Prof. A's class got better than a C–. In other words, $P(C|A') = \frac{1}{3}$.

Three quarters of Professor F's class consisted of former students of Professor A, so $P(A) = .75$.

We are asked to find what percentage of the students who got C– or worse were former students of Prof. A: $P(A|C)$.

$$P(A|C) = \frac{P(C|A)P(A)}{P(C|A)P(A) + P(C|A')P(A')} = \frac{(1)(.75)}{(1)(.75) + (1/3)(.25)} = .9$$

Therefore, we estimate that 9 of the 10 students in the delegation were former students of Prof. A.

29. Use the following events: H: Husband employed; W: wife employed.

$$P(H) = .95, \quad P(W|H) = .71$$

Since either the husband or wife in a couple with earnings had to be employed, $P(W|H') = 1$.

We are asked to find $P(H|W)$.

$$P(H|W) = \frac{P(W|H)P(H)}{P(W|H)P(H) + P(W|H')P(H')} = \frac{(.71)(.95)}{(.71)(.95) + (1)(.05)} = \frac{.6745}{.6745 + .05} \approx .9310$$

31. Use the following events: A: Arrested by age 14; C: Become a chronic offender.

$$P(C|A) = 17.9P(C|A'), \quad P(A) = .001$$

$$P(A|C) = \frac{P(C|A)P(A)}{P(C|A)P(A) + P(C|A')P(A')} = \frac{17.9P(C|A')(.001)}{17.9P(C|A')(.001) + P(C|A')(.999)}$$

Now cancel the $P(C|A')$ to get

$$\frac{(17.9)(.001)}{(17.9)(.001) + .999} = \frac{.0179}{.0179 + .999} \approx .0176, \text{ or } 1.76\%.$$

33. Use the following events: D: has diabetes; A: very active.

$$P(D|A) = .5P(D|A'), \quad P(A) = \frac{1}{3}$$

$$P(A|D) = \frac{P(D|A)P(A)}{P(D|A)P(A) + P(D|A')P(A')} = \frac{.5P(D|A')P(A)}{.5P(D|A')P(A) + P(D|A')P(A')}$$

Cancel the $P(D|A')$ to obtain

$$\frac{.5P(A)}{.5P(A) + P(A')} = \frac{(.5)(1/3)}{(.5)(1/3) + 2/3} = \frac{1/6}{1/6 + 2/3} = .2.$$

35. Use the following events: K: The child was killed; D: The air bag deployed.

$$P(K|D) = 1.31 P(K|D'), \ \ P(D) = .25$$

We are asked to compute $P(D|K)$.

$$P(D|K) = \frac{P(K|D)P(D)}{P(K|D)P(D) + P(K|D')P(D')} = \frac{1.31 P(K|D')P(D)}{1.31 P(K|D')P(D) + P(K|D')P(D')}$$

Cancel the terms $P(K|D')$ to obtain

$$\frac{1.31 P(D)}{1.31 P(D) + P(D')} = \frac{(1.31)(.25)}{(1.31)(.25) + .75} = \frac{.3275}{.3275 + .75} \approx .30.$$

Communication and reasoning exercises

37. Show him an example such as Example 1 of this section, where $P(T|A) = .95$ but $P(A|T) \approx .64$.

39. Suppose that the steroid test gives 10% false negatives and that only 0.1% of the tested population uses steroids. Then the probability that an athlete uses steroids, given that he or she has tested positive, is

$$\frac{(0.9)(0.001)}{(0.9)(0.001) + (0.01)(0.999)} \approx .083.$$

41. Draw a tree in which the first branching shows which of R_1, R_2, or R_3 occurred and the second branching shows which of T or T' then occurred. There are three final outcomes in which T occurs:

$$P(R_1 \cap T) = P(T|R_1)P(R_1)$$

$$P(R_2 \cap T) = P(T|R_2)P(R_2)$$

$$P(R_3 \cap T) = P(T|R_3)P(R_3).$$

In only one of these, the first, does R_1 occur. Thus,

$$P(R_1|T) = \frac{P(R_1 \cap T)}{P(T)} = \frac{P(T|R_1)P(R_1)}{P(T|R_1)P(R_1) + P(T|R_2)P(R_2) + P(T|R_3)P(R_3)}.$$

43. The reasoning is flawed. Let A be the event that a Democrat agrees with Safire's column, and let F and M be the events that a Democrat reader is female and male, respectively. Then A. D. makes the following argument:

$$P(M|A) = .9, \ \ P(F|A') = .9.$$

Therefore,

$$P(A|M) = .9.$$

According to Bayes' theorem, we cannot conclude anything about $P(A|M)$ unless we know $P(A)$, the percentage of all Democrats who agreed with Safire's column. This was not given.

Section 7.7

1. The transition matrix P is organized as follows:

$$P = \begin{bmatrix} 1 \to 1 & 1 \to 2 \\ 2 \to 1 & 2 \to 2 \end{bmatrix}.$$

Therefore, the diagram

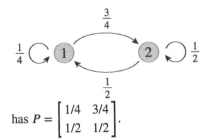

has $P = \begin{bmatrix} 1/4 & 3/4 \\ 1/2 & 1/2 \end{bmatrix}.$

3. The transition matrix P is organized as follows:

$$P = \begin{bmatrix} 1 \to 1 & 1 \to 2 \\ 2 \to 1 & 2 \to 2 \end{bmatrix}.$$

Therefore, the diagram

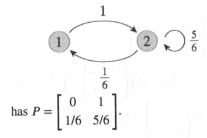

has $P = \begin{bmatrix} 0 & 1 \\ 1/6 & 5/6 \end{bmatrix}.$

5.

Diagram	Organization	Transition Matrix

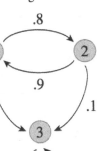

$$P = \begin{bmatrix} 1 \to 1 & 1 \to 2 & 1 \to 3 \\ 2 \to 1 & 2 \to 2 & 2 \to 3 \\ 3 \to 1 & 3 \to 2 & 3 \to 3 \end{bmatrix} \quad P = \begin{bmatrix} 0 & .8 & .2 \\ .9 & 0 & .1 \\ 0 & 0 & 1 \end{bmatrix}$$

7.

Diagram	Organization	Transition Matrix

$1 \circlearrowright 1 \qquad 2 \circlearrowright 1$

$$P = \begin{bmatrix} 1 \to 1 & 1 \to 2 & 1 \to 3 \\ 2 \to 1 & 2 \to 2 & 2 \to 3 \\ 3 \to 1 & 3 \to 2 & 3 \to 3 \end{bmatrix} \quad P = \begin{bmatrix} 1 & 0 & 0 \\ 0 & 1 & 0 \\ 0 & 0 & 1 \end{bmatrix}$$

$3 \circlearrowright 1$

9. The diagram

$$\begin{array}{cccccc} 1\circlearrowleft & \frac{1}{3} & \frac{1}{3} & \frac{1}{3} & \frac{1}{3} & \\ 1 & 2 & 3 & 4 & 5 & 6 \\ & \frac{2}{3} & \frac{2}{3} & \frac{2}{3} & \frac{2}{3} & 1 \end{array}$$

has $P = \begin{bmatrix} 1 & 0 & 0 & 0 & 0 & 0 \\ 2/3 & 0 & 1/3 & 0 & 0 & 0 \\ 0 & 2/3 & 0 & 1/3 & 0 & 0 \\ 0 & 0 & 2/3 & 0 & 1/3 & 0 \\ 0 & 0 & 0 & 2/3 & 0 & 1/3 \\ 0 & 0 & 0 & 0 & 1 & 0 \end{bmatrix}.$

11. a. $P^2 = \begin{bmatrix} .5 & .5 \\ 0 & 1 \end{bmatrix} \begin{bmatrix} .5 & .5 \\ 0 & 1 \end{bmatrix} = \begin{bmatrix} .25 & .75 \\ 0 & 1 \end{bmatrix}$

369

b. Distribution after one step: $vP = \begin{bmatrix} 1 & 0 \end{bmatrix} \begin{bmatrix} .5 & .5 \\ 0 & 1 \end{bmatrix} = \begin{bmatrix} .5 & .5 \end{bmatrix}$

After two steps: $vP^2 = \begin{bmatrix} .5 & .5 \end{bmatrix} \begin{bmatrix} .5 & .5 \\ 0 & 1 \end{bmatrix} = \begin{bmatrix} .25 & .75 \end{bmatrix}$

After three steps: $vP^3 = \begin{bmatrix} .25 & .75 \end{bmatrix} \begin{bmatrix} .5 & .5 \\ 0 & 1 \end{bmatrix} = \begin{bmatrix} .125 & .875 \end{bmatrix}$

13. a. $P^2 = \begin{bmatrix} .2 & .8 \\ .4 & .6 \end{bmatrix} \begin{bmatrix} .2 & .8 \\ .4 & .6 \end{bmatrix} = \begin{bmatrix} .36 & .64 \\ .32 & .68 \end{bmatrix}$

b. Distribution after one step: $vP = \begin{bmatrix} .5 & .5 \end{bmatrix} \begin{bmatrix} .2 & .8 \\ .4 & .6 \end{bmatrix} = \begin{bmatrix} .3 & .7 \end{bmatrix}$

After two steps: $vP^2 = \begin{bmatrix} .3 & .7 \end{bmatrix} \begin{bmatrix} .2 & .8 \\ .4 & .6 \end{bmatrix} = \begin{bmatrix} .34 & .66 \end{bmatrix}$

After three steps: $vP^3 = \begin{bmatrix} .34 & .66 \end{bmatrix} \begin{bmatrix} .2 & .8 \\ .4 & .6 \end{bmatrix} = \begin{bmatrix} .332 & .668 \end{bmatrix}$

15. a. $P^2 = \begin{bmatrix} 1/2 & 1/2 \\ 1 & 0 \end{bmatrix} \begin{bmatrix} 1/2 & 1/2 \\ 1 & 0 \end{bmatrix} = \begin{bmatrix} 3/4 & 1/4 \\ 1/2 & 1/2 \end{bmatrix}$

b. Distribution after one step: $vP = \begin{bmatrix} 2/3 & 1/3 \end{bmatrix} \begin{bmatrix} 1/2 & 1/2 \\ 1 & 0 \end{bmatrix} = \begin{bmatrix} 2/3 & 1/3 \end{bmatrix}$

After two and three steps, the result will remain the same: $\begin{bmatrix} 2/3 & 1/3 \end{bmatrix}$.

17. a. $P^2 = \begin{bmatrix} 3/4 & 1/4 \\ 3/4 & 1/4 \end{bmatrix} \begin{bmatrix} 3/4 & 1/4 \\ 3/4 & 1/4 \end{bmatrix} = \begin{bmatrix} 3/4 & 1/4 \\ 3/4 & 1/4 \end{bmatrix}$

b. Distribution after one step: $vP = \begin{bmatrix} 1/2 & 1/2 \end{bmatrix} \begin{bmatrix} 3/4 & 1/4 \\ 3/4 & 1/4 \end{bmatrix} = \begin{bmatrix} 3/4 & 1/4 \end{bmatrix}$

After two steps: $vP^2 = \begin{bmatrix} 3/4 & 1/4 \end{bmatrix} \begin{bmatrix} 3/4 & 1/4 \\ 3/4 & 1/4 \end{bmatrix} = \begin{bmatrix} 3/4 & 1/4 \end{bmatrix}$

After three steps, the result will remain the same: $\begin{bmatrix} 3/4 & 1/4 \end{bmatrix}$.

19. a. $P^2 = \begin{bmatrix} .5 & .5 & 0 \\ 0 & 1 & 0 \\ 0 & .5 & .5 \end{bmatrix} \begin{bmatrix} .5 & .5 & 0 \\ 0 & 1 & 0 \\ 0 & .5 & .5 \end{bmatrix} = \begin{bmatrix} .25 & .75 & 0 \\ 0 & 1 & 0 \\ 0 & .75 & .25 \end{bmatrix}$

b. Distribution after one step: $vP = \begin{bmatrix} 1 & 0 & 0 \end{bmatrix} \begin{bmatrix} .5 & .5 & 0 \\ 0 & 1 & 0 \\ 0 & .5 & .5 \end{bmatrix} = \begin{bmatrix} .5 & .5 & 0 \end{bmatrix}$

After two steps: $vP^2 = \begin{bmatrix} .5 & .5 & 0 \end{bmatrix} \begin{bmatrix} .5 & .5 & 0 \\ 0 & 1 & 0 \\ 0 & .5 & .5 \end{bmatrix} = \begin{bmatrix} .25 & .75 & 0 \end{bmatrix}$

After three steps: $vP^3 = \begin{bmatrix} .25 & .75 & 0 \end{bmatrix} \begin{bmatrix} .5 & .5 & 0 \\ 0 & 1 & 0 \\ 0 & .5 & .5 \end{bmatrix} = \begin{bmatrix} .125 & .875 & 0 \end{bmatrix}$

21. a. $P^2 = \begin{bmatrix} 0 & 1 & 0 \\ 1/3 & 1/3 & 1/3 \\ 1 & 0 & 0 \end{bmatrix} \begin{bmatrix} 0 & 1 & 0 \\ 1/3 & 1/3 & 1/3 \\ 1 & 0 & 0 \end{bmatrix} = \begin{bmatrix} 1/3 & 1/3 & 1/3 \\ 4/9 & 4/9 & 1/9 \\ 0 & 1 & 0 \end{bmatrix}$

b. Distribution after one step: $vP = \begin{bmatrix} 1/2 & 0 & 1/2 \end{bmatrix} \begin{bmatrix} 0 & 1 & 0 \\ 1/3 & 1/3 & 1/3 \\ 1 & 0 & 0 \end{bmatrix} = \begin{bmatrix} 1/2 & 1/2 & 0 \end{bmatrix}$

After two steps: $vP^2 = \begin{bmatrix} 1/2 & 1/2 & 0 \end{bmatrix} \begin{bmatrix} 0 & 1 & 0 \\ 1/3 & 1/3 & 1/3 \\ 1 & 0 & 0 \end{bmatrix} = \begin{bmatrix} 1/6 & 2/3 & 1/6 \end{bmatrix}$

After three steps: $vP^3 = \begin{bmatrix} 1/6 & 2/3 & 1/6 \end{bmatrix} \begin{bmatrix} 0 & 1 & 0 \\ 1/3 & 1/3 & 1/3 \\ 1 & 0 & 0 \end{bmatrix} = \begin{bmatrix} 7/18 & 7/18 & 2/9 \end{bmatrix}$

23. a. $P^2 = \begin{bmatrix} .1 & .9 & 0 \\ 0 & 1 & 0 \\ 0 & .2 & .8 \end{bmatrix} \begin{bmatrix} .1 & .9 & 0 \\ 0 & 1 & 0 \\ 0 & .2 & .8 \end{bmatrix} = \begin{bmatrix} .01 & .99 & 0 \\ 0 & 1 & 0 \\ 0 & .36 & .64 \end{bmatrix}$

B. Distribution after one step: $vP = \begin{bmatrix} .5 & 0 & .5 \end{bmatrix} \begin{bmatrix} .1 & .9 & 0 \\ 0 & 1 & 0 \\ 0 & .2 & .8 \end{bmatrix} = \begin{bmatrix} .05 & .55 & .4 \end{bmatrix}$

After two steps: $vP^2 = \begin{bmatrix} .05 & .55 & .4 \end{bmatrix} \begin{bmatrix} .1 & .9 & 0 \\ 0 & 1 & 0 \\ 0 & .2 & .8 \end{bmatrix} = \begin{bmatrix} .005 & .675 & .32 \end{bmatrix}$

After three steps: $vP^3 = \begin{bmatrix} .005 & .675 & .32 \end{bmatrix} \begin{bmatrix} .1 & .9 & 0 \\ 0 & 1 & 0 \\ 0 & .2 & .8 \end{bmatrix} = \begin{bmatrix} .0005 & .7435 & .256 \end{bmatrix}$

25. To find the steady-state vector $v_\infty = \begin{bmatrix} x & y \end{bmatrix}$, we solve

$x + y = 1$

$\begin{bmatrix} x & y \end{bmatrix} \begin{bmatrix} 1/2 & 1/2 \\ 1 & 0 \end{bmatrix} = \begin{bmatrix} x & y \end{bmatrix}$

The above equations give:.

$x + y = 1 \qquad x/2 + y = x \qquad x/2 = y$

Rewriting in standard form and dropping the last equation (which is the same as the next-to-last), we get

$x + y = 1 \qquad -x/2 + y = 0.$

Solving this system gives $x = 2/3, y = 1/3$.

So, $v_\infty = \begin{bmatrix} 2/3 & 1/3 \end{bmatrix}$

27. To find the steady-state vector $v_\infty = \begin{bmatrix} x & y \end{bmatrix}$, we solve

$x + y = 1$

$\begin{bmatrix} x & y \end{bmatrix} \begin{bmatrix} 1/3 & 2/3 \\ 1/2 & 1/2 \end{bmatrix} = \begin{bmatrix} x & y \end{bmatrix}$

The above equations give:

$x + y = 1 \qquad x/3 + y/2 = x \qquad 2x/3 + y/2 = y.$

Rewriting in standard form and dropping the last equation (which is the same as the next-to-last), we get

$x + y = 1 \qquad -2x/3 + y/2 = 0.$

Solving this system gives $x = 3/7, y = 4/7$.

So, $v_\infty = \begin{bmatrix} 3/7 & 4/7 \end{bmatrix}$

29. To find the steady-state vector $v_\infty = \begin{bmatrix} x & y \end{bmatrix}$, we solve

$x + y = 1$

$$\begin{bmatrix} x & y \end{bmatrix} \begin{bmatrix} .1 & .9 \\ .6 & .4 \end{bmatrix} = \begin{bmatrix} x & y \end{bmatrix}$$

The above equations give:

$x + y = 1 \qquad .1x + .6y = x \qquad .9x + .4y = y.$

Rewriting in standard form and dropping the last equation (which is the same as the next-to-last), we get

$x + y = 1 \qquad -.9x + .6y = 0.$

Solving this system gives $x = 2/5, y = 3/5$.

So, $v_\infty = \begin{bmatrix} 2/5 & 3/5 \end{bmatrix}$

31. To find the steady-state vector $v_\infty = \begin{bmatrix} x & y & z \end{bmatrix}$, we solve

$x + y + z = 1$

$$\begin{bmatrix} x & y & z \end{bmatrix} \begin{bmatrix} .5 & 0 & .5 \\ 1 & 0 & 0 \\ 0 & .5 & .5 \end{bmatrix} = \begin{bmatrix} x & y & z \end{bmatrix}$$

The above equations give:

$x + y = 1 \qquad .5x + y = x \qquad .5z = y \qquad .5x + .5z = z.$

Rewriting in standard form, we get

$x + y + z = 1 \qquad -.5x + y = 0 \qquad -y + .5z = 0 \qquad .5x - .5z = 0.$

Solving this system gives $x = 2/5, y = 1/5, z = 2/5$.

So, $v_\infty = \begin{bmatrix} 2/5 & 1/5 & 2/5 \end{bmatrix}$

33. To find the steady-state vector $v_\infty = \begin{bmatrix} x & y & z \end{bmatrix}$, we solve

$x + y + z = 1$

$$\begin{bmatrix} x & y & z \end{bmatrix} \begin{bmatrix} 0 & 1 & 0 \\ 1/3 & 1/3 & 1/3 \\ 1 & 0 & 0 \end{bmatrix} = \begin{bmatrix} x & y & z \end{bmatrix}$$

The above equations give:

$x + y = 1 \qquad y/3 + z = x \qquad x + y/3 = y \qquad y/3 = z.$

Rewriting in standard form, we get

$x + y + z = 1 \qquad -x + y/3 + z = 0 \qquad x - 2y/3 = 0 \qquad y/3 - z = 0.$

Solving this system gives $x = 1/3, y = 1/2, z = 1/6$.

So, $v_\infty = \begin{bmatrix} 1/3 & 1/2 & 1/6 \end{bmatrix}$

35. To find the steady-state vector $v_\infty = \begin{bmatrix} x & y & z \end{bmatrix}$, we solve

$x + y + z = 1$

$$\begin{bmatrix} x & y & z \end{bmatrix} \begin{bmatrix} .1 & .9 & 0 \\ 0 & 1 & 0 \\ 0 & .2 & .8 \end{bmatrix} = \begin{bmatrix} x & y & z \end{bmatrix}$$

The above equations give:

$x + y = 1 \qquad .1x = x \qquad .9x + y + .2z = y \qquad .8z = z.$

Rewriting in standard form, we get

$$x + y + z = 1 \qquad -.9x = 0 \qquad .9x + .2z = 0 \qquad -.2z = 0.$$

Solving this system gives $x = 0, y = 1, z = 0$.

So, $v_\infty = \begin{bmatrix} 0 & 1 & 0 \end{bmatrix}$

Applications

37. Take 1 = Sorey State, 2 = C&T.

$$P = \begin{bmatrix} 1/2 & 1/2 \\ 1/4 & 3/4 \end{bmatrix}$$

We are asked to find the (1, 1) entry of the 2-step transition probability matrix.

$$P^2 = \begin{bmatrix} 1/2 & 1/2 \\ 1/4 & 3/4 \end{bmatrix}\begin{bmatrix} 1/2 & 1/2 \\ 1/4 & 3/4 \end{bmatrix} = \begin{bmatrix} 3/8 & 5/8 \\ 5/16 & 11/16 \end{bmatrix}$$

The (1, 1) entry is $3/8 = .375$.

39. a. Take 1 = Not checked in, 2 = Checked in.

$$P = \begin{bmatrix} .4 & .6 \\ 0 & 1 \end{bmatrix}; P^2 = \begin{bmatrix} .4 & .6 \\ 0 & 1 \end{bmatrix}\begin{bmatrix} .4 & .6 \\ 0 & 1 \end{bmatrix} = \begin{bmatrix} .16 & .84 \\ 0 & 1 \end{bmatrix}$$

$$P^3 = \begin{bmatrix} .4 & .6 \\ 0 & 1 \end{bmatrix}\begin{bmatrix} .16 & .84 \\ 0 & 1 \end{bmatrix} = \begin{bmatrix} .064 & .936 \\ 0 & 1 \end{bmatrix}$$

b. 1 hour: $P_{12} = .6$ 2 hours: $(P^2)_{12} = .84$ 3 hours: $(P^3)_{12} = .936$

c. Eventually, all the roaches will have checked in.

41. Take 1 = High risk, 2 = Low risk.

$$P = \begin{bmatrix} .50 & .50 \\ .10 & .90 \end{bmatrix}$$

To find the steady-state vector $\begin{bmatrix} x & y \end{bmatrix}$, we solve

$$x + y = 1$$

$$\begin{bmatrix} x & y \end{bmatrix}\begin{bmatrix} .50 & .50 \\ .10 & .90 \end{bmatrix} = \begin{bmatrix} x & y \end{bmatrix}.$$

The above equations give:

$$x + y = 1 \qquad .5x + .1y = x \qquad .5x + .9y = y.$$

Rewriting in standard form and dropping the last equation, we get

$$x + y = 1 \qquad -.5x + .1y = 0.$$

Solving this system gives $x = 1/6, y = 5/6$.

So, $\begin{bmatrix} x & y \end{bmatrix} = \begin{bmatrix} 1/6 & 5/6 \end{bmatrix}$.

In the long term, $1/6 \approx 16.67\%$ fall into the high-risk category, and $5/6 \approx 83.33\%$ into the low-risk category.

43. a. Take 1 = User, 2 = Non-User.

To set up the transition matrix, note that the entries in each row have to add up to 1:

$$P = \begin{bmatrix} 2/3 & 1/3 \\ 1/10 & 9/10 \end{bmatrix}.$$

The 2-year transition matrix is

$$P^2 = \begin{bmatrix} 2/3 & 1/3 \\ 1/10 & 9/10 \end{bmatrix}\begin{bmatrix} 2/3 & 1/3 \\ 1/10 & 9/10 \end{bmatrix} = \begin{bmatrix} 43/90 & 47/90 \\ 47/300 & 253/300 \end{bmatrix}.$$

Non-user → User in 2 steps:

$(P^2)_{21} = 47/300 \approx .156667$

b. To find the steady-state vector $\begin{bmatrix} x & y \end{bmatrix}$, we solve

$x + y = 1$

$$\begin{bmatrix} x & y \end{bmatrix} \begin{bmatrix} 2/3 & 1/3 \\ 1/10 & 9/10 \end{bmatrix} = \begin{bmatrix} x & y \end{bmatrix}.$$

The above equations give:

$x + y = 1$ $(2/3)x + (1/10)y = x$ $(1/3)x + (9/10)y = y.$

Rewriting in standard form and dropping the last equation, we get

$x + y = 1$ $-(1/3)x + (1/10)y = 0.$

Solving this system gives $x = 3/13$, $y = 10/13$.

So, $\begin{bmatrix} x & y \end{bmatrix} = \begin{bmatrix} 3/13 & 10/13 \end{bmatrix}$.

In the long term, 3/13 of the college instructors will be users of this book.

45. a. Take 1 = Paid Up, 2 = 0–90 Days, 3 = Bad Debt.

$$P = \begin{bmatrix} .5 & .5 & 0 \\ .5 & .3 & .2 \\ 0 & .5 & .5 \end{bmatrix}$$

To find the steady-state vector $\begin{bmatrix} x & y & z \end{bmatrix}$, we solve

$x + y + z = 1$

$$\begin{bmatrix} x & y & z \end{bmatrix} \begin{bmatrix} .5 & .5 & 0 \\ .5 & .3 & .2 \\ 0 & .5 & .5 \end{bmatrix} = \begin{bmatrix} x & y & z \end{bmatrix}.$$

The above equations give:

$x + y + z = 1$ $.5x + .5y = x$ $.5x + .3y + .5z = y$ $.2y + .5z = z.$

Rewriting in standard form and dropping the last equation, we get

$x + y + z = 1$ $-.5x + .5y = 0$ $.5x - .7y + .5z = 0.$

Solving this system gives $x = 5/12$, $y = 5/12$, $z = 1/6$.

So, $\begin{bmatrix} x & y \end{bmatrix} = \begin{bmatrix} 5/12 & 5/12 & 1/6 \end{bmatrix}$.

5/12 or approximately 41.67% of the customers will be in the Paid Up category, 5/12 or approximately 41.67% in the 0–90 Days category, and 1/6 or approximately 16.67% in the Bad Debt category.

47. a. Take 1 = Affluent, 2 = Middle class, 3 = Poor.

$p_{11} = 1 - .729 = .271$, $p_{12} = .729$, $p_{13} = 0$,

$p_{21} = .075$, $p_{22} = 1 - (.075 + .085) = .84$, $p_{23} = .085$

$p_{31} = 0$, $p_{32} = .304$, $p_{33} = 1 - .304 = .696$

$$P = \begin{bmatrix} .729 & .271 & 0 \\ .075 & .84 & .085 \\ 0 & .304 & .696 \end{bmatrix}$$

b. The period 1980–2002 is a 22-year period, corresponding to two 11-year transition steps.

$$P^2 = \begin{bmatrix} .729 & .271 & 0 \\ .075 & .84 & .085 \\ 0 & .304 & .696 \end{bmatrix} \begin{bmatrix} .729 & .271 & 0 \\ .075 & .84 & .085 \\ 0 & .304 & .696 \end{bmatrix} = \begin{bmatrix} .552 & .425 & .023 \\ .118 & .752 & .131 \\ .023 & .467 & .51 \end{bmatrix}$$

Affluent → Poor in 2 steps:

$(P^2)_{13} = .023$, or 2.3%

c. Let us use technology to compute a high enough power of P directly so that the rows are approximately the same:

$$P^\infty \approx P^{64} \approx \begin{bmatrix} .178 & .643 & .18 \\ .178 & .643 & .18 \\ .178 & .643 & .18 \end{bmatrix}.$$

(We used the 64th power because powers of 2 are computed most efficiently in the online Matrix Algebra Tool.)
Answer:
Affluent: 17.8%; Middle class: 64.3%; Poor: 18.0%

49. Take 1 = Bottom 10%, 2 = 10–50%, 3 = 50–90%, 4 = Top 10%.
The transition matrix is given by the table:

$$P = \begin{bmatrix} .3 & .52 & .17 & .01 \\ .1 & .48 & .38 & .04 \\ .04 & .38 & .48 & .1 \\ .01 & .17 & .52 & .3 \end{bmatrix}.$$

Let us use technology to compute a high enough power of P directly so that the rows are approximately the same:

$$P^\infty \approx P^{64} \approx \begin{bmatrix} .0843 & .4157 & .4157 & .0843 \\ .0843 & .4157 & .4157 & .0843 \\ .0843 & .4157 & .4157 & .0843 \\ .0843 & .4157 & .4157 & .0843 \end{bmatrix}.$$

(We used the 64th power because powers of 2 are computed most efficiently in the online Matrix Algebra Tool.)
Thus, the percentages in each category are given by $\begin{bmatrix} 8.43 & 41.57 & 41.57 & 8.43 \end{bmatrix}$.
The reason they are not the "expected" 10%, 40%, 405, and 10% is that the movement between the income groups changes their relative sizes.

51. a. Number the states as follows: 1: Verizon, 2: Cingular, 3: AT&T, 4: Other. The figures next to the arrows show all the transition probabilities (obtained by dividing by 100) except the ones from each state to itself. To compute the missing probabilities, use the fact that the entries in each row add to 1. This leads to:

$$P = \begin{bmatrix} .981 & .005 & .005 & .009 \\ .01 & .972 & .006 & .012 \\ .01 & .006 & .973 & .011 \\ .008 & .006 & .005 & .981 \end{bmatrix}.$$

b. Each time-step is one quarter. At the end of the third quarter, the distribution is given as

$$w = \begin{bmatrix} 0.297 & 0.193 & 0.181 & 0.329 \end{bmatrix}.$$

We are asked to find the distribution v at the beginning of that quarter: one time step earlier. We know that

Distribution after one quarter $= vP = \begin{bmatrix} 0.297 & 0.193 & 0.181 & 0.329 \end{bmatrix}$

To obtain v, multiply both sides on the right by P^{-1}: $v = \begin{bmatrix} 0.297 & 0.193 & 0.181 & 0.329 \end{bmatrix} P^{-1}$.
On the Matrix Algebra Tool, use the format $w*P\verb|^|-1$: We obtain (rounding to 3 decimal places):

$$v = \begin{bmatrix} 0.296 & 0.194 & 0.182 & 0.328 \end{bmatrix}.$$

Thus, the market shares at the beginning of the quarter were:
    Verizon: 29.6%, Cingular: 19.4%, AT&T: 18.2%, Other: 32.8%.

c. The end of 2005 is 9 quarters from the end of the third quarter in 2003. Therefore, the distribution is predicted as

$$wP^9 = \begin{bmatrix} 0.303 & 0.186 & 0.176 & 0.335 \end{bmatrix}$$

Verizon: 30.3%, Cingular: 18.6%, AT&T: 17.6%, Other: 33.5%. (Technology format: $w*P\verb|^|9$) The biggest gainers are Verizon and Other, each gaining 0.6%.

53. From the diagram, the transition matrix is

$$P = \begin{bmatrix} 1/2 & 1/2 & 0 & 0 & 0 \\ 1/2 & 0 & 1/2 & 0 & 0 \\ 0 & 1/2 & 0 & 1/2 & 0 \\ 0 & 0 & 1/2 & 0 & 1/2 \\ 0 & 0 & 0 & 1/2 & 1/2 \end{bmatrix}.$$

For the steady-state vector, we solve:

$x + y + z + u + v = 1$

$(1/2)x + (1/2)y = x$

$(1/2)x + (1/2)z = y$

$(1/2)y + (1/2)u = z$

$(1/2)z + (1/2)v = u$

$(1/2)u + (1/2)v = v.$

Rewriting in standard form and dropping the last equation, we get the system

$x + y + z + u + v = 1$

$-(1/2)x + (1/2)y = 0$

$(1/2)x - y + (1/2)z = 0$

$(1/2)y - z + (1/2)u = 0$

$(1/2)z - u + (1/2)v = 0.$

Solving gives $x = 1/5, y = 1/5, z = 1/5, u = 1/5, v = 1/5.$

Hence the steady-state vector is $\begin{bmatrix} 1/5 & 1/5 & 1/5 & 1/5 & 1/5 \end{bmatrix}$. The system spends an average of 1/5 of the time in each state.

Communication and reasoning exercises

55. Answers will vary.

57. There are two assumptions made by Markov systems that may not be true about the stock market: the assumption that the transition probabilities do not change over time and the assumption that the transition probability depends only on the current state.

59. If q is a row of Q, then by assumption, $qP = q$. Thus, when we multiply the rows of Q by P, nothing changes, and $QP = Q$.

61. At each step only 0.4 of the population in state 1 remains there, and nothing enters from any other state. Thus, when the first entry in the steady-state distribution vector is multiplied by 0.4, it must remain unchanged. The only number for which this is true is 0.

63. An example is

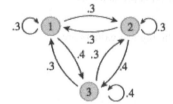

Its transition matrix is $P = \begin{bmatrix} .3 & .3 & .4 \\ .3 & .3 & .4 \\ .3 & .3 & .4 \end{bmatrix}$ and is therefore already in a steady state!

65. If $vP = v$ and $wP = w$, then

$$\frac{1}{2}(v+w)P = \frac{1}{2}vP + \frac{1}{2}wP = \frac{1}{2}v + \frac{1}{2}w = \frac{1}{2}(v+w).$$

Further, if the entries of v and w add up to 1, then so do the entries of $(v+w)/2$.

Chapter 7 Review

1. $n(S) = 2 \times 2 \times 2 = 8$;

$E = \{HHT, HTH, HTT, THH, THT, TTH, TTT\}$; $P(E) = \dfrac{n(E)}{n(S)} = \dfrac{7}{8}$

3. $n(S) = 6 \times 6 = 36$;

$E = \{(1,6),(2,5),(3,4),(4,3),(5,2),(6,1)\}$; $P(E) = \dfrac{n(E)}{n(S)} = \dfrac{6}{36} = \dfrac{1}{6}$

5. $n(S) = 6$ and $E = \{2\}$. However, the outcomes are not equally likely, so we need the probability distribution. Start with

Outcome	1	2	3	4	5	6
Probability	$2x$	x	x	x	x	$2x$

Since the sum of the probabilities of the outcomes must be 1, we get

$8x = 1$, so $x = \dfrac{1}{8} = .125$.

This gives:

Outcome	1	2	3	4	5	6
Probability	.25	.125	.125	.125	.125	.25

$P(E) = P(\{2\}) = .125 \text{ (or } 1/8)$

7. $P(2 \text{ heads}) = \dfrac{fr(2 \text{ heads})}{N} = \dfrac{12}{50}$

$P(\text{At least 1 tail}) = 1 - P(2 \text{ heads}) = 1 - \dfrac{12}{50} = \dfrac{38}{50} = .76$

9. We use the following events for a randomly selected novel:

U: You have read the novel. R: Roslyn has read the novel.

$P(U \cup R) = P(U) + P(R) - P(U \cap R) = \dfrac{150}{400} + \dfrac{200}{400} - \dfrac{50}{400} = \dfrac{300}{400} = .75$

Therefore, the probability that a novel has been read by neither you nor your sister is

$P[(U \cup R)'] = 1 - P(U \cup R) = 1 - .75 = .25$.

11. Let us use the following events: A: A student is in category A; B: A student is in category B.

$P(A \cup B) = 1$, $P(A) = P(B) = \dfrac{24}{32} = .75$

We are asked to find $P(A \cap B)$.

$P(A \cup B) = P(A) + P(B) - P(A \cap B)$

$1 = .75 + .75 - P(A \cap B) \Rightarrow P(A \cap B) = .75 + .75 - 1 = .5$

13. Use the following events: A: Is a Model A; R: Is orange.

$$P(A \cup R) = P(A) + P(R) - P(A \cap R) = \frac{1}{3} + \frac{1}{5} - \frac{1}{15} = \frac{7}{15}$$

15. $n(S) = C(12, 5) = 792$; $P(E) = \dfrac{n(E)}{n(S)} = \dfrac{C(4, 4)C(8, 1)}{792} = \dfrac{8}{792}$

17. $n(S) = C(12, 5) = 792$; $P(E) = \dfrac{n(E)}{n(S)} = \dfrac{C(4, 1)C(2, 1)C(1, 1)C(3, 1)C(2, 1)}{792} = \dfrac{48}{792}$

19. $n(S) = C(12, 5) = 792$

Decision algorithm for constructing a set with at least 2 yellow:

Alternative 1: 2 yellow: $C(3, 2)C(9, 3) = 252$ choices

Alternative 2: 3 yellow: $C(3, 3)C(9, 2) = 36$ choices

This gives a total of $252 + 36 = 288$ choices, so $P(E) = \dfrac{n(E)}{n(S)} = \dfrac{288}{792}$.

21. $n(S) = C(52, 5)$; $P(E) = \dfrac{n(E)}{n(S)} = \dfrac{C(8, 5)}{C(52, 5)}$

23. $n(S) = C(52, 5)$; $P(E) = \dfrac{n(E)}{n(S)} = \dfrac{C(4, 3)C(1, 1)C(3, 1)}{C(52, 5)}$

25. $n(S) = C(52, 5)$

Decision algorithm for constructing a full house of commons:

Step 1: Select the denomination for the triple $(2, 3, 4, 5, 6, 7, 8, 9, 10)$: $C(9, 1)$ choices.

Step 2: Select one of the remaining 8 denominations for the double: $C(8, 1)$ choices.

Step 3: Select 3 cards of the chosen denomination for the triple: $C(4, 3)$ choices.

Step 4: Select 2 cards of the chosen denomination for the double: $C(4, 2)$ choices.

$$P(E) = \frac{n(E)}{n(S)} = \frac{C(9, 1)C(8, 1)C(4, 3)C(4, 2)}{C(52, 5)}$$

27. A: The sum is 5. B: The green one is not a 1 and the yellow one is 1.

$A \cap B$: The sum is 5, the green one is not 1, and the yellow one is 1.

$B = \{(2, 1), (3, 1), (4, 1), (5, 1), (6, 1)\}$; $A \cap B = \{(4, 1)\}$

$$P(A \cap B) = \frac{1}{36}, \quad P(B) = \frac{5}{36}$$

$$P(A|B) = \frac{P(A \cap B)}{P(B)} = \frac{1/36}{5/36} = \frac{1}{5}$$

$$P(A) = \frac{4}{36} = \frac{1}{9}$$

Since $P(A|B) \neq P(A)$, the events A and B are dependent.

29. A: The yellow one is 4. B: The green one is 4.

$A \cap B$: Both the yellow and green dice are 4.

$$P(A \cap B) = \frac{1}{36}, \ P(B) = \frac{1}{6}$$

$$P(A|B) = \frac{P(A \cap B)}{P(B)} = \frac{1/36}{1/6} = \frac{1}{6}$$

$$P(A) = \frac{1}{6} = P(A|B).$$

Therefore, the events A and B are independent.

31. A: The dice have the same parity. B: Both dice are odd.

$A \cap B$: The dice have the same parity and are both odd.

Note that $A \cap B = B$.

$$P(A \cap B) = \frac{9}{36}, \ P(B) = \frac{9}{36}$$

$$P(A|B) = \frac{P(A \cap B)}{P(B)} = \frac{9/36}{9/36} = 1$$

$$P(A) = \frac{18}{36} = \frac{1}{2}$$

Since $P(A|B) \neq P(A)$, the events A and B are dependent.

33. Take 1 = Brand A, 2 = Brand B. $P = \begin{bmatrix} 1/2 & 1/2 \\ 1/4 & 3/4 \end{bmatrix}$

35. From Exercise 33, $P = \begin{bmatrix} 1/2 & 1/2 \\ 1/4 & 3/4 \end{bmatrix}$.

$$P^2 = \begin{bmatrix} 1/2 & 1/2 \\ 1/4 & 3/4 \end{bmatrix}\begin{bmatrix} 1/2 & 1/2 \\ 1/4 & 3/4 \end{bmatrix} = \begin{bmatrix} 3/8 & 5/8 \\ 5/16 & 11/16 \end{bmatrix}$$

$$P^3 = \begin{bmatrix} 1/2 & 1/2 \\ 1/4 & 3/4 \end{bmatrix}\begin{bmatrix} 3/8 & 5/8 \\ 5/16 & 11/16 \end{bmatrix} = \begin{bmatrix} 11/32 & 21/32 \\ 21/64 & 43/64 \end{bmatrix}$$

Take $v = \begin{bmatrix} 2/3 & 1/3 \end{bmatrix}$

Distribution after 3 years is $vP^3 = \begin{bmatrix} 2/3 & 1/3 \end{bmatrix}\begin{bmatrix} 11/32 & 21/32 \\ 21/64 & 43/64 \end{bmatrix} = \begin{bmatrix} 65/192 & 127/192 \end{bmatrix}$.

Brand A: $65/192 \approx .339$, Brand B: $127/192 \approx .661$

37. Take the first letter of each category shown in the table to stand for the corresponding event: S: The book is Sci Fi;

W: The book is stored in Washington, and so on.

$$P(S \cup T) = \frac{94 + 33 - 15}{200} = \frac{112}{200} = \frac{14}{25}$$

	S	H	R	O	Total
W	10	12	12	30	64
C	8	12	6	16	42
T	15	15	20	44	94
Total	33	39	38	90	200

39. Take the first letter of each category shown in the table to stand for the corresponding event: S: The book is Sci Fi;

W: The book is stored in Washington, and so on.

$$P(S|T) = \frac{n(S \cap T)}{n(T)} = \frac{15}{94}$$

	S	H	R	O	Total
W	10	12	12	30	64
C	8	12	6	16	42
T	15	15	20	44	94
Total	33	39	38	90	200

41. Take the first letter of each category shown in the table to stand for the corresponding event: S: The book is Sci Fi;

W: The book is stored in Washington, and so on.

$$P(T|S') = \frac{n(T \cap S')}{n(S')} = \frac{79}{167}$$

	S	H	R	O	Total
W	10	12	12	30	64
C	8	12	6	16	42
T	15	15	20	44	94
Total	33	39	38	90	200

43. Take H: Visited OHaganBooks.com; $P(H) = .02$.

$P(H') = 1 - P(H) = 1 - .02 = .98$, or 98%

45. Take H: Visited OHaganBooks.com; $P(H) = .02$; C: Visited a competitor; $P(C) = .05$.

Since H and C are independent, $P(H \cap C) = P(H)P(C) = .02 \times .05 = .001$.
Therefore,

$P(H \cup C) = P(H) + P(C) - P(H \cap C) = .02 + .05 - .001 = .069$, or 6.9%.

47. Take H: Visited OHaganBooks.com; $P(H) = .02$; C: Visited a competitor; $P(C) = .05$.

$P[(H \cup C)'] = 1 - P(H \cup C) = 1 - .069 = .931$

49. We are told that an online shopper visiting a competitor was more likely to visit OHaganBooks.com that a

381

randomly selected online shopper. That is, $P(H|C) > P(H)$.

Multiplying both sides by $P(C)$ gives $P(H|C)P(C) > P(H)P(C)$.

That is, $P(H \cap C) > P(H)P(C)$.

Therefore, $P(H \cap C)$ is greater.

51. Take H: Visited OHaganBooks.com; B: Purchased books.

$P(B|H) = .08$, $P(H) = .02$, $P(B|H') = .005$

We are asked to compute $P(H|B)$. Using Bayes' theorem,

$$P(H|B) = \frac{P(B|H)P(H)}{P(B|H)P(H) + P(B|H')P(H')} = \frac{(.08)(.02)}{(.08)(.02) + (.005)(.98)} = \frac{.0016}{.0016 + .0049} \approx .246.$$

53. Use the following events: A: Admitted; I: In-state.

$P(I) = .72$, $P(A|I) = .56$, $P(A|I') = .15$

We are asked to compute $P(I|A)$.

$$P(I|A) = \frac{P(A|I)P(I)}{P(A|I)P(I) + P(A|I')P(I')} = \frac{(.56)(.72)}{(.56)(.72) + (.15)(.28)} = \frac{.4032}{.4452} \approx .91$$

55. Number the states in the order shown in the table, so $P = \begin{bmatrix} .8 & .1 & .1 \\ .4 & .6 & 0 \\ .2 & 0 & .8 \end{bmatrix}$.

Starting distribution (July 1): $v = \begin{bmatrix} .2 & .4 & .4 \end{bmatrix}$

Distribution after 1 month:

$$vP = \begin{bmatrix} .2 & .4 & .4 \end{bmatrix} \begin{bmatrix} .8 & .1 & .1 \\ .4 & .6 & 0 \\ .2 & 0 & .8 \end{bmatrix} = \begin{bmatrix} .40 & .26 & .34 \end{bmatrix}$$

40% for OHaganBooks.com, 26% for JungleBooks.com, and 34% for FarmerBooks.com

57. Here are three factors that the Markov model does not take into account:
(1) It is possible for someone to be a customer at two different enterprises.
(2) Some customers may stop using all three of the companies.
(3) New customers can enter the field.

Section 8.1

1. $X =$ the sum of the numbers facing up when you roll two dice. The possible sums are $2, 3, ..., 12$. Thus, X is finite and has the set $\{2, 3, ..., 12\}$ of possible values.

3. X is the profit, to the nearest dollar, earned in a year if you purchase one share of a stock selected at random. The possible values of X are $0, \pm 1, \pm 2, ...$ (negative numbers indicate a loss). This gives an infinite number of discrete possible values for X. Therefore, X is a discrete infinite random variable with the set of possible values $\{0, 1, -1, 2, -2, ...\}$.

5. X is the time the second hand of your watch reads in seconds. This can be any real number between 0 and 60. Therefore, X is a continuous random variable that can assume any value between 0 and 60 (including 0).

7. The total number of goals that can be scored, up to a maximum of 10, is $1, 2, 3, ..., 9$, or 10. Therefore, X is a finite random variable with values in the set $\{0, 1, 2, ..., 10\}$.

9. The possible energies of an electron in a hydrogen atom are $k/1$, $k/4$, $k/9$, $k/16$, ... Thus, X is a discrete infinite random variable with set of possible values $\{k/1, k/4, k/9, k/16, ...\}$.

11. a. $S = \{HH, HT, TH, TT\}$

b. X is the rule that assigns to each outcome the number of tails.

c. Counting the number of tails in each outcome gives us the following values of X:

Outcome	HH	HT	TH	TT
Value of X	0	1	1	2

13. a. $S = \{(1, 1), (1, 2), ..., (1, 6), (2, 1), (2, 2), ..., (6, 6)\}$

b. X is the rule that assigns to each outcome the sum of the two numbers.

c. Computing the sum of the numbers in each outcome gives us the following values of X:

Outcome	$(1, 1)$	$(1, 2)$	$(1, 3)$	...	$(6, 6)$
Value of X	2	3	4	...	12

15. a. Take each outcome to be a pair of numbers (# of red marbles, # of green ones). Thus, $S = \{(4, 0), (3, 1), (2, 2)\}$. (The pairs $(1, 3)$ and $(0, 4)$ are impossible since there are only 2 green marbles.)

b. X is the rule that assigns to each outcome the number of red marbles.

c. Writing down the number of red marbles (the first coordinate) in each outcome gives the following values of X.

Outcome	(4, 0)	(3, 1)	(2, 2)
Value of X	4	3	2

17. a. $S =$ the set of students in the study group.

b. X is the rule that assigns to each student his or her final exam score.

c. The values of X, in the order given, are $89\%, 85\%, 95\%, 63\%, 92\%, 80\%$.

19. a. Assign letters to the missing values (we use the same letter for both since they are given to be equal):

x	2	4	6	8	10
$P(X = x)$	.1	.2	x	x	.1

Since the probabilities add to 1, we get

$$.1 + .2 + 2x + .1 = 1 \quad \Rightarrow \quad .4 + 2x = 1 \quad \Rightarrow \quad 2x = .6 \quad \Rightarrow \quad x = .3.$$

The completed table is:

x	2	4	6	8	10
$P(X = x)$	.1	.2	.3	.3	.1

b. $P(X \geq 6) = P(X = 6) + P(X = 8) + P(X = 10) = .3 + .3 + .1 = .7$

$P(2 < X < 8) = P(X = 4) + P(X = 6) = .2 + .3 = .5$

21. Since the probability that any specific number faces up is 1/6, the probability distribution for X is given by the following table and histogram:

x	1	2	3	4	5	6
$P(X = x)$	$\frac{1}{6}$	$\frac{1}{6}$	$\frac{1}{6}$	$\frac{1}{6}$	$\frac{1}{6}$	$\frac{1}{6}$

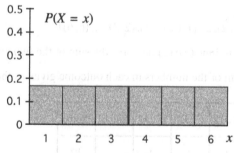

From the distribution, $P(X < 5) = P(X = 1) + P(X = 2) + P(X = 3) + P(X = 4) = \dfrac{4}{6} = \dfrac{2}{3}$.

23. The number of heads showing when you toss 3 fair coins is 0, 1, 2, or 3. The corresponding probabilities are 1/8, 3/8, 3/8, and 1/8 (see Example 3). The corresponding values of X are the squares of 0, 1, 2, and 3; that is, 0, 1, 4, and 9. This gives us the following distribution and histogram:

x	0	1	4	9
$P(X = x)$	$\frac{1}{8}$	$\frac{3}{8}$	$\frac{3}{8}$	$\frac{1}{8}$

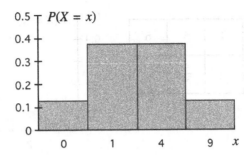

From the distribution, $P(1 \leq X \leq 9) = P(X = 1) + P(X = 4) + P(X = 9) = \frac{7}{8}$.

25. When two distinguishable dice are thrown, there are 36 possible outcomes. The values of X are the possible sums of the numbers facing up: $2, 3, \ldots, 11, 12$. To calculate their probabilities, we can use

$$P(X = 2) = P\{(1, 1)\} = \frac{1}{36}$$

$$P(X = 3) = P\{(1, 2), (2, 1)\} = \frac{2}{36}$$

$$P(X = 4) = P\{(1, 3), (2, 2), (3, 1)\} = \frac{3}{36}$$

and so on. The completed probability distribution is as follows:

x	2	3	4	5	6	7	8	9	10	11	12
$P(X = x)$	$\frac{1}{36}$	$\frac{2}{36}$	$\frac{3}{36}$	$\frac{4}{36}$	$\frac{5}{36}$	$\frac{6}{36}$	$\frac{5}{36}$	$\frac{4}{36}$	$\frac{3}{36}$	$\frac{2}{36}$	$\frac{1}{36}$

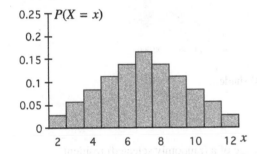

From the distribution, $P(X \neq 7) = 1 - P(X = 7) = 1 - \frac{6}{36} = \frac{30}{36} = \frac{5}{6}$.

27. The possible values of X are $1, 2, \ldots, 6$ (the larger of the two numbers facing up). We calculate their probabilities as follows:

$$P(X = 1) = P\{(1, 1)\} = \frac{1}{36}$$

$$P(X = 2) = P\{(1, 2), (2, 2), (2, 1)\} = \frac{3}{36}$$

$$P(X = 3) = P\{(1, 3), (2, 3), (3, 3), (3, 2), (3, 1)\} = \frac{5}{36}$$

and so on. The completed probability distribution is as follows:

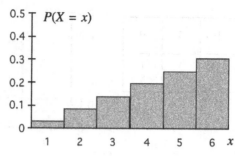

x	1	2	3	4	5	6
$P(X = x)$	$\frac{1}{36}$	$\frac{3}{36}$	$\frac{5}{36}$	$\frac{7}{36}$	$\frac{9}{36}$	$\frac{11}{36}$

From the distribution, $P(X \le 3) = P(X = 1) + P(X = 2) + P(X = 3) = \frac{9}{36} = \frac{1}{4}$.

Applications

29. a. For the values of X, use the rounded midpoints of the measurement classes given:

$(19{,}999 - 0)/2 \approx 10{,}000$ $\qquad$ $(39{,}999 - 20{,}000)/2 \approx 30{,}000$

$(59{,}999 - 40{,}000)/2 \approx 50{,}000$ $\qquad$ $(79{,}999 - 60{,}000)/2 \approx 70{,}000$

$(99{,}999 - 80{,}000)/2 \approx 90{,}000$

To obtain the probabilities, divide each frequency by the sum of all the frequencies, 1,000:

x	10,000	30,000	50,000	70,000	90,000
Freq	240	290	180	170	120
$P(X = x)$	.24	.29	.18	.17	.12

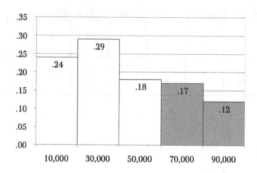

b. Above on the right is the histogram with the area corresponding to $X > 50{,}000$ shaded.

$P(X > 50{,}000) = .17 + .12 = .29$

31. The associated random variable is shown on the x axis of the histogram: $X =$ age of a (randomly selected) resident in Mexico.

To determine the probability distribution of X, use a frequency table. For the values of X, use the rounded midpoints of the measurement classes given:

$(0 + 14)/2 = 7$ $\quad$ $(15 + 29)/2 = 22$ $\quad$ $(30 + 64)/2 = 47$ $\quad$ $(65 + 74)/2 \approx 70$.

To obtain the relative frequencies, divide each frequency by the sum of all the frequencies, 250:

x	7	22	47	70
Freq	75	68	92	15
$P(X = x)$	.300	.272	.368	.060

33. a. The values of X are the possible tow ratings: 2,000, 3,000, 4,000, 5,000, 6,000, 7,000, 8,000 (7,000 is optional).

b. The following table shows the frequency (number of models with each tow rating) and the resulting probabilities (divide each frequency by the sum of all the frequencies):

x	2,000	3,000	4,000	5,000	6,000	7,000	8,000
Freq.	2	1	1	1	2	0	3
$P(X = x)$	.2	.1	.1	.1	.2	.0	.3

c. $P(X \leq 5{,}000) = .2 + .1 + .1 + .1 = .5$ (Add the probabilities in the first row of the above table.)

35. a. The values of X are the (rounded) changes of the Dow: $-700, -600, -500, -400, -300, -200, -100, 0, 100,$

$200, 300, 400, 500, 600, 700, 800, 900$ ($-600, 0, 100, 300, 500, 600, 700,$ and 800 are optional).

b. The following table shows the frequency (number of days with each specified change in the Dow) and the resulting probabilities (divide each frequency by the sum of all the frequencies, 20):

x	−700	−600	−500	−400	−300	−200	−100	0
Freq.	2	0	2	1	1	4	4	0
$P(X = x)$	.1	0	.1	.05	.05	.2	.2	0

x	100	200	300	400	500	600	700	800	900
Freq.	0	2	0	2	0	0	0	0	2
$P(X = x)$	0	.1	0	.1	0	0	0	0	.1

c. $P(X < -200) = .05 + .05 + .1 + .1 = .3$

37. To obtain the frequency table, count how many of the scores fall in each measurement class:

Class	1.1–2.0	2.1–3.0	3.1–4.0
Freq.	4	7	9

To obtain the probability distribution, divide each frequency by the sum of all the frequencies, 20:

x	1.5	2.5	3.5
$P(X = x)$	.20	.35	.45

39. First, construct the probability distribution:

x	0	1	2	3	4	5	6	7	8	9	10
Freq	140	350	450	650	200	140	50	10	5	15	10
$P(X = x)$	.07	.175	.225	.325	.1	.06	.025	.005	.0025	.0075	.005

$P(X < 6) = .07 + .175 + .225 + .325 + .1 + .06 = .955.$ Therefore, 95.5% of cars are newer than 6 years old.

41. The frequency distribution for X is obtained from the first row (small cars) of the given table, and we can use that to compute the probability distribution (divide each frequency by the sum of the frequencies, 16):

x	3	2	1	0
$P(X = x)$	.0625	.6875	.125	.125

43. $P(X \geq 2) = P(X = 2) + P(X = 3) = .6875 + .0625 = .75$

The relative frequency that a randomly selected small car is rated Good or Acceptable is .75.

45. The following tables give the frequency distributions for Y and Z:

y	3	2	1	0
Freq	1	4	4	1
$P(Y = y)$	.1	.4	.4	.1

z	3	2	1	0
Freq	3	5	3	4
$P(Z = z)$	.2	.3333	.2	.3333

$P(Y \geq 2) = P(Y = 2) + P(Y = 3) = .4 + .1 = .5$

$P(Z \geq 2) = P(Z = 2) + P(Z = 3) \approx .333 + .2 = .533$

Since a crash rating of at least 2 indicates "acceptable" or "good," the data suggest that medium SUVs are safer than small SUVs in frontal crashes.

47. $P(X = 3) = \dfrac{1}{16} = .0625$ $P(Y = 3) = \dfrac{1}{10} = .1$ $P(Z = 3) = \dfrac{3}{15} = .2$

$P(U = 3) = \dfrac{3}{13} \approx .2308$ $P(V = 3) = \dfrac{3}{15} = .2$ $P(W = 3) = \dfrac{9}{19} \approx .4737$

The lowest is $P(Y = 3)$, for small cars.

49. From Exercise 41, the probability that a randomly selected small car will be rated at least 2 is $P(X \geq 2) = .75$.

From Exercise 43, the probability that a randomly selected small SUV will be rated at least 2 is $P(Y \geq 2) = .5$.
Since these two events are independent (both vehicles are selected at random), the probability that both will be rated at least 2 is $P(X \geq 2) \times P(Y \geq 2) = .75 \times .5 = .375$.

51. The sample space is the set of all sets of 4 tents selected from 7; $n(S) = C(7, 4) = 35$. The possible values of X are 1, 2, 3, and 4. (X cannot equal 0, since that would require 4 green tents, and there are only 3.)

$$P(X = 1) = \frac{C(4, 1)C(3, 3)}{35} = \frac{4}{35} \qquad P(X = 2) = \frac{C(4, 2)C(3, 2)}{35} = \frac{18}{35}$$

$$P(X = 3) = \frac{C(4, 3)C(3, 1)}{35} = \frac{12}{35} \qquad P(X = 4) = \frac{C(4, 4)C(3, 0)}{35} = \frac{1}{35}$$

Probability distribution:

x	1	2	3	4
$P(X = x)$	$\dfrac{4}{35}$	$\dfrac{18}{35}$	$\dfrac{12}{35}$	$\dfrac{1}{35}$

$$P(X \geq 2) = 1 - P(X = 1) = 1 - \frac{4}{35} = \frac{31}{35} \approx .886$$

53. Answers will vary.

Communication and reasoning exercises

55. No; for instance, if X is the number of times you must toss a coin until heads comes up, then X is infinite but not continuous.

57. By measuring the values of X for a large number of outcomes and then using the estimated probability (relative frequency).

59. Here are two examples:

(1) Let X be the number of times you have read a randomly selected book.

(2) Let X be the number of days a diligent student waits before beginning to study for an exam scheduled in 10 days' time.

61. The bars should be 1 unit wide so that their height is numerically equal to their area.

63. Answers may vary. If we are interested in exact page counts, then the number of possible values is very large, and the values are (relatively speaking) close together, so using a continuous random variable might be advantageous. In general, the finer and more numerous the measurement classes, the more likely it becomes that a continuous random variable could be advantageous.

Section 8.2

1. $n = 5$, $p = .1$, $q = .9$; $P(X = 2) = C(5, 2)(.1)^2(.9)^3 = .0729$

3. $n = 5$, $p = .1$, $q = .9$; $P(X = 0) = C(5, 0)(.1)^0(.9)^5 = .59049$

5. $n = 5$, $p = .1$, $q = .9$; $P(X = 5) = C(5, 5)(.1)^5(.9)^0 = .00001$

7. $n = 5$, $p = .1$, $q = .9$; $P(X \leq 2) = P(X = 0) + P(X = 1) + P(X = 2)$

$P(X = 0) = C(5, 0)(.1)^0(.9)^5 = .59049$

$P(X = 1) = C(5, 1)(.1)^1(.9)^4 = .32805$

$P(X = 2) = C(5, 2)(.1)^2(.9)^3 = .0729$

Therefore, $P(X \leq 2) = .59049 + .32805 + .0729 = .99144$.

9. $n = 5$, $p = .1$, $q = .9$; $P(X \geq 3) = P(X = 3) + P(X = 4) + P(X = 5)$

$P(X = 3) = C(5, 3)(.1)^3(.9)^2 = .0081$

$P(X = 4) = C(5, 4)(.1)^4(.9)^1 = .00045$

$P(X = 5) = C(5, 5)(.1)^5(.9)^0 = .00001$

Therefore, $P(X \geq 3) = .0081 + .00045 + .00001 = .00856$.

11. $n = 6$, $p = .4$, $q = 1 - p = .6$; $P(X = 3) = C(6, 3)(.4)^3(.6)^3 = .27648$

13. $n = 6$, $p = .4$, $q = 1 - p = .6$; $P(X \leq 2) = P(X = 0) + P(X = 1) + P(X = 2)$

$P(X = 0) = C(6, 0)(.4)^0(.6)^6 = .046656$

$P(X = 1) = C(6, 1)(.4)^1(.6)^5 = .18662$

$P(X = 2) = C(6, 2)(.4)^2(.6)^4 = .31104$

Therefore, $P(X \leq 2) = .046656 + .18662 + .31104 = .54432$.

15. $n = 6$, $p = .4$, $q = 1 - p = .6$; $P(X \geq 5) = P(X = 5) + P(X = 6)$

$P(X = 5) = C(6, 5)(.4)^5(.6)^1 = .036864$ $P(X = 6) = C(6, 6)(.4)^6(.6)^0 = .004096$

Therefore, $P(X \geq 5) = .036864 + .004096 = .04096$.

17. $n = 6$, $p = .4$, $q = 1 - p = .6$; $P(1 \leq X \leq 3) = P(X = 1) + P(X = 2) + P(X = 3)$

$P(X = 1) = C(6, 1)(.4)^1(.6)^5 = .18662$

$P(X = 2) = C(6, 2)(.4)^2(.6)^4 = .31104$

$P(X = 3) = C(6, 3)(.4)^3(.6)^3 = .27648$

Therefore, $P(1 \leq X \leq 3) = .18662 + .31104 + .27648 = .77414$.

19. $n = 5$, $p = \dfrac{1}{4}$, $q = \dfrac{3}{4}$

We used Excel to generate the distribution. Format: BINOMDIST(x,n,p,0)

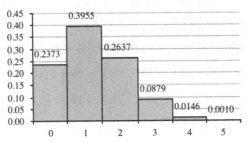

TI-83/84 Plus: binompdf(5,0.25,x)

21. $n = 4$, $p = \dfrac{1}{3}$, $q = \dfrac{2}{3}$; $P(X \le 2)$

We used Excel to generate the distribution.
Format: BINOMDIST(x,n,p,0).   (TI-83/84 Plus: binompdf(4,1/3,x))

The resulting values are shown on the histogram, with the portion corresponding to $P(X \le 2)$ shaded.

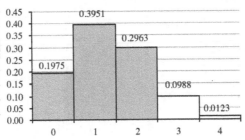

$P(X \le 2) \approx .1975 + .3951 + .2963 = .8889$

Applications

23. Take "success" = connect to the Internet immediately upon waking.

$n = 5$, $p = .25$, $q = 1 - .25 = .75$; $P(X = 2) = C(5,2)(.25)^2(.75)^3 \approx .2637$

25. Take "success" = a stock market success.

    $n = 10$, $p = .2$, $q = 1 - .2 = .8$; $P(X \ge 1) = 1 - P(X = 0)$

    $P(X = 0) = C(10,0)(.2)^0(.8)^{10} \approx .10737$

Therefore, $P(X \ge 1) \approx 1 - 10737 = .8926$.

27. Take "success" = selecting a male. $n = 3$, $p = .5$, $q = 1 - .5 = .5$

    $P(X \ge 1) = 1 - P(X = 0)$

    $P(X = 0) = C(3,0)(.5)^0(.5)^3 = .125$

Therefore, $P(X \ge 1) = 1 - 125 = .875$.

29. Take "success" = selecting a defective bag. $n = 5$, $p = .1$, $q = 1 - .1 = .9$

a. $P(X = 3) = C(5,3)(.1)^3(.9)^2 = .0081$

b. $P(X \ge 2) = 1 - P(X \le 1)$

$$P(X = 0) = C(5, 0)(.1)^0(.9)^5 = .59049$$

$$P(X = 1) = C(5, 1)(.1)^1(.9)^4 = .32805$$

Therefore, $P(X \geq 2) = 1 - (.59049 + .32805) = .08146$.

31. Take "success" = watching a rented video at least once. $n = 10$, $p = .71$, $q = 1 - .71 = .29$

$P(X \geq 8) = P(X = 8) + P(X = 9) + P(X = 10)$ $P(X = 8) = C(10, 8)(.71)8(.29)2 \approx .244$

$P(X = 9) = C(10, 9)(.71)9(.29)1 \approx .133$ $P(X = 10) = C(10, 10)(.71)10(.29)0 \approx .033$

Therefore, $P(X \geq 8) \approx .244 + .133 + .033 \approx .41$.

33. a. Take "success" = in foreclosure. $n = 10$, $p = .24$, $q = 1 - .24 = .76$

$$P(X = 5) = C(10, 5)(.24)^5(.76)^5 \approx .0509$$

b. Following is the distribution generated by the utility at Website   →   On-line Utilities   →   Binomial Distribution Utility:

(Number of trials) n = 10	(Probability of success) p = .24	
P(	≤X≤	)
	P(0) = 0.06428889	
	P(1) = 0.2030175	
	P(2) = 0.2884986	
	P(3) = 0.2429462	
	P(4) = 0.1342597	
	P(5) = 0.05087738	
	P(6) = 0.01338878	
	P(7) = 0.002416021	
	P(8) = 0.0002861078	
	P(9) = 0.00002007774	
	P(10) = 6.340338e-7	

c. The value of X with the largest probability is $X = 2$. So, the most likely number of homes to have been in foreclosure was __2__.

35. Take "success" = computer malfunction. $n = 3$, $p = .01$, $q = 1 - .01 = .99$

$$P(X \geq 2) = P(X = 2) + P(X = 3)$$

$$P(X = 2) = C(3, 2)(.01)^2(.99)^1 \approx .000297$$

$$P(X = 3) = C(3, 3)(.01)^3(.99)^0 \approx .000001$$

Therefore, $P(X \geq 2) \approx .000297 + .000001 = .000298$.

37. Take "success" = answering a question correctly. $n = 100$, $p = .80$

We use technology to compute $P(75 \leq X \leq 85) = P(X \leq 85) - P(X \leq 74)$

TI-83/84 Plus: `binomcdf(100,0.8,85)-binomcdf(100,0.8,74)`
Spreadsheet: `BINOMDIST(85,100,0.8,1)-BINOMDIST(74,100,0.8,1)`

Answer: $P(75 \leq X \leq 85) \approx .8321$

39. Take "success" = containing more than 10 grams of fat. $n = 50$, $p = .43$

We use technology to generate the probability distribution:
Spreadsheet: `BINOMDIST(x,50,0.43,0)`   TI-83/84 Plus: `binompdf(50,0.43,x)`

	A	B
1	x	P(X=x)
2	0	6.21932E-13
3	1	2.34588E-11
4	2	4.33577E-10
5		
49		
50	48	1.01478E-15
51	49	3.12463E-17
52	50	4.71436E-19

a. Since 43% of all the burgers contain more than 10 grams of fat, we can expect about $0.43 \times 50 = 21$ of them to contain more than 10 grams of fat.

b. The probability that k or more patties contain more than 10 grams of fat is $P(X \geq k) = 1 - P(X \leq k - 1)$
We want this to equal approximately .71:

$$1 - P(X \leq k - 1) \approx .71 \quad \Rightarrow \quad P(X \leq k - 1) \approx .29.$$

To answer this question, use the cumulative probability distribution:
　　Spreadsheet: `BINOMDIST(x,50,0.43,1)`　　TI-83/84 Plus: `binomcdf(50,0.43,x)`

	A	B
1	k	P(X ≤ k)
2	0	6.21932E-13
3	1	2.40808E-11
4	2	4.57658E-10
5		
19		
20	18	0.19634337
21	19	0.285669736
22	20	0.390118898
23	21	0.502683159
24	22	0.61461907

From the table, $k - 1 = 19$, so $k = 20$.
There is approximately a 71% chance that a batch of 50 ZeroFat patties contains _20_ or more patties with at least 10 grams of fat.
c. Graphs

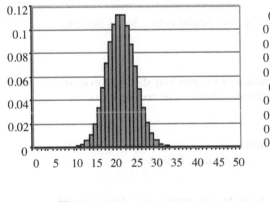

$n = 50$

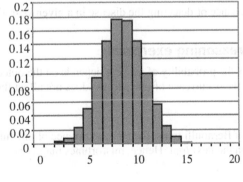

$n = 20$

The graph for $n = 50$ trials is more widely distributed than the graph for $n = 20$.

41. Take "success" = bad bulb. $p = .01$, $n = ?$

We want $P(X \geq 1) \geq .5$ $\Rightarrow$ $1 - P(X = 0) \geq .5$ $\Rightarrow$ $P(X = 0) \leq .5$

$C(n, 0)(.01)^0(.99)^n \leq .5$

$1 \times 1 \times .99^n \leq .5$ $\Rightarrow$ $.99^n \leq .5$

Using technology, we compute the values of $.99^n$ for $n = 0, 1, 2, ...$:

n	$.99^n$
0	1
1	0.99
...	...
68	0.50488589
69	0.49983703
...	...

The probability first dips below .5 when $n = 69$ trials.

43. The estimated probability of an accident in a given mile is

$$p = \frac{\text{Number of accidents}}{\text{Number of miles}} = \frac{562}{100,000,000} = .562 \times 10^{-5}$$

Since this is the estimated probability of a success, it is the probability that a male driver will have an accident in a 1-mile trip.

45. Let X be the number of cases of mad cow disease found among the 243,000 tested. Then X is a binomial random

variable with $p = \frac{5}{45,000,000} \approx 1.1111 \times 10^{-7}$ and $n = 243,000$.

We are asked to compute the probability of at least one "success": $P(X \geq 1)$.

$$P(X \geq 1) = 1 - P(X = 0) = 1 - C(n, 0)p^0(1 - p)^{n-0} = 1 - (1)(1)(1 - 1.1111 \times 10^{-7})^{243,000}$$

$$= 1 - (.999999889)^{243,000} \approx 1 - .9734 = .0266$$

Since there is only a 2.66% chance of detecting the disease in a given year, the government's claim seems dubious.

Communication and reasoning exercises

47. No; in the given scenario, the probability of success depends on the outcome of the previous shot. However, in a sequence of Bernoulli trials, the occurrence of one success does not affect the probability of success on the next attempt.

49. No; if life is a sequence of Bernoulli trials, then the occurrence of one misfortune ("success") does not affect the probability of a misfortune on the next trial. Hence, misfortunes may very well not "occur in threes."

51. Think of performing the experiment as a Bernoulli trial with "success" being the occurrence of E. Performing the experiment n times independently in succession would then be a sequence of n Bernoulli trials.

53. The probability of selecting a red marble changes after each selection, as the number of marbles left in the bag decreases. This violates the requirement that, in a sequence of Bernoulli trials, the probability of "success" is contant.

Section 8.3

1. $\bar{x} = \dfrac{-1+5+5+7+14}{5} = 6$

To compute the median, arrange the scores in order and take the (average of the) middle score(s).

    −1, 5, 5, 7, 14

Median = middle score = 5; Mode = most frequent score(s) = 5

3. $\bar{x} = \dfrac{2+5+6+7-1-1}{6} = 3$

To compute the median, arrange the scores in order and take the (average of the) middle score(s).

    −1, −1, 2, 5, 6, 7

Median = average of middle scores = $\dfrac{2+5}{2} = 3.5$; Mode = most frequent score(s) = −1

5. In decimal notation, the given scores are: 0.5, 1.5, −4, 1.25

$\bar{x} = \dfrac{0.5+1.5-4+1.25}{4} = -0.1875$

To compute the median, arrange the scores in order and take the (average of the) middle score(s).

    −4, 0.5, 1.25, 1.5

Median = average of middle scores = $\dfrac{0.5+1.25}{2} = 0.875$; Mode = most frequent score(s) = Every value

7. $\bar{x} = \dfrac{2.5-5.4+4.1-0.1-0.1}{5} = 0.2$

To compute the median, arrange the scores in order and take the (average of the) middle score(s).

    −5.4, −0.1, −0.1, 2.5, 4.1

Median = middle score = −0.1; Mode = most frequent score(s) = −0.1

9. Answers may vary. Two examples are: 0, 0, 0, 0, 0, 6 and 0, 0, 0, 1, 2, 3.

    $\bar{x} = 1$, Median = 0 for each sample.

11. We use the tabular method described in Example 3 in the textbook:

x	0	1	2	3
$P(X = x)$	.5	.2	.2	.1
$xP(X = x)$	0	0.2	0.4	0.3

$E(X)$ = Sum of entries in the bottom row = 0.9

13. We first convert the fractions into decimals, and then use the tabular method described in Example 3 in the textbook:

x	10	20	30	40
$P(X = x)$	.3	.4	.2	.1
$xP(X = x)$	3	8	6	4

$E(X)$ = Sum of entries in the bottom row = 21

15. We use the tabular method described in Example 3 in the textbook:

x	−5	−1	0	2	5	10
$P(X = x)$	.2	.3	.2	.1	.2	0
$xP(X = x)$	-1	-0.3	0	0.2	1	0

$E(X)$ = Sum of entries in the bottom row = −0.1

17. The probability distribution of X is

x	1	2	3	4	5	6
$P(X = x)$	1/6	1/6	1/6	1/6	1/6	1/6

Using the tabular method described in Example 3, we get

x	1	2	3	4	5	6
$P(X = x)$	1/6	1/6	1/6	1/6	1/6	1/6
$xP(X = x)$	1/6	2/6	3/6	4/6	5/6	6/6

$E(X)$ = sum of numbers in bottom row = 3.5

19. X is a binomial random variable with $n = 2$ and $p = .5$. Therefore,

$E(X) = np = 2(.5) = 1.$

21. We compute the probability distribution of X (see the solution to Exercise 27 in Section 8.1) and then use the tabular method of Example 3 to compute $E(X)$:

x	1	2	3	4	5	6
Freq	1	3	5	7	9	11
$P(X = x)$	$\dfrac{1}{36}$	$\dfrac{3}{36}$	$\dfrac{5}{36}$	$\dfrac{7}{36}$	$\dfrac{9}{36}$	$\dfrac{11}{36}$
$xP(X = x)$	$\dfrac{1}{36}$	$\dfrac{6}{36}$	$\dfrac{15}{36}$	$\dfrac{28}{36}$	$\dfrac{45}{36}$	$\dfrac{66}{36}$

$$E(X) = \text{Sum of entries in the bottom row} = \frac{161}{36} \approx 4.4722$$

23. Number of sets of 4 marbles $= C(6, 4) = 15$. If X is the number of red marbles, then the possible values of X are 2, 3, 4, and

$$P(X = x) = \frac{C(4, x)C(2, 4 - x)}{15}.$$

x	2	3	4
$P(X = x)$	$\dfrac{6}{15}$	$\dfrac{8}{15}$	$\dfrac{1}{15}$
$xP(X = x)$	$\dfrac{12}{15}$	$\dfrac{24}{15}$	$\dfrac{4}{15}$

$$E(X) = \text{Sum of entries in the bottom row} = \frac{40}{15} \approx 2.6667$$

25. X is a binomial random variable with $n = 20$ and $p = .1$. Therefore,

$$E(X) = np = 20(.1) = 2.$$

27. The number of possible hands of 5 cards is $C(52, 5) = 2,598,960$. The possible values of X are 0, 1, 2, 3, 4, and

$$P(X = x) = \frac{C(4, x)C(48, 5 - x)}{2,598,960}.$$

x	0	1	2	3	4
$P(X = x)$	.6588	.2995	.0399	.0017	.0000
$xP(X = x)$	0.0000	0.2995	0.0799	0.0052	0.0001

$$E(X) = \text{Sum of entries } xP(X = x) \approx 0.3846$$

Applications

29. The sum of the given Dow changes is $-1,500$. Therefore, $\overline{x} = \dfrac{-1,500}{10} = -150$.

If we arrange the Dow changes in order, we get

$$-700, -700, -500, -400, -200, -100, -100, -100, 400, 900.$$

The two middle scores are -200 and -100. Therefore, Median $= \dfrac{-200 + (-100)}{2} = -150.$

There were as many days with a change in the Dow above $\underline{-150}$ points as there were with changes below that. (See the definition of the median.)

31. The sum of the scores is 10,688. Therefore, $\bar{x} = \dfrac{10{,}688}{10} = 1{,}068.8.$

Arranging the scores in order gives:

$$1{,}049, 1{,}062, 1{,}062, 1{,}068, 1{,}071, 1{,}072, 1{,}072, 1{,}075, 1{,}076, 1{,}081.$$

The two middle scores are 1,071 and 1,072. Therefore, Median $= \dfrac{1{,}071 + 1{,}072}{2} = 1{,}071.5.$

The modes are the most frequently occurring scores: 1,062 and 1,072, which appear twice each.
Over the 10-business-day period sampled, the price of gold averaged \$1,068.80 per ounce. It was above \$1,071.50 as many times as it was below that price, and stood at \$1,062 and \$1,072 per ounce more often than at any other price.

33. a. We use the tabular method described in Example 3 in the textbook:

x	1	2	3	4	5	6	7	8	9	10
$P(X = x)$	.01	.04	.04	.08	.10	.15	.25	.20	.08	.05
$xP(X = x)$	0.01	0.08	0.12	0.32	0.5	0.9	1.75	1.6	0.72	0.5

$\mu = E(X) = $ sum of values $xP(X = x) = 6.5.$

There were, on average, 6.5 checkout lanes in a supermarket that was surveyed.

b. $P(X < \mu) = P(X < 6.5) = .01 + .04 + .04 + .08 + .10 + .15 = .42$

$P(X > \mu) = P(X > 6.5) = .25 + .20 + .08 + .05 = .58,$

and is thus larger. Most supermarkets have more than the average number of checkout lanes.

35. Using the rounded midpoints of the given measurement classes, we get the following table. (The probabilities are obtained by dividing the given frequencies by their sum, 72 and then rounding to 2 decimal places.)

x	5	10	15	20	25	35
$P(X = x)$	.17	.33	.21	.19	.03	.07
$xP(X = x)$	0.85	3.3	3.15	3.8	0.75	2.45

$E(X) = $ sum of values $xP(X = x) = 14.3$

Interpretation: The average age of a student in 1998 was 14.3.

37. For the values of X, use the rounded midpoints of the measurement classes given:

$$(0 + 14)/2 = 7 \qquad (15 + 29)/2 = 22 \qquad (30 + 64)/2 = 47 \qquad (65 + 74)/2 \approx 70$$

To estimate the average age, we use the tabular method described in Example 3 in the textbook:

x	7	22	47	70
Freq	75	68	92	15
$P(X = x)$	.300	.272	.368	.060
$xP(X = x)$	2.1	5.984	17.296	4.2

Average age = expected value of X = sum of values $xP(X = x) = 29.58 \approx 29.6$.

39. Using the rounded midpoints of the measurement classes, we get the following table:

x	10,000	30,000	50,000	70,000	90,000
$P(X = x)$	.24	.29	.18	.17	.12
$xP(X = x)$	2,400	8,700	9,000	11,900	10,800

The sum of the entries in the bottom row is 42,800, representing an average income of $42,800 \approx \$43,000$.

41. The probabilities in the tables below are obtained by dividing the given frequencies by the sum, 16 for X and 10 for Y:

x	3	2	1	0
$P(X = x)$	.0625	.6875	.125	.125
$xP(X = x)$	0.1875	1.375	0.125	0

y	3	2	1	0
$P(Y = y)$	.1	.4	.4	.1
$yP(Y = y)$	0.3	0.8	0.4	0

$E(X)$ = sum of values in the bottom row = 1.6875 $E(Y)$ = sum of values in the bottom row = 1.5

Because small cars (X) have a higher average rating, small cars performed better in frontal crashes.

43. The probabilities in the tables below are obtained by dividing the given frequencies by the sum, 16 for X and 10 for Y:

Small Cars:

x	3	2	1	0
$P(X = x)$	.0625	.6875	.125	.125
$xP(X = x)$	0.1875	1.375	0.125	0

Midsize Cars:

v	3	2	1	0
$P(V = v)$	.2	.333	0	.467
$vP(V = v)$	0.6	0.667	0	0

$E(X) = 1.6875$

$E(V) \approx 1.267$

Large Cars:

w	3	2	1	0
$P(W = w)$	.474	.263	.158	.105
$wP(W = w)$	1.421	0.526	0.158	0

$E(W) \approx 2.105$

Of the three, large cars (W) performed best.

45. Either you lose \$1 ($X = -1$) or win \$1 ($X = 1$).

There are 20 losing numbers out of 38, so $P(X = -1) = \dfrac{20}{38}$.

There are 18 winning numbers out of 38, so $P(X = 1) = \dfrac{18}{38}$.

x	-1	1
$P(X = x)$	$\dfrac{20}{38}$	$\dfrac{18}{38}$
$xP(X = x)$	$-\dfrac{20}{38}$	$\dfrac{18}{38}$

$E(X) = -\dfrac{20}{38} + \dfrac{18}{38} = -\dfrac{2}{38} \approx -0.53$ Expect to lose 53¢.

47. The given experiment consists of $n = 40$ Bernoulli trials with $p = .63$. Thus, the expected number of students that will shop at a mall during the next week is $\mu = np = 40 \times .63 = 25.2$ students.

49. a. The given experiment consists of $n = 20$ Bernoulli trials with $p = .10$. Therefore, the expected number of defective air bags is $\mu = np = 20(.10) = 2$.

b. $p = .10$, $\mu = 12$ and n is unknown.

$\mu = np \quad \Rightarrow \quad 12 = .10n \quad \Rightarrow \quad n = \dfrac{12}{.10} = 120$ air bags

51. The probability distribution for X = number of red tents is derived in the solution to Exercise 51 in Section 1 of this chapter. Here, we add a new row to compute the expected value:

x	1	2	3	4
$P(X = x)$	$\dfrac{4}{35}$	$\dfrac{18}{35}$	$\dfrac{12}{35}$	$\dfrac{1}{35}$
$xP(X = x)$	$\dfrac{4}{35}$	$\dfrac{36}{35}$	$\dfrac{36}{35}$	$\dfrac{4}{35}$

$E(X) = $ Sum of bottom row entries $= \dfrac{80}{35} \approx 2.2857$ tents

53. According to Exercise 37 in the section on Probability and counting techniques, the probability of winning this bet is $(8! \times 2^8)/16!$. Your expected winnings are therefore

$$1{,}000{,}000 \times \frac{8! \times 2^8}{16!} - 1 \times \left(1 - \frac{8! \times 2^8}{16!}\right) \approx -0.51.$$

So, you expect to lose about 51¢ on this bet, on average.

55. According to Exercise 39 in the section on Probability and counting techniques, the probability of someone picking all the correct winners at random was $1/2^{63}$. If X was the expected number of winners, then X was a binomial variable with $p = 1/2^{63}$ and $n = 50{,}000{,}000$, hence the expected value of X was

$$np = \frac{50{,}000{,}000}{2^{63}}.$$

The expected payout was therefore

$$1{,}000{,}000{,}000 \times \frac{50{,}000{,}000}{2^{63}} \approx 0.005.$$

In other words, Quicken Loans expected to pay out about half a cent. Put another way, they really didn't expect to have to pay out anything.

57. Let $X = $ rate of return of Fastforward Funds, and let $Y = $ rate of return of SolidState Securities.

The following worksheets show the computation of the expected values of X and Y:

	A	B	C
1	x	P(X = x)	xP(X = x)
2	-0.4	0.015	-0.006
3	-0.3	0.025	-0.0075
4	-0.2	0.043	-0.0086
5	-0.1	0.132	-0.0132
6	0	0.289	0
7	0.1	0.323	0.0323
8	0.2	0.111	0.0222
9	0.3	0.043	0.0129
10	0.4	0.019	0.0076
11		Total:	0.0397

	E	F	G
1	y	P(Y = y)	yP(Y = y)
2	-0.4	0.012	-0.0048
3	-0.3	0.023	-0.0069
4	-0.2	0.05	-0.01
5	-0.1	0.131	-0.0131
6	0	0.207	0
7	0.1	0.33	0.033
8	0.2	0.188	0.0376
9	0.3	0.043	0.0129
10	0.4	0.016	0.0064
11		Total:	0.0551

From the worksheets, we read off the following expected rates of return:
    FastForward: 3.97%; SolidState: 5.51%.
SolidState gives the higher expected return.

59. If a driver wrecks a car, the net cost to the insurance company is $\$100{,}000 - \$5{,}000 = \$95{,}000$,

so $X = -95{,}000$.

If a driver does not wreck a car, the net profit for the insurance company is the premium: $\$5{,}000$, so $X = 5{,}000$.
To compute the probability distribution, use the following tree:

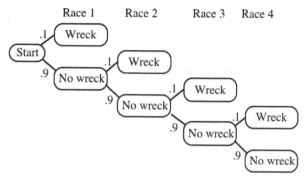

Race 1   Race 2   Race 3   Race 4

Probability of no wreck is $.9^4 = .6561$,

Probability of a wreck is $1 - .9^4 = .3439$,

x	-95,000	5000
$P(X = x)$	.3439	.6561
$xP(X = x)$	-32670.5	3280.5

$E(X) = $ sum of entries in bottom row $= -29,390$. The company can expect to lose \$29,390 per driver.

Communication and reasoning exercises

61. Since there are as many scores above the median as below the median, the correct choice is (A): The median and mean are equal.

63. He is wrong; for example, the collection 0, 0, 300 has mean 100 and median 0.

65. No; the expected number is the average number of times you will hit the bull's-eye per 50 shots; the average of a set of whole numbers need not be a whole number.

67. Not necessarily; it might be the case that only a small fraction of people in the class scored better than you but received exceptionally high scores that raised the class average. Suppose, for instance, that there are 10 people in the class. Four received 100%, you received 80%, and the rest received 70%. Then the class average is 83%, 5 people have lower scores that you, but only 4 have higher scores.

69. No; the mean of a very large sample is only an *estimate* of the population mean. The means of larger and larger samples *approach* the population mean as the sample size increases.

71. Wrong; the statement attributed to President Bush asserts that the mean tax saving would be \$1,000, whereas the statements referred to as "The Truth" suggest that the median tax saving would be close to \$100 (and that the 31st percentile would be zero).

73. Select a U.S. household at random, and let X be the income of that household. The expected value of X is then the population mean of all U.S. household incomes.

Section 8.4

1. For ease of computation, we arrange the data in a table. The first row lists the values of X, the second row lists the numbers $(x_i - \bar{x})$, and the third row lists their squares. The totals are shown in the last column.

x	−1	5	5	7	14	30
$(x - \bar{x})$	−7	−1	−1	1	8	0
$(x - \bar{x})^2$	49	1	1	1	64	116

$$\bar{x} = \frac{30}{5} = 6 \qquad \sum_i (x - \bar{x})^2 = 116 \qquad s^2 = \frac{\sum_i (x - \bar{x})^2}{n - 1} = \frac{116}{5 - 1} = 29 \qquad s = \sqrt{x^2} \approx 5.39$$

3. Below are the data arranged in a table, with the totals in the last column.

x	2	5	6	7	−1	−1	18
$(x - \bar{x})$	−1	2	3	4	−4	−4	0
$(x - \bar{x})^2$	1	4	9	16	16	16	62

$$\bar{x} = \frac{18}{6} = 3 \qquad \sum_i (x - \bar{x})^2 = 62 \qquad s^2 = \frac{\sum_i (x - \bar{x})^2}{n - 1} = \frac{62}{6 - 1} = 12.4 \qquad s = \sqrt{x^2} \approx 3.52$$

5. In the following table, we first converted all the fractions to decimals.

x	0.5	1.5	−4	1.25	−0.75
$(x - \bar{x})$	0.6875	1.6875	−3.8125	1.4375	0
$(x - \bar{x})^2$	0.4727	2.8477	14.5352	2.0664	19.9219

$$\bar{x} = \frac{-0.75}{4} = -0.1875 \qquad \sum_i (x - \bar{x})^2 \approx 19.9219 \qquad s^2 = \frac{\sum_i (x - \bar{x})^2}{n - 1} \approx \frac{19.9219}{4 - 1} \approx 6.64 \qquad s = \sqrt{x^2} \approx 2.58$$

7. Below are the data arranged in a table, with the totals in the last column.

x	2.5	−5.4	4.1	−0.1	−0.1	1
$(x - \bar{x})$	2.3	−5.6	3.9	−0.3	−0.3	0
$(x - \bar{x})^2$	5.29	31.36	15.21	0.09	0.09	52.04

$$\bar{x} = \frac{1}{5} = 0.2 \qquad \sum_i (x - \bar{x})^2 \approx 52.04 \qquad s^2 = \frac{\sum_i (x - \bar{x})^2}{n - 1} = \frac{52.04}{5 - 1} = 13.01 \qquad s = \sqrt{x^2} \approx 3.61$$

9. We use the tabular method described in Example 3 in the textbook:

x	0	1	2	3
$P(X = x)$	.5	.2	.2	.1
$x P(X = x)$	0	0.2	0.4	0.3

μ = Sum of entries in the bottom row = 0.9

$x - \mu$	−0.9	0.1	1.1	2.1
$(x - \mu)^2$	0.81	0.01	1.21	4.41
$(x - \mu)^2 P(X = x)$	0.405	0.002	0.242	0.441

σ^2 = Sum of entries in bottom row = 1.09 $\qquad \sigma = \sqrt{\sigma^2} \approx 1.04$

11. We first convert the fractions into decimals, and then use the tabular method described in Example 3 in the textbook:

x	10	20	30	40
$P(X = x)$	.3	.4	.2	.1
$x P(X = x)$	3	8	6	4

μ = Sum of entries in the bottom row = 21

$x - \mu$	−11	−1	9	19
$(x - \mu)^2$	121	1	81	361
$(x - \mu)^2 P(X = x)$	36.3	0.4	16.2	36.1

σ^2 = Sum of entries in bottom row = 89 $\qquad \sigma = \sqrt{\sigma^2} \approx 9.43$

13. We use the tabular method described in Example 3 in the textbook:

x	−5	−1	0	2	5	10
$P(X = x)$	.2	.3	.2	.1	.2	0
$xP(X = x)$	−1	−0.3	0	0.2	1	0

μ = Sum of entries in the bottom row = −0.1

$x - \mu$	−4.9	−0.9	0.1	2.1	5.1	10.1
$(x - \mu)^2$	24.01	0.81	0.01	4.41	26.01	102.01
$(x - \mu)^2 P(X = x)$	4.802	0.243	0.002	0.441	5.202	0

σ^2 = Sum of entries in bottom row = 10.69 $\sigma = \sqrt{\sigma^2} \approx 3.27$

15. The probability distribution and expected value calculated in Exercise 17 of Section 8.3:

x	1	2	3	4	5	6
$P(X = x)$	1/6	1/6	1/6	1/6	1/6	1/6
$xP(X = x)$	1/6	2/6	3/6	4/6	5/6	6/6

μ = sum of numbers in bottom row = 3.5

$x - \mu$	−2.5	−1.5	−0.5	0.5	1.5	2.5
$(x - \mu)^2$	6.25	2.25	0.25	0.25	2.25	6.25
$(x - \mu)^2 P(X = x)$	1.042	0.375	0.042	0.042	0.375	1.042

σ^2 = Sum of entries in bottom row ≈ 2.92 $\sigma = \sqrt{\sigma^2} \approx 1.71$

17. X is a binomial random variable with $n = 2$ and $p = .5$. Therefore,

$$\mu = E(X) = np = 2(.5) = 1 \qquad \sigma^2 = npq = 2(.5)(.5) = 0.5 \qquad \sigma = \sqrt{\sigma^2} \approx 0.71.$$

19. The probability distribution was calculated in Exercise 27 of Section 8.1. To continue the calculation, we use decimal approximations of the fractions:

x	1	2	3	4	5	6
$P(X = x)$	.0278	.0833	.1389	.1944	.25	.3056
$xP(X = x)$	0.0278	0.1667	0.4167	0.7778	1.25	1.8333

μ = Sum of entries in the bottom row ≈ 4.47

$x - \mu$	−3.4722	−2.4722	−1.4722	−0.4722	0.5278	1.5278
$(x - \mu)^2$	12.0563	6.1119	2.1674	0.223	0.2785	2.3341
$(x - \mu)^2 P(X = x)$	0.3349	0.5093	0.301	0.0434	0.0696	0.7132

σ^2 = Sum of entries in bottom row ≈ 1.9715 $\sigma = \sqrt{\sigma^2} \approx 1.40$

21. The probability distribution was calculated in Exercise 23 of Section 8.3. To continue the calculation, we use decimal approximations of the fractions:

x	2	3	4
$P(X = x)$	.4	.5333	.0667
$xP(X = x)$	0.8	1.6	0.2667

μ = Sum of entries in the bottom row $= \dfrac{40}{15} \approx 2.67$

$x - \mu$	−0.6667	0.3333	1.3333
$(x - \mu)^2$	0.4444	0.1111	1.7778
$(x - \mu)^2 P(X = x)$	0.1778	0.0593	0.1185

σ^2 = Sum of entries in bottom row ≈ 0.36 $\sigma = \sqrt{\sigma^2} \approx 0.60$

23. X is a binomial random variable with $n = 20$ and $p = .1$. Therefore,

$$\mu = E(X) = np = 20(.1) = 2 \qquad \sigma^2 = npq = 20(.1)(.9) = 1.8 \qquad \sigma = \sqrt{\sigma^2} \approx 1.34$$

Applications

25. a. Below are the data arranged in a table, with the totals in the last column.

x	3	2	0	9	1	15
$(x - \bar{x})$	0	−1	−3	6	−2	0
$(x - \bar{x})^2$	0	1	9	36	4	50

$$\bar{x} = \frac{15}{5} = 3 \qquad \sum_i (x - \bar{x})^2 = 50 \qquad s^2 = \frac{50}{5-1} \approx 12.5 \qquad s = \sqrt{x^2} \approx 3.54$$

b. The empirical rule states that approximately 68% of the class will rank you between

$$\bar{x} - s = 3 - 3.54 = -0.54 \quad \text{and} \quad \bar{x} + s = 3 + 3.54 = 6.54.$$

That is, in the interval [0, 6.54] (We replaced the negative score by 0, since no rankings can be negative.)
We must assume that the population distribution is bell shaped and symmetric.

27. a. Below are the data arranged in a table, with the totals in the last column.

x	4.2	4.7	5.4	5.8	4.9	25
$(x - \bar{x})$	−0.8	−0.3	0.4	0.8	−0.1	0
$(x - \bar{x})^2$	0.06	0.09	0.16	0.64	0.01	1.54

$$\bar{x} = \frac{25}{5} = 5.0 \qquad \sum_i (x - \bar{x})^2 = 1.54 \qquad s^2 = \frac{1.54}{5-1} = 0.385 \qquad s = \sqrt{x^2} \approx 0.6$$

b. The empirical rule states that approximately 95% of the data will fall between

$$\bar{x} - 2s \approx 5.0 - 1.2 = 3.8 \quad \text{and} \quad \bar{x} + 2s \approx 5.0 + 1.2 = 6.2.$$

29. a. Below are the data arranged in a table, with the totals in the last column.

x	−400	−500	−200	−700	−100	900	−100	−700	400	−100	−1,500
$(x - \bar{x})$	−250	−350	−50	−550	50	1050	50	−550	550	50	0
$(x - \bar{x})^2$	62,500	12,2500	2,500	302,500	2,500	1,102,500	2,500	302,500	302,500	2,500	2,205,000

$$\bar{x} = \frac{-1,500}{10} = -150 \qquad \sum_i (x - \bar{x})^2 = 2,205,000 \qquad s^2 = \frac{2,205,000}{10-1} = 245,000 \qquad s = \sqrt{x^2} \approx 495$$

b. The empirical rule states that approximately 68% of the data will fall between

$$\bar{x} - s \approx -150 - 495 \quad \text{and} \quad \bar{x} + s \approx -150 + 495 = 345.$$

Thus, $100 - 68 = 32\%$ of the time the data fall outside that range, and so, by symmetry, 16% of the data will be below

−645. That is, 16% of the time, the Dow will drop by more than 645 points.

To obtain the actual percentage of times the Dow fell by more than 645 points, count how many times this actually happened: twice.
Since 2 scores is 20% of the original 10, we conclude that the Dow actually fell by more than 645 points 20% of the time.

31. a. Below is a worksheet computation of the variance:

	A	B
1	x	(x - mu)^2
2	17	0.5625
3	18	0.0625
4	17	0.5625
5	18	0.0625
6	21	10.5625
7	16	3.0625
8	21	10.5625
9	18	0.0625
10	16	3.0625
11	14	14.0625
12	15	7.5625
13	22	18.0625
14	17	0.5625
15	19	1.5625
16	17	0.5625
17	18	0.0625
18	mean (mu)	variance
19	17.75	4.733333

From the worksheet,
$$s^2 \approx 4.7333 \qquad s = \sqrt{x^2} \approx 2.1756 \approx 2.18.$$

b. Chebyshev's inequality predicts that at least 8/9 of the scores will fall between
$$\bar{x} - 3s \approx 17.75 - 3(2.1756) \approx 11.22$$
and
$$\bar{x} + 3s \approx 17.75 + 3(2.1756) \approx 24.28.$$

That is, they will fall in the interval $[11.22, 24.28]$.

c. Every single score, or <u>100%</u> of the scores fall in the range $[11.22, 24.28]$. Since the empirical rule predicts that 99.7% of the scores will fall in the given range, the empirical rule is a more accurate predictor than Chebyshev's inequality.

33. We use the tabular method of arranging the data described in Example 3.

x	0	1	2	3	4
$P(X = x)$	.4	.1	.2	.2	.1
$xP(X = x)$	0	0.1	0.4	0.6	0.4

$\mu =$ Sum of entries in the bottom row $= 1.5$

$x - \mu$	−1.5	−0.5	0.5	1.5	2.5
$(x - \mu)^2$	2.25	0.25	0.25	2.25	6.25
$(x - \mu)^2 P(X = x)$	0.9	0.025	0.05	0.45	0.625

$\sigma^2 =$ Sum of entries in bottom row $= 2.05 \qquad \sigma = \sqrt{\sigma^2} \approx 1.43$

The range of X within two standard deviations of the mean is given by

Lower limit $= \mu - 2\sigma \approx 2.05 - 2(1.43) = -0.81 \qquad$ Upper limit $= \mu + 2\sigma \approx 2.05 + 2(1.43) = 4.91$

All (100%) of the values of X are within the interval $[-0.81, 4.91]$. Therefore, 100% of malls have a number of movie theater screens within two standard deviations of μ.

35. The probabilities for the distribution are computed by dividing the frequencies by the total: 1,000.

x	10	30	50	70	90
$P(X = x)$	.24	.29	.18	.17	.12
$xP(X = x)$	2.4	8.7	9.0	11.9	10.8

μ = Sum of entries in the bottom row = 42.8 $\approx$ 43, representing an average income of $43,000.

$x - \mu$	−32.8	−12.8	7.2	27.2	47.2
$(x - \mu)^2$	1,075.84	163.84	51.84	739.84	2,227.84
$(x - \mu)^2 P(X = x)$	258.2016	47.5136	9.3312	125.7728	267.3408

σ^2 = Sum of entries in bottom row = 708.18 $\sigma = \sqrt{\sigma^2} \approx 26.6$, representing about $27,000.

The range of X within one standard deviation of the mean is given by

Lower limit = $\mu - \sigma \approx 42.8 - 26.6 = 16.2$ Upper limit = $\mu + \sigma \approx 42.8 + 26.6 = 68.8$.

The difference is $68.8 - 16.2 = 52.6$ (two standard deviations), representing an income gap of about $53,000.

37. a. The rounded midpoints of the measurement classes (ages) are:

$(15 + 24.9)/2 \approx 20$ $(25 + 54.9)/2 \approx 40$ $(55 + 64.9)/2 \approx 60$.

In the following table, all intermediate calculations have been rounded to two decimal places.

x	20	40	60	Total
N	16,000	13,000	1,600	30,600
$P(X = x)$	.52	.42	.05	
$xP(X = x)$	10.4	16.8	3	30.2
$x - \mu$	−10.2	9.8	29.8	
$(x - \mu)^2$	104.04	96.04	888.04	
$(x - \mu)^2 P(X = x)$	54.1	40.34	44.4	138.84

Expected value = μ = Sum of entries in 4th row $\approx$ 30.2 yrs old

Variance = σ^2 = Sum of entries in bottom row $\approx$ 138.84

St. deviation = $\sigma = \sqrt{\sigma^2} \approx 11.78$ years

b. According to the empirical rule approximately 68% of all male Hispanic workers fall in the interval

$[\mu - \sigma, \mu + \sigma] = [30.2 - 11.78, 30.2 + 11.78] = [18.42, 41.98] \approx [18, 42]$.

So, approximately 68% of all male Hispanic workers are 18–42 years old.

39. Since the probability distribution is highly skewed, we need to use Chebyshev's rule.

$\mu = 2,\ \sigma = 0.15$

Following is a representation of the mean ± several standard deviations (each division is one standard deviation in width):

The range $[1.4, 2.6]$ represents the interval $\mu \pm 4\sigma$, so by Chebyshev's rule, at least 15/16 of companies have a lifespan in this range. Therefore, *at most* 1/16 have a lifespan outside this range (the gray regions above). Since the distribution is not symmetric, we cannot conclude that half of the 1/16 is in the range on the right (companies at least as old as yours).

Therefore, all we can say is that at most 1/16, or 6.25%, of all companies are in the range on the right (at least as old as yours).

41. Since the probability distribution is not known to be bell shaped, we need to use Chebyshev's rule.

$\mu = 9,\ \sigma = 2$

Following is a representation of the mean ± several standard deviations (each division is one standard deviation in width):

The range $[5, 13]$ represents the interval $\mu \pm 2\sigma$, so by Chebyshev's rule, at least 3/4 of all Batmobiles have a lifespan in this range. Therefore, *at most* 1/4 have a lifespan outside this range (the gray regions above). Since the distribution is symmetric, at most half of these, 1/8 of all Batmobiles, have life spans more than 13 years (the gray region on the right).

Therefore, there is *at most* (Choice B) a 12.5% chance that your new Batmobile will last 13 years or more.

43. a. Take "success" to mean shopping at a mall. The given distribution is a binomial distribution with $n = 40$,

$p = .63,\ q = 1 - .63 = .37$

$\mu = np = 40 \times .63 = 25.2$ teenagers

$\sigma = \sqrt{npq} = \sqrt{40 \times .63 \times .37} \approx 3.05$

b. Since the binomial distribution is symmetric and bell shaped, we can use the empirical rule, which says there is a 95% chance that between $\mu - 2\sigma = 25.2 - 2(3.05) \approx 19$ and $\mu + 2\sigma = 25.2 + 2(3.05) \approx 31$ teenagers in the sample will shop at a mall during the next week.

Therefore, there is a 5% chance of this not happening—that is, a 5% chance that either 19 or fewer teenagers will shop at a mall, or that 31 or more will. Since the distribution is symmetric, there is a 2.5% chance that <u>31</u> or more teenagers in the group will shop at a mall during the next week.

45. a. Take "success" to mean *not* having a checking account. The given distribution is a binomial distribution with $n = 1{,}000,\ p = 1 - .22 = .78,\ q = .22$

$\mu = np = 1{,}000 \times .78 = 780$ teenagers

$\sigma = \sqrt{npq} = \sqrt{1{,}000 \times .78 \times .22} \approx 13.1$

b. Since the binomial distribution is symmetric and bell shaped, we can use the empirical rule, which says there is a 95% chance that that between $\mu - 2\sigma = 780 - 2(13.1) \approx 754$ and $\mu + 2\sigma = 780 + 2(13.1) \approx 806$ teenagers in the sample will not have checking accounts.

47. a. Below is a worksheet computation of the variance (the bottom row gives the column sums):

	A	B	C	D
1	x	P(x)	x*P(x)	(x-mu)^2*P(x)
2	1	0.01	0.01	0.3025
3	2	0.04	0.08	0.81
4	3	0.04	0.12	0.49
5	4	0.08	0.32	0.5
6	5	0.1	0.5	0.225
7	6	0.15	0.9	0.0375
8	7	0.25	1.75	0.0625
9	8	0.2	1.6	0.45
10	9	0.08	0.72	0.5
11	10	0.05	0.5	0.6125
12	55	1	6.5	3.99

From the worksheet,
$$\mu = 6.5 \qquad \sigma^2 = 3.99 \approx 4.0 \qquad \sigma = \sqrt{\sigma^2} \approx 2.0.$$
b. According to Chebyshev's inequality, at least 3/4 or 75% of all supermarkets will have between
$$\mu - 2\sigma \approx 6.5 - 2(2.0) = 2.5$$
and
$$\mu + 2\sigma \approx 6.5 + 2(2.0) = 10.5$$
checkout lanes.

The smallest (whole) number of checkout lanes in this range is 3.

49. The household income of a poor family in the U.S. is $38,000 - 1.3(21,000) = \$10,700$ or less.

51. The household income of a rich family in the U.S. is $38,000 + 1.3(21,000) = \$65,300$ or more.

53. Cutoffs for poor families (1.3 standard deviations):

U.S.: $38,000 - 1.3(21,000) = \$10,700$ Canada: $35,000 - 1.3(17,000) = \$12,900$

Switzerland: $39,000 - 1.3(16,000) = \$18,200$ Germany: $34,000 - 1.3(14,000) = \$15,800$

Sweden: $32,000 - 1.3(11,000) = \$17,700$

The U.S. has the poorest households.

55. The gap between rich and poor is measured by $2 \times 1.3 = 2.6$ standard deviations. Since the U.S. has the largest standard deviation listed, it has the largest gap between rich and poor.

57. An income of $17,000 is 1 standard deviation below the U.S. mean income. By the empirical rule, approximately 68% earned within 1 standard deviation, so that approximately 32% earn outside the one-standard-deviation interval. Half of that, approximately 16%, earn less.

59. By the empirical rule, approximately 99.7% earned within three standard deviations of the mean. This is the range
$$34,000 - 3(14,000) = -8,000 \quad \text{to} \quad 34,000 + 3(14,000) = 76,000$$
Since income can't be negative, the answer is 0–$76,000.

61. Using technology: TI: Enter the data in L_1 and then `1-Var Stats`   Spreadsheet: `=STDEVP()`

2000 data: $\mu = 12.56\%$, $\sigma \approx 1.8885\%$ 2010 data: $\mu = 13.30\%$, $\sigma \approx 1.6643\%$

63. The mean for of the aging populations in 2010 is larger than that for 2000, suggesting that the population was older in 2010.
The standard deviation for 2010 is less than that for 2000, so that the variation across states of percentage of the aging population is lower, suggesting that the population was less diverse with respect to age in 2010. (Choice (B))

65. The empirical rule predicts approximately 68%. The one-standard-deviation interval based on the calculations in Exercise 61 is
$$\mu - \sigma = 13.30 - 1.6643 \approx 11.64 \qquad \mu + \sigma = 13.30 + 1.6643 \approx 14.96$$
The scores in that range are shown in bold:
8, 9, 10, 11, **11, 11, 12, 12, 12, 12, 12, 12, 12, 13, 13, 13, 13, 13, 13, 13, 13, 13, 14, 14, 14, 14, 14, 14, 14, 14, 14, 14, 14, 14, 14, 14, 14, 14, 14, 14, 14, 14**, 15, 15, 15, 15, 15, 16, 16, 17
These are 36 states, representing 36/50 = 72% of all states. This differs substantially from the empirical rule

prediction. One reason for the discrepancy is that the associated probability distribution is roughly bell shaped but not symmetric, as shown in the following figure:

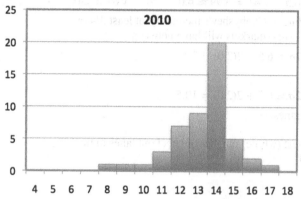

67. The two-standard-deviation interval based on the calculations in Exercise 61 is

$$\mu - 2\sigma = 13.30 - 2(1.6643) \approx 9.971 \qquad \mu + 2\sigma = 13.30 + 2(1.6643) \approx 16.63.$$

The scores in that range are shown in bold:
8, 9, **10, 11, 11, 11, 12, 12, 12, 12, 12, 12, 12, 13, 13, 13, 13, 13, 13, 13, 13, 13, 14, 14, 14, 14, 14, 14, 14, 14, 14, 14,
14, 14, 14, 14, 14, 14, 14, 14, 14, 14, 15, 15, 15, 15, 15, 16, 16,** 17
These are 47 states, representing 47/50 = 94% of all states. Chebyshev's rule is valid, since it predicts that *at least* 75% of the scores are in this range. This is not surprising, as Chebyshev's rule is *always* valid.

69. (A) The graph shows standard deviations, not the actual power grid frequency, so we cannot conclude (A).
(B) The standard deviation indicates the variability in the power supply frequency. Since it was lower in mid-1999 than in 1995, this indicates greater stability in mid-1999 than in 1995, so the assertion is true.
(C) The standard deviation indicates the variability in the power supply frequency. Since it was higher in mid-2002 than in mid-1995, this indicates *less* stability in mid-2002, so we cannot conclude (C).
(D) The standard deviation was greatest in 2001–2002, indicating that the greatest fluctuations in the power grid frequency occurred during that period, so the assertion is true.
(E) Around January 1995, the standard deviation was closest to its average of 0.9 but was lower around January 1999, so the power grid was more stable around January 1999 than around January 1995. Thus, (E) is false.

71. (A), (B), (C) From 2002 on, the mean is definitely trending upward, but the fluctuations appear to be trending downward, so the standard deviation is not increasing (choice (B)).
(D) False; what was greater in 2003 was the mean. The fluctuations in 2001 were significantly greater
(E) The mean of the monthly means in 2002 was greater than 4.0, whereas the mean of the 2000 monthly means was considerably lower. On the other hand, those means fluctuated more in 2000 than in 2002, so the statement is true.

Communication and reasoning exercises

73. The sample standard deviation is bigger; the formula for sample standard deviation involves division by the smaller term $n - 1$ instead of n, which makes the resulting number larger.

75. The grades in the first class were clustered fairly close to 75. By Chebyshev's inequality, at least 88% of the class had grades in the range 60–90. On the other hand, the grades in the second class were widely dispersed. The second class had a much wider spread of ability than did the first class.

77. Since the standard deviation is 0, there is no variability at all; that is, the variable must be constant. Therefore, the variable must take on only the value 10, with probability 1.

79. Since there are 2 data points, the mean is midway between them, at a distance of $(y - x)/2$ from each. Summing the square of this distance twice gives

$$\sum_i (x_i - \mu)^2 = \frac{(y-x)^2}{4} + \frac{(y-x)^2}{4} = \frac{(y-x)^2}{2}.$$

Dividing by 2 gives the variance:

$$\sigma^2 = \frac{(y-x)^2}{4}.$$

Therefore, $\sigma = \frac{y-x}{2}$.

Section 8.5

1. Note: Answers for this section were computed by using the 4-digit table in the Appendix, and may differ slightly from the more accurate answers generated by using technology.

We use the table to find $P(0 \leq Z \leq 0.5) = .1915$.

Z	0.00	0.01	0.02
0.4	.1554	.1591	.1628
0.5	.1915	.1950	.1985
0.6	.2257	.2291	.2324

3. The table gives us $P(0 \leq Z \leq 0.71) = .2611$.

Z	0.00	0.01	0.02
0.6	.2257	.2291	.2324
0.7	.2580	.2611	.2642
0.8	.2881	.2910	.2939

$P(-0.71 \leq Z \leq 0.71)$ is twice this area, as shown:

$P(-0.71 \leq Z \leq 0.71) = 2(.2611) = .5222$

5. The table gives us $P(0 \leq Z \leq 1.34) = .4099$.

Z	0.03	0.04	0.05
1.2	.3907	.3925	.3944
1.3	.4082	.4099	.4115
1.4	.4236	.4251	.4265

$P(0 \leq Z \leq 0.71) = .2611$

Z	0.00	0.01	0.02
0.6	.2257	.2291	.2324
0.7	.2580	.2611	.2642
0.8	.2881	.2910	.2939

$P(-0.71 \leq Z \leq 1.34)$ is obtained by adding these areas (see figure).

7. From the table,

$P(0 \leq Z \leq 1.5) = .4332$

$P(0 \leq Z \leq 0.5) = .1915$.

To obtain $P(0.5 \leq Z \leq 1.5)$, subtract the smaller from the larger (see figure).

$P(0.5 \leq Z \leq 1.5) = .4332 - .1915 = .2417$

$$P(-0.71 \leq Z \leq 1.34) = .4099 + .2611 = .6710$$

9. $\mu = 50$, $\sigma = 10$. Standardize the given problem:

$$P(a \leq X \leq b) = P\left(\frac{a - \mu}{\sigma} \leq Z \leq \frac{b - \mu}{\sigma}\right)$$

$$P(35 \leq X \leq 65) = P\left(\frac{35 - 50}{10} \leq Z \leq \frac{65 - 50}{10}\right) = P(-1.5 \leq Z \leq 1.5) = 2(.4332) = .8664$$

11. $\mu = 50$, $\sigma = 10$. Standardize the given problem:

$$P(a \leq X \leq b) = P\left(\frac{a - \mu}{\sigma} \leq Z \leq \frac{b - \mu}{\sigma}\right)$$

$$P(30 \leq X \leq 62) = P\left(\frac{30 - 50}{10} \leq Z \leq \frac{62 - 50}{10}\right) = P(-2 \leq Z \leq 1.2) = .3849 + .4772 = .8621$$

13. $\mu = 100$, $\sigma = 15$. Standardize the given problem:

$$P(a \leq X \leq b) = P\left(\frac{a - \mu}{\sigma} \leq Z \leq \frac{b - \mu}{\sigma}\right)$$

$$P(110 \leq X \leq 130) = P\left(\frac{110 - 100}{15} \leq Z \leq \frac{130 - 100}{15}\right) \approx P(0.67 \leq Z \leq 2) = .4772 - .2486 = .2286$$

15. The Z-value measures the number of standard deviations form the mean. Therefore, the given problem translates to

$$P(-0.5 \leq Z \leq 0.5) = 2(.1915) = .3830.$$

17. This is the probability that Z is either $> \frac{2}{3}$ or $< -\frac{2}{3}$. The complement of this event is the event that $-\frac{2}{3} \leq Z \leq \frac{2}{3}$.

$$P(-0.67 \leq Z \leq 0.67) = 2(.2486) = .4972$$

Therefore, the desired probability is

$$1 - .4972 = .5028.$$

19. Normalizing, we have

$$P(100 \leq X \leq b) = P\left(0 \leq Z \leq \frac{b - 100}{10}\right) = .3.$$

Looking inside the table, we see that $P(0 \leq Z \leq 0.84) \approx .3$, so

415

$$\frac{b - 100}{10} = 0.84$$

$$b = 10 \times 0.84 + 100 = 108.4.$$

21. Normalizing, we have

$$P(X \geq a) = P\left(Z \geq \frac{a - 100}{10}\right) = .04.$$

Because $P(Z \geq b) = .5 - P(0 \leq Z \leq b)$, we look inside the table to find $P(0 \leq Z \leq 1.75) \approx .46$, hence

$P(Z \geq 1.75) \approx .04.$ So,

$$\frac{a - 100}{10} = 1.75$$

$$a = 10 \times 1.75 + 100 = 117.5.$$

23. $P(10 \leq X \leq 15) = P(9.5 \leq Y \leq 15.5)$, where Y has a mean of $\mu = np = 100 \times \frac{1}{6} \approx 16.6667$ and a standard

deviation of $\sigma = \sqrt{npq} = \sqrt{100 \times \frac{1}{6} \times \frac{5}{6}} \approx 3.7268$. We now standardize Y:

$$P(9.5 \leq Y \leq 15.5) = P\left(\frac{9.5 - 16.6667}{3.7268} \leq Z \leq \frac{15.5 - 16.6667}{3.7268}\right)$$

$$\approx P(-1.92 \leq Z \leq -0.31) = .4726 - .1217 = .3509 \approx .35.$$

25. $P(X < 25) = P(0 \leq X \leq 24) = P(-0.5 \leq Y \leq 24.5)$, where Y has a mean of $\mu = np = 200 \times \frac{1}{6} \approx 33.3333$ and a

standard deviation of $\sigma = \sqrt{npq} = \sqrt{200 \times \frac{1}{6} \times \frac{5}{6}} \approx 5.2705$. We now standardize Y:

$$P(-0.5 \leq Y \leq 24.5) = P\left(\frac{-0.5 - 33.3333}{5.2705} \leq Z \leq \frac{24.5 - 33.3333}{5.2705}\right)$$

$$\approx P(-6.42 \leq Z \leq -1.68) = .5000 - .4535 = .0465 \approx .05.$$

Applications

27. $\mu = 500$, $\sigma = 100$

$$P(450 \leq X \leq 550) = P\left(\frac{450 - 500}{100} \leq Z \leq \frac{550 - 500}{100}\right) = P(-0.5 \leq Z \leq 0.5) = .1915 + .1915 = .3830$$

29. $\mu = 151$, $\sigma = 7$

$$P(137 \leq X \leq 158) = P\left(\frac{137 - 151}{7} \leq Z \leq \frac{158 - 151}{7}\right) = P(-2 \leq Z \leq 1) = .4772 + .3413 = .8185$$

31. $\mu = 100$, $\sigma = 16$

$$P(110 \leq X \leq 140) = P\left(\frac{110 - 100}{16} \leq Z \leq \frac{140 - 100}{16}\right) \approx P(0.63 \leq Z \leq 2.5) = .4938 - .2357 = .2581,$$

approximately 26%

33. $\mu = 100$, $\sigma = 16$

$$P(X \geq 120) = P\left(Z \geq \frac{120 - 100}{16}\right) = P(Z \geq 1.25) = .5 - .3944 = .1056$$

The total number of such people in the United States is $.1056 \times 323{,}000{,}000 \approx 34{,}100{,}000$.

35. $\mu = 500$, $\sigma = 100$. We seek a such that $P(X \geq a) = .05$. Normalizing, we have

$$P(X \geq a) = P\left(Z \geq \frac{a - 500}{100}\right) = .05.$$

Because $P(Z \geq b) = .5 - P(0 \leq Z \leq b)$, we look inside the table to find $P(0 \leq Z \leq 1.645) \approx .45$ (it appears to be halfway between 1.64 and 1.65), hence $P(Z \geq 1.645) \approx .05$. So,

$$\frac{a - 500}{100} = 1.645$$

$$a = 100 \times 1.645 + 500 \approx 665.$$

37. $\mu = 0.250$, $\sigma = 0.03$

$$P(X \geq 0.400) = P\left(Z \geq \frac{0.400 - 0.250}{0.03}\right) = P(Z \geq 5) = .5 - .5000 = .0000$$

to 4 decimal places. The total number of such batters expected is therefore 0.

39. $\mu = 7.5$, $\sigma = 1$

$$P(X \geq 9) = P\left(Z \geq \frac{9 - 7.5}{1}\right) = P(Z \geq 1.5) = .5 - .4332 = .0668$$

The total number of jars is therefore $.0668 \times 100{,}000 \approx 6{,}680$.

41. $\mu = 38$, $\sigma = 21$ (in thousands of dollars)

$$P(X \geq 50) = P\left(Z \geq \frac{50 - 38}{21}\right) \approx P(Z \geq 0.57) = .5 - .2157 = .2843;$$

approximately 28%

43. $\mu = 39$, $\sigma = 16$ (in thousands of dollars)

Very rich: $P(X \geq 100) = P\left(Z \geq \frac{100 - 39}{16}\right) \approx P(Z \geq 3.81) = .5 - .5000 = 0$

Very poor: $P(X \leq 12) = P\left(Z \leq \frac{12 - 39}{16}\right) \approx P(Z \leq -1.69) = .5 - .4545 = .0455;$

approximately 5%

45. United States: $\mu = 38$, $\sigma = 21$ (in thousands of dollars)

$$P(X \leq 12) = P\left(Z \leq \frac{12 - 38}{21}\right) \approx P(Z \leq -1.24) = .5 - .3925 = .1075$$

Canada: $\mu = 35$, $\sigma = 17$ (in thousands of dollars)

$$P(X \leq 12) = P\left(Z \leq \frac{12 - 35}{17}\right) \approx P(Z \leq -1.35) = .5 - .4115 = .0885$$

The United States has a higher proportion of very poor people.

47. Wechsler; as this test has a smaller standard deviation, a greater percentage of scores fall within 20 points of the mean.

49. $\mu = 6$, $\sigma = 1$. The Z-value corresponding to $X = 1$ month is

$$Z = \frac{1-6}{1} = -5.$$

This is surprising, because the time between failures was more than 5 standard deviations away from the mean, which happens with an extremely small probability.

51. Task 1: $\mu = 11.4$, $\sigma = 5.0$

$$P(X \geq 10) = P\left(Z \geq \frac{10 - 11.4}{5}\right) \approx P(Z \geq -0.28) = .5 + .1103 = .6103$$

53. Task 1: By Exercise 51, $P(X_1 \geq 10) = .6103$.

Task 2: $\mu = 11.9$, $\sigma = 9$

$$P(X_2 \geq 10) = P\left(Z \geq \frac{10 - 11.9}{9}\right) \approx P(Z \geq -0.21) = .5 + .0832 = .5832$$

Because the times taken to complete the tasks are independent,

$$P(X_1 \geq 10 \text{ and } X_2 \geq 10) = P(X_1 \geq 10) \times P(X_2 \geq 10) = .6103 \times .5832 \approx .3559$$

55. The total time it takes to complete tasks 1 and 2 is $X = X_1 + X_2$, which is normal with

$$\mu = 11.4 + 11.9 = 23.3 \text{ and } \sigma = \sqrt{5^2 + 9^2} \approx 10.2956$$

$$P(X \geq 20) = P\left(Z \geq \frac{20 - 23.3}{10.2956}\right) \approx P(Z \geq -0.32) = .5 + .1255 = .6255.$$

57. $\mu = np = 1,200 \times .71 = 852$

$$\sigma = \sqrt{npq} = \sqrt{1,200 \times 0.71 \times 0.29} \approx 15.719$$

We compute $\mu - 3\sigma \approx 805$ and $\mu + 3\sigma \approx 899$. Because these are between 0 and $n = 1,200$, the normal approximation of the binomial distribution is valid.

$$P(840 \leq X) \approx P(839.5 \leq Y) = P\left(\frac{839.5 - 852}{15.719} \leq Z\right) \approx P(-0.80 \leq Z) = .5 + .2881 = .7881$$

59. $n = 10,000,000$, $p = .00000276$

$$\mu = np = 27.6; \sigma = \sqrt{npq} = \sqrt{27.6 \times .99999724} \approx 5.254$$

We compute $\mu - 3\sigma \approx 12$ and $\mu + 3\sigma \approx 43$. Because these are between 0 and $n = 10,000,000$, the normal approximation of the binomial distribution is valid.

$$P(X < 35) = P(X \leq 34) = P(Y \leq 34.5) = P\left(Z \leq \frac{34.5 - 27.6}{5.254}\right) \approx P(Z \leq 1.31) = .4049 + .5 = .9049$$

61. $n = 10,000,000$, $p = .00000276$

$$\mu = np = 27.6; \sigma = \sqrt{npq} = \sqrt{276 \times .99999724} \approx 5.254$$

We compute $\mu - 3\sigma \approx 12$ and $\mu + 3\sigma \approx 43$. Because these are between 0 and $n = 10,000,000$, the normal approximation of the binomial distribution is valid.

Suppose there are X crashes. Since 10 people buy insurance, the payout is $10 \times 1,000,000 = \$10,000,000$ per crash; that is, $10,000,000X$. On the other hand, the premium you receive is

$$10 \times 3 \times 10,000,000 = \$300,000,000.$$

For breakeven you must have

$$10,000,000X = 300,000,000$$

so $X = 30$ flights. To lose money, $X > 30$:

$$P(X > 30) = P(X \geq 31) = P(Y \geq 30.5) = P\left(Z \geq \frac{30.5 - 27.6}{5.254}\right) \approx P(Z \geq 0.55) = .5 - .2088 = .2912.$$

63. Let "success" = a person polled says that he or she prefers Goode.

$$p = .9 \times .55 + .1 \times .45 = .54$$

$$n = 1,000$$

$$\mu = np = 540$$

$$\sigma = \sqrt{npq} = \sqrt{1,000 \times .54 \times .46} \approx 15.761$$

We compute $\mu - 3\sigma \approx 493$ and $\mu + 3\sigma \approx 587$. Because these are between 0 and $n = 1,000$, the normal approximation of the binomial distribution is valid.

$$P(X > 520) = P(X \geq 521) = P(Y \geq 520.5) = P\left(Z \geq \frac{520.5 - 540}{15.761}\right) \approx P(Z \geq -1.25) = .5 + .3925 = .8925$$

65. $\mu = 100$, σ unknown. Let the Z-score of someone who just barely qualifies be k. To be in the top 2%,

$P(Z \geq k) = .02$. Therefore,

$$P(0 \leq Z \leq k) = .5 - .02 = .48.$$

Looking in the table for the closest score to .48 (which is .4798) gives $k \approx 2.05$. Since $Z = \dfrac{148 - 100}{\sigma}$, we have

$$k = 2.05 = \frac{148 - 100}{\sigma} = \frac{48}{\sigma}$$

$$2.05\sigma = 48,$$

so

$$\sigma = \frac{48}{2.05} \approx 23.4.$$

Note: A more accurate guess at k would be midway between 2.05 and 2.06. Using $k = 2.055$ gives the same answer to one decimal place.

Communication and reasoning exercises

67. Since the empirical rule is based on the normal distribution (see the remarks before Example 3 in the textbook), the empirical rule gives the exact results when the distribution is exactly normal.

69. Neither. They are equal, because they differ by $P(X = a)$, which is zero for a continuous random variable.

71. The total area under the curve must be equal to 1, and the area is a rectangle with width $(b - a)$. Therefore, its height must be $\dfrac{1}{b - a}$.

73. A normal distribution with standard deviation 0.5, because it is narrower near the mean but must enclose the same amount of area as the standard curve, so it must be higher.

Chapter 8 Review

1. X = the number of boys. X is a binomial random variable with $p = .5$, $n = 2$.

$$P(X = 0) = C(2, 0)(.5)^0(.5)^2 = 1/4$$

$$P(X = 1) = C(2, 1)(.5)^1(.5)^1 = 1/2$$

$$P(X = 2) = C(2, 2)(.5)^2(.5)^0 = 1/4$$

Probability distribution:

x	0	1	2
$P(X = x)$	1/4	1/2	1/4

Histogram:

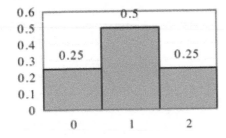

3. X = the sum of the two numbers when a four-sided dice is rolled twice. There are $4 \times 4 = 16$ possible outcomes:

$X = 2$: Outcomes $\{(1, 1)\}$ $\qquad$ $P(X = 2) = 1/16$

$X = 3$: Outcomes $\{(1, 2), (2, 1)\}$ $\qquad$ $P(X = 3) = 2/16$

$X = 4$: Outcomes $\{(1, 3), (2, 2), (3, 1)\}$ $\qquad$ $P(X = 4) = 3/16$

$X = 5$: Outcomes $\{(1, 4), (2, 3), (3, 2), (4, 1)\}$ $\quad$ $P(X = 5) = 4/16$

$X = 6$: Outcomes $\{(2, 4), (3, 3), (4, 2)\}$ $\qquad$ $P(X = 6) = 3/16$

$X = 7$: Outcomes $\{(3, 4), (4, 3)\}$ $\qquad$ $P(X = 7) = 2/16$

$X = 8$: Outcomes $\{(4, 4)\}$ $\qquad$ $P(X = 2) = 1/16$

Probability distribution:

x	2	3	4	5	6	7	8
$P(X = x)$	1/16	2/16	3/16	4/16	3/16	2/16	1/16

Histogram:

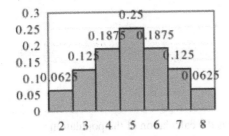

5. $X =$ number of defective joysticks chosen when 3 are selected from a bin that contains 20 defective joysticks and 30 good ones.

$$P(X = 0) = \frac{C(20, 0)C(30, 3)}{C(50, 3)} \approx .2071$$

$$P(X = 1) = \frac{C(20, 1)C(30, 2)}{C(50, 3)} \approx .4439$$

$$P(X = 2) = \frac{C(20, 2)C(30, 1)}{C(50, 3)} \approx .2908$$

$$P(X = 3) = \frac{C(20, 3)C(30, 1)}{C(50, 3)} \approx .0582$$

Probability distribution:

x	0	1	2	3
$P(X = x)$	.2071	.4439	.2908	.0582

Histogram:

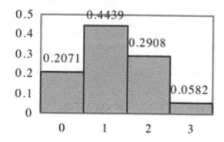

7. The first row lists the values of X, giving us the mean $\bar{x}$. The second row lists the numbers $(x_i - \bar{x})$, and the third row lists their squares.

x	−1	2	0	3	6	10
$(x - \bar{x})$	−3	0	−2	1	4	0
$(x - \bar{x})^2$	9	0	4	1	16	30

The right column shows the sums.

$$\bar{x} = \frac{10}{5} = 2 \qquad \sum_i (x_i - \bar{x})^2 = 30 \qquad s^2 = \frac{\sum_i (x_i - \bar{x})^2}{n - 1} = \frac{30}{5 - 1} = 7.5 \qquad s = \sqrt{s^2} \approx 2.7386$$

For the median m, arrange the scores in order and select the middle score:

−1, 0, **2**, 3, 6 $m = 2$.

9. Two examples are: 0, 0, 0, 4 and −1, −1, 1, 5.

11. An example is −1, −1, −1, 1, 1, 1. (Also see Exercise 78 in Section 9.4.) Here is the calculation of the population standard deviation (the right column shows the sums):

x	−1	−1	−1	1	1	1	0
$(x - \bar{x})$	−1	−1	−1	1	1	1	0
$(x - \bar{x})^2$	1	1	1	1	1	1	6

$$\bar{x} = \frac{0}{6} = 0 \qquad \sum_i (x_i - \bar{x})^2 = 6 \qquad \sigma^2 = \frac{\sum_i (x_i - \bar{x})^2}{n} = \frac{6}{6} = 1 \qquad \sigma = \sqrt{\sigma^2} = 1$$

13. For each of the following 8 solutions: To construct the probability distribution for the weighted die, take p to be the probability of a 1. The given information implies that the probability distribution for the die is

x	1	2	3	4	5	6
$P(X = x)$	p	p	p	p	p	$2p$

Since $7p = 1$, we have $p = 1/7$, so the probability distribution for a single die is

x	1	2	3	4	5	6
$P(X = x)$	$\frac{1}{7}$	$\frac{1}{7}$	$\frac{1}{7}$	$\frac{1}{7}$	$\frac{1}{7}$	$\frac{2}{7}$

Throwing the die 4 times is a sequence of 4 Bernoulli trials with $p = 2/7$ and $q = 1 - 2/7 = 5/7$.

$$P(X = 1) = C(4, 1)\left(\frac{2}{7}\right)^1 \left(\frac{5}{7}\right)^3 \approx .4165$$

15. The probability that 6 comes up at most twice is $P(X \leq 2) = P(X = 0) + P(X = 1) + P(X = 2)$.

$$P(X = 0) = C(4, 0)\left(\frac{2}{7}\right)^0 \left(\frac{5}{7}\right)^4 \approx .2603$$

$$P(X = 1) = C(4, 1)\left(\frac{2}{7}\right)^1 \left(\frac{5}{7}\right)^3 \approx .4165$$

$$P(X = 2) = C(4, 2)\left(\frac{2}{7}\right)^2 \left(\frac{5}{7}\right)^2 \approx .2499$$

Therefore,
$$P(X \leq 2) \approx .2603 + .4165 + .2499 = .9267.$$

17. The probability that X is more than 3 is $P(X > 3) = P(X = 4) = C(4, 4)\left(\frac{2}{7}\right)^4 \left(\frac{5}{7}\right)^0 \approx .0067$.

19. $P(1 \leq X \leq 3) = P(X = 1) + P(X = 2) + P(X = 3)$. We computed $P(X = 1)$ and $P(X = 2)$ in Exercise 15.

$$P(X = 3) = C(4, 3)\left(\frac{2}{7}\right)^3 \left(\frac{5}{7}\right)^1 \approx .0666$$

Therefore,

$$P(1 \leq X \leq 3) = P(X = 1) + P(X = 2) + P(X = 3) \approx .4165 + .2499 + .0666 = .7330.$$

21. Think of the experiment as a sequence of 3 Bernoulli trials with "success" = girl, $n = 3$ and $p = .5$.

$$\mu = np = 3(.5) = 1.5$$

$$\sigma = \sqrt{npq} = \sqrt{3 \times .5 \times .5} \approx 0.8660$$

To answer the last part,

$$[\mu - \sigma, \mu + \sigma] = [0.634, 2.366]$$

This interval does not include all 3 values.

$$[\mu - 2\sigma, \mu + 2\sigma] = [-0.232, 3.232]$$

This interval includes all the scores, so all values of X lie within __2__ standard deviations of the expected value.

23. The frequencies add up to 16, so we obtain the probabilities by dividing the frequencies by 16:

x	-3	-2	-1	0	1	2	3
$P(X = x)$	0.0625	0.125	0.1875	0.25	0.1875	0.125	0.0625
$xP(X = x)$	-0.1875	-0.25	-0.1875	0	0.1875	0.25	0.1875

$\mu =$ sum of entries in right-hand column $= 0$

$x - \mu$	-3	-2	-1	0	1	2	3
$(x - \mu)^2$	9	4	1	0	1	4	9
$P(X = x) \times (x - \mu)^2$	0.5625	0.5	0.1875	0	0.1875	0.5	0.5625

$\sigma^2 =$ sum of entries in right-hand column $= 2.5$

$$\sigma = \sqrt{\sigma^2} \approx 1.5811$$

For the last part, note that 14 of the 16 scores in the frequency table (exclude the first and last value) are in the interval $[-2, 2]$; that is, in the interval

$$[\mu - 2, \mu + 2]$$

because $\mu = 0$. The number of standard deviations that 2 represents is approximately

$$\frac{2}{1.5811} \approx 1.3 \text{ standard deviations.}$$

Therefore, 87.5% (or 14/16) of the time, X is within __1.3__ standard deviations of the expected value.

25. By Chebyshev's rule, X is guaranteed to lie within k standard deviations with of μ with a probability of at least $1 - \dfrac{1}{k^2}$. Thus,

$$1 - \frac{1}{k^2} = .90$$

$$\frac{1}{k^2} = 1 - .90 = .10$$

$$k^2 = \frac{1}{.10} = 10$$

$$k = \sqrt{10} \approx 3.162.$$

The associated interval is therefore

$$[\mu - k\sigma, \mu + k\sigma]$$

$$\approx [100 - 3.162(16), 100 + 3.162(16)]$$

$$\approx [49.4, 150.6].$$

27. Since X is bell-shaped and symmetric, the empirical rule applies so approximately 99.7% of samples of X are within the interval

$$[\mu - 3\sigma, \mu + 3\sigma].$$

This means that approximately $0.3/2 = 0.15\%$ of the samples of X are greater than

$$\mu + 3\sigma = 200 + 3(20) = 260.$$

29. $X = Z$; From the table, $P(0 \leq Z \leq 1.5) = .4332.$

31. $X = Z$; $P(|Z| \geq 2.1) = 2(.5 - .4821) = .0358$

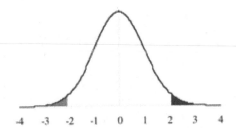

33. Standardize the given problem:

$$P(X \leq b) = P\left(Z \leq \frac{b - \mu}{\sigma}\right)$$

$$P(X \leq -1) = P\left(Z \leq \frac{-1 - 0}{2}\right) = P(Z \leq -0.5) = .5 - .1915 = .3085.$$

35. The frequency distribution for the price is

x	5.50	10	12	15
$fr(X = x)$	1	2	3	4

Dividing the frequencies by the sum, 10, gives the probability distribution. We also add a row for the computation of $E(X)$:

x	5.50	10	12	15
$P(X = x)$	.1	.2	.3	.4
$xP(X = x)$	0	0.2	0.4	0.3

$E(X) = $ sum of entries in bottom row $= \$12.15$

37. Revenue = Price per copy sold × Number of copies sold. The values of the weekly revenue are obtained by multiplying the prices by the weekly sales:

x **(Revenue)**	34,100	35,000	36,000	15,000
$P(X = x)$	.1	.2	.3	.4
$xP(X = x)$	3,410	7,000	10,800	6,000

$E(X) = $ sum of entries in bottom row $= \$27{,}210$

39. False; let $X = $ price and $Y = $ weekly sales. Then weekly revenue $= XY$. However, $27{,}210 \neq 12.15 \times 2{,}620$. In other words, $E(XY) \neq E(X)E(Y)$.